Holt Meyer, Susanne Rau, Katharina Waldner (Eds. / Hrsg.)

SpaceTime of the Imperial

SpatioTemporality / RaumZeitlichkeit

Practices – Concepts – Media /
Praktiken – Konzepte – Medien

Edited by / Herausgegeben von
Sebastian Dorsch, Bärbel Frischmann, Holt Meyer,
Susanne Rau, Sabine Schmolinsky,
Katharina Waldner

Volume / Band 1

SpaceTime of the Imperial

Edited by
Holt Meyer, Susanne Rau, Katharina Waldner

DE GRUYTER
OLDENBOURG

The realization of the book was financially supported by Sciencefunding Erfurt gGmbH.

ISBN 978-3-11-041973-3
e-ISBN (PDF) 978-3-11-041875-0
e-ISBN (EPUB) 978-3-11-041885-9
ISSN 2365-3221

Library of Congress Cataloging-in-Publication Data
A CIP catalog record for this book has been applied for at the Library of Congress.

Bibliografische Information der Deutschen Nationalbibliothek
The Deutsche Nationalbibliothek lists this publication in the Deutsche Nationalbibliographie; detailed bibliographic data are available in the Internet at http://dnb.dnb.de.

Cover Image: Holt Meyer, Erfurt
Typesetting: Dr. Rainer Ostermann, München
Printing: CPI books GmbH, Leck
♾ Printed on acid free paper
Printed in Germany

www.degruyter.com
www.magnespress.co.il

MIX
Papier aus verantwortungsvollen Quellen
FSC www.fsc.org FSC® C083411

Contents

Europe, Spatiotemporal Orientation, and the Imperial

God(s) in the Empire: Mapping Imperial Religion

Cartographies of the Imperial Age

Media Narratives on Königsberg/Kaliningrad: Spatiotemporalities of the Displaced

Holt Meyer, Susanne Rau and Katharina Waldner

"… this smooth space of Empire … "

Introduction to Six Spatiotemporal 'Stabs' at Analyzing the Imperial

When we speak of *spatiotemporality* – and we say this keeping in mind that this is the first volume of a series on that very topic[1] – we have in mind that spatiality and temporality are treated as constructs in inextricable correlation with each other in contexts both historical and contemporary. Theories in and after the "spatial turn" have taught us that spaces are not considered any more as something simply physically given, but rather as something socially created. Spaces are results of negotiations and designs, and these take place in time. The focus on the genesis, the practices and processes connected to space also brings to mind Michel Foucault's insight that not only 'actual' places or buildings such as the hospital or the prison, but just as much social processes and the regulation or administration of the body in general can enter into reflection on spatiality. Already here we have an inseparability of space and time.[2] The role of space and time in people's sociocultural and life-world concepts of themselves and in media representations is the core of our concern in this volume and the whole series. In the respective context of the series with its disciplinary or thematic orientation, different theories of spatiotemporality, e.g. Lefebvre's "rhythmanalysis", musicological or versological concepts, Soja's "thirdspace", Bakhtin's "chronotope" (more on this later in this introduction), or even the spacetime of Einstein's relativity theory come to the fore. Empire seems to us to be a topic which, in launching the series, has substantial temporal and spatial elements which historians have extremely often conceived of separately, but not in terms of a correlation in which the one is not thought without the other and both are considered explicitly. Often spatial elements are to be found in phenomena which seem to be purely temporal or vice versa. Some of the most important examples for this in the book involve

1 The editors of the series entitled *Spatiotemporality: Practices – Concepts – Media* are Sebastian Dorsch, Bärbel Frischmann, Holt Meyer, Susanne Rau, Sabine Schmolinsky, Katharina Waldner. They are all members of the *Erfurt Spatiotemporal Studies Group*: https://www.uni-erfurt.de/philosophische-fakultaet/raumzeit-forschung/ (21.07.2016). We especially thank Bettina Neuhoff from De Gruyter for her inspiring and always friendly and professional support for starting the book series and producing its first volume.

2 See Susanne Rau. *Räume. Konzepte – Wahrnehmungen – Nutzungen*. 2d ed. Frankfurt/Main: Campus, 2014, esp. 47, 153.

DOI 10.1515/9783110418750-001

maps: A particular treatment of the geographical map is one model of representation which at first seems purely spatially oriented, but at a second glance displays many layers of temporality which condition all visualization and verbalization of space. The verbalization contains the aspect of the state and the historical status of the language itself, which, in turn, is an effect of temporality.

***

Our concentration on the spatiotemporal aspect of empire takes a 'stab' at analyzing the imperial – in the sense of reaching into a dark area and trying to literally grasp it, but also, of course, in the sense of calling it radically into question, i.e. taking a literal 'stab' at the imperial itself as an ideology.[3] One example of a radical questioning of the imperial in our days is the book *Empire* (2000) by Michael Hardt and Antonio Negri, which is addressed in this collection less as an unquestioned authority than as a constructive provocation of a debate on the imperial beyond the classic colonial (and thus 'post-colonial') paradigm.[4]

---

3 This collection of essays is the result of a conference held at the University of Erfurt from October 8th to 11th, 2014. We would like to thank the Thuringian Ministry for Economic Affairs, Science and Digital Society, the Ernst-Abbe-Stiftung Jena as well as the University of Erfurt for funding this scientific encounter and exchange. We also thank Florian Heintze, Robert Fischer, René Smolarski, and Charlotte Krause for their editorial support in preparing this volume.

4 Michael Hardt and Antonio Negri. *Empire.* Cambridge, Mass.: Harvard University Press, 2000. The book came out in its fourth printing in 2001 already (at the moment of writing this edition was freely accessible online). For a critical approach to the book see Manfred B. Steger's review in *The American Political Science Review* 96:1 (Mar., 2002): 264–265. Hardt and Negri followed up on this publication with several publications in the same vein, above all Michael Hardt and Antonio Negri, *Multitude. War and democracy in the age of Empire*. New York: The Penguin Press, 2004 responding to the Bush administration reactions to 9/11 (Francis Fukuyama in a *New York Times* article on July 25, 2004, called it an "antidote to Empire", referring mainly to the category "multitude" itself), but also with *Commonwealth.* Cambridge, Mass.: Belnap, 2009. In *Declaration*, Kindle edition epub 2012, they reacted to the occupy movements. Since the category of empire, particularly from an historical perspective, gets its most thorough treatment in *Empire* of 2000, while the other, much smaller, books are much more calls to political action, most of our references are to this first and most extensive study of Hardt and Negri. There have been other important contributions in recent years to the subject, particularly Jane Burbank and Frederick Cooper. *Empires in World History: Power and the Politics of Difference*. Princeton, NJ: Princeton University Press, 2010. But their approach is a more traditional historical one, which is of course very valuable. Burbank and Cooper's book, however, has little to do with Hardt and Negri's 'unified field theory', and complements it. No text in Burbank and Cooper refers to *Empire*. Herfried Münkler. *Imperien. Die Logik der Weltherrschaft – vom Alten Rom bis zu den Vereinigten Staaten*. 2d ed. Berlin: Rowohlt Verlag, 2005 refers to Hardt and Negri in passing in two footnotes (p. 306f.).

Getting a grasp on the imperial is difficult, because, as Hardt and Negri note, the object of study in their book *Empire* is a "smooth space". Empire in Hardt and Negri's understanding is "smooth", among other things, because it is a kind of Proteus: it is everywhere and nowhere and takes endless new forms. It seems almost impossible to even take a stab at it, for this is like stabbing at the air we breathe. Above all, it is not an era of the past – we live today, according to Hardt and Negri, perhaps in the post-colonial, but by no means in the post-imperial. For many post-imperial claims turn out to be clandestine and thus disingenuous imperial ones. The first move in this game may have been made by the USA, and the game was certainly continued by the USSR.

We have chosen the approach of Hardt and Negri as an important reference also because it doesn't make life easy for itself by locating the imperial in some other time, or in some other place, but rather in the entirely of the contemporary world. This does not mean that the authors of all the contributions of this collection or even all the authors of this introduction necessarily completely share this view in the same way. It does mean that we view one of the most challenging views of the imperial as a position to be reckoned with.

This introduction explains how and why this book, despite these extreme difficulties, takes six 'stabs' at analyzing empire. These 'stabs', or attempts, are the collection's six chapters, each of which takes up a specific aspect of the imperial. These six aspects are conceptual locations from which an attempt is made to bring the Proteus known – or not known – as empire to the ground and thus to grasp it. Looking at the issue (the history, the systematic nature, the types) of the imperial and/or colonialism from the point of view of spatiotemporality seems logical, almost trivial. As such it is, however, a new approach. Spatiotemporal analysis with the version of empire in mind which Hardt and Negri outlined at the end of the 20th century is all the more new, especially since the book *Empire* does not explicitly put spatiotemporality into the foreground.

In a conventional contemporary view of the *space* of empire, empire is an expanse of land or a group of expanses of land belonging to a strong (in classic cases European) nation. The expanses of land are militarily controlled and the inhabitants subjugated, and that is the end of the story – until the "liberation". But before modernity there were, indeed, many different political, social and cultural ways to organize power over vast territories (spaces!), conventionally called "empires" ("Reiche").[5] Bringing in the perspective of spatiotemporality also

5 See e.g. for the ancient Mediterranean and Near Eastern Empires: Susan E. Alcock et al. (eds.). *Empires: Perspectives from Archaeology and History*. Cambridge: University Press 2001; for the Roman Empire: Richard Hingley. *Globalizing Roman Culture. Unity, Diversity, and Empire*. Lon-

means to explore the spatial dimension of such formations without projecting contemporary views of Empire upon them.

In a conventional contemporary view of the *time* of empire, one often encounters the comforting conviction that there was a colonial age in the past which had a beginning and an end. This narrative claims that the epoch ended with the various decolonialization processes in the Africa, Asia and Americas, which, in turn were at the core of historical development in the three decades following World War II (although of course the breaking away from empire is the stuff of the history of the USA and a number of South American countries in the late 18th and 19th century). Our spatiotemporal approach tries to go beyond these conventional assumptions. The question of the imperial in the sense treated here is historically to a great degree a question of Europe, i.e. the "West". It is, in any case, this "western" model of the imperial which Hardt and Negri are addressing and which is arguably the only one which could be seen to making the global ideological claim which is described in *Empire*.[6] At the same time, it is recognized at the latest since Edward Said's *Orientalism* in the 1970s that Europe is defined by the Non-European, just as the Non-European 'orientalizes' itself by linking itself to Europe according to an imperial agenda. This is why it is essential to have a section on "Europe" itself.[7]

***

One of the most well known models of spatiotemporality is the concept of the "chronotope", derived from the thought of Mikhail Bakhtin of the 1940s[8], which strives to find a modelling of space and time which is characteristic for a particular epoch or formation. Would one consider something like a chronotope of empire, one could assemble characteristic links of space and time. We will take as an example Joseph Conrad's 1899 novella *Heart of Darkness* (and also Francis Ford Coppola's *Apocalypse Now* of 1979 as a 'film version' which marks

don: Routledge 2005; for medieval and early modern times: Christine Lebeau (ed.). *L'espace du Saint-Empire: du Moyen-Âge à l'époque moderne*. Strasbourg: Presses Univ. de Strasbourg, 2004.

6 In this sense they are a step beyond Niall Ferguson. *Empire: How Britain Made the Modern World*. London: Penguin 2004, an apology for the British Empire, as the subtitle suggests, and also Herfried Münkler's *Imperien* of 2005, which is an apology for the European model which is similarly skeptical towards the USA but mentions Hardt and Negri only twice (see above note 3).

7 See below, 169.

8 Mikhail Bakhtin. Forms of time and of the chronotope in the novel. In *The Dialogic Imagination: Four Essays*. Michael Holquist, Caryl Emmerson (ed. and transl.), 84–258. Austin: Univ. Texas Press.

the 'post-colonial' era relevant for Hardt and Negri's *Empire*).[9] Conrad's text is considered by many to be a salient narrative working out of the imperial chronotope. Edward Said began by no means coincidentally as a Conrad scholar – and, thus, as a literary scholar. Also in the 1970s, Chinua Achebe's impassioned intervention into the debate on the text brought it front and center in the negotiation of empire.[10]

Taking Conrad's text as a starting point, one might test the claim that the chronotope of empire is the transport up a river and/or across a body of water with the purpose of someone on the vessel engaging an imperially relevant subject on some shore or embankment – in this case it would be the narrator Marlow engaging the object of his concerns and of his African mission, Kurtz. Marlow is instructed to travel to Africa and engage Kurtz, since the activities of the latter are deemed in their extremity to push the imperial project too far and thus to endanger it as a whole. The plot of Conrad's classic text is Marlow's own account of that engagement. Thus, one might say that the imperial as a whole is negotiated in and by the text. A sail from London up the Thames provides the framework for Marlow's telling the tale to the others on the boat. Since this tale-telling comprises most of the narrative and, opposing Rome to Britain as negative and positive imperial models respectively, this fact could be seen as driving the status of the chronotope all the more home. In this sense, in looking for a narratively oriented visual representation of the classic age of colonialism, the location on a ship, be it a 17th century Dutch representation (e.g. by Willem van de Velde de Jonge) or a 19th century English one (e.g. by Joseph Mallord William Turner) in a typical temporal arrangement or occasion would be a logical place to fix the imperial chronotope, and Marlow's double river passage (in the narrative frame in London and in the narrative itself in the unnamed African country which might be Congo, a country never actually mentioned in the text) at the turn of the 20th century might be the chronotope coming to itself, perhaps even deconstructing itself.

Taking up the global aspect of Empire, we need to keep in mind the words of the great contemporary historian of the ordinary, Alf Lüdtke, who addresses "the colonial/ imperial as the dark side of the West"[11], and indeed when we speak of

9 Joseph Conrad, Heart of Darkness. In *The Portable Conrad*, Michael Gorra (ed.), 277–364. London: Penguin, 2007.

10 Chinua Achebe. An Image of Africa: Racism in Conrad's 'Heart of Darkness'. In *Heart of Darkness, An Authoritative Text, background and Sources Criticism*. Robert Kimbrough (ed.). 3rd ed. London: W. W Norton and Co., 1988, 251–261 (originally Massachusetts Review. 18. 1977).

11 See the introduction to the first chapter "Narrating the Other" by Robert Fischer (see below, 21–24, quotation: 21).

the imperial in the context at hand, we are again speaking of 'Western empire'. Would one want to bring the "East" into the mix one might claim that Peter the Great's entrance onto the military scene in the Baltic at the end of the 17th century raises for the first time in the modern age the question of the 'Western empire' and its being challenged both from the 'Eastern' outside or from its Eastern edge (this difference would on closer consideration be decisive). Pushkin's verse narrative *The Bronze Horseman (1833)* representing Peter the Great presiding over the horrifying image of the river Neva's waters flooding the city and destroying lives would be a working through of this "chronotope" as seen from the land and/as the East.

The "chronotope" as an instrument for determining how configurations of time and space are represented in language and discourse is an indicator of a certain discursive formation. It is just one possibility of conceiving of spatiotemporality in depictions (in the broadest sense) of empire.

The concern with "mapping" (also in the broad sense) is another option. The production of the map is – we will return to this thought later in the specific context – a media activity which is highly conditioned by "geopolitics". The term itself is as problematic as it is spatiotemporal, for "politics" necessarily take place in time.

Turning again to Conrad's *Heart of Darkness*, Marlow's famous recollection of his observation of a map with large blank spaces turns the product of cartography into an operational plane, but also into a further spatiotemporal model:

> Now when I was a little chap I had a passion for maps. I would look for hours at South America, or Africa, or Australia, and lose myself in all the glories of exploration. At that time there were many blank spaces on the earth, and when I saw one that looked particularly inviting on a map (but they all look that) I would put my finger on it and say, "When I grow up I will go there." The North Pole was one of these places, I remember.[12]

The two texts on the arctic in our collection (Walsh, Holtorf) could be brought to bear on this point here, but we will stay with the general argumentation. A child bathing in the "glories of exploration", and the "hours" spent doing this, are also an example of an imperial chronotope – the chronotope of the time and place of imperial education or self-education. They are perhaps analogous to the travels of the actual "explorer" on the ships which take him to the prized coasts of trade and exploitation.

When one considers the "double mapping" which Christopher GoGwilt suggests as a reading of Conrad's colonial narratives (especially *Heart of Darkness*,

12 Conrad, Heart of Darkness, 282.

but also memoirs) in his book *The Invention of the West*[13], an East-West dynamic again enters into the discussion, for the Russian colonialization of a good part of Conrad's native Poland becomes the 'actual' field of reference of Conrad's text. *Heart of Darkness* becomes a huge trope, just as the intervention of Peter the Great into the world of "Western" ship-building and navies creates a foil for 'Western' imperialism which defines its specific qualities and indeed represents an "Invention of the West" in the first place.[14]

In the alternative history of the 'Western'-based imperial project represented by Hardt and Negri's *Empire*, the 'East-West'-divide is downplayed, for the book relativizes traditional political and cultural borders in order to attain a global vision. Indeed, the "West" as such appears not a single time in the book, but we do read things about the "Western" such as:

> The legacy of modernity is a legacy of fratricidal wars, devastating "development," cruel "civilization," and previously unimagined violence. Erich Auerbach once wrote that tragedy is the only genre that can properly claim realism in *Western* literature, and perhaps this is true precisely because of the tragedy *Western* modernity has imposed on the world[15]

and

> Considered simply in cultural terms, Islamic fundamentalism is a paradoxical kind of postmodernist theory – postmodern only because it chronologically follows and opposes Islamic modernism. It is more properly postmodernist, however, when considered in geopolitical terms. Rahman writes: "The current postmodernist fundamentalism, in an important way, is novel because its basic elan is *anti-Western* ... Hence its condemnation of classical modernism as a purely *Westernizing* force." Certainly, powerful segments of Islam have been in some sense "*anti-Western*" since the religion's inception. What is novel in the contemporary resurgence of fundamentalism is really the refusal of the powers that are emerging in the new imperial order. From this perspective, then, insofar as the Iranian revolution was a powerful rejection of the world market, we might think of it as the first postmodernist revolution.[16]

---

13 Christopher GoGwilt. *The Invention of the West. Joseph Conrad and the Double-mapping of Europe and Empire*, Stanford: Stanford Univ. Press 1995.

14 Christopher GoGwilt. True West: the Changing Idea of the West from the 1880s to the 1920s. In *Enduring Western Civilization*, Silvia Federici (ed.), 37–62. Westort, Conn.: Praeger, 1995. GoGwilt's study demonstrates this 'invention' on the basis of material from the self-description of the British imperial configuration and from Russian thought of the 19th century which ever and ever again comes back to the assessment of Peter.

15 Hardt and Negri, *Empire*, 46 (our italics).

16 Hardt and Negri, *Empire*, 149 (our italics).

Despite this, it is fair to say that for Hardt and Negri, the source of the imperial is the "West", materializing first as Europe, and then as Europe overtaken by the USA, the executive in the translational post-nation-state of Empire, and the heir of the classic colonial world order. That does and does not answer the question of the place and the time of empire and Empire (as the political state of the contemporaneity of the 21st century), the latter of which is, as we have already said, eerily omnipresent and without place:

> In this sense, the clearly defined crisis of modernity gives way to an omnicrisis in the imperial world. In this smooth space of Empire, there is no place of power – it is both everywhere and nowhere. Empire is an ou-topia, or really a non-place.[17]

Here, the "West" is geographically 'de-westernized' – one might even say 'deterritorialized'. This is of course reminiscent of the thought of Gilles Deleuze and Félix Guattari. In a political and economic sense it is, on the other hand, maximally 'westernized'. It is a "smooth space", not a "serrated" one. That is also a central observation in a spatiotemporal view. Their reference to Deleuze and Guattari (they are mentioned over 50 times in Hardt and Negri's book) is constitutive. They say explicitly, for instance: "Since no one point in the network is necessary for communication among others, it is difficult for it to regulate or prohibit their communication."[18] This "democratic" model is what Deleuze and Guattari call a rhizome, a nonhierarchical and noncentered network structure. Opposed to this is the "oligopolistic network model", which is

> characterized by broadcast systems. According to this model, for example in television or radio systems, there is a unique and relatively fixed point of emission, but the points of reception are potentially infinite and territorially indefinite, although developments such as cable television networks fix these paths to a certain extent. The broadcast network is defined by its centralized production, mass distribution, and one-way communication. The entire culture industry – from the distribution of newspapers and books to films and video cassettes – has traditionally operated along this model. A relatively small number of corporations (or in some regions a single entrepreneur, such as Rupert Murdoch, Silvio Berlusconi, or Ted Turner) can effectively dominate all of these networks. This oligopolistic model is not a rhizome but a tree structure that subordinates all of the branches to the central root.[19]

17 Hardt and Negri, *Empire*, 190.
18 Hardt and Negri, *Empire*, 299.
19 Hardt and Negri, *Empire*, 300.

The temporality of both is not explicitly addressed, but is apparent: the simultaneity of the "democratic" rhizome is opposed to the succession of steps down the hierarchy of the imperial "oligopoly". One might even consider this a medialized version of the classical colonial chronotope, with the river crossings being replaced by the passages through media hierarchies to the entrepreneur as the new Kurtz. The imperial en route to Empire in this spatiotemporal de-bordering, in this extreme generalization of the concept (or regulator of practices) known as the "frontier" which can be translated into the assemblage of "oligopolistic" networks which are combatted by the "rhizomatic" multitude: that would be a first approach to Hardt and Negri (in primarily Deleuzian formulation) from a spatiotemporal point of view.

***

In dividing up the treatment of empire (and Empire) into six separate and at the same time interwoven themes, this book addresses the divisibility of the issue itself on the basis of the criteria chosen: criteria of 'technical' media (e.g. maps and/or cartography), 'conceptual' geography (Europe, Königsberg/Kaliningrad), societal divides (gender), textual activity or genre (narrative othering), institutional-philosophical discourses (religion), we implicitly ask if empire (and Empire) has these categories at all, and how this breaking down into categoriality informs the spatiotemporal view of the whole volume.

The six 'stabs' at an analysis of the imperial, the six chapters of this book, have the following concerns: 1. narrative othering, 2. gender, 3. Europe, 4. God(s) respectively religion, 5. cartographies and 6. displacement (in the case of East Prussia). This surely does not sound like a systematically complete treatment of every significant aspect of the imperial from a spatiotemporal point of view, and it is not meant to be. What it is meant to be is an inductive demonstration of the usefulness of spatiotemporality as a way of grasping empire(s)/Empire, of wrestling it to six types of ground in order to get a good look at it from each of the six sides. With this, we introduce a systematic methodology which augments Hardt and Negri's often essayistic approach, and also adds in aspects which are not specifically addressed in the treatise.

We will briefly run through the basic aspects of each of these six concerns, the six sides of the cube which contains the ensemble of our concerns, reminding the reader at the same time that each is laid out in detail in separate introductions to the individual chapters.

1 Narrative Othering

In beginning with the issue of "narrative othering through imperial time spaces", we are in a way dealing with the very 'stuff' of the imperial. It is nowhere if it is not in narratives (of course also that of *Heart of Darkness*), indeed can be defined as a complex narrative (comprised of myriad textual and media sources such as *Heart of Darkness, Apocalypse Now, Orientalism, Homeland*, etc.). The basic narrative is formulated by Robert Fischer in the introduction to the chapter, who notes that the "others" in the imperial configuration: "were connected to Western colonial/ imperial powers through established trade networks or settlements, conquered exotic lands or attempts to missionize and educate the timeless other (*mission civilisatrice, white man's burden*)".[20] Isolating "othering" as a fundamental operation in imperial practices is self-evident, and the characterization of the "other" as "timeless" gives a first hint at how this operation can be grasped from a spatiotemporal point of view. The imperial power and aggressor as the "beacon of civilization contrary to the dark barbaric "other" represents time equated with progress and history, defined by an overcoming the "barbarian's" 'timelessness'. As far as space goes, in the context of alternative dealings with this side of spatiotemporality, the reference to "scales" points to a key feature of the narrative. "Zooming in" as opposed to implementing the grand narrative of the colonizer reveals what actually happens 'on the ground' – "happens", of course in the sense of the passage of time. Examining "othering" thus involves reclaiming time for a space which the imperial discourse determined – precisely in its narrative strategies – to have 'not yet entered history'. Getting a 'foothold' on this supposedly 'a-historical' ground is a method in making headway in the terrain in question.

2 Gender

Narratives of women being treated as 'historiographically irrelevant' is an analogous challenge facing a focus on gender and the imperial. It is clear for this and many other reasons that "the concept of 'Empire' cannot do without the analytical category of gender".[21] Even if it is correct that Hardt and Negri's category of the "multitude" includes gender aspects, the lack of a sufficient focus on

20 All quotes heretofore in this section are from Robert Fischer's chapter introduction "Reflecting on Narrative Othering through Imperial Time Spaces" (see below, 21–24).

21 All quotes heretofore in this section are from Sabine Schmolinsky's chapter introduction "Gender in the Empire" (see below, 95–99).

gender in their account of Empire's history needs to be not only abstractly taken seriously, but must lead to concrete analytical work, two examples of which are contained in the corresponding chapter.

While it is true that the issue of gender is addressed in a more concerted manner in Hardt and Negri's later books, particularly the 2004 book *Multitude*, precisely the role of gender in the writing of a history of empire and Empire is a question which the later books do not answer, because they are concerned with the contemporary. This is a deficit of the work, which subordinates gender to other issues while retracing empire's path into the present. This goes for both the role of women and for the issue of masculinity. The two articles in this chapter try to shed light on both sides of the gender divide. Joanna de Groot expresses constructive critique of the place of the category of gender at the core of most theories of empire and imperialism (as opposed to the thematization of the place of individuals or groups of women in imperial practices, be they victims or perpetrators, which is not unrelated, but still another issue). She addresses this issue frontally and speaks even of a "gender blindness" of empire theories, including that of Hardt and Negri, which needs to be rectified.

Silvan Niedermeier's study on "male white self-representation at war abroad as it is performed through the medium of the soldiers' private photography or of professional photography by male photographers" brings up significant gender issues, and the matters are "determined by spatiotemporality in that they directly referred to the soldiers as they inscribed themselves in the USA's becoming an overseas empire in particular spaces around 1900". This is a further example of the concretization of the imperial in the analysis of theories and of practices. Both issues treated and both perspectives drive home the extreme significance of the category of gender precisely for a spatiotemporal treatment of empire.

3 Europe

Since this collection was completed in the weeks and months immediately following the 'Brexit vote' of June 2016 in the United Kingdom, a focus on the European Union and its predecessors going as far back as the Early Modern period, and their relation to the imperial, would appear to be more than topical. One might think in this context of Marlow in Conrad's *Heart of Darkness* as he sails up the Thames and valorizes the English imperial model over others. One might also think of present apologetics for the British Empire such as those of Niall Ferguson.[22]

We ask in this chapter whether the constituting of "Europe" already contains imperial aspirations which remain throughout its existence (Olaf Asbach). We get more specific in the late 18th and early 19th century by examining the exploring activity of and the discourses which it gave rise to (Elizabeth Millán). We examine the "projection of the most diverse interests and fantasies" onto the North Pole as a general European concern, and the "acquisition of knowledge, symbol systems, technological possibilities, interests, and consequent actions" which can be seen in the general European conceptualizations of the arctic region (Stephen Walsh) this will come up again in the section on cartography). All this allows us to witness the emergence of spatiotemporalities from 'above' which automatically provoke the question of whether what is going on on the ground in various regions of Europe might call Europe as a whole and the spatiotemporalities of its implicit and explicit narratives about itself into question.[23]

4 God(s)

"Mapping religion in a very concrete spatiotemporal sense"[24]: this is the issue with which the introduction to the chapter "God(s) in the Empire" ends, and indeed a view of religion in empire(s) based on Hardt and Negri does require serious historical concretization by using a spatiotemporal perspective which is what the chapter achieves in its three articles. At the same time it analyzes Hardt and Negri's use of "religion", most specifically of Christianity.

By taking two examples from antiquity and one from the contemporary USA, the authors of the chapter "God(s) in the Empire" choose material which links the beginning and the end of the imperial narrative addressed in this book and also in *Empire*, particularly considering the fact that the US project is described by Hardt and Negri as the working through of two versions of Rome.[25] As in the case of gender, it seems that religion is in the shadows of thinking about empire and

22 See above, note 6.

23 All quotes heretofore in this section are from Bärbel Frischmann's chapter introduction "Europe, Spatiotemporal Orientation, and the Imperial" (see below, 171–173).

24 All quotes heretofore in this section are from Katharina Waldner's chapter introduction "Empire and Religion" (see below, 247–253).

25 See the chapter "The American Revolution and the Model of Two Romes", Hardt and Negri. *Empire*, 161f.

needs to be brought out of those shadows. In a different way than in the case of gender, a view to this topic reveals interesting, probably unconscious tendencies within the approach to empire in the book *Empire*, particularly as concerns Christian millenarism, which Hardt and Negri in an odd way ironically mirror – precisely because the authors of *Empire* do not take religion head on (it is noted that in the index of the book neither God nor religion are mentioned). The arguments in all three texts thus note a strong oscillation of Hardt and Negri themselves on the issue of religion. The concretization of the three cases brings the connection of religion and empire down to earth, and the perspective of spatiotemporality brings a certain amount of order into this issue. Harry Maier analyzes early Christian martyrdom accounts and takes issue with the comprehending of empire as a "smooth space", taking a much more differentiated view. He finds much that is "serrated" on the ground of the actual practice of ancient Roman imperial religion and politics. We are dealing in the final analysis with "texts of resistance" that use the Thirdspace (Soja) to construct and narrate a new Christian identity.

The chapter on religion brings furthermore a media and genre perspective into the debate, something which remains in the book's contributions through the next two chapters on cartography and displacement. For Alfred Schmid, the literal "Weltbild" (image of the world) for imperial Rome plays a big role, something "based on the imagination of the world as perfect globe". Maier argues strongly with narrative devices of his "martyrdom acts". And in his study on American civil religion, Jens Kugele maps the place of religion in contemporary Empire using a strongly medialized example by analyzing the example of the memorial service for Neil Armstrong in 2013, in the National Cathedral in Washington DC Kugele shows how national memory and sacred space are "performed" by "architectural and liturgical devices" which "communicate a certain version of US imperialism"[26]. Religion and media enter into a close alliance and become almost interchangeable, which for a spatiotemporal link between religion and empire is of significance of the highest order. In the end it turns out that Hardt and Negri's perspective on and use of religion is versatile and somehow contradictory: Following Macchiavelli they state that Christian religion destroyed the civic religion of the Roman Empire by reconstructing the past from the Renaissance onwards, they oscillate between ascribing to religion an antimodernist slant as well as a kind of postmodern revolutionary potential. In their vision of the world to come religions tends to be reduced or hypostasized to a "communicative matrix that compromises the interhuman world itself".[27]

26 Taken again from Katharina Waldner's chapter introduction (see below, 252)

27 Chris Fox. From Representation to Constituent Power. Religion, or something like it, in Hardt and Negri's *Empire*. *Journal for Cultural and Religious Theory* 9.2 (2008): 30–42, quote: 40.

5 Cartographies

The notion of "looking more closely into the many colonial situations on the ground where we might potentially discover the collective character of cartographic knowledge production" brings once again to the fore the thought that one needs to get beyond or behind constructions of official imperial spatiotemporality in order to find other models coming from those colonially 'generalized'.[28] Keeping in mind, as stated at the outset of this introduction, that space is socially and temporally formed, one can state that the 'actual' ground on which the map is 'based' produces signs and signals in danger of being erased as soon as they are 'organized'. This applies to scholarly accounts of the imperial themselves and the call to "see the bricolage character of the imperial endeavor on the whole" goes a step further in addressing not only imperial projects, but also later historiographies. The notion of "Western science" itself as an empire put forth by Steven Walsh in his article is important here, particularly since to this day we are loath to question on ideas of "objective geographical knowledge".

How time and space are actually arranged "on the ground", for instance in the case of the medialization of arctic islands, in this case by the Habsburg Empire, can be retraced by "taking the localized and historicized actors of mapping into account" and thus bringing the making of the map itself as imperial activity into focus – beyond the often noted fact that cartography is basically born as and in empire.

The representation of a fixed space on the two-dimensional plane of the map suggests a moment of 'objective geographical truth' which becomes particularly apparent when spatiotemporality becomes the focus. The cultural limitations of this become apparent when a map modelled on colonial Africa is produced in Japan, and can not "meet the daily practical demands of its expected consumers" in this case the "contract foreigners" operating on Japanese territory. In other words, the temporal processes into which maps are meant to enter there in order to cope with the space in question. The "smoothed", i.e. colonialistically organized map laid of the ground of Africa, of course already there deeply problematic, contains notations which get bogged down in the serrated space it is meant to grasp. This, in turn, helps us grasp the spatiotemporality of the imperial as a whole.

28 All quotes heretofore in this section are from the chapter introduction "Spatiotemporalities of Cartographic Empire-Building" by Sebastian Dorsch and Iris Schröder (see below, 333–337).

6 Displacement/East Prussia

Maps appear also in the last section – literally in the form of totalitarian projects of naming, renaming, reshaping and rebuilding at various points in the ferocious tearing up of the territory known until 1945 as "East Prussia". In taking this special geographical case at the end of the collection and connecting the three studies in question with the central idea of "displacement", the three authors looking both from the 'Russian side' and the 'German side' note strategies of the depletion of actual historical time in the territory and an attempt to fix it in an eternally identical national spatiality.[29]

The transformation of the 'most Eastern' territory of the "West" (if one looks at Imperial Weimar or Nazi Germany in that way, which is not unproblematic) into the 'most Western' territory of the "East", which, as the Russian Federation, is now (once again) self-defined as an anti-Western political entity: This process brings up once again the fate of 'Western empire' in the period of the emergence of 'Empire'. And when Payne describes the balancing act of West Germany unambiguously entering the "West" (NATO, etc.) and at the same time accounting for mythical post-Nazi and concrete empirical accounts of the former inhabitants of Ostpreußen, he is getting to the core of the 'clashes of spatiotemporalities' which condition the post-war "order", this in turn being the object of Hardt and Negri's study. One learns here that Hardt and Negri's somewhat simplifying treatment of the totalitarianisms of Central and East Europe as extreme cases of capitalist nationalism might, as Meyer suggests, itself be accused of casting an 'anti-imperial imperial' eye on a field which looks quite different when taking account of a models 'on the ground' in the immediate post-war period. Hoppe notes similar discrepancies in the area of architectural plans and realizations. In this case, as in the whole collection, a constructive critical application of Hardt and Negri necessarily leads to a transcending, indeed in some cases to an outright refutation of a number of their claims. We are certain that, at least in principle, the authors of Empire themselves as critical thinkers would enthusiastically accept such constructive critical endeavors as ours.

---

29 Here, too, we refer to the chapter introduction "Temporal and Spatial Displacement: German and Russian Persons, Names and Cities in Königsberg/Kaliningrad" by Holt Meyer and Charlton Payne (see below, 415–421).

Conclusion

> The brown current ran swiftly out of the heart of darkness, bearing us down towards the sea with twice the speed of our upward progress; and Kurtz's life was running swiftly, too, ebbing, ebbing out of his heart into the sea of inexorable time. [...] It is strange how I accepted this unforeseen partnership, this choice of nightmares forced upon me in the tenebrous land invaded by these mean and greedy phantoms.[30]

Conrad's *Heart of Darkness* at the end of the 19th century (and retracings of it from Said to GoGwilt), like Hardt and Negri at the end of the 20th, shows that there are texts from within the imperial which reveal its uncanniness, elusiveness and at the same time its omnipresence. They are texts in which the imperial deconstructs itself.

Some, like Lenin (the ideological lord and master of Kaliningrad, to pick up on the last section of the book) in his famous treatise about "Imperialism as the Highest Form of Capitalism" still claim to have found the perfect recipe against the imperial and just need to follow that recipe, and others have declared our age to have been "post-colonial" for quite a while and have transported the imperial to another time and a different place than ours (e.g in the case of countries claiming to have not or only minimally participated in colonialism), washing their hands of the infringements and crimes connected to it. We, on the other hand, hope with this volume to show that the traces of empire are very much present, as is Empire itself. Both have many manifestations in need of analysis for which the recognition and comprehension of the imperial in the first place is a prerequisite. This volume will show that picking up on these traces with a view to their spatiotemporal aspects leads to essential insights, and that the cross-section that we have taken is plausibly representative. Of course, ten times as many categories and 'stabs' as chosen here are thinkable, even necessary in order to comprehensively cover anything close to all the bases.

Coming back to the outset of our argumentation, we express our confidence that reaching into the dark area of the history and current presence of the imperial in traces and in practices has opened up new perspectives, also by implementing, going beyond, or in some cases working against Hardt and Negri. We express our hope that by taking these 'stabs' at a spatiotemporal account of empire from a 21st century perspective (with a view both to the present and – mainly – to the past, less with the political view of the future, a concern of Hardt and Negri which these studies in general do not pursue explicitly) will be enlightening for the

30 Conrad, Heart of Darkness, 314.

reader, and that the six approaches can inspire others in further areas of spatio-temporal empire studies to take a 'stab' at accounting for this central category of social sciences, history, sign and media studies and further areas.

References

Achebe, Chinua. An Image of Africa: Racism in Conrad's 'Heart of Darkness'. In *Heart of Darkness: An Authoritative Text, background and Sources Criticism*, Robert Kimbrough (ed.), 251–261. 3rd ed. London: W. W Norton and Co., 1988 (originally Massachusetts Review. 18. 1977).

Alcock, Susan E. et al. (eds.). *Empires: Perspectives from Archaeology and History.* Cambridge: Cambridge University Press 2001.

Bakhtin, Mikhail. Forms of time and of the chronotope in the novel. In *The Dialogic Imagination: Four Essays*. Michael Holquist, Caryl Emmerson (ed. and transl.), 84–258. Austin: Univ. Texas Press.

Burbank, Jane and Frederick Cooper. *Empires in World History: Power and the Politics of Difference*. Princeton, NJ: Princeton University Press 2010.

Conrad Joseph. Heart of Darkness. In *The Portable Conrad*, Michael Gorra (ed.), 277–364. London: Penguin, 2007.

Ferguson, Niall. *Empire: How Britain Made the Modern World.* London: Penguin 2004.

Fox, Chris. From Representation to Constituent Power. Religion, or something like it, in Hardt and Negri's *Empire. Journal for Cultural and Religious Theory* 9.2 (2008): 30–42.

GoGwilt, Christopher. True West: the Changing Idea of the West from the 1880s to the 1920s. In *Enduring Western Civilization*, Silvia Federici (ed.), 37–62. Westort/Conn.: Praeger 1995.

GoGwilt, Christopher. *The Invention of the West. Joseph Conrad and the Double-mapping of Europe and Empire*, Stanford: Stanford Univ. Press 1995.

Hardt, Michael and Antonio Negri, *Multitude. War and democracy in the age of Empire.* New York: The Penguin Press, 2004.

Hardt, Michael and Antonio Negri, *Empire*. Cambridge, MA: Harvard University Press, 2001.

Hardt, Michael and Antonio Negri. *Commonwealth*. Cambridge, MA: Belnap, 2009.

Hardt, Michael and Antonio Negri. *Declaration*. Kindle Edition (epub) 2012.

Hingley, Richard. *Images of Rome. Perceptions of ancient Rome in Europe and the United States in the modern age*. Ann Arbor, Michigan: Cushing-Malloy, 2001.

Lebeau, Christine (ed.). *L'espace du Saint-Empire: du Moyen-Âge à l'époque moderne.* Strasbourg: Presses Univ. de Strasbourg, 2004.

Münkler, Herfried. *Imperien. Die Logik der Weltherrschaft – vom Alten Rom bis zu den Vereinigten Staaten*. 2d ed. Berlin: Rowohlt Verlag, 2005.

Rau, Susanne. *Räume. Konzepte – Wahrnehmungen – Nutzungen*. 2d ed. Frankfurt/Main: Campus, 2014.

Reflecting on Narrative Othering through Imperial TimeSpaces

Robert Fischer

Introduction
Reflecting on Narrative Othering through Imperial TimeSpaces

The 'empire' or the 'imperial' derives from the Latin words 'im' (in, lat. in, within) and 'parare' (paro, lat. supply, obtain, order) which entail the self-understanding of the power asymmetries that are inherent to imperial thinking – to create order where there supposedly is none and to create this order in one's own understanding. The European explorers and colonizers were confronted with a diverse and inexplicable new world that could not be seen as the outside anymore but had to be incorporated in order to situate oneself in the cacophony of other voices. Here narratives emerged that were characterized by European/ Western dominance: The others were depicted as far away in a spatial and/ or temporal matter – at the periphery of the globe or a few steps below on the ladder of societal evolution.[1] They were connected to Western colonial/ imperial powers through established trade networks or settlements, conquered exotic lands or attempts to missionize and educate the timeless other (*mission civilisatrice*, *white man's burden*). In this form the other was subject to daily practices of control that often included subtle and open forms of force. But the other could also be excluded and be exposed to massive forms of violence once the dominators were confronted with resistance. This could result in the annihilation of whole tribes/ populations realizing the idea of the so-called "uninhabited" lands. The imperial setting was embedded in a hierarchical relationship that rested on a binary understanding of the colonizers as the superior self and the colonized as the inferior other.

Consequently the Westerners understood themselves as the beacon of civilization contrary to the dark barbaric other.[2] Multiple dichotomies between the self and the other were created in imperial settings in order to produce difference:[3] the noble savage, the hypersexual black woman, the lazy and dirty Mexican or the deceitful Chinese to name but a few. Westerners however, constructed themselves as the modest white rational being, which was only possi-

1 Walter D. Mignolo. *The Darker Side of Western Modernity. Global Futures, Decolonial Options*. Durham: Duke University Press, 2011; Edward W. Said. *Orientalism*. London: Routledge, 1978.

2 Ann Laura Stoler. *Carnal Knowledge and Imperial Power. Race and the Intimate in Colonial Rule*. Berkeley: University of California Press, 2002, 144.

3 Jane Burbank and Frederick Cooper. *Empires in World History. Power and the Politics of Difference*. Princeton: Princeton University Press, 2010.

DOI 10.1515/9783110418750-002

ble through said distinction from the other. This process that excludes groups as being different from the Western norm systems can be understood in a dialectic as the encounter produced the self and the other simultaneously.[4] Spivak first describes this coeval process in a colonial setting by analyzing the construction of the white emancipated European woman contrary to the colonized woman.[5] In her concept the Other/ other is attributed as the desired as well as the mastered by the different processes of othering whereas the multiplicity of processes hint towards a heterogeneity of the imperial project.[6]

The processes of othering are still to be found in colonial archives, present day imperial settings[7] and in the modern concept of writing about the past itself. The idea of writing history in the nineteenth century originated as a national project that was created in order to develop a national self that could be differentiated from other nation-states. Writing about a distant other meant the reproduction of othering through the discipline of history. Since then scholars have shown in various ways that the writing of history can break the imperial/ eurocentric gaze. So how can a perspective through the study of time and/ or space contribute to evading the reproduction of these narratives? The stated difference can as a starting point be well described in a spatial or temporal way with concepts like Foucault's heterotopy or Fabian's denial of coevalness.[8] Both concepts situate the other outside of a supposedly universal modern spacetime. But there are certain ways to mellow the rigidity of this distinction: For example by moving in and out of time and/ or space scales, by following the daily and the extraordinary life of locals, or by describing the flexibility of the limits.

Thus, borders between supposedly homogeneous entities could be characterized as rather fluid and penetrable by individual actors leading to asymmet-

---

4 Gayatri Chakravorty Spivak. *A Critique of Postcolonial Reason. Toward a History of the Vanishing Present.* Cambridge, Mass.: Harvard University Press, 1999.

5 Gayatri Chakravorty Spivak. Three Women's Texts and a Critique of Imperialism. *Critical Inquiry*, 12:1 (1985), 243–261.

6 Othering. In *Post-Colonial Studies. The Key Concepts*, Bill Ashcroft, Gareth Griffiths and Helen Tiffin (eds.), London: Routledge, 1998, 188–190.

7 Although Hardt and Negri (Michael Hardt and Antonio Negri. *Empire*. Cambridge, Mass.: Harvard University Press, 2000.) argue for homogenizing processes by corporations building a rather subtle notion of empire this does not contradict the concept of Burbank and Cooper's "politics of difference" per se but rather broadens it and opens a way of thinking about empire in a postmodern world. See also concept of "soft power". (Joseph S. Nye. *Soft Power. The Means to Success in World Politics.* New York: Public Affairs, 2004.)

8 Michel Foucault. Of Other Spaces, Jay Miskowiec (trans.). *Diacritics* 16 (1986): 22–27; Johannes Fabian. *Time and the Other. How Anthropology Makes its Object*. New York: Columbia University Press, 1983.

rical interaction as well as resistance and parallel reaffirmation towards their construction. Concepts of alternate behavior towards time or space could be described in a process of interaction making it possible to think spaces and times differently: as in-between, overlapping, intermingling or of parallel existence. The analysis towards a historical anthropology of modern spacetime in Saurabh Dube's contribution reveals that imperial binaries are not to be seen as absolute but they are rather set in a network of asymmetrical power relations making all participants subject to the encompassing setting – not as modern subjects but as subjects of modernity.

Aside from overlapping times and spaces and its porous borders, scales can be of great importance. Zooming into the daily life of colonized communities can show much about the actual conditions of a colonial regime contrary to elements of an official colonial discourse. This way eurocentric perspectives could be evaded as the researcher is capable of considering the Western actors as being the others – being capable of switching between the two perspectives and even holding in-between positions. Ishita Banerjee-Dube shows in her article that the analysis of time on a daily level uncovers the coexistence of different rhythms and time concepts and thus demonstrates that othering is often times produced to enforce difference and (re)establish imperial power relations.

The zooming out also enables one to gain broader perspectives. The study of the rise and the fall of other empires around the world helps to put the Western empires in perspective. But also the consideration of phenomena on a large temporal scale reveals the small time scales we are usually thinking in. Over a span of thousands of years the territorial conquest of the world by Western imperial powers seems rather short and less significant. Following concepts of property as Gesa Mackenthun elaborates in her contribution may enable us to unmask the ephemerality of European dominance and deconstruct the alleged existence of Western universalities and subsequent legitimation. The uncovered voices of the subaltern are capable of breaking a supposed European dominance.

As Alf Lüdtke suggested in his comment at the conference a perspective on the colonial/ imperial as the dark side of the West should not be omitted when referring to the concept of/ the practices within the empire and modernity. Lüdtke stressed that in order to rupture the progressive narrative of civilization[9] it is of utmost importance to identify the thresholds of difference. The distinction between the dominated and the dominating or the daily and the exceptional may be a blurred line yet it is worth examining when trying to go beyond the reflection of the past conditions.

---

9 Norbert Elias. *The Civilizing Process. Sociogenetic and Psychogenetic Investigations*. 1939. Oxford: Blackwell, 2000.

Thus, considerations about the past from a spatiotemporal perspective may focus on imperial and counter-imperial practices as both are products and production sites of the asymmetrical power settings. The masked and changing lines of domination and counter-narratives may be uncovered for example by elaborating on imperial contradictions in its legitimizing and civilizing discourse and its actual barbarism on site. Here Chakrabarty's paradox of European thought as "both indispensable and inadequate"[10] may be disentangled and deconstructed by breaking dichotomies, by changing the perspective, by listening to the unheard, or by following the diversity in center and periphery. By trying to overcome the inadequate the analysis however should not return to the starting point of ordering the other but rather pursue these new ways of analyzing and describing highly asymmetrical interactions.

References

Burbank, Jane and Frederick Cooper. *Empires in World History. Power and the Politics of Difference*. Princeton: Princeton University Press, 2010.

Chakrabarty, Dipesh. *Provincializing Europe. Postcolonial Thought and Historical Difference*. Princeton: Princeton University Press, 2000.

Elias, Norbert. *The Civilizing Process. Sociogenetic and Psychogenetic Investigations*. Oxford: Blackwell, 2000 [1939].

Fabian, Johannes. *Time and the Other. How Anthropology Makes its Object*. New York: Columbia University Press, 1983.

Foucault, Michel. Of Other Spaces, Jay Miskowiec (trans.). *Diacritics* 16 (1986): 22–27.

Hardt, Michael and Antonio Negri. *Empire*. Cambridge, Mass.: Harvard University Press, 2000.

Mignolo, Walter D. *The Darker Side of Western Modernity. Global Futures, Decolonial Options*. Durham: Duke University Press, 2011.

Nye, Joseph S. *Soft Power. The Means to Success in World Politics*. New York: Public Affairs, 2004.

Othering. In *Post-Colonial Studies. The Key Concepts*, Bill Ashcroft, Gareth Griffiths and Helen Tiffin (eds.), 188–190. London: Routledge, 1998.

Said, Edward W. *Orientalism*. London: Routledge, 1978.

Spivak, Gayatri Chakravorty. *A Critique of Postcolonial Reason. Toward a History of the Vanishing Present*. Cambridge, Mass.: Harvard University Press, 1999.

Spivak, Gayatri Chakravorty. Three Women's Texts and a Critique of Imperialism. *Critical Inquiry* 12:1 (1985): 243–261.

Stoler, Ann Laura. *Carnal Knowledge and Imperial Power. Race and the Intimate in Colonial Rule*. Berkeley: University of California Press, 2002.

10 Dipesh Chakrabarty. *Provincializing Europe. Postcolonial Thought and Historical Difference*. Princeton: Princeton University Press, 2000, 16.

Saurabh Dube

Enchantments and Incitements: Modernity, Time/Space, Margins

This paper explores issues of modernity and history, turning on critical considerations of time, space, and their enmeshments. Drawing in socio-spatial subjects and tousled temporalities, it thinks through the enchantments (and oppositions) – as well as incitements (and ambivalences), especially towards pasts and communities – spawned under modernity. Here are to be found pervasive procedures of the temporalization of space and spatialization of time. These rest on routine portrayals of homogeneous time (that is yet founded on inaugural, spatial ruptures) and antinomian blueprints of social space (which nonetheless entail a singular, temporal hierarchy), each one binding the other.[1]

Overture

The idea of modernity rests on rupture. It brings into view a monumental narrative – the breaching of magical covenants, the surpassing of medieval superstitions, and the undoing of hierarchical traditions. The advent of modernity, then, insinuates the disenchantment of the world: the progressive control of nature through scientific procedures of technology; and the inexorable demystification of enchantments through powerful techniques of reason. Indeed, it is possible to argue – along with Martin Heidegger, for example – that the privileged dispensation of legislative reason within regimes of modernity gathers together nature and humanity as conjoint attributes of a disenchanted world.

Yet processes of modernity create their own enchantments. Here are to be found enchantments that extend from the immaculately imagined origins and ends of modernity through to the dense magic of money and markets; and from novel mythologies of nation and empire through to hierarchical oppositions between myth and history, emotion and reason, ritual and rationality, East and West, and tradition and modernity. Intensely spectral but concretely palpable, forming tangible representations and informing forceful practices, the one bound

---

1 This chapter addresses themes that are to find fuller elaboration in Saurabh Dube. *Subjects of Modernity. Time-Space, Disciplines, Margins.* Manchester: Manchester University Press, forthcoming 2017.

DOI 10.1515/9783110418750-003

to the other, such enticements stalk the worlds of modernity's doing and undoing. The enchantments of modernity give shape to the past and the present by ordering and orchestrating these terrains. It is through the "oblique" perspective of the enchantments of modernity – rather than the dominant motif of its disenchantment – that this paper approaches the terms of this volume.[2]

Unsurprisingly, a part of my effort is to track the critical overlaps yet mutual distinctions between modernity, modernization, and modernism. Here, I explore how the developmental idea of a surpassing of the past is central to modern imaginaries – of academic and everyday natures as well as their entwined expressions. At the same time, it should become clear that such segregation of the past from the present, although assumed to be principally temporal, nonetheless embodies profoundly spatial attributes. Thus, the place-holding presumption of a homogeneous history allows an imaginary yet palpable West – its singular temporal trajectory working in tandem with its exclusive spatial location – to become the horizon for the present and posterity of other cultures, which are seen as succeeding or failing to meet their destiny. Yet, historical ruptures also insinuate stubborn knots, which once again irreducibly braid the temporal and the spatial. This is to say that prior places/times, at once anachronistic yet entirely coeval, appear enmeshed with contemporary stages/spaces, thus intimating the tangles, tatters, and textures of the past and the present, the spatial and the temporal. Taken together, these overlapping measures reveal that routine representations of historical, temporal ruptures alongside their hierarchical, spatial distinctions being discussed underlie homologous oppositions between tradition and modernity, ritual and rationality, myth and history, the magical/medieval and the modern, community and state, and East and West. Such matrices require understanding as the enduring enticements of modernity. But here are to be exactly found narratives, oppositions, and enchantments that should not be treated as mere objects of knowledge, which can then be readily discarded or easily overcome. Rather, these stories, antimonies, and seductions need to be approached as key conditions of knowing under modernity.

2 Indeed, precisely recognizing the worldly (or ontological) dimensions of the enchantments (and disenchantments) of modernity, I do not offer here a consideration of the different theories of modernity. Rather, I provide a provocative account of modern disenchantments and their enchantments, raising critical questions around these categories-entities as well as highlighting wider issues of the intricate interplay of history and modernity, time and space, and pasts and communities.

1 Abiding Antinomies

There is something uncannily haunting, unerringly close to home, about modernity's enchantments, now drawing in and now reaching beyond scholarly understandings. Consider the manner in which the term "medieval" bears enormous import for delineations of modernity, an issue that I have discussed elsewhere in relation to imperatives of contemporary politics.[3] The point here is that specters of the medieval – think of the Taliban, of ISIS, of Islam at large, among other examples – darkly delineate practices, beliefs, cultures, faiths, and histories as at once a prior presence *and* an ongoing horror in the mirrors of modernity. They hover in the present in ominous ways.

Why should this be the case? I began by noting that as an idea, ideal, and ideology modernity and the modern appear as premised upon fundamental ruptures: a surpassing of tradition, a break with the medieval.[4] Time after time, in this vision of the past, present, and posterity, an exclusive, imaginary, and bloated West has morphed into history, modernity, and destiny – for each society, any culture, and every people.[5] Even more widely, assiduously plotted against the

3 Saurabh Dube. *After Conversion. Cultural Histories of Modern India*. New Delhi: Yoda Press, 2010, pp. 178–186.

4 This is not deny the complex pasts of the term "modern", whose "conceptual history" in Western Europe, for example, Hans Ulrich Gumbrecht traces in interesting ways – a history that makes clear the articulations of the "modern" with the "ancient", the "classical", and the "romantic." Rather, it is to stay longer with the moment of Gumbrecht's understanding where the concept "modern" yields to the category "modernity", while recognizing that a purely "internal" account of a concept can elide its multiple hierarchies, played out on distinct registers. See Hans Ulrich Gumbrecht's chapter "A history of the concept 'modern'" in his book: Hans Ulrich Gumbrecht. *Making Sense in Life and Literature*, Glen Burns (trans.). Minneapolis: University of Minnesota Press, 1992.

5 This pervasive, "metageographical" projection appears elaborated in several ways, from the evidently aggressive to the seemingly benign, embedded of course in "modernization" theory, yet also long lodged within the interstices of Western social and political thought. The way all this might come together is evident in the following statement of Gumbrecht: "From our perspective at least, modernization in the underdeveloped countries is ... taking place somewhere between decolonization and our own present." The "stagist" presumptions of time and space here are not so far apart from the wide-ranging elisions of authoritative accounts – for example, by Anthony Giddens and Jürgen Habermas – that see modernity as a self-generated, European phenomenon. As I discuss later, the projection also finds contradictory articulations within discrete expressions of "tradition" that question "modernity" by reversing the moral import of its constitutive hierarchies and oppositions. To consider the enchantments of modernity is to think through such oppositions, hierarchies, and elisions. Gumbrecht, *Making Sense*, 108; Anthony Giddens. *Consequences of Modernity*. Stanford: Stanford University Press, 1990; and

horizon of a singular modernity, distinct meanings, practices, and institutions appear as primitive or progressive, lost or redeemable, savage or civilized, barbaric or exotic, ever-behind or nearly-there, medieval or modern.

These peoples have missed the bus of universal history, or they hang precariously from one of its symmetrical sides. Patiently or impatiently, they still wait for the next vehicle plying the road of modernity. Comfortably or uncomfortably, they now sit within this transportation of historical time, this vehicle traversing social space. Their distance from the modern registers redemptive virtue or their falling behind on this route reflects abject failure.[6] Rather more than ideological errors, awaiting their inexorable exorcism through superior knowledge, such mappings circulate as structures of feeling, instituted as categorical entities, intimating the measures and the means of the modern – which is to say, they are abiding enchantments of modernity.

From where do such hierarchal oppositions and their immense enchantments arise? For a long time now, formidable antimonies between static, traditional communities and dynamic, modern societies have played an important role in understandings of history and culture.[7] At first, the duality might seem to be little more than an ideological plank of modernization theory, counter-posing (primarily non-Western) tradition with (chiefly Western) modernity. But the antinomy has wider implications and deeper underpinnings.[8] It is not only that the duality has animated and articulated other enduring oppositions, such as those between ritual and rationality, myth and history, community and state,

Jürgen Habermas. *The Philosophical Discourse of Modernity. Twelve Lectures*, Frederick Lawrence (trans.). Cambridge, Mass.: MIT Press, 1987.

6 Even as prior enchantments can appear as an antidote to a disenchanted modernity, so too logics of "exclusion" and terms of "inclusion" bind each other within the temporal and spatial hierarchies of modernity. Keeping this in view, works that initially influenced my own emphases here include Uday Mehta. *Liberalism and Empire. A Study in Nineteenth-Century British Liberal Thought*. Chicago: University of Chicago Press, 1999; Dipesh Chakrabarty. *Provincializing Europe. Postcolonial Thought and Historical Difference*. Princeton: Princeton University Press, 2000; and John Comaroff and Jean Comaroff. *Ethnography and the Historical Imagination*. Boulder: Westview Press, 1992. See also, Michael Taussig. *Shamanism, Colonialism, and the Wild Man. A Study in Terror and Healing*. Chicago: University of Chicago Press, 1987.

7 This section draws upon yet also develops further arguments first presented in Saurabh Dube, "Anthropology, history, historical anthropology", in Saurabh Dube. Anthropology, history, historical anthropology. In *Historical Anthropology*, Saurabh Dube (ed.), 1–73. New Delhi and New York: Oxford University Press, 2007.

8 For a wider discussion of the mappings of the traditional and the modern see Saurabh Dube. *Stitches on Time. Colonial Textures and Postcolonial Tangles*. Durham and London: Duke University Press, 2004.

magic and the modern, and emotion and reason. It is also that as a lasting legacy of the developmental idea of universal, natural history and an aggrandizing representation of an exclusive, Western modernity, such oppositions have found varied expressions among the distinct subjects that they have named, described, and objectified since at least the eighteenth century.[9] At stake, indeed, are mappings of time and orderings of space, which substantialize both (space and time) in antinomian ways.

Representations emanating from the European Enlightenment have played a key role here. Now, it would be hasty and erroneous to see the European Enlightenment of the seventeenth and eighteenth centuries as all of a piece. From contending strains of rationalism in France and of empiricism in Britain through to different conceptions of universal and natural history, it is more useful to speak in the plural of *Enlightenments*.[10] Here were to be found, too, challenges to rationalist procedures through varieties of *Counter-Enlightenments*, which shaped the Enlightenment.[11] Despite such plural procedures, it has been generally accepted that the period of the Enlightenment was accompanied and marked by ideas and processes of the secularization of Judeo-Christian time.[12] Actually, such secularization of Judeo-Christian time during the Enlightenment was an emergent and consequential idea, but a circumscribed and limited process.[13]

In this context, discrete yet overlaying developmental schemes underwrote grand designs of human history, from the rationalist claims of Voltaire and Kant through to the historicist frames of Giambattista Vico and Johann Gottfried von Herder. There was profound contention among such schemas, yet in different

9 This is not to deny prior formations of the modernity of the Renaissance and the New World, issues discussed, for example, in Saurabh Dube and Ishita Banerjee-Dube (eds.). *Unbecoming Modern. Colonialism, Modernity, Colonial Modernities.* New Delhi: Social Science Press, 2006.

10 Roy Porter. *The Creation of the Modern World. The Untold Story of the British Enlightenment.* New York: Norton, 2001; John G. A. Pocock. *Barbarism and Religion.* Vol. Two, *Narratives of Civil Government.* Cambridge: Cambridge University Press, 1999; and Donald R. Kelley. *Faces of History. Historical Inquiry from Herodotus to Herder.* New Haven: Yale University Press, 1998. See also, Sankar Muthu. *Enlightenment against Empire.* Princeton: Princeton University Press, 2003.

11 Isaiah Berlin. *Against the Current. Essays in the History of Ideas.* Princeton: Princeton University Press, 2001, 1–24; and Darrin M. McMahon. *Enemies of the Enlightenment. The French Counter-Enlightenment and the Making of Modernity.* New York: Oxford University Press, 2002.

12 See, for example, Johannes Fabian. *Time and the Other. How Anthropology makes its Object.* New York: Columbia University Press, 1983, especially 26–7, 146–7.

13 I discuss these issues in much greater detail in Dube, Anthropology, history, historical anthropology. Although even in terms purely of eighteenth-century Enlightenment thought consider the emphases of Carl L. Becker. *The Heavenly City of the Eighteenth-century Philosophers.* New Haven: Yale University Press, 1932.

ways they each projected developmental blueprints of universal history, turning on space and time.[14] Such contrary strains and convergent emphases were bound to the fact, many times overlooked, that the Enlightenment was as much historical as philosophical, as much about the rewriting of history as about the rethinking of philosophy. The consequences were limited yet significant. On the one hand, throughout the nineteenth century but also afterwards Judeo-Christian and Messianic time, temporality, and *telos* – and the spatial imaginaries on which they rested – did not lose their influence in Western worlds.[15] On the other hand, by the second half of the nineteenth century, at the very least in the Protestant West, secularized time could acquire a naturalized aura and developmental thought was distilled (uncertainly yet potently) as historical progress, each seeking to transform spatially segregated worlds in their image and wake.[16]

It followed that time and space, articulated in tandem, came to be increasingly mapped in hierarchical ways to plot peoples and cultures in the movement of history that was primarily projected as the passage of progress. Frequently articulated by the *Ur*-opposition between the primitive and the civilized, in place here nonetheless was neither a singular Western "self" nor an exclusive non-Western "other". Rather, at play in this terrain were the cultural severalty of Western selves and the historical hierarchies of non-Western otherness. In this scenario, many peoples (for example, Africans, African-Americans, and indigenous groups in the Americas and across the world) were still stuck in the stage of barbarism and savagery with few prospects of advancement. Other societies (for example, those of India and China) had reached the ascending steps of civilization yet lacked the critical foundations of reason. Still other people (chiefly of Western and Northern European stock) had evolved to the higher reaches of humanity through advantages of race and rationality and propensities of history and nationality. Indeed, it was the past and the present of this last set of people, comprising the enlightened European elect, that was seized on and rendered as a looking glass at large. In this mirror was envisioned the universal history of

14 Put differently, not only rationalist, analytical schemes but contending historicist, hermeneutic traditions articulated in distinct ways the terms of developmental, universal history, its projections of time and space. Kelley, *Faces of History*, 211–262.

15 Concerning only the United States and its Puritan model of the secularization of the world, consider R. Laurence Moore. *Touchdown Jesus. The Mixing of Sacred and Secular in American History.* Louisville, Kentucky: Westminster John Know Press, 2003; and Vincent Crapanzano. *Serving the Word. Literalism in America from the Pulpit to the Bench.* New York: New Press, 2000.

16 I am neither attributing an inexorable quality to these developments nor denying that such processes have possessed contradictory and critical pasts. See, for example, Fabian, *Time and the Other*, 12–16.

human destiny – a destiny represented as groups and societies, caught in warps of space and vortexes of time, either failing before or rising to the stage of modernity, cast as spatial and temporal apex.

It was registered earlier that the notion and narrative of modernity involve a break with the past, a carving up of space: here, stories of modernity ever intimate ruptures with ritual and magic and breaches with enchantment and tradition, setting to work procedures of the temporalization of space and the spatialization of time. Following authoritative understandings, as an epochal concept, modernity has been seen as embodying a distinct and new status from preceding periods. Two immensely influential, contemporary discussions explicating the critical attributes of modernity should suffice here.

The philosopher Jürgen Habermas has suggested that under modernity the notion of the "new" or the "modern" world loses a "merely chronological meaning" to take on instead "the oppositional significance of an emphatically 'new' age." It follows from this that the normative order of modernity has to be grounded out of itself, rather than drawing its dispositions from models offered by other, obviously earlier, epochs.[17] Similarly, the historian Reinhart Koselleck has argued that, starting in the eighteenth century the regimes of historicity under modernity have entailed a series of homologous disjunctions between the past and the present, prophecy and prediction, and eschatological imaginings and secular visions. This is to say that modernity innately insinuates novel orientations to the past, present, and future.[18]

Now, these are persuasive arguments that carry their own truths. But they also principally present modernity in idealized terms, overlooking their own implicit articulations of time and space. At the same time, precisely for these reasons, the understandings are acutely representative. None of this should be surprising. For the persuasions and truths of such arguments and their presentation of modernity in idealized terms are inextricably entwined with each other. Indeed, at stake here is nothing less than the abiding enchantments of modernity.

First, influential and commonplace explications of modernity have for a very long time now proceeded by locating its constitutive terms as being entirely internal to an imaginary yet tangible space-time called Europe/West. This is to say that they have understood modernity as phenomena generated purely internally within the West. Produced within this spectral yet palpable Europe, it was only later that modernity was variously exported to other parts of humanity. Now,

17 Habermas, *Philosophical Discourse of Modernity*, 5.

18 Reinhart Koselleck. *Futures Past. On the Semantics of Historical Time*, Keith Tribe (trans.). Cambridge: MIT Press, 1985, especially 3–20.

precisely this measure serves to override dynamics of colonizer and colonized, race and reason, Enlightenment and empire, and indeed space and time. It is these dynamics that been constitutive of the terms and textures of modernity as history.[19]

Second, processes of modernity have been all too frequently understood by being sieved modernity's through resolutely overarching filters the necessarily uneven historical processes that have attended emergence(s) and development(s). From history to sociology and philosophy, modular designs of modernity are assumed in place more or less *a priori*, which then provide the means with which to approach, analyze, and apprehend the causes, characteristics, and consequences – as well as the terms, terrains, and trajectories – of the phenomena, spreading across the world in pre-determined ways.[20] This has served to subordinate the everyday manifestations and critical margins of modernity – entailing of course key co-ordinates of space, time, and their regular and irregular productions – further underplaying its contentions and contradictions, in Western and non-Western worlds.

Third and finally, representations and definitions of modernity – and its attendant processes such as secularization as well as its cognate concepts such as liberty – have entailed a ceaseless interplay between their ideal attributes and their actual manifestations. This has meant not only that the actual has been apprehended in terms of the ideal, but that even when a gap is recognized between the two the actual is seen as tending toward the ideal with each shoring up the other. At stake are more than simple errors of understanding, since it is exactly the admixtures of the actual articulations and the idealized projections of modernity that have defined its worldly dimension.[21] Taken together, these pro-

---

19 Consider that in some ways this problem continues to characterize Charles Taylor's sensitive and suggestive, truly remarkable recent opening up of the terms, terrains, and trajectories of modernity. Charles Taylor. *Modern Social Imaginaries*. Durham: Duke University Press, 2005.

20 Instead of providing instances of the seemingly endless writings that embody such procedures in straightforward ways, let me take up a more critical example. In their influential text on history-writing, Joyce Appleby, Lynn Hunt, and Margaret Jacob adopt a dryly-ironic, gently-mocking tone toward what they call "a heroic model of science" of the Enlightenment – one they see as shaping and structuring modern knowledge, especially history. Yet, not only do these authors insufficiently probe the plurality of Enlightenment traditions but for the most part their own understanding of knowledge schemes rests upon the heroic model that they pillory. Here are to be found exactly the enchantments of modernity under discussion. Joyce Appleby, Lynn Hunt, and Margaret Jacob. *Telling the Truth About History*. New York: W. W. Norton, 1995. Such difficulties extend from history through to philosophy, and I discuss the place of modular assumptions of modernity as shoring up the important, critical work of Habermas in Saurabh Dube. *After Conversion. Cultural Histories of Modern India*. New Delhi: Yoda Press, 2010.

21 Elsewhere, I have discussed the active interchange between the "ought" and the "is", the "ideal" and the "real", especially in relation to propositions of the secularization of the world.

cedures announce salient registers of hierarchical mappings of time and space. In both conscious and inadvertent ways, such registers entail two simultaneous measures. Rehearsing the West as modernity, they equally stage modernity "*as* the West."[22]

The idea of modernity as a coming apart from the past rests on the imagination of ruptures within Western history, that terrain of the pre-modern intimating different, diminished co-ordinates of time and space. But such an idea cannot help also turning on the importance of disjunctions of the West with non-Western worlds, a categorically distinct, lower space-time, whether explicitly or implicitly. On the one hand, the caesura defined by modernity as the new beginning is shifted into the past, "precisely to the start of modern times" in Europe.[23] It is ahead of this threshold that the present is seen as being renewed in its vitality and novelty under modernity. On the other hand, exactly when the modern is privileged as the most recent period, the novelty and vitality of modernity confront specters of the "medieval", the "superstitious", the "prophetic", and the "spiritual" meandering in their midst. These spirits are a prior presence *and* an ongoing process. Each attempt to engage them in the present entails marking them as an attribute of the past. My reference is to the ways in which in dominant representations, the Taliban, Al-Qaeda, Hamas, or ISIS are simultaneously "coeval" and "medieval"; and the manner in which in pervasive understandings the importance today of "indigenous spirituality" and "local tradition" is at once contemporary yet anachronistic.

I am suggesting, then, that the meanings, understandings, and actions that fall outside the disenchantment-driven horizons of modernity have to be plotted as lagging behind this novel stage. Here spatial mappings and temporal measurements of the West and the non-West come to rest on the trajectory of time, an axis that claims to be normatively neutral but is in fact profoundly hierarchical. This is to say that the precise notion of modernity as a rupture with the past carves up social and historical worlds into the traditional and the modern, further naming

---

The point is that instead of considering either as a mere straw-figure, it is important to track how the interlacing of these propositional forms has under-girded not only academic and everyday understandings but equally the social worlds that they seek to explain. Saurabh Dube, Modernity and its Enchantments. An Introduction. In *Enchantments of Modernity. Empire, Nation, Globalization*, Saurabh Dube (ed.), 1–43. London: Routledge, 2010.

22 Timothy Mitchell. The stage of modernity. In *Questions of Modernity,* Timothy Mitchell (ed.), 1–34. Minneapolis: University of Minnesota Press, 2000, 15, emphasis in the original.

23 Habermas, *Philosophical Discourse of Modernity*, 5. Saliently, Habermas is summarizing significant historical and philosophical writings here.

and animating other oppositions such as those between ritual and rationality, myth and history, and magic and modernity.

Why should the antinomies of modernity have played an important role in the mapping and making of social worlds? These oppositions emerged embedded within formidable projects of power and knowledge, turning on Enlightenment, empire, and nation as well as within the challenges to these projects. These have been motivated if diverse projects "not simply of looking and recording but of recording and remaking" the world, as Talal Asad tells us.[24] Unsurprisingly, the oppositions themselves assumed persuasive analytical authority *and* acquired pervasive worldly attributes, variously articulated with representations of modernity and its trajectory as a self-realizing project of progress and a self-evident embodiment of history. As worldly knowledge, then, these neat proposals, abiding oppositions, and their formative presumptions entered the lives of historical subjects – formidably disseminated as ways of approaching and modes of apprehending social worlds, they have appeared equally instituted as tissues of affect and textures of experience within everyday arenas. Unsurprisingly, their critical questioning notwithstanding, these oppositions continue to beguile and seduce.[25]

---

24 Talal Asad. *Genealogies of Religion. Discipline and Reasons of Power in Christianity and Islam.* Baltimore: The Johns Hopkins University Press, 1993, 269.

25 A cautious clarification is in order at this point. While it is entirely salient to register the place of the abiding oppositions of modernity in the molding of social worlds, it is equally important to attend to the contending elaborations of the analytical, ideological, and everyday separation between enchanted or traditional cultures and disenchanted or modern societies, turning on space and time. The contentions are present at the core of post-Enlightenment thought and non-Western scholarship, each including critiques of the West in the past and the present. Indeed, the actual elaborations of the hierarchical oppositions of modernity have imbued them with contradictory value and contrary salience. Here are to be found ambivalences, ambiguities, and excesses of meaning and authority. All of this is registered by the *particular* unraveling of divergent traditions of understanding and explanation at the heart of modernity as ideology and history. My reference is to the opposed tendencies that have been described as those of rationalism and historicism, of the analytical and the hermeneutical, and of the progressivist and the romantic. It is critical to track the frequent combination in intellectual practice of these tendencies in order to trace the contradictions and contentions and ambivalences and excesses of modern knowledge(s). Amidst else, such measures allow us to guard against what the Haitian historical anthropologist Michel Rolph-Trouillot has described as our "contemporary arrogance" to overplay the uniqueness of our own times and knowledge. Michel Rolph-Trouillot. North Atlantic Universals: Analytical Fictions, 1492–1945. In *Enchantments of Modernity. Empire, Nation, Globalization*, Saurabh Dube (ed.), 45–66, quotation on p. 46. London: Routledge, 2010.

2 Untangling Modernity

The pervasive presence of modern oppositions, including in intellectual arenas, derives in no small measure from the manner in which modernity is often elided with modernization, and at other times folded into modernism. As is generally known, the notion of modernization, as expressed by its different theorists/theories, refers to modular, temporal-spatial, projections of material, organizational, and technological – as well as economic, political, and cultural – transformation(s), principally envisioned in the look-glass of Western development. Here, different, often hierarchically ordered, societies are seen as succeeding (or failing) to evolve from their traditional (or pre-modern) states through linear stages of succession to become modernized (or capitalist) arenas.[26] Now, the simplistic, step-by-step, spatial schemas and the reductive, totalizing, temporal templates of modernization theories have always been far too tendentious.[27] And so, too, have they been decisively questioned and firmly rejected by critical scholarship for some time now. Yet, motifs of modernization have also crucially carried wide resonance, easily elided with mappings of modernity, such that each shores up the other.

Why should this be the case? To begin with, as was just discussed, a crucial characteristic of dominant discourses concerning Western modernity has hinged on their positing of the phenomena as marked by a carving up of space and time, a break with the past, a rupture with tradition, a surpassing of the medieval. In this scenario, the blueprints of modernization have actually distilled the designs of modernity, the aggressive spatial assumption of the latter holding in place the schematic temporal prognosis of the former. Taken together, modernity's discourses and modernization theories, inextricably entwined, the one with the other, have articulated an imaginary but palpable, distended and aggrandizing West/Europe as history, modernity, and destiny, the telos of time and space – for each society, culture, and people.

Yet, there is more to the picture. Beyond routine representations, in artistic, intellectual, and aesthetic arenas, each understood broadly, modernity has often appeared in intimate association with its cognate (or conceptual cousin),

---

26 Here, especially influential statements included, W. W. Rostow. *The Stages of Economic Growth. A Non-Communist Manifesto.* Cambridge: Cambridge University Press, 1960; David E. Apter. *The Politics of Modernization.* Chicago: University of Chicago Press, 1965.

27 I acknowledge that reassessments of modernization have emphasized the place of "tradition" in elaborations of "development", for example. But such understandings continue to be based on the enduring oppositions – and teleological templates of space-time – of discourses of modernity.

modernism. Now, modernism is also an enormously contentious term that necessarily follows from the contested and contradictory character of the tendencies it describes. Here are to be found cultural movements, styles, and representations, going back to the mid-nineteenth century and extending into our own times, which have been diversely expressed and performed in different parts of the world. Following Theodor Adorno, modernism has been a principally "qualitative" rather than a merely "chronological" category:[28] but it is also the case that the internal endeavors within modernisms to surpass the past, articulate the present, and envision the future have been intrinsically heterogeneous ones. They have variously engaged and interrogated, accessed and exceeded Enlightenment thought and Romantic tradition, abstract reason and religious truth, surface coherence and tonal depth, Western representations and pre-colonial narratives, the certainties of science and the presence of God, and governmental authority and popular politics.[29] All of this raises intriguing issues of the configurations of time and space within the ideational articulations and aesthetic practices of modernism(s).

On the one hand, from Charles Baudelaire's avowal of the "transitory" and the "fleeting" through to modernist rejections of realism and replication in favor of discontinuity and disruption, and from Ezra Pound's invitation to art to "make new" through to the many manifestations of modernisms flowing from the mid-twentieth century (and before), a key characteristic of these cultural tendencies has been to emphasize the difference of the contemporary present from past epochs. On the other hand, as Peter Childs has argued, modernism has always involved "paradoxical if not opposed trends towards revolutionary and reactionary positions, fear of the new and delight at the disappearance of the old, nihilism and fanatical enthusiasm, creativity and despair."[30] Now, to hold together the discourses of modernity and the articulations of modernism is not only to trace the interleaving yet distinct ways in which they each offer a cessation and overcoming of the past. It is also to register that the constitutive contradictions and contentions of modernism(s) can hold a mirror up to the characteristics, contingencies, contentions, and co-ordinates, especially of space and time, of that acutely authoritative universal – modernity.

---

28 Theodor Adorno. *Minima Moralia*. London: Verso, 2005, 218.

29 A parallel here is the manner in which opposed tendencies of modern knowledge – defined as those of rationalism and historicism, of the analytical and the hermeneutical, and of the progressivist and the romantic – can frequently combine in intellectual practice, leading to contradictions and contentions and ambivalences and excesses. Dube, Anthropology, History, Historical Anthropology, 10–15; see also, Dube, *After Conversion*.

30 Peter Childs. *Modernism. The New Cultural Idiom*. New York: Routledge, 2000, 17.

To approach the entanglements between modernity, modernism, and modernization in this manner, where the one is not simply folded into the other yet their mutual linkages are adequately acknowledged, might have critical consequences. Building on my own prior proposals, modernity is now understood not only as a forceful idea and ideology but as also entailing heterogeneous histories and plural processes.[31] These imaginings and procedures extend back to the last five centuries and interlock in critical ways, such that both models of modernization and movements of modernism appear as crucial components yet small parts in the broader articulation of modernity. There are at least two faces to the phenomena, each insinuated in the other. Intrinsic to each is a querying of pervasive procedures of the temporalization of space and the spatialization of time.

On the one hand, as part of a familiar picture, constitutive of modernity are processes of reason and science, industry and technology, commerce and consumption, nation-state and citizen-subject, public spheres and private spaces, and secularized religion(s) and disenchanted knowledge(s). It warrants emphasis that vigilance is required regarding the endless unfolding of these developments as heroic histories. Indeed, instead of teleological tales of the march of modernization/modernity, such stories require being unraveled as rather more checkered narratives, even as models of modernization are registered as part of the protocols of modernity.

On the other hand, although this is often overlooked, at the core of modernity are also processes of empires and colonies, race and genocide, resurgent faiths and reified traditions, disciplinary regimes and subaltern subjects, and seductions of the state and enchantments of the modern. Lessons learned from the split, Janus-faced nature of modernism assume salience here. This is to say simultaneously that the ceaseless portrayals of modernity as embodying a singular, seamless trajectory are actually shored up by hierarchical presumptions and antinomian projections of space and time; and that, unsurprisingly, procedures of modernity have been contradictory, contingent, and contested – protocols that are incessantly articulated yet also critically out of joint with themselves.[32]

It is precisely these procedures that emerge expressed by subjects of modernity. Here, my reference is to historical actors who have been active participants in processes of modernity: social actors who have been both *subject to* these processes but also *subjects shaping* these processes. Over the past few centuries, the subjects of modernity have included, to take just a few instances, peasants,

---

31 See, for example, Dube, *Stitches on Time*; Dube, *After Conversion*; and Dube (ed.), *Enchantments of Modernity*.

32 Dube, *Stitches on Time*, particularly 11.

artisans, and workers in South Asia that have diversely articulated processes of colony and post-colony; indigenous communities in the Americas under colonial and national rule; peoples of African descent not only on that continent but in different Diasporas across the world; and, indeed, subaltern, marginal, and elite women and men in non-Western and Western theatres. Unsurprisingly, these subjects have registered within their measures and meanings the formative contradictions, contentions, and contingencies of modernity.[33]

I am suggesting that at stake in this discussion of subjects of modernity are key questions of heterogeneous yet coeval temporalities and overlapping but contending productions of space. First, it is well known that conceptions of modernity generally proceed by envisioning the phenomenon in the image of the European and Euro-American (frequently, implicitly, male) modern subject.[34] On the contrary, I am indicating the inadequacy of conflating *the modern subject* with the *subject of modernity*. Is it perhaps the case then that my articulation of subjects of modernity productively widens the range of address of modernity and its participants? And that it does this by querying the hierarchies and antinomies of time and space that underlie formidable projections of routinely timeless traditions and an endlessly dynamic modernity? Moreover, mine is not a chronological claim that everyone living in the modern age counts as a modern subject. For, subjects of modernity have revealed, again and again, that there are different ways of being modern, now accessing and now exceeding the stipulations of the Western modern subject. Yet, all too often, in fashioning themselves, subjects of modernity have also barely bothered about the Western modern subject exactly while articulating the enduring terms of modernity. What are the implications of such recognition for weaving in distinct textures and transformations of affects and subjectivities – including inherently plural experiences, articulations, and elaborations of time, space, and their enmeshments – in considerations of modernity?[35] Finally, it bears emphasis that there are other modern subjects besides Western ones, embodying formidable heterogeneity yet coevalness –of the temporal and the spatial, the affective and the subjective. Does this not suggest the need in discussions of modernity to rethink exclusive images of the modern

---

33 Saurabh Dube. Modernity and its Enchantments. An Introduction. In *Enchantments of Modernity. Empire, Nation, Globalization*, Saurabh Dube (ed.), 1–43. London: Routledge, 2010.

34 I am developing here ideas that were first initiated in Dube, *Stitches on Time*.

35 Indeed, all of this is to emphasize, too, the importance of affect and subjectivity – long privileged within modernism(s) – in explorations of modernity. Yet, it is to do so while refusing to approach affect(s) as the repressed other of the modern as well as eschewing an understanding of subject(s) as sovereign of modernity.

subject, both in non-Western arenas and in Western ones?[36] This also brings me to the next step of the paper, involving issues of history and community.

3 Pasts and Communities

My bid now is to concretize some of what I have been saying about the temporal and the spatial, their mutual hierarchies and denials, their shared containments and constitution by turning to the critical rethinking of history and community at the core of recent scholarship, especially historical anthropology. On the one hand, members of the South Asian subaltern studies collective[37] – alongside other intellectuals, focusing on diverse geo-political areas[38] – have pointed to the place of power in the production of the past. This has served to underscore the inherently political character of history writing while putting a question mark on the very nature of the academic-historical archive. On the other, scholars of anthropology, history, and other disciplines have emphasized the precise plurality of cultural pasts, the manner in which history and temporality are differently approached and understood and seized upon and set to work by distinct social groups in conversation with their identities.[39]

---

36 These various modern subjects in the West and the non-West are also subjects of modernity. But, once more, not all subjects of modernity are modern subjects, of course. At any rate, I hope it is clear that the dispositions to modernity that I am outlining do not claim to comprehensively define this category, entity, and process. Rather, my bid is to open up spaces and suggest resources for discussing procedures of modernity and their many persuasions.

37 Ranajit Guha. *Elementary Aspects of Peasant Insurgency in Colonial India.* Delhi: Oxford University Press, 1983; Ranajit Guha. *Dominance without Hegemony. History and Power in Colonial India.* Cambridge, Mass.: Harvard University Press, 1997; Chakrabarty. *Provincializing Europe*; Gyanendra Pandey. *Remembering Partition. Violence, Nationalism and History in India.* Cambridge: Cambridge University Press, 2001.

38 Michel-Rolph Trouillot. *Silencing the Past: Power and the Production of History.* Boston: Beacon Press, 1995; David William Cohen. *The Combing of History.* Chicago: University of Chicago Press, 1994; Kerwin Lee Klein. *Frontiers of Historical Imagination. Narrating the European Conquest of Native America, 1890–1990.* Berkeley: University Of California Press, 1999. Also see, Saidiya H. Hartman. *Scenes of Subjection. Terror, Slavery, and Self-Making in Nineteenth-Century America.* New York: Oxford University Press, 1997; Saidiya H. Hartman. *Lose Your Mother. A Journey Along the Atlantic Slave Route.* New York: Farrar, Straus and Giroux, 2007.

39 For example, Renato Rosaldo. *Ilongot Headhunting 1883–1974. A Study in Society and History.* Stanford: Stanford University Press, 1980; Joanne Rappaport. *Cumbe Reborn. An Andean Ethnography of History.* Chicago: University of Chicago Press, 1994; Nancy Florida. *Writing the Past, Inscribing the Future. History as Prophecy in Colonial Java.* Durham: Duke University Press, 1995;

Three overlaying emphases have played a salient part in such considerations.[40] To begin with, it has been diversely admitted that forms of historical consciousness vary in their degree of symbolic elaboration, their ability to pervade multiple contexts, and their capacity to capture people's imaginations – between and across cultural groups and their identities. Second, it has been increasingly noted that history does not just refer to events and processes out there, but that it exists as a negotiated resource at the core of shifting configurations of historical worlds and social identities. Third and finally, there has been an opening up of critical questions considering the coupling of history writing with the modern nation and of the haunting presence of a reified "West" in widespread beliefs in historical progress. (The importance of these moves for the ways in which I approach configurations of time and space should be clear, and I shall not belabor the points.)

Together, in approaching the past and the present, such efforts toward critical history writing have often bound the impulse to cautiously probe and affirm social worlds with the desire to carefully narrate and describe them. The endeavors have taken truly seriously requirements of evidence and fidelity to facts. Yet they have also sieved historical evidence through critical filters and construed facts unexpected – facts which speak in the uneasy echoes of limiting doubt rather than deal in dead certainties.[41] It only follows that the emphases outlined above have not resorted to oppositions involving cyclical notions of the past as characteristic of the East and linear conceptions of history as constitutive of the West. Nor have they approached the assertive appropriations and enunciations of the past in historical and contemporary worlds by submitting to views that each of these visions is equally true. Rather, they have precisely probed such overwrought blueprints and solipsistic schemes by tracking expressions of history as made up of interleaving, conflict-ridden processes of meaning and authority, ever entailing identity and alterity.[42]

David William Cohen and E.S. Atieno Odhiambo. *Siaya. Historical Anthropology of an African Landscape.* Cleveland: Ohio University Press, 1989; Richard Price. *First-Time. The Historical Vision of an Afro-American People.* Baltimore: Johns Hopkins University Press, 1983; Richard Price. *Alabi's World.* Baltimore: Johns Hopkins University Press, 1990; Luise White. *Speaking with Vampires. Rumor and History in Colonial Africa.* Berkeley: University of California Press, 2000.

40 Dube, Anthropology, history, historical anthropology.

41 Peter Redfield. *Space in the Tropics. From Convicts to Rockets in French Guiana.* Berkeley: University of California Press, 2000; Dube, *Stitches on Time.*

42 Vinayak Chaturvedi. *Peasant Pasts. History and Memory in Western India.* Berkeley: University of California Press, 2007; Prachi Deshpande. *Creative Pasts. Historical Memory and Identity in Western India, 1700–1960.* New York: Columbia University Press, 2007; Shail Mayaram. *Against*

In this terrain, the explorations have extended from tracing the variability and mutability that can inhere in the perceptions and practices of the past of historical communities through to tracking the uses of the past and their contending validities in the making of identities, especially the play of power in the production of history. In elaborations of these conjoint emphases, particularly pertinent are Shahid Amin's innovative account of the interplay between governmental demands and subaltern desires in the remembering and monumentalizing of a critical event of Indian nationalism in north India across the twentieth century;[43] Ajay Skaria's thickly textured study of wildness, environment, gender, and politics among the Dangis of western India, especially as based on these people's historical narratives of "colonial" and "extra-colonial" times;[44] and Ishita Banerjee-Dube's imaginative enquiry into the unfolding of oral and written histories and sectarian and ascetic formations – each inflected by the presence of the law and the state – within a popular religious formation in eastern India from the mid-nineteenth century through to the present.[45]

All of these writings have variously combined historical fieldwork and ethnographic archival research. Unsurprisingly, they have been accompanied by analyses that have not only unraveled the persistence of oppositions between myth and history in authoritative projections but precisely placed question marks on pervasive projections of the West and nation as history, modernity, and destiny – for all people and every identity, as was noted earlier. Important examples of such work reside in the challenges posed by Dipesh Chakrabarty's forceful, philosophical critique of the developmental premises of "historicist" thinking as well as by Gyanendra Pandey's critical considerations of the formidable violence that is at once embodied and ignored, made routine and glossed over, by the modern

---

History, Against State. Counterperspectives from the Margins. Delhi: Permanent Black, 2004; Richard Price. *The Convict and the Colonel. A Story of Colonialism and Resistance in the Caribbean.* Boston: Beacon Press, 1998; Ann Gold and Bhoju Ram Gujar. *In the Time of Trees and Sorrows. Nature, Power, and Memory in Rajasthan.* Durham: Duke University Press, 2002; Saurabh Dube. *Untouchable Pasts. Religion, Identity, and Power among a Central Indian Community, 1780–1950.* Albany: State University of New York Press, 1998; Trouillot, *Silencing the Past.*

43 Shahid Amin. *Event, Metaphor, Memory. Chauri Chaura 1922–1992.* Berkeley: University of California Press, 1995.

44 Ajay Skaria. *Hybrid Histories. Forest, Frontiers and Wildness in Western India.* New Delhi: Oxford University Press, 1999.

45 Ishita Banerjee-Dube. *Religion, Law and Power. Tales of Time in Eastern India, 1860–2000.* London: Anthem Press, 2007.

coupling of nation and history.[46] Clearly, all of this raises critical questions for our deliberations in this volume of time, space, and their enmeshments.

No less than history, the acute rethinking of community has been at the core of recent critical scholarship. Here, too, there has been a braiding of two apparently incommensurable yet actually complementary emphases. On the one hand, several scholars associated with subaltern studies, particularly Partha Chatterjee,[47] have underscored the key role of the community as an ethical formation in questioning and challenging projects of power – of colony and empire and nation and history. On the other hand, distinct strands of critical scholarship have queried persistent portrayals of the community as an ineluctably anachronistic, tightly bounded entity, caught in curious warps of space and time – community as tending toward consensus in its expression, entailing allegiance to primordial tradition, and as broadly opposed to modernity. Together, communities have come to be understood as active participants in wider processes of colonialism and empire, nation and nationalism, state and citizen, and modernity and globalization, participants that imbue such processes – themselves made up of diverse relationships of meaning and power, time and space – with their own terms and textures.[48]

Work in historical anthropology has explored the many meanings of community construed by its members, especially their symbolization and elaboration of boundaries as providing substance to their differences and identities. To start off, this has involved examinations of the constitutive location of community within wide-ranging processes of power as well as of its internal divisions as expressed in terms of property, gender, law, and office.[49] Moreover, such efforts

---

46 Chakrabarty. *Provincializing Europe*; Dipesh Chakrabarty. *Habitations of Modernity. Essays in the Wake of Subaltern Studies.* Chicago: University of Chicago Press, 2002; Gyanendra Pandey. *Remembering Partition*; Gyanendra Pandey. *Routine Violence. Nations, Fragments, Histories.* Stanford: Stanford University Press, 2006; Ashis Nandy. History's Forgotten Doubles, *History and Theory*, 34:1 (1995): 44–66.

47 Partha Chatterjee. *The Nation and its Fragments. Colonial and Postcolonial Histories.* Princeton: Princeton University Press, 1993.

48 For a wider discussion see, Dube, Anthropology, history, historical anthropology.

49 Veena Das. *Critical Events. An Anthropological Perspective on Contemporary India.* Delhi: Oxford University Press, 1995; Prem Chowdhry. *The Veiled Woman: Shifting Gender Equations in Rural Haryana 1880*–1980. Delhi: Oxford University Press, 1994; Malavika Kasturi. *Embattled Identities. Rajput Lineages and the Colonial State in Nineteenth-Century North India.* New Delhi: Oxford University Press, 2002; Dube, *Stitches on Time*; Charu Gupta. *Sexuality, Obscenity, and Community. Women, Muslims, and the Hindu Public in Colonial India.* Delhi: Permanent Black, 2002; see also, Shalini Puri. *The Caribbean Postcolonial. Social Equality, Post-Nationalism, and Cultural Hybridity.* New York: Palgrave Macmillan, 2004.

have been fortified by incisive accounts of communities as questioning and contesting dominant projects of meaning and power, including those turning on empire and nation and religion and race, unraveling their challenge to authority in a historically and ethnographically layered manner.[50] Finally, there have been diverse endeavors to write greater heterogeneity into the concept of community. Indeed, recent reconfigurations of the category have derived further support from the thinking through of the antinomy between community and state, moves that have queried the analytical binaries of modern disciplines, which are closely tied to totalizing templates of a universal history and exclusive blueprints of a Western modernity. Once more, key consideration of the spatial and the temporal are at stake here.

Some studies have combined these overlapping emphases. In addition to the work of Skaria on the Dangs in western India and of Banerjee-Dube on Orissa in eastern India that has been referred to above,[51] this is evident in my historical and anthropological exploration of an untouchable and heretical caste-sect formation of Chhattisgarh in central India over the past two centuries.[52] This account focuses on a large, internally-differentiated community in order to: trace the endeavors of its members within changing relations of power and property under pre-colonial regimes and colonial rule in the region; explore the group's negotiation and reproduction of ritual authority and gender hierarchies, and track its articulations of caste and Hinduism, evangelism and empire, and state and nation, especially as these were played out in everyday arenas. These writings suggest that prudent procedures are afoot today in the rethinking not only of community and history but also of colony and empire, nation and state, modernity and globalization as shaped by innately heterogeneous yet variously overlapping construal of space and production of time, each often accessing and exceeding the other.

---

50 Anupama Rao. *The Caste Question. Untouchable Struggles for Rights and Recognition.* Berkeley: University of California Press, 2009; Shail Mayaram. *Resisting Regimes. Myth, Memory and the Shaping of a Muslim Identity.* Delhi: Oxford University Press, 1997; David Hardiman. *The Coming of the Devi. Adivasi Assertion in Western India.* Delhi: Oxford University Press, 1987; Guha. *Elementary Aspects.*

51 Skaria, *Hybrid Histories*; Banerjee-Dube, *Religion, Law and Power.*

52 Saurabh Dube, *Untouchable Pasts.*

Coda

At the end, let me point to how this discussion suggests key dispositions toward intellectual endeavor, bids that critically query the privileging of theory as the primary object of enquiry.[53] Here, to start off, it is salient to think through projections that render modernity and history (as well as time and space) as mere objects of knowledge, awaiting their ineluctable endorsement, inevitable refinement, or irrevocable exorcism at the hands of prescient knowledge(s). Instead, it is crucial to approach these as acutely intimating *conditions of knowing*: to explore them as entities, concepts, and coordinates that shore up our worlds, demanding critical articulation *and* careful affirmation.

Next, it only follows that such procedures – of a simultaneous querying and affirmation – can be usefully understood as entwining hermeneutic impulses and critical considerations. This is to say as protocols entailing the interplay of prudent questionings of social worlds and their academic apprehensions *with* intimate accounts of the diversity and distinction of these terrains. Here, there is neither the excision of the details by their being assimilated to the endless analytics of unpicking and unmasking nor is there the privileging of the particulars by their being presented as innate embodiments of alterity and difference.

Further, it is critical to query the pervasive antinomies between the universal and the particular and power and difference. After all, it is much too easy to rail against the claims of universality and power in order to simply celebrate particularity and alterity. Instead, the more challenging task involves exploring the articulation of historical worlds and their social subjects as expressing the shared entailments and mutual productions of power and difference, as interleaving the founding exclusions and constitutive contradictions of authority and alterity.

This means, finally, that the productive possibilities of our efforts inhere in constant vigilance against our self-projections as always subversive, already known modes of scholarly knowledge and political criticism. Rather, it is through the self-questioning of our formative presumptions and formidable limitations that we can more adequately explore modernity, history, time, space, and their

---

53 The proposals in this section derive from the protocols and procedures of what I have called a "history with warranty", developed in Dube, *Stitches on Time* and Dube, *After Conversion*. Now, while this chapter has implicitly engaged throughout with Hardt and Negri's *Empire*, the dispositions that I now describe explicitly, critically think through a crucial, latent aspect of their work: namely, its formative neo-vitalism that gestures toward a present that is a break with the past, albeit with a difference. It signals theory as object-cum-subject of enquiry, a sort of (Gnostic) return of the repressed. Clearly, my own articulations of time, space, and modernity are rather more modest, all too worldly, representing distinct articulations of the subject, theory, and immanence.

interplay – as shaped by the concatenations of distinct yet coeval temporalities-spatiality and of overlapping yet heterogeneous histories-contemporaneity. After all, such distinct but coeval times, overlapping yet heterogeneous spaces have been the soul, stuff, and substance of modernity and its subjects, of subjects of modernity and modern subjects, which is to say of all of us reading this text.

References

Adorno, Theodor. *Minima Moralia*. London: Verso, 2005.

Amin, Shahid. *Event, Metaphor, Memory. Chauri Chaura 1922–1992*. Berkeley: University of California Press, 1995.

Appleby, Joyce, Lynn Hunt, and Margaret Jacob. *Telling the Truth About History*. New York: W. W. Norton, 1995.

Apter, David E. *The Politics of Modernization*. Chicago: University of Chicago Press, 1965.

Asad, Talal. *Genealogies of Religion. Discipline and Reasons of Power in Christianity and Islam*. Baltimore: The Johns Hopkins University Press, 1993.

Banerjee-Dube, Ishita. *Religion, Law and Power. Tales of Time in Eastern India, 1860–2000*. London: Anthem Press, 2007.

Becker, Carl L. *The Heavenly City of the Eighteenth-century Philosophers*. New Haven: Yale University Press, 1932.

Berlin, Isaiah. *Against the Current. Essays in the History of Ideas*. Princeton: Princeton University Press, 2001.

Chakrabarty, Dipesh. *Habitations of Modernity. Essays in the Wake of Subaltern Studies*. Chicago: University of Chicago Press, 2002.

Chakrabarty, Dipesh. *Provincializing Europe. Postcolonial Thought and Historical Difference*. Princeton: Princeton University Press, 2000.

Chatterjee, Partha. *The Nation and its Fragments. Colonial and Postcolonial Histories*. Princeton: Princeton University Press, 1993.

Chaturvedi, Vinayak. *Peasant Pasts. History and Memory in Western India*. Berkeley: University of California Press, 2007.

Chowdhry, Prem. *The Veiled Woman: Shifting Gender Equations in Rural Haryana 1880–1980*. Delhi: Oxford University Press, 1994.

Cohen, David William and E.S. Atieno Odhiambo. *Siaya. Historical Anthropology of an African Landscape*. Cleveland: Ohio University Press, 1989.

Cohen, David William. *The Combing of History*. Chicago: University of Chicago Press, 1994.

Childs, Peter. *Modernism. The New Cultural Idiom*. New York: Routledge, 2000.

Comaroff, John and Jean Comaroff. *Ethnography and the Historical Imagination*. Boulder: Westview Press, 1992.

Crapanzano, Vincent. *Serving the Word. Literalism in America from the Pulpit to the Bench*. New York: New Press, 2000.

Das, Veena. *Critical Events. An Anthropological Perspective on Contemporary India*. Delhi: Oxford University Press, 1995.

Deshpande, Prachi. *Creative Pasts. Historical Memory and Identity in Western India, 1700–1960*. New York: Columbia University Press, 2007.

Dube, Saurabh. *Subjects of Modernity. Time/Space, Disciplines, Margins.* Manchester: Manchester University Press, forthcoming 2017.

Dube, Saurabh. *After Conversion. Cultural Histories of Modern India.* New Delhi: Yoda Press, 2010.

Dube, Saurabh. Anthropology, history, historical anthropology. In *Historical Anthropology*, Saurabh Dube (ed.), 1–73. New Delhi and New York: Oxford University Press, 2007.

Dube, Saurabh. Modernity and its Enchantments. An Introduction. In *Enchantments of Modernity. Empire, Nation, Globalization*, Saurabh Dube (ed.), 1–43. London: Routledge, 2010.

Dube, Saurabh. *Stitches on Time. Colonial Textures and Postcolonial Tangles.* Durham and London: Duke University Press, 2004.

Dube, Saurabh. *Untouchable Pasts. Religion, Identity, and Power among a Central Indian Community, 1780–1950.* Albany: State University of New York Press, 1998.

Dube, Saurabh and Ishita Banerjee-Dube (eds.). *Unbecoming Modern. Colonialism, Modernity, Colonial Modernities.* New Delhi: Social Science Press, 2006.

Fabian, Johannes. *Time and the Other. How Anthropology makes its Object.* New York: Columbia University Press, 1983.

Florida, Nancy. *Writing the Past, Inscribing the Future. History as Prophecy in Colonial Java.* Durham: Duke University Press, 1995.

Giddens, Anthony. *Consequences of Modernity*. Stanford: Stanford University Press, 1990.

Gold, Ann and Bhoju Ram Gujar. *In the Time of Trees and Sorrows. Nature, Power, and Memory in Rajasthan.* Durham: Duke University Press, 2002.

Guha, Ranajit. *Dominance without Hegemony. History and Power in Colonial India.* Cambridge, Mass.: Harvard University Press, 1997.

Guha, Ranajit. *Elementary Aspects of Peasant Insurgency in Colonial India.* Delhi: Oxford University Press, 1983.

Gupta, Charu. *Sexuality, Obscenity, and Community. Women, Muslims, and the Hindu Public in Colonial India.* Delhi: Permanent Black, 2002.

Gumbrecht, Hans Ulrich. *Making Sense in Life and Literature*, Glen Burns (trans.). Minneapolis: University of Minnesota Press, 1992.

Habermas, Jürgen. *The Philosophical Discourse of Modernity. Twelve Lectures*, Frederick Lawrence (trans.). Cambridge, Mass.: MIT Press, 1987.

Hardiman, David. *The Coming of the Devi. Adivasi Assertion in Western India.* Delhi: Oxford University Press, 1987.

Hartman, Saidiya H. *Lose Your Mother. A Journey Along the Atlantic Slave Route.* New York: Farrar, Straus and Giroux, 2007.

Hartman, Saidiya H. *Scenes of Subjection. Terror, Slavery, and Self-Making in Nineteenth-Century America.* New York: Oxford University Press, 1997.

Kasturi, Malavika. *Embattled Identities. Rajput Lineages and the Colonial State in Nineteenth-Century North India.* New Delhi: Oxford University Press, 2002.

Kelley, Donald R. *Faces of History. Historical Inquiry from Herodotus to Herder.* New Haven: Yale University Press, 1998.

Klein, Kerwin Lee. *Frontiers of Historical Imagination. Narrating the European Conquest of Native America, 1890–1990.* Berkeley: University Of California Press, 1999.

Koselleck, Reinhart. *Futures Past. On the Semantics of Historical Time*, Keith Tribe (trans.). Cambridge: MIT Press, 1985.

Mayaram, Shail. *Against History, Against State. Counterperspectives from the Margins.* Delhi: Permanent Black, 2004.

Mayaram, Shail. *Resisting Regimes. Myth, Memory and the Shaping of a Muslim Identity.* Delhi: Oxford University Press, 1997.
McMahon, Darrin M. *Enemies of the Enlightenment. The French Counter-Enlightenment and the Making of Modernity.* New York: Oxford University Press, 2002.
Mehta, Uday. *Liberalism and Empire. A Study in Nineteenth-Century British Liberal Thought.* Chicago: University of Chicago Press, 1999.
Mitchell, Timothy. The stage of modernity. In *Questions of Modernity,* Timothy Mitchell (ed.), 1–34. Minneapolis: University of Minnesota Press, 2000.
Moore, R. Laurence. *Touchdown Jesus. The Mixing of Sacred and Secular in American History.* Louisville, Kentucky: Westminster John Know Press, 2003.
Muthu, Sankar. *Enlightenment against Empire.* Princeton: Princeton University Press, 2003.
Nandy, Ashis. History's Forgotten Doubles, *History and Theory*, 34:1 (1995): 44–66.
Pandey, Gyanendra. *Remembering Partition. Violence, Nationalism and History in India.* Cambridge: Cambridge University Press, 2001.
Pandey, Gyanendra. *Routine Violence. Nations, Fragments, Histories.* Stanford: Stanford University Press, 2006.
Pocock, John G. A. *Barbarism and Religion*. Vol. 2, *Narratives of Civil Government.* Cambridge: Cambridge University Press, 1999.
Porter, Roy. *The Creation of the Modern World. The Untold Story of the British Enlightenment.* New York: Norton, 2001.
Price, Richard. *Alabi's World.* Baltimore: Johns Hopkins University Press, 1990.
Price, Richard. *First-Time. The Historical Vision of an Afro-American People.* Baltimore: Johns Hopkins University Press, 1983.
Price, Richard. *The Convict and the Colonel. A Story of Colonialism and Resistance in the Caribbean.* Boston: Beacon Press, 1998.
Puri, Shalini. *The Caribbean Postcolonial. Social Equality, Post-Nationalism, and Cultural Hybridity.* New York: Palgrave Macmillan, 2004.
Rao, Anupama. *The Caste Question. Untouchable Struggles for Rights and Recognition.* Berkeley: University of California Press, 2009.
Rappaport, Joanne. *Cumbe Reborn. An Andean Ethnography of History.* Chicago: University of Chicago Press, 1994.
Redfield, Peter. *Space in the Tropics. From Convicts to Rockets in French Guiana.* Berkeley: University of California Press, 2000.
Rosaldo, Renato. *Ilongot Headhunting 1883–1974. A Study in Society and History.* Stanford: Stanford University Press, 1980.
Rostow, W. W. *The Stages of Economic Growth. A Non-Communist Manifesto.* Cambridge: Cambridge, 1960.
Skaria, Ajay. *Hybrid Histories. Forest, Frontiers and Wildness in Western India.* New Delhi: Oxford University Press, 1999.
Taussig, Michael. *Shamanism, Colonialism, and the Wild Man. A Study in Terror and Healing.* Chicago: University of Chicago Press, 1987.
Taylor, Charles. *Modern Social Imaginaries*. Durham: Duke University Press, 2005.
Trouillot, Michel-Rolph. *Silencing the Past: Power and the Production of History.* Boston: Beacon Press, 1995.
White, Luise. *Speaking with Vampires. Rumor and History in Colonial Africa.* Berkeley: University of California Press, 2000.

Gesa Mackenthun

Imperiality, Deep Time, and Indigenous Landmark Epistemologies in North America

The concept of imperial time-spaces is open toward various interpretations and moral assessments of imperiality. The following essay tries to look at the imperial history of Europe and the West by including the perspective of those whose communities and cultures were violently displaced and transformed by western imperialism in the Americas. Its point of departure is Walter Mignolo and Aníbal Quintano's important insight that economic and political expansion in the American Hemisphere as well as into other parts of the world now referred to as the global south was accompanied by an epistemology, as well as scientific practices, which frequently functioned to legitimize those processes.[1] Part of that legitimizing function consisted in suppressing and effacing other histories, other epistemologies, and other time-spaces – which amounts to a massive "theft of history," as Jack Goody called it.[2] Notions of time and space, too, were fundamentally affected by this coloniality of knowledge; it is worthwhile to try to 'provincialize'[3] them to a certain degree by showing how their universalist claims are in fact indebted to very particular historical and geographical points of view. The postcolonial paradigm has produced a whole series of contrapuntal readings of imperial historiography, and the present essay may be seen to contribute its tiny share to that growing critical corpus.[4]

After suggesting Mikhail Bakhtin's term "chronotope" as an alternative for "TimeSpace," this essay will shortly discuss the American career of the well-known concept of *translatio imperii et studi* as it has been used in colonial discourse in America. Visualized in the powerful image of the torch (ex oriente lux), this self-legitimizing metaphor continues to shed its light on various assumedly 'dark' corners of the earth (to use Conradian imagery).

---

1 Walter Mignolo. *The Darker Side of Western Modernity. Global Futures, Decolonial Options.* Durham: Duke University Press, 2011.

2 Jack Goody. *The Theft of History*. Cambridge: Cambridge University Press, 2006.

3 Dipesh Chakrabarty. *Provincializing Europe. Postcolonial Thought and Historical Difference.* Reissue. Princeton: Princeton University Press, 2008.

4 "Contrapuntal reading" is the method suggested by Edward Said for tickling out the resistant sub-texts of colonial texts, for showing up inconsistencies, ruptures, and hauntings that disturb the seemingly neat surface of colonial and imperial discourse. For one of the best readings of this kind with reference to colonial discourse in America, see Peter Hulme. *Colonial Encounters. Europe and the Native Caribbean, 1492–1797.* London: Methuen, 1986.

DOI 10.1515/9783110418750-004

Empire, I will furthermore argue, increasingly finds one of its cultural expressions in the universalist discourse and practice of world heritage whose sites are themselves chronotopical in their museal compression of historical time into what Bakhtin calls "castle time." This will be followed by a section that discusses how the heroic discourse of *translatio imperii,* with its specific construction of an imperial past and its projection of an imperial future, underwent an epistemic crisis in the nineteenth century due to the growing knowledge about the age of the earth and the sudden expansion of time way beyond the beginnings of empires and of human life itself. 'Fossil crisis' will be succeeded by a section that looks at non-western chronotopes to suggest what we can learn from the observation of seemingly insignificant creatures like ants and lizards. What unites "Torch," "Fossil," and "Ant" is my interest in looking at TimeSpaces both imperially and transculturally. Approaching as we do the point in history where humanity will prove or disprove its own legitimacy on earth by the way it deals with the challenges of energy and food production, as well as with global climate change, an inclusive and transcultural attitude to time and space seems more necessary than ever.

1 Torch: The Imperial Chronotope

Mikhail Bakhtin introduced the concept of chronotope – which literally means "time-space" – to refer to the way in which, in literary texts, time "thickens, takes on flesh, becomes artistically visible" while "space becomes charged and responsive to the movements of time, plot, and history." The term refers to the potential of literary texts to produce a "simultaneous existence [...] of phenomena taken from widely separate periods of time."[5] Although Bakhtin is primarily interested in the artistic creation of chronotopes, he notes in passing that of course chronotopes also exist in the real world and that they are transformed in the process of representation.[6]

The incredible career of the ideologeme of *translatio imperii et studi* is a good example of non-literary chronotoping. Anthony Pagden has described the emergence and intellectual route of this powerful narrative used as an instrument of self-legitimization by many successive empires:

---

5 Mikhail Bakhtin. *The Dialogic Imagination. Four Essays.* Austin: University of Texas Press, 1981, 84, 85.

6 Bakhtin, *The Dialogic Imagination,* 252–253.

> The continuity between the ancient and the modern *imperia* was guaranteed as much by the translation of power from Augustus to Constantine the Great and from Constantine to Charles V, the *alter Karolus*, via Charlemagne, as it was by the objectives which all these monarchs had pursued.[7]

The translatio trope was adopted by expanding Elizabethan England (see the work of Frances Yates)[8] from where it travelled to the British colonies of America and was built into the nationalist discourse of the early United States. The Irish Bishop Berkeley perhaps most famously gave it poetic expression in the last stanza of his poem "On the Prospect of Planting Arts and Learning in America" (1752) which reads:

> Westward the course of empire takes its way;
> The four first acts already past,
> A fifth shall close the drama with the day:
> Time's noblest offspring is the last.[9]

The early US national poets Philip Freneau, Joel Barlow, and Timothy Dwight readily included the *translatio* concept in their modern epics of Europe's massive movement to the New World. As Eric Cheyfitz argues, world empire translates itself through civilizing activities:

> The *translatio* [...] is inseparably connected with a 'civilizing' mission, the bearing of Christianity and Western letters to the barbarians, literally [...] those who do not speak the language of the empire. From its beginnings the imperialist mission is, in short, one of translation: the translation of the 'other' into the terms of the empire [...] Within the dynamics of the *translatio* it is the eloquent orator who is figured as the prime agent of this mission of translation.[10]

In American political discourse, the chronotope of the translation of empire continued its career as the guiding narrative of westward expansion after the acqui-

7 Anthony Pagden. *Lords of All the World*. Oxford: Oxford University Press, 1995.

8 Frances A. Yates. *Astraea. The Imperial Theme in the Sixteenth Century*. Harmondsworth: Penguin, 1975.

9 George Berkeley. On the Prospect of Planting Arts and Learning in America. In *Key Concepts in American Cultural History. From the Colonial Period to the End of the 19th Century*, Bernd Engler und Oliver Scheiding (eds.), 32. Trier: WVT, 2005 [1752]. See also Gesa Mackenthun. Expansionism. In: *A Companion to American Cultural History. From the Colonial Period to the End of the 19th Century*, Bernd Engler and Oliver Scheiding (eds.), 247–273. Trier: WVT, 2009.

10 Eric Cheyfitz. *The Poetics of Imperialism. Translation and Colonization from "The Tempest" to "Tarzan"*. Oxford: Oxford University Press, 1991, 112.

sition of Louisiana in 1803 – in other words, when it became clear that the whole continent was to come under the dominion of Anglo-Saxon settlers. It translated into the concept of Manifest Destiny, first spelled out by the journalist John L. O'Sullivan, editor of the *Democratic Review*, who in 1845 coined the term in an essay propagating the annexation of Texas. Like many of his contemporaries, he argues that Texas must be 'saved' from being taken by rival nations whose intention it was to limit and check "the fulfilment of our manifest destiny to overspread the continent allotted by Providence for the free development of our yearly multiplying millions."[11] Herman Melville condenses all the elements of the trope through the voice of White Jacket in his novel by that title (1850):

> we Americans are the peculiar, chosen people – the Israel of our time; we bear the ark of the liberties of the world. Seventy years ago we escaped from thrall; and, besides our first birthright – embracing one continent of earth – God has given to us, for a future inheritance, the broad domains of the political pagans. [...] The rest of the nations must soon be in our rear. We are the pioneers of the world; the advance-guard, sent on through the wilderness of untried things, to break a new path in the New World that is ours.[12]

The quotation fuses the providential thinking inherited from Puritanism with the aggressive jingoism of the period after the defeat of Mexico (1848). White Jacket regards the process of expansion by purchase, war and annexation itself as a "dispensation," that is as divine signs that the continent has been set apart for settlement by the descendants of revolutionary Americans. Others enhanced the trope with racial overtones. It is difficult to tell how much Melville himself identified with this position – he had severely criticized both European and American imperialism before in *Typee* and would do so again in *Moby-Dick*. However that may be, the mid-nineteenth century witnessed the emergence of the expansionist *Frontier* myth according to which the incorporation into the United States of all adjacent lands was legitimate as it was the fulfilment of a historical destiny devised long ago by Providence itself.[13] It finds visual expression in the famous painting "American Progress" by John Gast (1872/1874) which gives a panoramic view of the north American continent where various groups of people (Indians, settlers, stage coach) and symbols of civilizational progress (railway, telegraph)

---

11 John L. O'Sullivan. Annexation. In *Key Concepts in American Cultural History. From the Colonial Period to the End of the 19th Century*, Bernd Engler und Oliver Scheiding (eds.), 389. Trier: WVT, 2005 [1845].

12 Herman Melville. *White Jacket*. New York: Grove Press, 1956 [1850], 151.

13 Albert Weinberg. *Manifest Destiny. A Study of Nationalist Expansionism in American History*. Chicago: Quadrangle, 1963, 1–2.

can be seen to travel from East to West, bringing light into darkness. Hovering above the continent is the figure of Lady Columbia, allegorizing the translation of knowledge and civilized virtues. She is shown with a star on her forehead, a schoolbook under her right arm, and unfolding a telegraph wire with her left hand. Lady Columbia could be encountered again at the end of the century, as a gigantic statue dominating the World's Columbian Exposition at the White City in Chicago in 1893. Officially representing the Republic, she holds the insignia of empire in her hands: an imperial orb on which an eagle sprawls its wings in one hand, and in the other a staff topped by a Jacobin's cap, the symbol of republicanism. But by far the more famous statue is of course the Statue of Liberty which had been erected in New York in 1885, since then greeting the tired and poor immigrants from across the Atlantic with the torch of freedom raised high above her head.

During the Chicago Exposition, which was a global fair dedicated to the presentation of the latest technological inventions and a general exchange of knowledge on a global scale, Frederick Jackson Turner gave his famous speech "The Significance of the Frontier in American History" in which he announced that the frontier was closed, America had come into full possession of the United States, and that Americans, in order to retain their specific freedom-loving and republican character, would have to look for new frontiers. In the same decade the US would launch its official imperial phase by annexing Hawaii and entering the Spanish-American War. The journey of empire and knowledge, begun in Antiquity, would be continued beyond the continental confines of the United States.

Such, in all brevity, is the narrative of the translation of empire – the imperial chronotope which serves to legitimize imperial expansion through its in-built assumption of the inevitability of the westward movement of power and knowledge. The temporal-spatial movement acquires a kind of self-legitimacy through the power of narrative – from darkness to light, from ignorance to knowledge, from childhood to maturity, from barbarism to civilizational perfection. It is a narrative that bypasses the tortuous attempts of early colonists to legitimize their actions of conquest with reference to international legal principles – from the right of discovery inherent in the trope of vacuum domicilium, the right of civil tenure and government formulated in the light of Locke, all the way to the right by voluntary submission of America's indigenous inhabitants expressed in many a romantic tale of returning white leaders and enamoured princesses. One of the most successful tropes was, and still is, the trope of indigenous nomadism – a non-sedentary lifestyle that presumably deprives Native Americans of the right to territorial ownership. The ideologeme of native nomadism has reached mythical dimension in Hollywood-generated narratives of Cowboys and Indians whose power often overshadows the knowledge of the fact that the majority of America's indigenous inhabitants practiced various forms of agriculture and horticulture and were far less given to aimless wandering than imperial armchair theorists were able to understand. This insistence on the nomadic territoriality of indigenous peoples – which was used, for example, in legitimating the removal policy of the 1830s – may be regarded symptomatically as a 'screen trope' disguising the restless geographical mobility of European-Western imperial expansion itself; it strangely replicates the guiding expansionist metaphor of the moving Frontier.

In the twentieth century, the discourse of World Heritage can be seen to compensate for this frequent lack of territorial groundedness and territorial legitimacy, although it can of course not be reduced to this function and primarily fulfils the need for nationalist self-assertion (in spite of its claims to universal relevance). World Heritage sites frequently provide that kind of territorialized historical identity that is often missing from the ever-mobile activities of empire. It provides the possibility to lay *cultural* claims to certain places to make up for lacking *legal* evidence of land ownership. World Heritage is a method of freezing time – or various periods of time – in particular places within and outside the empire.[14] Al Quaida, ISIS and other jihadist militia groups involved in the spectacular destruction of World Heritage sites have understood the extreme symbolic power of such museal places – their function as material confirmations of the imperial master narrative, perhaps even as the cultural expression of global economic hegemony.

Using another of Bakhtin's categories, the logic of World Heritage can be described with the concept of "castle time" – the castle being a particular literary chronotope suggested by Bakhtin analogously to the castles of gothic novels. Bakhtin refers to the castle as both a novelistic chronotope and a new "territory" invented by the gothic novel: "The castle is saturated through and through with a time that is historical in the narrow sense of the word, that is, the time of the historical past."[15] But of course "castle time" is more than a condensation of cultural treasures in particular places. It should be extended to the plots of gothic novels which are disturbing the archival complacency of the castle's homogeneous

14 I'm here using "empire" not with reference to any particular nation but rather, somewhat analogously to Hardt and Negri, as a mobile ideological-political-economic formation irreducible to any particular state. Modern heritage discourse is at once a reassertion of specific national traditions and an assertion of a supranational cultural formation, represented by the UNESCO. In cultural terms, perhaps, the UNESCO could be seen as what Negri and Hardt describe as "a single power that overdetermines them all [i.e. all earlier distinct imperialist powers], structures them in a unitary way, and treats them under one common notion of right that is decidedly postcolonial and postimperialist." (Michael Hardt and Antonio Negri. *Empire*. Cambridge: Harvard University Press, 2000, 9.)

15 Bakhtin, *Dialogical Imagination*, 245–246.

and final temporality by showing that the place radiates with energies of a suppressed past, energies that cut through time and space to produce havoc among the present members of society. The gothic castle is, in Benjaminian terms, a chronotope shot through with the energies it only seemingly controls.

David Lowenthal draws a direct connection between the massive growth of the heritage sector and the mass migrations of the twentieth century. Displacement and loss of homeland, he argues, produce a nostalgic craving for heritage sites.[16] As Lowenthal clarifies in various books, and other scholars do in the wake of his work, heritage is not to be confused with history:

> In fact, heritage is not history at all: while it borrows from and enlivens historical study, heritage is not an inquiry into the past but a celebration of it, not an effort to know what actually happened but a profession of faith in a past tailored to present-day purposes.[17]

One of these purposes is to "console" us in the face of a growing fear of the future.[18] The recent addition of Natural Heritage sites to the UNESCO catalogue seems to confirm this analysis: specific natural sites are set apart to be enjoyed and consumed in a historically sanitized way while all around, beyond the borders of the heritage zone, the destruction of the environment continues. In this fatal ambivalence, I suggest, world heritage is one of the cultural expressions of empire.

2 Fossil: The Crisis of Deep Time

Since the nineteenth century, the imperial chronotope has come to be confronted with its own historical insignificance. While the heritage industry adamantly holds on to the human importance of man-made structures and artifacts – an activity that consoles not only migrants but also reconciles many more sedentary people with the uncertainties of modernity and the unpredictability of the future – the discovery of the real age of the earth and of deep time have shaken the foundations of imperial historiography. The eighteenth-century geologist John Hutton had formulated in 1788:

> Time, which measures every thing in our idea, and is often deficient to our schemes, is to nature endless and as nothing. […] If the succession of worlds is established in the system of

16 David Lowenthal. *The Heritage Crusade and the Spoils of History*. Cambridge: Cambridge University Press, 1998, 9.

17 Lowenthal, *Heritage Crusade,* X.

18 Lowenthal, *Heritage Crusade,* XIII.

> nature, *it is in vain to look for anything higher in the origin of the earth.* The result, therefore, of our present enquiry is, that *we find no vestige of a beginning, – no prospect of an end.*[19]

The gradual dissemination of this insight was suitable to disclose the logical flaws of imperial historiography: the fact that reference to the ancient past, primarily that of Rome, Greece, and the so-called Holy Land, is not really helpful in making territorial claims to areas far away from both those antique places and from the homelands of the nations making such claims. Old World antiquity provides no logical evidence for territorial proprietorship as the sites of that antiquity are located within other sovereign nations, from Italy and Greece to Egypt, Ethiopia, Iran and Syria. In fact, the flaw is inherent in the concept of *translatio imperii* itself, which rests on the notion of spatial mobility – of an absence of a fixed form of territoriality – but is then paradoxically used to lay claims to fixed territories.

With the new discoveries by Lyell in geology and Darwin in biology, the imperial world, especially imperial America, had to learn that, as Johannes Fabian writes, "Time was no longer the vehicle of a continuous, meaningful story." Rather, it was now reduced to an attempt "to order an essentially discontinuous and fragmentary geological and paleontological record."[20] What became exposed, in other words, was the constructivist role of history, as a meaning-making narrative machine – a cultural necessity but no longer a legitimate basis for territorial claims. The discovery of geological time exposed to man the "*vastness*" of the past in which "the history of mankind, recorded or reconstructed, occupied a negligible span on the scale of natural evolution." For the imperial mind, it was just as hard to accept the "stark *meaninglessness* of mere physical duration." Until then, imperial ideology had been far too infused with the "conviction that Time 'accomplished' or brought about things in the course of evolution." Too much, in other words, had it held on to a belief that events themselves happened as the events of a narrative, transforming that narrative into a script for the future. Among the convictions now to be shed was also the Western belief in a universal succession of stages of civilization, from a Savage stage, through an Arcadian and Agricultural stage, all the way to modern metropolitan civilization, each of which were regarded to be "as meaningful as a sentence leading toward the conclusion of a story."[21]

---

19 Quoted in Stephen Jay Gould. *Time's Arrow Time's Cycle. Myth and Metaphor in the Discovery of Geological Time.* Cambridge: Harvard University Press, 1987, 64–65, emphasis added.

20 Johannes Fabian. *Time and the Other. How Anthropology Makes its Object.* 2d ed. New York: Columbia University Press, 2002, 14

21 Fabian, *Time,* 14–15, emphases added. While my subject of reflection is "empire," Fabian's are the "social evolutionists" of the nineteenth century.

Next to the dawning understanding that the history of the earth was much older than announced in the Bible, the structure of its temporality was called into question by the discoveries of geology and evolutionary theory. Such insights were theoretically not irreconcilable with the imperial chronotope of the Westward Course of Empire but it mitigated that chronotope's temporal scope. Fossils of extinct species were found all over the Americas, some of them mixed with human bones, testifying to the longevity of man's existence on the continent before the arrival of Columbus. Frequently the reactions to these finds consisted in denial, mixed with horror and scorn. A typical case of denial is the reaction to the discovery of petrified human bones, mixed with the bones of extinct species, by the Danish naturalist Peter Wilhelm Lund in the caves of Lagoa Santa, Minas Gerais, Brazil, in 1843. The sensation of Lund's find should have been its proof of the great antiquity of the human presence in the Americas, but in the process of reception it was reduced to Lund's skull measurements which, most scholars agreed, were identical with those of all other skulls of indigenous Americans.[22] Lund's extraordinary find was thus translated into familiar categories; what was 'unthinkable' (the extreme age of the bones and what this meant for the legitimacy of Euroamerica's presence on the continent) was domesticated into the dehistoricizing categories of craniology and scientific racism.

But the response to 'fossil time' was not unanimous. A recently discovered panorama of the Mississippi River by the Irish immigrant John Egan includes images of Native American burial mounds, fossil bones of extinct animals, and scenes of Indian life before Columbus. It shows Mississippi steamers traveling through landscapes strewn with mounds as well as the living descendants of the dead buried in them. Around the same time the panorama was created (in the 1840s), the German immigrant Albert Koch toured the US displaying recomposed fossil skeletons together with Native American stories about ancient monsters. American spiritualists like Hudson Tuttle likewise had no problem imagining a native presence even in primeval times – maybe owing to spiritualism's general rejection of reading the Bible literally. Herman Melville, on the other hand, was

22 For more information on Lund and his reception, see Gesa Mackenthun. Night of First Ages. Deep Time and the Colonial Denial of Temporal Coevalness. In *Crossroads in American Studies. Transnational and Biocultural Encounters,* Frederike Offizier, Marc Priewe, Ariane Schröder (eds.). Heidelberg: Winter, 2015; Gesa Mackenthun. Fossils and Immortality. Geological Time and Spiritual Crisis in Nineteenth-Century America. In *Deutungsmacht. Religion und Belief Systems in Deutungsmachtkonflikten,* Philipp Stoellger (ed.), 259–283. Tübingen: Mohr Siebeck, 2014. See also Pedro de Luna. The Man Who Faced the Saber-Toothed Cat. In *Fugitive Knowledge. The Preservation and Loss of Knowledge in Cultural Contact Zones,* Andreas Beer and Gesa Mackenthun (eds.), 163–179. Münster: Waxmann, 2015.

filled with horror when considering the old age of the world and its pagan inhabitants, whether whales (in his chapters on the fossil whale in *Moby-Dick*) or ancient Egyptians (in his diary of his journey to Egypt).[23] Comedy mixed with scorn was the register chosen by Mark Twain for expressing existential uncertainty. In his essay "Was the World Made for Man?" (1909), he satirically revises the imperial chronotope in comparing the span of man's existence on earth with the skin of paint on the summit of the Eiffel Tower while forcefully arguing that, without any doubt, the world was created exclusively for Man.

These stories and representations refer to the heteroglossy of the imperial episteme which was far from homogenous. Nevertheless, certain branches of archaeology nervously hold on to imperial chronotopes and master narratives until our own time, now trying to project them into the more distant past.[24]

By the twenty-first century, the anxiety inspired by the idea of deep antiquity has found a new location in the cultural field of horror fiction, especially in the sub-genre of archaeological horror thrillers. In late-imperial gothic thrillers such as *Indiana Jones*, *The Exorcist*, or the thrillers of Preston and Child, the archaeological dig or the archaeological museum are the chronotopes where the history of empire "thickens," in Bakhtin's words: where undead mummies, murderous creatures from the rainforest, and Native American skinwalkers violently voice (or rather mumble) their territorial claims.[25] Here the "castle time" of classical gothic fiction mutates into the multilayerd temporality of the dig, forming what we may call 'dig time'.

The cultural discourse of archaeological horror, then, may be seen as an articulation of imperial anxiety.[26] It is an anxiety about time, but even more so

---

23 For a more extended discussion see Mackenthun, Fossils and Immortality.

24 I'm thinking of attempts by Smithsonian archaeologists to prove transatlantic Ice Age migrations that would suggest that at least Atlantic America has been first peopled from Europe after all. The conspicuous aggressiveness with which some archaeologists try to find evidence that "Kennewick Man" – or "The Ancient One" – was not of Native American descent is explicable as a symptom of a long-lasting imperial pathology about the unsolved question of land rights in North America. A recent Ancient DNA analysis has shown that "The Ancient One" has indeed the strongest match with the DNA of the local tribes (Umatilla, Colville) who had laid claim to the right of reburying the remains right from the beginning.

25 See Gesa Mackenthun. Hidden Cities in the American Wilderness. The Cultural Work of a Romantic Trope. *Miranda* [Online], 11 (2015), miranda.revues.org/6954 (06 December 2015).

26 The beginnings of the discourse of imperial horror can be traced to the peak periods of imperialism – to the works of Haggard, Kipling, Conrad, Stevenson, Blackwood, and Stoker in the UK (see Patrick Brantlinger. *Rule of Darkness. British Literature and Imperialism, 1830–1914*. Ithaca: Cornell University Press, 1988) and to that of Poe for early US imperialism (i.e., domestic slavery and territorial dispossession; see Teresa A. Goddu. *Gothic America. Narrative, History, and*

about place. After all, imperial historiography nervously evokes its temporal genealogies, tracing its origins back to biblical times, as a mere surrogate for its incapacity to legitimize itself spatially. The persistent identification of indigenous peoples as 'nomads' in American savagist discourse – as people who ranged and wandered up and down the land rather than inhabiting it – serves as a projection of the fact that it is indeed imperialism itself whose agents are constantly on the move.[27] What distinguishes indigenous from imperial 'wanderers' is the fact that the latter migrants possess a narrative of a larger historical purpose or 'mission'.[28]

In addition to this, imperialist chronologies are the victims of empire's own cultural arrogance: they become uncertain whenever they leave the safe grounds offered by written documentation and enter the slippery terrain of material-archaeological evidence where their claims must be measured with the same instruments and methodologies as those of non-literate cultures or cultures whose acquisition of literacy is of a more recent date. While Hegelian historiography and science have persistently disarticulated the oral traditions of non-Western cultures and disqualified them as alternative archives, literate and nonliterate cultures become equals in the archaeological dig or on the site of accidental geological-archaeological finds, such as the recent discoveries of human remains in underwater caves in Yucatan,[29] or the 12,700 year-old bones of the so-called

Nation. New York: Columbia University Press, 1997) and H.P. Lovecraft in the early twentieth century (where the problem of the territorial claims of antiquity is joined with anti-immigration sentiment).

27 The trope of nomadism entered international legal discourse through the work of Emmerich de Vattel who, in his *Law of Nations* (1758), indiscriminately declares all indigenous inhabitants of North America to be nomads with an inferior right to territorial possession: "though the conquest of the civilized empires of Peru and Mexico was a notorious usurpation, the establishment of many colonies on the continent of North America might, on their confining themselves within just bounds, be extremely lawful. The people of those extensive tracts rather ranged through than inhabited them." Emmerich de Vattel. *The Law of Nations*. London: Newberry, 1759. Book I, Capter VII, § 81, http://www.constitution.org/vattel/vattel_01.htm (06 December 2015).

28 One of the best studies of savagism is still Roy Harvey Pearce. *Savagism and Civilization. A Study of the Indian and the American Mind*. Rev ed. Berkeley: University of California Press, 1988. On the 'nomadic' quality of imperialism and colonialism, see Peter Hulme. Tales of Distinction. European Ethnography and the Caribbean. In *Implicit Understandings. Europeans and Other Peoples in the Early Modern Era*. Stuart Schwartz (ed.), 16–34. Cambridge: Cambridge University Press, 1994.

29 The text from the *Nature* article on the discovery in 2007 of the so-called Naia girl in an underwater cave in Yucatán unwittingly betrays the temporal problems I'm discussing: "The near-intact skeleton of a delicately built teenage girl, *who died more than 12,000 years ago* in what is today's Mexico, could help to solve the riddle of how the Americas were first populated. – Cave divers discovered the skeleton seven years ago in a complex of flooded caverns known as Hoyo

Anzick Boy found in Montana in 1968 but DNA-sequenced only in 2014.[30] Where the excavations relate to colonial times, imperial and Indigenous forms of memory meet at such sites as the recently discovered remains of Sir John Franklin's ships in the Arctic ice near King William Island in the Canadian province Nunavut on 7 September, 2014 – precisely in that place at which local tribes had been pointing their fingers for the last 130 years.[31]

Negro, in the jungles of the Yucatan Peninsula. *They called her Naia, after the naiads, the water nymphs of Greek mythology*. She lies in a collapsed chamber *together with the remains of 26 other large mammals, including a sabre-toothed tiger*, 600 metres from the nearest sinkhole. *Most of the mammals became extinct around 13,000 years ago*". (Alison Abbott. Mexican skeleton gives clue to American ancestry. Genetic signature from cave remains matches that of modern Native Americans. *Nature*, 14 May 2014, www.nature.com/news/mexican-skeleton-gives-clue-to-american-ancestry-1.15226 (1 December 2015); emphasis added.) Abbott is merely repeating the information given her by James Chatters, who in 1996 had claimed that Kennewick Man had a 'Caucasian' skull, thus firing theories about ice age migrations from Europe across the Atlantic. "12,000 years ago" still fits into the Clovis consensus about man's first arrival in America across Bering's Strait; to consider a synchronicity between the human remains and those of the animals who became extinct "around 13,000 years ago" would cause problems for adherents of Clovis First. The other two interesting features about the short text are the classical European reference for the skeleton's name and the fact that the "girl" is being subsumed under the category of animals. This may be a trans-species symbolic gesture in favor of the "other large mammals," or it may be a residue of scientific racism. For the Clovis debates, see Charles Mann. *1491. New Revelations of the Americas before Columbus.* New York: Knopf, 2015. Modern methods of ascertaining ancient DNA likewise serves ill the defenders of imperial chronotopes. Abbott also writes that "Naia's mitochondrial DNA reveals genetic signatures in common with modern Native Americans, despite her very different skull shape." Maybe the craniological method, invented at the peak of Manifest Destiny, is in need of serious revision. See Stephen Jay Gould, *The Mismeasure of Man.* 1981. New York: W.W. Norton & Company, 1996; Fabian, Ann. *The Skull Collectors. Race, Science, and America's Unburied Dead.* Chicago: University of Chicago Press, 2010.

30 Michael Balter. Native Americans Descend From Ancient Montana Boy. *Science*, 12 February 2014. news.sciencemag.org/archaeology/2014/02/native-americans-descend-ancient-montana-boy (06 December 2015).

31 "Search parties later recorded Inuit testimony in the late 1840s that claimed one ship sank in deep water west of King William Island, and one ship went perhaps as far south as Queen Maud Gulf or into Wilmot and Crampton Bay. The location of this wreck backs up that testimony." (Lost Franklin expedition ship found in the Arctic. *CBC News*, 9 September 2014, www.cbc.ca/news/politics/lost-franklin-expedition-ship-found-in-the-arctic-1.2760311 (06 December 2015).)

3 Ant: Indigenous Landmark Epistemologies

Although the imperial formation crucially depends on spatial possessions, it consistently seeks to legitimize these with reference to its old age, not by virtue of a particular intellectual or emotional rapport with the places whose possession it claims. The impossibility of doing so, I argued above, is one of the sources of the anxiety of imperial discourse. Within the episteme of empire, concepts of space and place have, with a few exceptions, only recently begun to be seriously theorized.[32] Following in the tracks of empire, the novel (as Bakhtin, Firdous Azim, and Edward Said have variously argued)[33] is particularly outspoken in laying out the spatial and geographical coordinates of its actions – a fact that is probably owing to its emergence from the genre of travel narrative. Frequently, the domestic dramas set in the homeland are 'haunted' by contexts of colonial dispossession and exploitation mentioned only in their margins yet of decisive relevance for the novelistic action. These spatial coordinates, though only half-visible in the classical novels of empire, have been brought forth in recent years by way of contrapuntal readings in the wake of Said's important critical impulses.

Following the same impulse to foreground space and place in the analysis of culture, a few scholars, like Lee Schweninger, Shari Huhndorf, and Joy Porter, have recently analyzed Native American literature for its fictionalization of spatial relations.[34] Concurrently, the field of anthropology, especially in the work of Julie Cruikshank and in works on Indigenous astronomical knowledge (Cajete, Hamilton, Lankford), has recently opened new paths of reassessing Indigenous oral traditions by mining them for their information about cultural relationships to space and place.[35] Alternative chronotopes come into view.

---

32 I'm thinking of the work of Said, Soja, and Lefebvre. (Edward W. Said. History, Literature, and Geography. In *Reflections on Exile and Other Literary and Cultural Essays*, Edward W. Said (ed.), 453–473. London: Granta, 2000; Edward W. Soja. *Postmodern Geographies. The Reassertion of Space in Critical Social Theory*. London: Verso, 1989; Henri Lefebvre. *The Production of Space*, Donald Nicholson-Smith (trans.). Oxford: Blackwell, 1991 [1974].)

33 Bakhtin, *The Dialogic Imagination* 11–12; Firdous Azim. *The Colonial Rise of the Novel*. London: Routledge, 1991, 30; Said, "Jane Austen and Empire." *Culture and Imperialism*. London: Chatto & Windus, 1993. 95–116.

34 Lee Schweninger. *Listening to the Land. Native American Literary Responses to the Landscape*. Athens: University of Georgia Press, 2008; Shari Huhndorf. *Mapping the Americas. The Transnational Politics of Contemporary Native Culture*. Ithaca: Cornell University Press, 2009; Joy Porter. *Native American Environmentalism. Land, Spirit, and the Idea of Wilderness*. Lincoln: University of Nebraska Press, 2014.

35 Julie Cruikshank. *The Social Life of Stories. Narrative and Knowledge in the Yukon Territory*. Lincoln: University of Nebraska Press, 1998; Greg Cajete. Native Astronomy. An Indigenous View

A crucial medium of Indigenous epistemologies – next to petroglyphs and different forms of writing – are orally transmitted narratives. Traditional stories, as Colin Calloway points out, "connected the people to an ancient world whose lessons they must not forget and to the natural world in which they could survive only by maintaining proper relations with other forms of life." He adds that many of these stories transporting important historical and moral lessons are tied to specific landscape features or "storied places" which "reinforced [their] continuity and accuracy."[36] Before the break-up of traditional culture due to colonialism, Indigenous stories were passed on by specially trained experts and the act of transmission included non-verbal aspects like song, dance, and gesture. Often, these stories were highly codified forms of passing on knowledge (in the form of narrative, allegory, and personification). In American Indigenous epistemologies, specific places have acquired their meaning over many centuries by being preserved in oral and material texts. Space, place, and land are 'owned' by virtue of being thoroughly 'storied': knowledge about them exists in the form of song, ritual, and orature.

Meanwhile, the introduction of geological and evolutionary 'time' has also shifted attention from time to place. But because, as Johannes Fabian and more recently Lynn Hunt have suggested, our understanding of the past is still very much indebted to teleological and providential narratives, evolutionary theory was itself quickly incorporated into such a providentialist structure.[37] Evolutionary and geological chronotoping needs more serious attention as an alternative way of conceptualizing planetary events, without being integrated in the timeless taxonomies of scientific racism or the double-edged discourse about the anthropocene.[38] Indigenous oral traditions offer an alternative historical archive to which scholarship has paid but little attention. One of their essential features is their relationship to particular geographical sites that are considered sacred

Dancing the Heavens.In *The Pari Dialogues. Essays in Indigenous Knowledge and Western Science*, F. David Peat (ed.), 41–84. Pari (Italy): Pari Publishing, 2013; Ross Hamilton. *Star Mounds. Legacy of a Native American Mystery*. Berkeley: CA: North Atlantic Books, 2012; George E. Lankford. *Reachable Stars. Patterns in the Ethnoastronomy of Eastern North America*. Tuscaloosa: University of Alabama Press, 2007.

36 Colin G. Calloway. *One Vast Winter Count. The Native American West before Lewis and Clark*. Lincoln: University of Nebraska Press, 2003, 7.

37 Fabian, *Time;* Lynn Hunt. *Measuring Time, Making History*. Budapest and New York: Central European University Press, 2008.

38 ... a discourse that frequently serves to critique man's dominion over the earth but whose claim to a discrete geological epoch also contains a certain pride about man's powerful impact on the history of the earth.

or culturally significant. Orally transmitted narratives express a collective sense of belonging to the land. As the website of the Colville Confederacy (whose DNA turned out to be closest to that of Kennewick Man) states on its website:

> These legends are not considered 'myths' by the people but rather are our spiritual and religious beliefs that connect us with the land. Our elders passed down these legends through the generations to their children and grandchildren. [...] Each legend tells a story, provides historic information, relates to resource management and provides moral and spiritual guidance. *These stories and legends witness our ancestral claim to the land, to the mighty Columbia River and its tributaries, and to the resources found here. These landmark legends define our home.* (emphasis added)[39]

There is an abundant archive of Indigenous place-related stories in North America, especially stories dedicated to cultural heroes or connected to the emergence of a social group as a unified cultural group (unified by language, custom, form of government, etc.). Sometimes confirmation of the significance of these places can be found graven into the landscape features themselves, as in the case of the petroglyphs found at the site where the mythical Hero Twins of the Navaho are said to have dwelled.[40] The Great Smoky Mountains have a similar function in the oral tradition of the Cherokee.[41] But the tie that binds story to place is easily cut through forced migration (as in the case of the Cherokee), more or less voluntary migration (Navaho), removal to reservations and prohibition to leave,[42] and indeed through the forced interruption of the oral tradition itself by way of colonial reeducation. Often the landscapes of origin are destroyed, as effected by the Navaho Dam in the 1950s by whose erection the petroglyphs referring to the story of the Hero Twins became submerged.[43] Sometimes the stories are taken along and assigned to new landscapes.[44]

---

39 Jennifer K. Ferguson and Matilda George (ed.). *Upper Columbia River Book of Legends.* Nespelam, WA: Confederated Tribes of the Colville Reservation, 2007, 6, www.colvilletribes.com/book_of_legends.php (06 December 2015).

40 Polly Schaafsma and Will Tsosie. Xeroxed on Stone. Times of Origin and the Navajo Holy People in Canyon Landscapes. In *Landscapes of Origin in the Americas* Jessica Joyce Christie (ed.), 15–31. Tuscaloosa: University of Alabama Press, 2009, 17.

41 Christopher Arris Oakley. The Center of the World. The Principle People and the Great Smoky Mountains. In *Landscapes of Origin in the Americas.* Jessica Joyce Christie (ed.), 3–14. Tuscaloosa: University of Alabama Press, 2009.

42 Oakley, Center, 29.

43 Oakley, Center, 17.

44 Oakley, Center, 26.

But there are a few cases where tribes had the good luck to stay in the same area which had already been inhabited by their ancestors. The Northwestern Rockies and British Columbia are such an area, which offers astounding concurrences between Indigenous oral traditions about certain cataclysmic events and Western geological reconstructions of these events.[45] As a consequence of their economic success (compared to other tribes who barely survived), tribes in the Northern Rockies were threatened with termination and some of them were actually terminated, like the Klamath in California and Oregon. This included the end of Klamath sovereignty, and hence the end of treaty rights; it caused dependency on welfare and territorial dispossession, including loss of the right to preserve their scared lands.[46] As the Five Civilized Tribes which were removed from their ancestral lands in the 1830s, the Klamath were simply too 'civilized' and too much of an economic rival to American farmers and the logging industry. It took them a long legal struggle (with the help of the Native American Rights Fund, a legal NGO) to regain some of their rights of self-government (Klamath Restoration Act, 1986). Today, they continue to regain some of their lands as well. In legal battles such as this, it is crucial that the tribes can prove their 'continuous occupation' of the land. Place-related oral traditions – so-called geomyths – have in the past been helpful to achieve such partial successes.

The other scientific area in which Indigenous topological knowledge is used is the large field of ecological resource management. In spite of ongoing attempts to denounce Native Americans as ecological villains, there is abun-

45 See for example Gesa Mackenthun. 'Unhallowed Mysteries' in the Colonial Archive. Competing Epistemologies in North America. In *Concurrences in Colonial and Postcolonial Studies*. Diana Bryden, Peter Forsgren and Gunlög Fur (eds.). Amsterdam: Rodopi, 2015. See also Vine Deloria. *Red Earth White Lies. Native Americans and the Myth of Scientific Fact*. Golden: Fulcrum, 1997.

46 "Termination" was a federal policy adopted by the United States Congress in 1953. The thrust of the policy, in its simplest terms, was to force the assimilation of Indian people into the mainstream American culture. There were, of course, a whole array of other purposes, including

- The unilateral abrogation of all treaty and other agreements between the governments of the various tribes and the United States.
- The abolition of tribal governments.
- The eradication of reservations and all tribal holdings of lands and assets.
- The unilateral abrogation of all federal responsibilities owed to the tribes and their citizens under the treaties and other tribal-federal agreements.
- The elimination of the federal bureaucracy dedicated to the support of Indian programs and fulfillment of treaty guarantees.
- The consequent reduction of the federal budget.

All information taken from: Allen Foreman. Termination, www.klamathtribes.org/background/termination.html (14 December 2015).

dant proof that Indigenous cultures possess a vast knowledge about ecological issues and relationships.[47] This Traditional Ecological Knowledge (TEK) – the "local understandings of plant, animal, and habitat relations held by Indigenous peoples"[48] – is now being used in applied resource management, partly to resolve "a myriad of problems created by industrial resource extraction and intensive factory-style agriculture" and the spectacular collapses that this system periodically creates.

Charles Menzies distinguishes between the historical evidence of Indigenous ecological knowledge as preserved in stories and the knowledge gained from "the active use of the landscape."[49] Indigenous scholars argue that, as ecological knowledge continuously has to adapt to ever changing circumstances, historical knowledge in the traditional sense is often insufficient if not accompanied by constant observation; historical knowledge preserved in oral tradition is often at an advantage in providing a "cultural framework within which knowledge of the environment is transmitted." But in order to "understand actual ecological knowledge one must participate in the real processes of hunting, fishing, and gathering. This is a form of pragmatic, tactile knowledge that ultimately is dynamic and responsive to change within the material environment."[50] Yet, as Caroline Butler warns in the same volume, it would be dangerous to rely exclusively on the effectiveness of Indigenous knowledge for our ecological survival. After all, the transmission of Indigenous epistemologies was brutally disrupted as a result of colonialism such as forced reeducation in boarding schools and outright genocide.[51] Under these circumstances it would be disingenuous to look to Native Americans as the ones to save us from the global crisis produced by industrial capitalism.[52] Yet much knowledge of local topographies, of fauna and flora, has been reinvigorated and is now being used in resource management (e.g. in the regulation of the Salmon fisheries in the Pacific Northwest). In addition to its practical uses, Indigenous knowledge is increasingly becoming a "political

47 Michael E. Harkin and David Rich Lewis (eds.). *Native Americans and the Environment. Perspectives on the Ecological Indian*. Lincoln: University of Nebraska Press, 2007.

48 Charles R. Menzies. Ecological Knowledge, Subsistence, and Livelihood Practices. In *Traditional Ecological Knowledge and Natural Resource Management*. Charles R. Menzies (ed.), 87–104. Lincoln: University of Nebraska Press, 2006, 87.

49 Menzies, Ecological Knowledge, 101.

50 Menzies, Ecological Knowledge, 101.

51 Caroline Butler. Historicizing Indigenous Knowledge. In *Traditional Ecological Knowledge and Natural Resource Management*, Charles R. Menzies (ed.), 107–126. Lincoln: University of Nebraska Press, 2006, 112.

52 Butler, Historicizing, 107.

resource."[53] Its use "is a political act – it is a claim of Aboriginality, an assertion of land and resource rights, and a demand for management power" (i.e. stewardship).[54] But perhaps most importantly, traditional ecological and topological knowledge has proven useful in the legal field. In recent court trials conducted to clarify native land and resource rights, Indigenous oral traditions were used as evidence in settling claims to sovereignty and resource ownership.

Orally transmitted stories may be old but we cannot tell just how old. They tend to telescope various historical events into one and the same story, largely ignoring historical chronology except in oral accounts of the lineage of leaders. While imperial culture crucially depends on chronological and linear constructions of time because its spatial constellations are continuously changing (its spatial memory culture is as global as its economic reach), Indigenous cultures have stronger ties to particular local places than to time. Of course Native Americans feel responsible for keeping a memory of the deeds and wisdom of the ancestors (and for burying their bones). And of course they feel a responsibility for the future – for the coming "seven generations." But they have little understanding for imperial cultures' restless (and nervous) search for true origins: the first immigrant, the oldest artifact, endlessly imagined in science and science fiction and scientific horror fiction. From an Indigenous perspective, the incontestable fact is that they were in America many generations before the newcomers. If we accept a Native American logic of ownership, matters of primacy lose their importance. Ownership, they suggest, is gained not from proving millennia-old ancestry but from familiarity with the place you inhabit – with the land, the flora and fauna of the region where you live and for whose future prosperity you adopt responsibility. "If this is your land, where are your stories," is the quote from a tribal elder which Edward Chamberlin takes for the title of a book that reflects on the relevance of stories as proof of residence and on the significance of topographical stories as evidence of cultural belonging.[55] This requires a knowledge of how all of these things are related to one another. And it requires a certain attachment to these things – a tactile knowledge that Western culture and science have only just begun to seriously pursue.

---

53 Butler, Historicizing, 119, quoting Mark Nuttall.

54 Butler, Historicizing, 119.

55 Edward Chamberlin. *If This Is Your Land, Where Are Your Stories? Finding Common Ground.* Toronto: Vintage, 2004.

Coda: Transcultural Chronotopes

In Indigenous chronotopes, the topical is paramount. Native American stories suggest that knowledge of place, in addition to knowledge of time, may prove indispensible for survival. As Leslie Marmon Silko writes when contemplating the southwestern desert:

> The bare vastness of the Hopi landscape emphasizes the *visual* impact of every plant, every rock, every arroyo. *Nothing is overlooked or taken for granted.* Each ant, each lizard, each lark is imbued with great value simply because the creature is there, simply because the creature is alive in *a place where any life at all is precious.* Stand on the mesa edge at Walpai and look west over the bare distances toward the pale blue outlines of the San Francisco peaks where the ka'tsina spirits reside. So little lies between you and the sky. So little lies between you and the earth. One look and you know that simply to survive is a great triumph, that every possible resource is needed, every possible ally – even the most humble insect or reptile. You realize you will be speaking with all of them if you intend to last out the year. Thus it is that the Hopi elders are grateful to the landscape for aiding them in their quest as spiritual people.[56]

While we may all need the knowledge of ants and lizards for our own survival, human beings will certainly need each others' skills and knowledge – a truly transcultural constellation. Bakhtin's chronotope of "threshold" comes close to such a transcultural view of chronotoping.[57] The threshold is traditionally the site where 'inside' and 'outside' meet and form contact zones, middle grounds, mixed places, perhaps heterotopia in one of the meanings of that term provided by Foucault.[58]

First offered by Fernando Ortiz in 1940 and then widely forgotten until the time was finally ripe for a revival of the concept in the early 1990s (see Mary Pratt's *Imperial Eyes*),[59] "transculturation" emphasizes a certain equilibrium between the cultural materials exchanged during processes of cultural encounter.[60] In his

56 Leslie Marmon Silko. Landscape, History, and the Pueblo Imagination. In *At Home on the Earth. Becoming a Native to Our Place*, David Landis Barnhill (ed.), 30–42. Berkeley: UC California, 1999, 42, emphasis added.

57 Bakhtin, *Dialogical Imagination*, 248–249.

58 In "Of Other Spaces," Foucault mentions the colony as an example for heterotope. He does not discuss the transcultural aspect of colonies, though. (Michel Foucault. Of Other Spaces. *Diacritics* 16:1 (1986): 22–27.) For "middle ground" see Richard White. *The Middle Ground. Indians, Empires, and Republics in the Great Lakes Region, 1650–1815*. Cambridge: Cambridge University Press, 1991; for "mixed space" see Lefebvre, *Production*.

59 Mary Louise Pratt. *Imperial Eyes. Travel and Transculturation*. London and New York: Routledge, 1992.

60 Fernando Ortiz. *Cuban Counterpoint. Tobacco and Sugar*, Fernando Coronil (ed.), Harriet de

introduction to the English edition of *Contrapunteo Cubano* (1940), Bronislaw Malinowski stresses the aspects of processuality and innovativeness in his own reformulation of the term coined by Ortiz:

> [T]ransculturation is a process [...] in which both parts of the equation are modified, a process from which a new reality emerges, transformed and complex, a reality that is not a mechanical agglomeration of traits, nor even a mosaic, but a new phenomenon, original and independent.[61]

The economic and ecological development of the world suggests that the imperial lion may have to bend to such offers of mutuality and reciprocity. It may actually have to listen to the humble desert ant to learn its lessons about survival. The coloniality of knowledge inherent in the imperial chronotope may thus give way to a transcultural episteme, to intellectual dialogue and equal and fair exchange.

References

Abbott, Alison. Mexican skeleton gives clue to American ancestry. Genetic signature from cave remains matches that of modern Native Americans. *Nature,* 14 May 2014, www.nature.com/news/mexican-skeleton-gives-clue-to-american-ancestry-1.15226 (1 December 2015).

Azim, Firdous. *The Colonial Rise of the Novel.* London: Routledge, 1991.

Bakhtin, Mikhail. *The Dialogic Imagination. Four Essays.* Austin: University of Texas Press, 1981.

Balter, Michael. Native Americans Descend From Ancient Montana Boy. *Science*, news.sciencemag.org/archaeology/2014/02/native-americans-descend-ancient-montana-boy (06 December 2015).

Berkeley, George. On the Prospect of Planting Arts and Learning in America. In *Key Concepts in American Cultural History. From the Colonial Period to the End of the 19th Century,* Bernd Engler und Oliver Scheiding (eds.), 32. Trier: WVT, 2005 [1752].

Brantlinger, Patrick. *Rule of Darkness. British Literature and Imperialism, 1830–1914.* Ithaca: Cornell University Press, 1988.

Butler, Caroline. Historicizing Indigenous Knowledge. In *Traditional Ecological Knowledge and Natural Resource Management,* Charles R. Menzies (ed.), 107–126. Lincoln: University of Nebraska Press, 2006.

Cajete, Greg. Native Astronomy. An Indigenous View Dancing the Heavens... In *The Pari Dialogues. Essays in Indigenous Knowledge and Western Science,* F. David Peat (ed.), 41–84. Pari (Italy): Pari Publishing, 2013.

Onís (trans.). Durham: Duke University Press, 1995 [1940].

61 Bronislaw Malinowski. Introduction. In Fernando Ortiz. *Cuban Counterpoint. Tobacco and Sugar*, Fernando Coronil (ed.), Harriet de Onís (trans.), lvii–lxiv. Durham: Duke University Press, 1995 [1940], lviii–lix.

Calloway, Colin G. *One Vast Winter Count. The Native American West before Lewis and Clark.* Lincoln: University of Nebraska Press, 2003.

Chakrabarty, Dipesh. *Provincializing Europe. Postcolonial Thought and Historical Difference.* Reissue. Princeton: Princeton University Press, 2008.

Chamberlin, Edward. *If This Is Your Land, Where Are Your Stories? Finding Common Ground.* Toronto: Vintage, 2004.

Cheyfitz, Eric. *The Poetics of Imperialism. Translation and Colonization from "The Tempest" to "Tarzan".* Oxford: Oxford University Press, 1991.

Cruikshank, Julie. *The Social Life of Stories. Narrative and Knowledge in the Yukon Territory.* Lincoln: University of Nebraska Press, 1998.

Cruikshank, Julie. *Do Glaciers Listen? Local Knowledge, Colonial Encounters, and Social Imagination.* Vancouver: British Columbia University Press, 2005.

Deloria, Vine. *Red Earth White Lies. Native Americans and the Myth of Scientific Fact.* Golden: Fulcrum, 1997.

Fabian, Ann. *The Skull Collectors. Race, Science, and America's Unburied Dead.* Chicago: University of Chicago Press, 2010.

Fabian, Johannes. *Time and the Other. How Anthropology Makes its Object.* 2d ed. New York: Columbia University Press, 2002.

Ferguson, Jennifer K. and Matilda George (ed.). *Upper Columbia River Book of Legends.* Nespelam, WA: Confederated Tribes of the Colville Reservation, 2007, www.colvilletribes.com/book_of_legends.php (06 December 2015).

Foreman, Allen. Termination, www.klamathtribes.org/background/termination.html (16 December 2015).

Foucault, Michel. Of Other Spaces. *Diacritics* 16:1 (1986): 22–27.

Goddu, Teresa A. *Gothic America. Narrative, History, and Nation.* New York: Columbia University Press, 1997.

Goody, Jack. *The Theft of History.* Cambridge: Cambridge University Press, 2006.

Gould, Stephen Jay. *Time's Arrow Time's Cycle. Myth and Metaphor in the Discovery of Geological Time.* Cambridge: Harvard University Press, 1987.

Gould, Stephen Jay. *The Mismeasure of Man.* 1981. New York: W.W. Norton & Company, 1996.

Gray, Robert. *A good speed to Virginia*, Wesley F. Craven (ed.). New York: Scholars' Facsimiles and Reprints, 1937 [1609].

Hamilton, Ross. *Star Mounds. Legacy of a Native American Mystery.* Berkeley: CA: North Atlantic Books, 2012.

Hardt, Michael and Antonio Negri. *Empire.* Cambridge: Harvard University Press, 2000.

Harkin, Michael E. and David Rich Lewis (eds.). *Native Americans and the Environment. Perspectives on the Ecological Indian.* Lincoln: University of Nebraska Press, 2007.

Huhndorf, Shari. *Mapping the Americas. The Transnational Politics of Contemporary Native Culture.* Ithaca: Cornell University Press, 2009.

Hulme, Peter. *Colonial Encounters. Europe and the Native Caribbean, 1492–1797.* London: Methuen, 1986.

Hulme, Peter. Tales of Distinction European Ethnography and the Caribbean. In *Implicit Understandings. Europeans and Other Peoples in the Early Modern Era.* Stuart Schwartz (ed.), 16–34. Cambridge: Cambridge University Press, 1994.

Hunt, Lynn. *Measuring Time, Making History.* Budapest and New York: Central European University Press, 2008.

Lankford, George E. *Reachable Stars. Patterns in the Ethnoastronomy of Eastern North America.* Tuscaloosa: University of Alabama Press, 2007.

Lefebvre, Henri. *The Production of Space*, Donald Nicholson-Smith (trans.). Oxford: Blackwell, 1991 [1974].

Lost Franklin expedition ship found in the Arctic. *CBC News*, 9 September 2014, www.cbc.ca/news/politics/lost-franklin-expedition-ship-found-in-the-arctic-1.2760311 (06 December 2015).

Lowenthal, David. *The Heritage Crusade and the Spoils of History*. Cambridge: Cambridge University Press, 1998.

Luna, Pedro de. The Man Who Faced the Saber-Toothed Cat. In *Fugitive Knowledge. The Preservation and Loss of Knowledge in Cultural Contact Zones,* Andreas Beer and Gesa Mackenthun (eds.), 163–179. Münster: Waxmann, 2015.

Mackenthun, Gesa. Expansionism. In: *A Companion to American Cultural History. From the Colonial Period to the End of the 19th Century,* Bernd Engler and Oliver Scheiding (eds.), 247–273. Trier: WVT, 2009.

Mackenthun, Gesa. Fossils and Immortality. Geological Time and Spiritual Crisis in Nineteenth-Century America. In *Deutungsmacht. Religion und Belief Systems in Deutungsmachtkonflikten,* Philipp Stoellger (ed.), 259–283. Tübingen: Mohr Siebeck, 2014.

Mackenthun, Gesa. Night of First Ages. Deep Time and the Colonial Denial of Temporal Coevalness. In *Crossroads in American Studies. Transnational and Biocultural Encounters,* Frederike Offizier, Marc Priewe, Ariane Schröder (eds.). Heidelberg: Winter, 2015. In print.

Mackenthun, Gesa. 'Unhallowed Mysteries' in the Colonial Archive. Competing Epistemologies in North America. In *Concurrences in Colonial and Postcolonial Studies*. Diana Bryden, Peter Forsgren and Gunlög Fur (eds.). Amsterdam: Rodopi, 2015. In print.

Mackenthun, Gesa. Hidden Cities in the American Wilderness. The Cultural Work of a Romantic Trope. *Miranda* [Online], 11 (2015), miranda.revues.org/6954 (06 December 2015).

Malinowski, Bronislaw. Introduction. In Fernando Ortiz. *Cuban Counterpoint. Tobacco and Sugar*, Bronislaw Malinowski (ed.), Harriet de Onís (trans.), lvii–lxiv. Durham: Duke University Press, 1995 [1940].

Mann, Charles. *1491. New Revelations of the Americas before Columbus.* New York: Knopf, 2015.

Melville, Herman. *White Jacket*. New York: Grove Press, 1956 [1850].

Menzies, Charles R. Ecological Knowledge, Subsistence, and Livelihood Practices. In *Traditional Ecological Knowledge and Natural Resource Management.* Charles R. Menzies (ed.), 87–104. Lincoln: University of Nebraska Press, 2006.

Mignolo, Walter. *The Darker Side of Western Modernity. Global Futures, Decolonial Options.* Durham: Duke University Press, 2011.

Oakley, Christopher Arris. The Center of the World. The Principle People and the Great Smoky Mountains. In *Landscapes of Origin in the Americas*. Jessica Joyce Christie (ed.), 3–14. Tuscaloosa: University of Alabama Press, 2009.

Ortiz, Fernando. *Cuban Counterpoint. Tobacco and Sugar*, Bronislaw Malinowski (ed.), Harriet de Onís (trans.). Durham: Duke University Press, 1995 [1940].

O'Sullivan, John L.. Annexation. In *Key Concepts in American Cultural History. From the Colonial Period to the End of the 19th Century,* Bernd Engler und Oliver Scheiding (eds.), 389. Trier: WVT, 2005 [1845].

Pagden, Anthony. *Lords of All the World.* Oxford: Oxford University Press, 1995.

Pearce, Roy Harvey. *Savagism and Civilization. A Study of the Indian and the American Mind.* Rev. reprint. Berkeley: University of California Press, 1988.

Porter, Joy. *Native American Environmentalism. Land, Spirit, and the Idea of Wilderness.* Lincoln: University of Nebraska Press, 2014.

Pratt, Mary Louise. *Imperial Eyes. Travel and Transculturation*. London and New York: Routledge, 1992.

Said, Edward W. History, Literature, and Geography. In *Reflections on Exile and Other Literary and Cultural Essays*, Edward W. Said (ed.), 453–473. London: Granta, 2000.

Schaafsma, Polly and Will Tsosie. Xeroxed on Stone. Times of Origin and the Navajo Holy People in Canyon Landscapes. In *Landscapes of Origin in the Americas* Jessica Joyce Christie (ed.), 15–31. Tuscaloosa: University of Alabama Press, 2009.

Schweninger, Lee. *Listening to the Land. Native American Literary Responses to the Landscape.* Athens: University of Georgia Press, 2008.

Silko, Leslie Marmon. Landscape, History, and the Pueblo Imagination. In *At Home on the Earth. Becoming a Native to Our Place,* David Landis Barnhill (ed.), 30–42. Berkeley: UC California, 1999.

Soja, Edward W. *Postmodern Geographies. The Reassertion of Space in Critical Social Theory.* London: Verso, 1989.

Vattel, Emmerich de. *The Law of Nations.* London: Newberry, 1759, www.constitution.org/vattel/vattel_01.htm (06 December 2015).

Weinberg, Albert. *Manifest Destiny. A Study of Nationalist Expansionism in American History.* 1935. Reprint. Chicago: Quadrangle, 1963.

Yates, Frances A. *Astraea. The Imperial Theme in the Sixteenth Century.* Harmondsworth: Penguin, 1975.

Ishita Banerjee-Dube

In Other Times: Apocalypse, Temporality, Spatiality in Eastern India

A few decades ago, Marc Bloch had defined history as the "science of men in time". Historical time, he had insisted, was a concrete and lived reality with "an irreversible onward rush".[1] Such time simultaneously embodied a continuum and perpetual change, making historical inquiry face serious problems with regard to the treatment of time. How was it to deal with the flow of time and focus on segregated time of a particular period?[2]

For Reinhart Koselleck, the renowned theorist of the concept (*Begriff*) of history, the discovery (in Europe, between 1750 and 1850) that historical time was different from the time of nature gave history its defining element and purpose. The singularity of "denaturalized" historical time lay in the fact that it was affected by human action and purposiveness in ways "natural time" was not. History, therefore, could "be made" and "suffered".[3] At the same time, this amenability to human action and aspiration made the idea of a singular historical time problematic; it was virtually impossible to think of time determined by "social and political units of action", acting and suffering human beings, and their organisations and institutions, as one.[4]

I draw upon Bloch's enunciation of the problem of historical inquiry, and his and Koselleck's affirmation that time in history is a concrete, lived reality to explore their implications for the actual, quotidian experience of time and space by ordinary peoples in a society where time and space are context-sensitive and have properties that affect human beings that dwell in them.[5] This distinct context-specificity shapes experience in discrete ways and opens up novel possibilities for the understanding of time and space. Historical and natural time flow into each other, as do divine and secular, cyclical and linear, segmentary and sequential time, inflecting thereby notions of natural and historical, abstract and universal, and concrete time in intricate ways.

1 Marc Bloch. *The Historian's Craft*. 7th ed. Manchester: Manchester University Press, 1992, 23.
2 Bloch, *Historian's Craft*, 24.
3 Hayden White. Foreword. In R. Koselleck. *The Practice of Conceptual History. Timing History: Spacing Concepts*, ix–xiv. California: Stanford University Press, 2002, xi.
4 Reinhart Koselleck. *The Practice of Conceptual History. Timing History: Spacing Concepts.* California: Stanford University Press, 2002, 110.
5 Attipat Krishnaswami Ramanujan. Is there and Indian Way of Thinking? An Informal Essay. *Contributions to Indian Sociology* (n.s.), 23:1 (1989): 41–58, 51.

DOI 10.1515/9783110418750-005

This chapter will explore distinct dimensions of time/space in the lived world of peoples and places far away from Europe. It will highlight how different qualities of concrete time/space and their varied perceptions shape apprehensions and experience of time, and lay bare different dispositions toward time/space that are not mere reflections of 'universal' time/space. In other words, it will try and open-up the notion of universal time/space as 'given' and 'provincialize' it as one among many constructions of time/space in lived experience. Taking aspirations and experience as crucial elements in configurations of time and space, and drawing inspiration from Bakhtin's 'extraordinary sensitivity to the plurality of experience',[6] the paper will try and unsettle the settled hierarchies of linear, homogeneous, 'rational' time, and cyclical (circular) mythical time, and interrogate the idea of a definite break between perceptions of time/space in the 'Middle Ages' and in 'modern' times. It will raise the issue of whether it is possible to grasp the *Begriff* of history among peoples and in places where plural, concrete time easily blends with the natural and the sacred. If yes, it is possible to uphold the singularity of historical time as totally distinct from natural and divine time?

To advance my arguments, I will focus on the adherents of Mahima Dharma, a radical, heterodox religious order of mid-nineteenth century Orissa, eastern India that exists till today, and *malikas*, a genre of apocalyptic texts current in Orissa for centuries.[7] My effort is two-fold. First, a quick analysis of the distinct notions and perceptions of time and temporality in Judaic, Christian and Hindu traditions will underscore the common issues and 'mysteries' that have shaped reflections on and orientations to time in the three traditions. This will be accompanied by a brief, critical reflection on influential scholarship (on India) that substantiated the spatial division and ordering of the world into orient and occident by means of scientific studies of language, cosmology and temporality. Such studies served to establish the orient as essentially different, albeit with common origins of peoples and languages. The two together will set the stage for an exploration of understandings and deployment of classical notions of time by the adherents of Mahima Dharma and other peoples of Orissa, deployments shaped by imaginative apprehensions of a genre of popular apocryphal texts. A focus on time and temporality – the length and duration, and beginning and end of time – together with its normative content, will enable a rethinking of divine and social, circular

6 Michael Holquist. Introduction. In *The Dialogic Imagination. Four Essays by M.M. Bakhtin*, M. Holquist (ed.), xv–xxxiii. Austin: University of Texas Press, 1981, xx.

7 The name of Orissa has been officially changed to Odisha and of the language to Odia in 2010. But since I am examining a period when the province was known as Orissa and all the works and reports I cite use Orissa, I use Orissa instead of Odisha and Oriya instead of Odia in the chapter.

and linear, empty and filled, and concrete historical time, and pose new questions about the qualitative and quantitative dimensions of time. This in turn will allow serious reflection on the nature of historical time.

1 Time of Gods and the Father

India, for long, has been seen to provide the best illustration of the circular/cyclical notion of time. Advocated by early British orientalists such as Horace Hayman Wilson in the introduction to the translation of the *Vishnu Purana*, the idea was forcefully perpetrated by Mircea Eliade in his *Cosmos and History. The Myth of the Eternal Return*.[8] Examining the ontology of pre-modern, traditional, archaic societies, of Asia, Europe and America, among which India figured prominently, Eliade argued that the conscious and continuous repetition of the "given paradigmatic gestures" by people of such societies revealed an original ontology.[9] The principal difference between the man of "archaic and traditional societies" and the man of "modern societies with their strong imprint of Judeo-Christianity", he stated in the Preface, lay in the fact that the archaic man feels himself indissolubly connected with the Cosmos and cosmic rhythms, while the modern man insists the he is connected only with History.[10] This accounted for the "immense importance" of "collective regeneration through repetition of the cosmogonic act", through "the abolition of concrete time", a fact that conferred on such rituals an "ahistorical intent".[11] Indian spirituality, in particular, was outstanding on account of its metaphysical and ahistorical structure revealed in the fact that Indians never "elaborated a cosmological New Year scenario".[12] This profound ahistorical intent integrated in endless repetition was then connected with an absence of a will to human action in order to change the course of life or history.[13] The sacred time of the religious man, wrote Eliade in *The Sacred and the Profane*, consisted of "a succession of eternities", periodically recoverable during the festivals that made up the sacred calendar. Since liturgical time of this calen-

8 Mircea Eliade. *Cosmos and History. The Myth of the Eternal Return*, trans. Williard R. Trusk. New York: Harper Torchbooks, 1959.
9 Eliade, *Cosmos and History*, 5.
10 Eliade, *Cosmos and History*, vii.
11 Eliade, *Cosmos and History*, 75, 85.
12 Eliade, *Cosmos and History*, 75.
13 Romila Thapar. Time Concepts, Social Identities and Historical Consciousness in Early India. In *Time in India: Concepts and Practices*, Angelika Malinar (ed.), 23–38. New Delhi: Manohar, 2007.

dar flowed in a closed circle, sacred time appeared to be cyclical, reversible and recoverable, "a sort of an eternal mythical present that is periodically reintegrated by means of rites."[14] The cycle extended to cover Hindu time in its entirety from the time of creation: a time divided into *yugas*, epochs that succeeded each other eternally, where the world was renewed periodically at the end of each epoch.

Hindu cyclical time was contrasted with "the Judeo-Christian eschatology of the Garden of Eden at the start of time and Judgement Day at the end."[15] It marked traditional ahistorical India's essential difference from the scientific, rational West with history. The secularization of Judeo-Christian time and space that entailed a changeover from time as the medium of divine/sacred events specific to a chosen people to a generalization and universalization of time that was now bereft of its eschatological content, was not seen to have occurred in archaic India. The secularization of Judeo-Christian time, in turn, enabled a break from "a conception of time/space in terms of a history of salvation" to "a secularization of Time as natural history".[16]

The striking similarity in the ways Jewish people and Hindus mark, relate to, and live in time notwithstanding, the lack of a sense of history or a will to human action in order change the course of history that supposedly characterizes the Indians (Hindus) is clearly absent in Judaism. In the words of Eliezer Berkovits, an orthodox Jewish theologian, "God conceals his face" so as to allow humanity to act. History, conceived as man's free and responsible activity, makes it necessary for God to conceal his face. When, as during the miracle of Exodus, he intervenes wonderfully in history, history in reality stands still.[17] At the same time, it is equally true that the history of Israel is posited on the idea of God's presence (and revelation) in spaces and time that makes the present eternal and time concrete and filled (*gefüllte Zeit*), a time filled with events.[18]

Biblical imagination of the New Testament, writes Peter Manchester, joins "historical and mythical time" – the "inside" of time and its "epic dimension" – in "a characteristic and philosophically influential way".[19] The unfolding of "sal-

14 Mircea Eliade. *The Sacred and the Profane. The Nature of Religion*. New York: Harcourt, Brace & World, 1959, 70.

15 Thapar, Time Concepts, 23.

16 Johannes Fabian. *Time and the Other. How Anthropology Makes its Object*. New York: Columbia University Press, 1984, 2.

17 Berkovits cited in Peter Steensgaard. Time in Judaism. In *Religion and Time*, A. Niyogi Balslev and J.N. Mohanty (eds.), 64–108. Leiden: E.J. Brill, 1993, 106.

18 Steensgaard, Time in Judaism, 64.

19 Peter Manchester. Time in Christianity. In *Religion and Time*, A. Niyogi Balslev and J.N. Mohanty (eds.), 109–137. Leiden: E.J. Brill, 1993, 126.

vation history" from Genesis through the Exodus to the giving of the Law and the coming of Christ as a fulfilment of the Law and the promises made to Abraham, to the Apocalypse and the second coming of Christ as eschatological judge, effects a smooth transition between mythical time and archeologically and philologically constructed chronology, with "eucharistic practice" allowing emergent Christianity to integrate the time of the dying and rising of Christ into the time of the rest of Christian history.[20] This smooth transition has meant, in Manchester's reckoning, that the "critical decoupling" of the two times demanded by "historical consciousness" remains "widely unattempted" in the cultures "shaped by biblical religion".[21]

We will have more to say on the "critical decoupling demanded by historical consciousness" later. For the present, it is important to note that notions of time and space are not as fundamentally distinct in Hinduism, Judaism and Christianity as it is taken to be. All of them show an engagement with questions of elapsed time (to the present), i.e., duration, and with the mystery of the end of time (*eschaton*). The two elements of chronological continuity (historical time) and end of time underpinned by a messianic hope of redemption and renewal appear in all three (as well as in Islam). What allegedly distinguishes Hindu time from Judeo-Christian time is the preponderance of "mythic" time embodied in the dissolution and renewal of the epochs.

Early India, it was been argued persuasively, showed evidence of the co-existence of linear and cyclical time, chronological and epic time.[22] Apart from the fact that texts of distinct genres dealt with distinct modes of time reckoning, the co-existence of linear and circular time is evident in the succession of the four classical *yugas*, cosmic cycles or epochs/eras – Satya (Kreta), Treta, Dwapara and Kali. These four yugas together make up the cycle of a *mahayuga* (a great yuga) with twilight periods in-between, all based on elaborate astronomical and numerical calculations.[23]

The sequence of the four yugas is marked by a progressive decline in the duration and the moral/normative content of time. Moreover, since time and space are endowed with qualities that affect the nature of human beings who live in them, the characteristics of people also vary in accordance with the yugas. 4,000 divine

20 Manchester, Time in Christianity, 126, 118.

21 Manchester, Time in Christianity, 126–127.

22 Romila Thapar. Society and Historical Consciousness. The Itihasa-Purana Tradition. In *Situating Indian History for Sarvepalli Gopal*, S. Bhattacharya, R. Thapar (eds.), 353–383. New Delhi: Oxford University Press, 1996; Romila Thapar. *Time as a Metaphor of History. Early India*. Oxford: Oxford University Press, 1997; Thapar, Time Concepts, for instance.

23 Thapar, Time Concepts, 25–27.

years make up *satyayuga*, with *treta*, *dwapara* and *kaliyuga* consisting respectively of 3,000, 2,000 and 1,000 divine years. Dharma, faith/right conduct/ morality (often depicted as a superior being) measures four feet in Satya and is reduced to only one in Kaliyuga. The four yugas, therefore, have to follow each other in a strict, linear sequence. Kaliyuga (the era of evil), the last and worst of the four, the current one, cannot come right after Satyayuga, the era of truth/virtue. And only after Kaliyuga comes to an end can a new cycle of mahayuga begin. So even if there is a "return" to Satyayuga at the end of Kaliyuga, it is not clear whether the return will be a mere repetition or a new beginning with time acquiring new normative content.

This notion of succession of yugas is intimately linked to that of the appearance of an *avatara*, an incarnation of the divine, particularly of Lord Vishnu, the preserver of the world, who comes down to earth to rid it of sin and re-establish *dharma*.

The end of the world (*eschaton*) and millenarian visions, we are aware, are current in all religious trends and involve similar understandings and apprehensions of time and temporality. Important works have demonstrated the currency of millenarianism in the "connected world" of the late fifteenth century; in particular, the influence of millenarian aspirations (and Franciscan apocalyptic thought) on Columbus's drive to set out on a westward voyage. Such notions in turn, found a curious and ironical parallel in the apocalyptic vision of some of the indigenous American peoples the Spaniards encountered in the aftermath of 1492.[24] New material conditions of the sixteenth century enabled millenarianism to propagate itself as a phenomenon that embraced a vast geographical space and yet had unique local manifestations.[25] If notions of the apocalypse predominated in the Mediterranean world, optimistic views about a possible re-ordering of the known world through the intercession of a *mujaddid* (Renewer) circulated in the Ottoman Empire, Iran and North Africa in the late sixteenth-early seventeenth century. The Mujaddid again, found a close parallel in the Imam Mahdi, the Concealed or the Expected One, who would emerge to reform the world in a radical fashion.[26] Both these notions were current in Mughal India and possibly have provenance till today.

24 Sanjay Subrahmanyam. Connected Histories. Notes Towards a Reconfiguration of Early-Modern Eurasia. *Modern Asian Studies* 31:3 (1997): 735–762, 749.

25 Subrahmanyam, Connected Histories, 749–750.

26 Subrahmanyam, Connected Histories, 749–751.

Eschaton, I shall argue, retains a lingering presence; it projects time in the horizon of the future and makes the future appear in the present, enabling thereby intricate apprehensions of current time/space in everyday social worlds.

The Book of Daniel (of the *Old Testament*) closely parallels descriptions of the dissolution of the world and the characteristics of the era prior to that, in the *Puranas*. *Puranas* are Hindu sacred texts that eulogize one of the three gods of the triumvirate (Brahma, Vishnu, Maheshwara/Shiva) and/or the Goddess, and reveal the story of the universe from its creation to destruction, divulging at the same time, cosmology, geography, philosophy as well as genealogies of kings, heroes, dynasties, sages and demigods in the form of a dialogue between two persons. According to the Book of Daniel, kings fight kings, some fall some rise, and the ones who rise are stronger and more cruel than the others (Book of Daniel, 10–12). In the *Vishnu Purana*, Kaliyuga is associated with the predominance of avaricious, low-caste (and foreign) rulers who exploit the subjects to the extreme, the reversal of social norms, decrease in the prevalence of dharma, and a rise in the number of heretical sects. All this makes Kalki(n), the tenth and last avatara of Vishnu come to earth toward the end of the yuga to destroy the *mlechhas* (foreigners/outcastes) and heretics and usher in the age of virtue, and, along with it, a new cycle of mahayuga.[27]

The Old Testament does not clearly predict what will happen after dissolution (a fact taken up seriously by Walter Benjamin in his discussion of Death as Salvation or Salvation only in and through Death, a point we will not be able to develop here). Apocalypse of the New Testament (the account of John the Apostle), on the other hand, is clear on what will come after the Apocalypse, which involves a series of disasters. Famine, disease, earthquake, lightning, thunder, death and destruction of human beings on a massive scale follow each other in regular sequence. There is, however, no mention of a second deluge. The end of old heaven and earth is followed by the emergence of new ones. "God's home is now with his people. He will live with them, and they will be his own. Yes, God will make his home among his people. He will wipe all tears from their eyes, and there will be no more death, suffering, crying, or pain. These things of the past are gone forever" (Book of Revelation, 21:3). A dazzling New Jerusalem, made bright by the glory of God, epitomizes this bright future. This idea of Redemption therefore, reinforces the notion of a cycle even as it speaks of linear, chronological time from the beginning to the end.

27 Thapar, *Time as a Metaphor*, 41.

2 Divine Time, Social Time

Different motifs characterise the end of Kaliyuga. It can be brought to an end by the appearance of an incarnation, or it can come to an end on account of its own sins and evils that brings forth the deluge, *pralaya*. Pralaya, which according to Monier-Williams' *Sanskrit-English Dictionary*, connotes to disappear, perish, die as well as dissolution, annihilation, reabsorption or the end of the world on account of a deluge, finds mention in post-Vedic literature such as *Shatapatha Brahmana*, and is developed further in the epic *Mahabharata* (and its part the *Bhagavad Gita*), and the *Matsya*, *Vishnu*, and *Bhagavata Purana*, i.e. texts related directly to the Vaishnavite tradition of Hinduism that advocates the theory of incarnation.

The earliest version of pralaya in the *Shatapatha Brahmana* (c. VII–VI B.C.) depicts cosmic dissolution as being produced by a flood of catastrophic proportions. It destroys all living beings on earth except Manu, the lawgiver, who saves himself by building a boat on the advice of Matsya (or Mina, a fish who is an avatara of Vishnu). In later versions of the legend, Manu saves other living beings by giving them refuge in the boat.[28]

It is important to note here that in the *Shatapatha Brahmana*, pralaya (deluge) is not connected with the idea of cyclical cosmic creation and destruction of the world. In a manner similar to the Book of Daniel, it refers to destruction of the world and all living beings with the exception of Manu. By the time of the *Puranas*, the notion of cosmic cycles had become prominent; hence, Manu is credited with saving other living beings of the world, and Kalki(n), the tenth incarnation of Vishnu is made to appear to bring Kaliyuga to an end. Kalki(n), interestingly, is a Brahmin who, in later versions of the legend, takes on the role of a warrior because a battle of great proportions is needed to destroy the evil Kali. Deluge and end are superseded by destruction caused by an incarnation and the re-establishment of Satyayuga signalling thereby the (in)famous beginning of the eternal cycle of periodic creation and destruction.

If the periodic regeneration and annulment of linear time characteristic of archaic societies fascinated Eliade, colonial scholar-administrators of the eighteenth century were troubled by yet another aspect of Hindu conception of time, namely the duration – elapsed time since Creation – of the yugas, and the extent of divine time (*aeon*). One day of Brahma (the Creator)'s time corresponds to one

28 *The Mahabharata*. Vol 2. Trans. and ed. J.A.B van Buitenen. Chicago: University of Chicago Press, 1975, 203–204; Wendy Doniger. *Hindu Myths. A Sourcebook Translated from the Sanskrit*. UK: Penguin Press, 2004, 180–181.

aeon: this makes the span of the epochs vast and takes them well beyond the time of Moses.

Once again, we are struck by the close parallel with Judaic time according to which the world is destined to last 6000 years, corresponding to six days of God's time. This means that a single day of God's time spans a 1000 years of human time.[29] The first two days, corresponding to 2000 years, compose the era of chaos, the second to the fourth days make up the 2000 years of the Torah and the last, fourth to sixth days, constitute the era of the Messiah. This long account of duration brought Judaic time in conflict with the Biblical account of Creation (conventionally dated to 2348/9 BC), just as the long duration of divine time and Hindu epochs caused serious concern by positing the beginning of the world well beyond Biblical chronology.

The findings of early orientalist scholars such as Nathaniel Brassey Halhed, in spite of an apparent conformity with Mosaic chronology, seemed to indicate that Hindu scriptures predated the deluge by about 900 years, which coincided with the beginning of Kaliyuga. It also explained why Hindu scriptures did not contain any "flood narrative",[30] which, as we have just seen, is not strictly true. If Kaliyuga, the last and the shortest, is believed to have begun in 3102 BC,[31] i.e. before Biblical time, the preceding three yugas took Indian antiquity back by several thousand years. The findings of "early orientalism", therefore, provided "ammunition to doubters of orthodox Christianity in France", and substantiated the conclusions of writers on ancient astronomy and mythology working in the sceptical tradition of Voltaire.[32]

William Jones, a renowned orientalist scholar and judge, and the founder of the Asiatic Society of Bengal (1784), conclusively settled the anxiety over the truth of the biblical account. Jones deployed his profound knowledge of Indian languages and texts to arrive at "a rational and scientific basis" on which to confirm Mosaic ethnography and Biblical chronology. He dismissed the first three yugas of extreme antiquity as "chiefly mythological" and dated the beginning of the current, *historical* era of Kaliyuga to no earlier than 2000 B.C.[33]

---

29 Talmud and the Epistle of Barnabas cited in Sylvie Anne Goldberg. Questions of Times. Conflicting Timescales in Historical Perspective. *Jewish History* 14 (2000): 267–286, 273.

30 Michael Dodson. *Orientalism, Empire and National Culture. India, 1770–1880*. New Delhi: Foundation Books, 2010, 29.

31 Heinrich von Stietencron. Kalkulierter Religionsverfall: Das Kaliyuga in Indien. In *Der Untergang von Religionen*, Hartmut Zinser (ed.), 135–150. Berlin: Dietrich Reimer Verlag, 1986, 135.

32 Michael Dodson. *Orientalism, Empire and National Culture*, 29.

33 Ibid, 30.

Once more, we are reminded of how, through prolonged and diverse debates, Jewish and non-Jewish scholars, historians, and specialists of the Bible and Antiquity, came to regard the Torah as an "essentially unhistorical entity",[34] since its chronology did not correspond to actual events enacted in real space. William Jones's new methodology allowed him to study the comparative grammatical structure and vocabulary of languages along with religious doctrines, mythology and archaeological evidence, in order to establish historical connections between languages and their speakers. In his several anniversary discourses at the Asiatic Society, he identified the Indians and Persians, together with the Greeks, Romans and Egyptians as descendants of Ham, the son of Noah, who originally spoke a common language. This "ethnological" study enabled Jones to "scientifically" substantiate the Mosaic account of creation as it appears in the Book of Genesis.[35] While rational and scientific knowledge established the validity of a divine account, the extreme antiquity of Indian classical time projected in sacred texts got discarded as mythical. An illustrative instance of how the 'scientific' basis of objective Western time is arrived at, and the historicity of the Biblical time upheld, at the cost of the 'ahistoricity' of 'mythical' time in India, enables serious reflection on practices of imperial 'othering' and an interrogation of the 'scientificity' and rationality of Western universals. With this in mind, let us turn to understandings of lived time/space by ordinary adherents of a radical religious order.

3 Preceptor's Time, Disciples' Time

At the beginning of the twenty-first century, followers of Mahima Dharma in the Sambalpur region of western Orissa surprised a German scholar with the daring statement that Kaliyuga had come to an end. It had ended in 1965, when Satyanarayan Baba, the deceased founder of this new group had appeared on earth. Satyanarayan Baba is believed to be an incarnation of Bhima Bhoi, the radical poet philosopher of Mahima Dharma. The fact that Kaliyuga had ended justified this group's preacher to look for and offer his disciples a new *mantra* (incantation) for this new era, since Bhima Bhoi's message was relevant only in Kaliyuga. The new era is not just satyayuga, the era of truth but *swadhin* (independent) satya (true/virtuous) and *bhogya* yuga (an era to be enjoyed). Interestingly,

34 Goldberg. Questions of Times, 279.

35 Thomas Trautmann. *Aryans and British India*. New Delhi: Yoda Press, 2004; Dodson, *Orientalism*, 28.

the incantation is an invocation of the eternal Vishnu and follows traditional Vishnu chants in rhythm and sequence. This blend of novelty and continuity, the proclamation of a break and the announcement of a new era premised on long prevalent ideas, graphically illustrates, not just the intermingling of the linear and the cyclical, and the new and the old in lived time, but also how aspirations mould experience of time and space, making them full and concrete.

Mahima Dharma, a heterodox religious order, emerged in the 1860s in Orissa at a time when a large part of the province was reeling from the effects of a devastating famine. Founded by Mahima Swami, an itinerant, abstemious ascetic who practised complete detachment, the faith advocated belief in a formless, all-pervasive, all-powerful Absolute who was equally accessible to all through devotion. This apparently simple message rendered redundant temples and rituals, priests and pilgrimage, and interrogated complex ritual and social hierarchies of caste and kingship, and the role of Brahmans as mediators between gods and men.[36]

Mahima Swami's simple but significant message, his austere lifestyle, constant mobility and practices that challenged authority and establishment in diverse ways, attracted a large number of subordinate men and women to him. The Swami initiated his large following into monastic and lay orders although he remained an itinerant ascetic throughout his life. He was deified during his lifetime as an incarnation of the indescribable Absolute whose *mahima* (radiance, glory) he preached. His faith *Alekh* (beyond writing/indescribable) Dharma or Mahima Dharma came to stand for the new, true faith of the new age in later traditions of Mahima Dharma.

At the same time, Mahima Swami's deification meant that his death shocked his followers. The divine incarnation was not meant to disappear from earth. The bunch of disparate followers, left with no nominated leader, no permanent structure and no written precepts, suffered a setback. At this time Bhima Bhoi, the "blind", unlettered Khondh (aboriginal) disciple of Mahima Swami emerged as a leader in western Orissa; he was bestowed with the "eyesight of knowledge" by the preceptor. Bhima Bhoi's inspired compositions in praise of the Absolute and Mahima Swami were noted down by Brahman scribes and came to constitute the theosophy-philosophy of the faith. There was friction between Bhima Bhoi who became a householder and the ascetic followers of Mahima Swami; Bhima Bhoi cut off his links with Joranda where the ascetics decided to set-up a memorial for the Guru; established in own *ashram* in Khaliapali in western Orissa and emerged as the leader of a large group of followers, who gradually came to regard him

36 For a detailed discussion see Ishita Banerjee-Dube. *Religion, Law, and Power. Tales of Time in Eastern India, 1860–2000*. London: Anthem Press, 2007.

as the founder of the faith and an incarnation of the Absolute. The reference to Satyanarayan Baba as an incarnation of Bhima Bhoi in the new era bears testimony to this.

I have briefly recounted the evolution of Mahima Dharma in order to highlight the constant and crucial presence of the metaphor of time (lived in space) in understandings of the faith and of their present by its devotees. Over and over again, the appearance of a preacher or the emergence of a new leader is associated with the descent to earth in human form of the divine to save the world and the devotees of the true faith from the evils of Kaliyuga. The bold step of declaring the end of Kaliyuga follows the same trend although one will have to carefully explore the perceptions that ground the statement because it gives an important twist to the ever-present metaphor of Kaliyuga as an era of evil characterised by general disorder. Given the fact that Kalki(n) is yet to appear, Kaliyuga lingers as a general metaphor of evil, and as a ready source of an appropriable past that gives meaning to the present and shapes the future.

What accounts for the continued presence of the classical notion of time in Orissa? Why do ordinary people view social time of their everyday lives through the prism of divine time lived in a particular space? In order to answer some of these questions, I now turn to the genre of *malika*s and their persistent prevalence in Orissa. *Malikas* conjoin time and space, sacred and social time, Bhima Bhoi and medieval mystics in important and innovative ways.

4 Time/Space in the *Malikas*

Malikas, apocryphal texts, have constituted an important part of Oriya literary tradition since the sixteenth century. Cast in the mode of Puranas, malikas deal with the theme of Kaliyuga, and prophesy its dissolution through the appearance of a redeemer and the establishment of true faith. In the dialogue of the malikas, the author barely makes an appearance: divinities discuss the future of the cosmos. If at all, the author interjects to emphasize the significance of the prophecies. If the absence of the strong imprint of a definite author and the general nature of the prophecies facilitate different readings of the malikas, they also enable their composition and interpolation at different moments.

The earliest malikas are ascribed to the *Panchasakhas*, Five Friends – eminent medieval mystics and exponents of a sixteenth century revival of Vaishnavism in Orissa; this lends great credibility to their prophecy. A close reading of malikas reveals the changing dimensions of the recurrent reference to Kaliyuga that form the basis for varied understandings and multiple appropriations of these texts.

The success of the malikas lies in their capacity to transform complex theological principles and esoteric religious ideas into an idiom grounded in the familiar and the everyday through the invocation of Kaliyuga. They serve as vehicles through which changing notions of time and history, good and evil, the enemy and the savior find expression, and cyclical and linear historical time come together to project a glorious past and a difficult present that pave the way for a promising future. The depiction of Kalki(n), the tenth incarnation of Vishnu in the form of a warrior gets prominence in the malikas, and the reversal of Brahmanical norms, a key feature of Kaliyuga, loses importance. Indeed, the heretical sects of the Puranas, come to represent true faith and their adherents join Kalki(n) and other mythical and legendary heroes to fight the evil forces of Kali who often embody real enemies. Time and space acquire force and dimension through the multiplicity of experience.

It is highly probable that the wide currency of malikas in Orissa is tied to Orissa's fame as the favorite abode of gods. Since the time of the *Mahabharata* and the Puranas, Orissa (or Odra *desh* or Utkal khand) has been has been hailed as an "uninterrupted *tirth*" (site of pilgrimage) in which distinct deities reside in different parts. Such acclaim probably accounts for the belief that the incarnation of the divine will appear in this sacred land and liberate the devout from the evils of Kali. The fullness and "thickness" of time represented by its inseparability from the past and the future, and by the unfolding of a (divine/social) plot makes space "charged and responsive";[37] like time it acquires "flesh" in which actual events occur.

Malikas are esoteric (yogic) texts written in a double language (*sandhya bhasha*): they convey one meaning to the uninitiated and another to the initiated. The description of actual places in Orissa as sites of battles, between the forces of Kali and the forces of Kalki, therefore, convey to the initiated the movement of the *kundalini* (source of energy that lies coiled up in the navel according to Yoga) to the head, a skill acquired only through serious practice of Yoga in order to make the body perfect, even immortal. The esoteric content of the malikas have been lost over time with the initiated dwindling in number; their dual language, however, gives an interesting twist to conceptions of time/space by intermingling divine, social, physical and bodily time/space in complex ways.

---

37 Mikhail M. Bakhtin. Forms of Time and of the Chronotrope in the Novel. In *The Dialogic Imagination. Four Essays by M. M. Bakhtin*, M. Holquist (ed.), 84–258. Austin: University of Texas Press, 1981, 84.

5 Malikas and Mahima Dharma

*Malika*s have been of crucial importance in the history of Mahima Dharma since its inception. Belief in Mahima Swami's divinity was, to a large extent, occasioned by the use of and belief in a malika by his disciples. Reports drawn up by colonial officials in the 1880s on the faith state that the followers of the sect were in possession of a book of predictions, which treated 'the incarnation of Alakh in the shape of Mahima Swami to save the world from the burden of sin and to pave the way towards salvation.' *Yasomati Malika*, ascribed to Yasovanta Das, one of the five medieval mystics, served a dual purpose. Trust in its prophecy served to highlight the inevitability of the appearance of an incarnation and the founding of true faith by him. It also established Mahima Swami as that *avatar* and Mahima Dharma as the true faith that was to bring Kaliyuga to an end. By the first decade of the twentieth century, when N. N. Vasu, a scholar serving in an official capacity in the feudatory state of Mayurbhanj, chanced upon *Yasomati Malika*, it had become one of the major scriptures of the sect. In the 1930s, the works of Biswanath Baba, the leading renouncer of the faith, quoted extensively from not one but several malikas and other works of the Panchasakhas to identify Mahima Swami with '*prabuddha narayan*' (the enlightened one) and '*iswar purush*' (god personified).

Bhima Bhoi, who played a major role in the propagation of the faith in western Orissa between the 1870s and the mid-1890s, did not only have his identity conflated with that of his divine preceptor; Baptist missionaries working in Orissa reproached him for claiming to be an incarnation (Annual Report 1886–1887).[38] The same missionaries, it is interesting to note, had commended Bhima Bhoi's success in getting adherents as a welcome sign that people were moving away "from the grosser forms of idolatry" (Annual Report 1881–1882).[39] Such people, however, did not take recourse in the precepts of the true faith, i.e. Christianity; they remained faithful to Bhima Bhoi instead. The missionaries now described him as an "impostor", "a master of blasphemous pretensions", who claimed to be a "new incarnation" that had appeared on earth to "inaugurate a new dispensation". Particularly irksome for the messengers of Christ was the fact that the false pretensions of "an ignorant man" could actually attract "hundreds, if not

38 *Indian Report of the Orissa Baptist Mission for 1887–88*, Angus Library, Baptist Mission Society Archives, Regent's Park College, Oxford, 46.

39 *Indian Report of the Orissa Baptist Mission for 1881–82*, Angus Library, Baptist Mission Society Archives, Regent's Park College, Oxford, 48.

thousands" of devotees (Ibid.).[40] It is difficult to gauge whether Bhima Bhoi gave himself out to be an incarnation. What is evident from N. N. Vasu's statement and the reports of the missionaries is that he had come to be regarded as a divine being. The malikas that had established Mahima Swami as the incarnation of the Absolute identified Bhima Bhoi as his divine devotee.

If Bhima Bhoi's identity got conflated with that of his divine preceptor resulting in his deification, the closeness in thought, ideas and language between his compositions and that of the Five Friends, set in motion novel interpretations of his works and teachings, and spurred the composition of malikas in his name. Of the three concepts that Bima Bhoi is believed to have taken from the Five Friends, that of "a future redeemer who will come and openly establish what is for the time being a secret doctrine"[41] is of crucial importance. The works of Bhima Bhoi do not only speak of the indescribable Absolute; they throb with the anguished cries of the poet to his Lord to save him and his fellow beings from the sufferings of Kaliyuga.

Malikas in the name of Bhima Bhoi appeared in the early decades of the twentieth century, a time of turmoil for the Dharma. Two groups of ascetics, distinguished by their garb, came to clash over control of the memorial of Mahima Swami. This overlapped with a tremendous boost given by Gandhi to the nationalist struggle for India's independence from British rule. His declaration in 1921 that Khilafat and Non-Cooperation together would result in *swaraj* (self-rule) within a year electrified the "masses". For the first time, ordinary men and women, peasants and workers, joined a political struggle led by the Indian National Congress in large numbers. Together, these intersecting battles inspired the composition of new malikas, now ascribed to Bhima Bhoi. They prophesied the reappearance of Mahima Swami in the new incarnation of a warrior. He was to lead the battle against the forces of Kali and cause their destruction.

Bhima Bhoi Malika ba Padmakalpa published by the Cuttack Dharma Grantha Store in 1971[42] is widely accepted by the followers as a creation of Bhima Bhoi. It has, however, not been given a place in *Bhima Bhoi Granthabali* (collected works of Bhima Bhoi)[43], published by the same store, as the ascetics of Mahima Dharma in charge at Joranda doubt its "authenticity". *Padmakalpa* was preceded

40 *Indian Report of the Orissa Baptist Mission for 1887–88*, Angus Library, Baptist Mission Society Archives, Regent's Park College, Oxford, 46.

41 Anncharlott Eschmann. Mahima Dharma and Tradition. In *Popular Religion and Ascetic Practices: New Studies on Mahima Dharma*, Ishita Banerjee-Dube and Johannes Beltz (eds.), 31–41. New Delhi: Manohar, 2008, 33.

42 Bhima Bhoi, *Bhima Bhoi Malika ba Padmakalpa*, Cuttack: Dharmagrantha Store, 1971.

43 *Bhaktakabi Bhimabhoinka Grathabali*, K.Sahu (comp.), Cuttack: Dharmagrantha Store, 1977.

by another short *malika* of Bhima Bhoi, which expressed Bhima Bhoi's anguish and confusion at the disappearance of the preceptor.[44] How could the Lord forget his devotees and leave them? What were the *bhaktas* (followers) to do? Whose commands were they to follow? Surely the Lord could not abandon his followers. He was to come back again, in the form of Kalki, the warrior, destroy Kaliyuga, and end the sufferings of the true devotees for all time to come. This belief is given greater finesse and flourish in *Padmakalpa*.

True to the tradition of malikas, *Padmakalpa* is composed in the form of a dialogue between Anadi (the eternal one) and Adi (Mata), mother of the world, who is also called Sati. It is to clear away Sati's doubts that Anadi unravels the mysteries of his divine play (*lila*) to her along with the events of Kaliyuga. Bhima Bhoi figures as the narrator of this dialogue. From time to time, he interjects his own comments, addressing the reader directly. The text is composed in a language of time and timelessness. There is a constant change of tense as the dialogue moves back and forth between what has happened, what is happening, and what will happen.

The text opens with Bhima Bhoi expressing his fear and helplessness at being surrounded by the evil ways of Kali and calling upon Sri Guru (the master), at whose feet he has taken shelter, to save him. This is followed by an account of Kaliyuga, and of its long line of cruel and exploitative rulers. The injustice of kings and Brahmans disgusts Lord Jagannath so much that he leaves his abode to dwell secretly for sixteen thousand years.

The *soma* (lunar) dynasty reigns over the world at the beginning of the age; the British, who follow the Gurkhas and Turks, come to power toward the end of the era and are meant to rule for 495 years. The British are the worst since there is a progressive degeneration in the nature of the rulers and in the condition of the earth. Their duplicity makes them particularly dangerous: with an outward show of speaking the truth they secretly oppress their subjects and destroy them gradually but thoroughly, in the manner of white ants destroying wood. This wile goes hand in hand with their dirty habits – they eat the flesh of cows and buffaloes and drink liquor every day. Is it surprising that they do not wash themselves after defecating?[45]

Exploitation by these extreme outcastes robs *prithvi devi* (mother earth) of her beauty and compels Lakshmi (goddess of sustenance, wealth and bounty)

44 Bhima Bhoi Malika, a hand-written version I found in the Eschmann Papers (*Nachlass*), *Sud-Asien Insitut*, Heidelberg in 1991. The papers are curently with Johannes Betz.

45 *Bhima Bhoi Malika ba Padmakalpa*, chapter 1, verses 5, 7 and 8.

to seek shelter in *patal* (the nether world). Scarcity is a constant feature: floods, droughts, and famines regular happenings.

Sinfulness comes to characterize humanity in general. True devotees waver. They fail to recognize the redeemer, the Brahma avatar (incarnation of the Absolute) when he appears and refuse to pay heed to his precepts. Under the influence of Kali they behave like "a host of blind men"[46] and, in a manner similar to the treatment meted out to the prophets in the Old Testament, the devotees ignore the Guru (preceptor)'s teachings.[47] He leaves the earth. Kaliyuga holds sway. Fear and despair prompt the gods to come together and pray to the Absolute in a gesture of abject surrender. In response to their prayers, Anadi (One without Beginning) comes to *yogaghara* (a place for meditation), sends his devotees to earth, enjoins saints who have attained self-realization to sit in meditation, and finally saves the earth by destroying Kali, lighting the Brahma *dhuni* (eternal flame), and establishing true faith.

A fierce battle has to be fought and great sufferings undergone before Kali is annihilated. Bhima trembles and his heart fills with sorrow as he narrates the events of the future. Kalki will play the game of non-violence in the era of evil. He will be called Gandhi-Sankarsana. His devotees, remarkably important leaders of the Congress, will be with him constantly. Some other followers, both men and women, will start an agitation. They will disobey the king and preach truth. A river of blood will flow during the 7th and 8th year of reign of Ram Chandra Deva. Sunya Guru will be fighting then, flying his sunya *bana* (flag).[48]

This agitation is a part of the war that paves the way for the beginning of a novel world and a new era by rooting out evil and effecting a complete break with the past. References to the war abound in the text, references which by virtue of their frequency, magnitude and detail create an intimidating picture of anarchy, destruction, devastation, and death.

The description is neither clear nor consistent. At times it approximates to a fight between Gandhi and his followers and the British, at times it assumes the massive proportions of the Kuruksetra war (of *Mahabharata*) for Kaliyuga. The human form of Anadi leads the forces of truth against the forces of Kaliyuga in this epic war. In course of his dialogue with Adi, he declares with remarkable clarity

---

46 Ibid, chapter 9, verse 12.

47 Johannes Beltz. Apocalyptic Predictions, Prophecies, and a New Beginning. Mahima Dharma and the arrival of Satyayuga. In *Popular Religion and Ascetic Practices. New Studies on Mahima Dharma*, Ishita Banerjee-Dube and Johannes Beltz (eds.), 79–101. New Delhi: Manohar, 2008, 87.

48 *Bhima Bhoi Malika*, chapter 10.

that he will appear during the reign of Ram Chandra Dev, in the year 1941.[49] He is Mahima Swami, Gandhi, and Kalki.

Anadi's soldiers are many: the gods and goddesses, *nagantis*, *yogantis*, and *risis* (super humans and saints), the great heroes of the past, the present, and the future – Ramchandra and Hanuman (of the *Ramayana*), the five Pandava brothers (of the *Mahabharata*), Gandhi and his followers, the followers of Anadi, troops from all over India and abroad. The forces of the evil age, that include the British, are powerful both in terms of numbers and the sophisticated arms and ammunition they command. It requires the combined strength of gods and men, of mythical, legendary, and historical figures, to overpower and destroy them.

The fierceness of the battle is difficult to convey in words. The earth trembles under the pressure of 700,000 marching soldiers, 400,000 horses and a million elephants. The battle continues for twelve years and devastates the world thoroughly. Humanity faces destruction; countries disappear from the face of the earth.

Mughals come to rule again in the newly engendered world and are assisted by brave men from the families of Dhruva and Prahlada, the favored devotees of Vishnu. People prosper under able rule; famines and scarcity are erased from memory. Anadi Brahma returns to his abode, after establishing satya dharma on earth.

Padmakalpa is a richly layered text; a spectacular example of ingenious blending of real and mythical time/space, and lived and utopic (messianic) experience and aspirations, inflected by particular characteristics of time/space that affect human lives. A text that draws upon a tradition that goes back to the *Puranas*, and uses a language that is at once simple and esoteric. Actual places of Orissa find mention as sites of battles; they inversely refer to the kundalini's journey upwards. Similarly, the battles are fought in real and mythic times, with legendary heroes and a political leader famed to be a saint. He is at once Mahima Swami (Anadi), Gandhi and Kalki. If Mahima Swami's identification with Kalki indicates the place he had secured for himself in the minds of his adherents, the prominence accorded to Gandhi is a good example of the hold of the "Mahatma" in popular imagination in a region that was never in the forefront of the nationalist struggle. This Mahatma, who plays the game of non-violence in Kaliyuga, also joins hands with Kalki and heroes of the epics to lead a fierce battle, to root out an evil and dangerously powerful enemy, the colonial masters.

The depiction of the British, the archetypal exploitative foreign ruler of Kaliyuga, shows originality even if it draws upon motifs current in the *Puranas*, and

49 *Bhima Bhoi Malika*, chapter 22.

yet demonstrates a familiarity with the rhetoric of the nationalist struggle. The cunning and duplicity of the British is combined with dirty habits to embellish the popular picture of the vile, powerful *mlechha*.

Remarkably, the memory of the revolt of 1857 is portrayed in a return to the rule of the Mughals after the destruction of Kaliyuga. The Mughals, however, have Dhruva and Prahlada – the devotees of Vishnu – as their lieutenants, symbolizing the harmony and amity that is to prevail at the end of the era of evil.

A similar coupling of the "mythic" and the "modern" characterizes the use of time in *Bhima Bhoi Malika ba Padmakalpa*. This *Malika*, as stated before, moves back and forth between the past, the present and the future; it records what happened in the past to make sense of the present. And in this process, the past is recreated as the present acquires new significance. Time is at once eternal, cyclical, linear, chronological and real. The events occur during the reigns of the four kings of Puri – Mukunda Dev, Ramchandra Dev, Birakishor Dev and Dibya Singha Dev. These are titles adopted by the kings of Puri when they ascend the throne. Temporally, then, their rule can cover centuries. At the same time, the year of appearance of Kalki is announced with remarkable precision. Moreover, Ramchandra Dev's rule features prominently in the discussion along with the activities of Mahima Swami, Gandhi, and their followers. This seems to suggest that the text was composed in the early twentieth century when Ram Chandra Dev was the king of Puri.

The text's resemblance to *Yasomati Malika* is striking. *Yasomati Malika* had announced the appearance of the incarnation in the shape of Mahima Swami and his divine disciple Bhima Bhoi. *Padmakalpa* announces the re-appearance of the redeemer. The first appearance of Mahima Swami had not brought about the end of suffering. Under the influence of Kaliyuga the devotees had failed to recognize the redeemer and had behaved like "a host of blind men". The immoral era had gained in strength and made true devotees "blind". Evil rulers had become specially powerful and dangerous. The fight against them – not imagined but real in the early decades of the twentieth century – thus came to assume the proportion of the *Mahabharata* war. On another front, "fake" yogis posing as the disciples of Mahima Swami had started to create dissension among the few faithful followers. The situation was ominous indeed. But all was not lost. The Preceptor was to appear again.

The *bhaktas* had to recognize him this time and then join him in the battle against the forces of evil. In other words, in direct opposition to Eliade's idea that cyclical time leaves the religious man bereft of any will to human action to change the course of history, *Padmakalpa* demands the true followers of Mahima Dharma to act. Their action is to change the course of the future. The corrupt ones will perish. Kaliyuga will be annihilated; truth will prevail.

Towards a conclusion

When a devastating cyclone struck Bhubaneswar, the capital of Orissa in 1999, *malikas* gained wide circulation. The end of the world and *kaliyuga* became imminent once more. Only this time natural calamity gained precedence over the appearance of an incarnation. Throughout this paper, I have tried to argue against clear demarcations of divine and social, cyclical and linear time tied to separation of myth and history, by underscoring their overlaps and inextricable blends. The ordering, separation and hierarchization of time/space are results of abstractions that only sustain the myth of modernity but do not correspond to social worlds. To substantiate my point, I have paid particular attention to a genre of texts and perceptions and beliefs of a religious order of Orissa. They graphically exhibit how a divine plan unfolds in everyday lives to bestow them with singular significance, and experience and aspiration come together to enable deployments of worldview in a way as to confer newer meanings and dimensions on time/space. Combinations of cyclical-linear, divine-social, normative-real time and lived, physical, bodily and utopic space bring precise historical years and real spaces to life; and jumbles of distinct time/space thicken the plot of a divine drama staged by human actors and mythical heroes.

My purpose, of course, is not to stress the difference or uniqueness of Orissa and (Hindu) India where the transition to modern time/space and history has been unsatisfactory and incomplete. Rather, I wish to bring into relief the existence of these muddled worlds in different parts of the globe and in distinct traditions to urge reflection on whether and how far universals work, and advocate the need to pay attention to the context,[50] in order to be more sensitive to experience and aspirations as well as dispositions and apprehensions of time/space. If we accept distinct modes of ordering time/space and past and future that continue to imbue temporality with normativity as valid and legitimate, we will be able to prise open the notion of a clearly separate and integral historical time, dislodging thereby European universals from their privileged position. It will allow us to guard against producing a history in stages that denies coevalness to large groups of peoples and places. An acknowledgement of different approaches and understandings of the past as valid articulations of historical consciousness will go a long way in shaking-up the imposed binaries of myth and history, magic and reason, orality and writing, and archaic (primitive) and modern. Does the divine, perhaps, ordain it?

---

50 Ramanujan, Indian Way of Thinking.

References

Archival Source

Baptist Missionary Society Archives, Angus Library, Regenet's Park College Oxford.

Literature

Bhima Bhoi, *Bhima Bhoi Malika ba Padmakalpa*. Cuttack: Dharmagrantha Store, 1971.
K. Sahu (comp.). *Bhaktakabi Bhimabhoinka* Granthabali. Cuttack: Dharmagrantha Store, 1977.
Books and Articles
Bakhtin, Mikhail M. Forms of Time and of the Chronotrope in the Novel. In *The Dialogic Imagination. Four Essays by M. M. Bakhtin*, M. Holquist (ed.), 84–258. Austin: University of Texas Press, 1981.
Banerjee-Dube, Ishita. *Religion, Law, and Power. Tales of Time in Eastern India, 1860–2000*. London: Anthem Press, 2007.
Beltz, Johannes. Apocalyptic Predictions, Prophecies, and a New Beginning. Mahima Dharma and the arrival of Satyayuga. In *Popular Religion and Ascetic Practices. New Studies on Mahima Dharma*, Ishita Banerjee-Dube and Johannes Beltz (eds.), 79–101. New Delhi: Manohar, 2008.
Berkovits, Eliezer. *Faith after the Holocaust*. New York: KTAV Pub. House, 1973.
Bloch, Marc. *The Historian's Craft*. 7th ed. Manchester: Manchester University Press, 1992 [1954].
Dodson, Michael. *Orientalism, Empire and National Culture. India, 1770–1880*. New Delhi: Foundation Books, 2010.
Doniger, Wendy. *Hindu Myths. A Sourcebook Translated from the Sanskrit*. UK: Penguin Press, 2004.
Eliade, Mircea. *Cosmos and History. The Myth of the Eternal Return*, trans. Williard R. Trusk. New York: Harper Torchbooks, 1959.
Eliade, Mircea. *The Sacred and the Profane. The Nature of Religion*. New York: Harcourt, Brace & World, 1959.
Eschmann, Anncharlott. Mahima Dharma and Tradition. In *Popular Religion and Ascetic Practices: New Studies on Mahima Dharma*, Ishita. Banerjee-Dube and Johannes Beltz (eds.), 31–41. New Delhi: Manohar, 2008.
Fabian, Johannes. *Time and the Other. How Anthropology Makes its Object*. New York: Columbia University Press, 1984.
Goldberg, Sylvie Anne. Questions of Times. Conflicting Timescales in Historical Perspective. *Jewish History* 14 (2000): 267–286.
Holquist, Michael. Introduction. In *The Dialogic Imagination. Four Essays by Mikhail M. Bakhtin*, M. Holquist (ed.), xv–xxxiii. Austin: University of Texas Press, 1981.
Koselleck, Reinhart. *The Practice of Conceptual History. Timing History: Spacing Concepts*. California: Stanford University Press, 2002.
Manchester, Peter. Time in Christianity. In *Religion and Time*, A. Niyogi Balslev and J.N. Mohanty (eds.), 109–137. Leiden: E.J. Brill, 1993.

Ramanujan, Attipat Krishnaswami. Is there and Indian Way of Thinking? An Informal Essay. *Contributions to Indian Sociology* (n.s.), 23:1 (1989): 41–58.

Steensgaard, Peter. Time in Judaism. In *Religion and Time*, A. Niyogi Balslev and J.N. Mohanty (eds.), 64–108. Leiden: E.J. Brill, 1993.

Stietencron, Heinrich von. Kalkulierter Relgionsverfall: Das Kaliyuga in Indien. In *Der Untergang von Religionen*, Hartmut Zinser (ed.), 135–150. Berlin: Dietrich Reimer Verlag, 1986.

Subrahmanyam, Sanjay. Connected Histories. Notes Towards a Reconfiguration of Early-Modern Eurasia. *Modern Asian Studies* 31:3 (1997): 735–762.

Thapar, Romila. Society and Historical Consciousness. The Itihasa-Purana Tradition. In *Situating Indian History for Sarvepalli Gopal*, S. Bhattacharya, R. Thapar (eds.), 353–383. New Delhi: Oxford University Press, 1996.

Thapar, Romila. *Time as a Metaphor of History. Early India.* Oxford: Oxford University Press, 1997.

Thapar, Romila. Time Concepts, Social Identities and Historical Consciousness in Early India. In *Time in India: Concepts and Practices*, Angelika Malinar (ed.), 23–38. New Delhi: Manohar, 2007.

Trautmann, Thomas. *Aryans and British India.* New Delhi: Yoda Press, 2004.

The Mahabharata, vol. 2. J.A.B van Buitenen (ed., trans.). Chicago: University of Chicago Press, 1975.

White, Hayden. Foreword. In R. Koselleck. *The Practice of Conceptual History. Timing History: Spacing Concepts*, ix–xiv. California: Stanford University Press, 2002.

Gender in the Empire

Sabine Schmolinsky

Introduction
Gender in the Empire

In their book on 'Empire' (2000) Michael Hardt and Antonio Negri emphasize that they use the term '"Empire"' "not as a *metaphor* [...] but rather as a *concept*, which primarily calls for a theoretical approach".[1] "Empire" appears to be marked by a spatio-temporal boundlessness which encompasses a totality of both time and space.[2] Moreover it regulates and even creates the entire social world and tries to rule over human nature.[3] Even when considering Empire's tendency to being ahistorical[4], a deficiency of the concept is to be noted at this point, for gender and gender differences are not included at its very core.

This may be realized, for instance, when Hardt and Negri in a chapter on "Resistance, Crisis, Transformation" (3.3) refer to "[f]eminist movements" of the 1960s and 1970s. These are described as having "raised the social value of what has traditionally been considered women's work, which involves a high content of affective or caring labour and centers on services necessary for social reproduction."[5] If "gender" is mentioned, it is referred to as an item of difference which occurs in the contemporary world,[6] but no specific explanatory power for the comprehension of the social and the cultural world is ascribed to it as fundamental.[7] Hardt and Negri, however, do not claim to deal with questions of gender, which seem not to be in the scope of their theoretical framework. This panel therefore proposes to engender the concept of "Empire" which needs to get more substance by being bound to the analytical category of gender.

1 Hardt, Michael and Antonio Negri. *Empire*. Cambridge, Mass., London: Harvard University Press, 2000, xiv.

2 It remains an open question, whether and if yes how far gendered digitalised spaces would contribute to the oppressive character of the Empire. Andrea Petö (Budapest) touched those topics in her conference talk on 'Imperial Imaginations Digitalised and Gendered: Imaginations of European Far Right'.

3 Hardt and Negri, *Empire*, xiv–xv.

4 Ibid.

5 Hardt and Negri, *Empire*, 274.

6 Hardt and Negri, *Empire*, 147, 153, 399 (gender in the world of labor).

7 For an exception see the list of categories of difference, which form part of the Empire's inclusive moment; cf. Hardt and Negri, *Empire*, 198. Gender in relation to the new body may not be discussed here; cf. Hardt and Negri, *Empire*, 215sq.

DOI 10.1515/9783110418750-006

To refer to gender as a category of analysis means to refer to Joan Scott's classical dictum.[8] It serves as a point of departure for Joanna de Groot's arguments on research evidence underscoring the crucial role of gender for historical and cultural analysis. In her contribution she focuses on the critics' adapting and developing the Eurocentric perspective of the notion into postcolonial, transnational, multi-cultural, and intersectional conceptualizations. These include numerous narrations of the history of empires combined with analytical narrations according to questions of race/ethnicity, class, gender, sexuality, and age in a cross-cultural sense. According to this method, Joanna de Groot expands an ample survey of research on past empires in a global perspective and its relation to the issue of gender.[9] She revisits the main conceptual tools within gender theory, which enable to transcend mere topicalization for the sake of the study of (past) conditions of empires. Emphasizing their analytical value, the notions of patriarchy, of masculinity and femininity, and of sexualities are discussed as instruments of analysis. Joanna de Groot comes to the conclusion that "there is a dense and varied body of gender and feminist scholarship which makes both analytical and empirical challenges to gender blind scholarship on empire."[10] Concerning biopolitics and biopower she points to the systematic lack of gender and gender theory in the conceptualization of the "imperial machine", which, according to Hardt and Negri, moves Empire.[11]

In the context of reviewing contemporary Italian Marxist theorists, however, Hardt and Negri conclude in a footnote that "[...], from a methodological point of view, we would say that the most profound and solid problematic complex that has yet been elaborated for the critique of biopolitics is found in feminist theory, particularly Marxist and socialist feminist theories that focus on women's work, affective labor, and the production of biopower."[12] In effect, this specification implies a narrowing of the perspective to female contributions to biopower, rather than widening the horizon to a gender-based perspective. The latter would include reflection on this type of singling out women's agency and activity by assuming men's agency and activity as a unquestioned starting point of all considerations. Analyzing the "new corporate culture" within capitalist organiza-

8 Scott, Joan W. Gender. A Useful Category of Analysis. *American Historical Review* 91:5 (1986): 1053–1075.

9 In her conference talk on 'Maria Theresa and the Habsburg-Dynasty: Aspects of Gender' Christine Lebeau (Paris) addressed these subjects.

10 See below, 146.

11 Hardt and Negri, *Empire*, 22 and passim.

12 Hardt and Negri, *Empire*, 422 sq., footnote 17, relating to page 29 quoted by de Groot in her foootnote 34.

tions configured in a postmodernist way, Hardt and Negri, however, note that "[t]he corporations seek to include difference within their realm and thus aim to maximize creativity, free play, and diversity in the corporate workplace. People of all different races, sexes, and sexual orientations should potentially be included in the corporation; [...]."[13] Thus by the brokerage of diversity and "[t]his project [...] aptly called 'diversity management'"[14], gender seems to be part of the essential principles underlying a set of cultural practices. This assumption, however, would imply to deconceptualize the category of gender and reduce its very essence.

So far, even in an undoubtedly decolonizing mode and context the reference to "[l]abels like Indian, or woman, or Muslim, or American", admittedly conceived as being "not more than starting-points", does not include a label "man".[15] This is no concept focused by Edward W. Said in his book "about the general relationship between culture and empire".[16] Gender discourse and gendered identities are at stake, however, when masculinity is envisaged as a factor in an empire's foreign policy. In the case of Robert D. Dean's notion of 'Imperial Brotherhood' "[t]he term 'ideology of masculinity' refers to the cultural system of prescription and proscription that organizes the 'performance' of an individual's role in society, that draws boundaries around the social category of manhood, and that can be used to legitimate power and privilege."[17] Here elite masculinities, politicians and high bureaucrats of the U.S. Cold War foreign policy are in the center of analysis.

A different perspective is proposed by Silvan Niedermeier in his contribution on the construction of imperial masculinities in the Philippine-American War (1899–1902). He focuses on male white self-representation at war abroad as it is performed through the medium of the soldiers' private photography or of professional photography by male photographers. The soldiers used war photographs "to create and retell a story of male imperial self-becoming that drew on spatiotemporal concepts of empire such as race, civilization, and modernity."[18] As documents of their becoming men during and by means of an imperial war, the soldiers shaped their personalized marks of authenticity as well as their

13 Hardt and Negri, *Empire*, 153. For a fundamental critique see Boron, Atilio A.: Empire *and Imperialism. A Critical Reading of Michael Hardt and Antonio Negri*, translated by Jessica Casiro. London and New York: Zed Books, 2005, 47–49.
14 Hardt and Negri, *Empire*, 153.
15 Said, Edward W. *Culture and Imperialism*. New York: Alfred A. Knopf, 1993, 336.
16 Said, *Culture*, xi.
17 Dean, Robert D. *Imperial Brotherhood. Gender and the Making of Cold War Foreign Policy*. Amherst: University of Massachusetts Press, 2001, 5.
18 See below, 103.

relics memorizing their participation in this war. The ensuing collections were connected to further practices of recollection as the soldiers and the veterans exchanged photographs in order to enlarge their individual collection; in addition, images formed part of the veterans' collective memory. The process of adopting a specific concept of masculinity was determined by spatiotemporality in that they directly referred to the soldiers as they inscribed themselves in the USA's becoming an overseas empire in particular spaces around 1900: in this period, white male middle-class Americans' self-conception was at stake and tended to be shaped by military ideals. Finally intersecting the two concepts of gender and race, Silvan Niedermeier emphasizes the deep immersion of imperial masculinization in the ongoing process of "'whitening'" it.[19]

Coming back to Hardt and Negri, the question arises whether gender awareness will be included in the "alternative powers" being geographically unlimited in the manner of Empire, since these contest and prefigure "an alternative global society" which is to be created by the multitude and its "resistances, struggles, and desires".[20] As globalization is conceived to have a second face providing "the possibility that, while remaining different, we discover the commonality that enables us to communicate and act together"[21] there is difference as a common feature. Hardt and Negri conceptualize the multitude as "an open and expansive network in which all differences can be expressed freely and equally, a network that provides the means of encounter so that we can work and live in common"[22], thus preparing a space for gender. Gender does appear as a category of difference, but it is destined to disappear as a destructive power: "When we say that we do not want a world without racial or gender difference but instead a world in which race and gender do not matter, that is, a world in which they do not determine hierarchies of power, a world in which differences express themselves freely, this is a desire for the multitude. And, of course, for the singularities that compose the multitude, in order to take away the limiting, negative, destructive character of differences and make differences our strength (gender differences, racial differences, differences of sexuality, and so forth) we must radically transform the world."[23]

In part 3 on "Democracy", Hardt and Negri emphasize their argument concerning gender difference: "The radicality of gender difference, for example, can

19 See below, 128.
20 Hardt and Negri, *Empire*, xvi, cf. xv.
21 Hardt and Negri, *Multitude*, xiii.
22 Hardt and Negri, *Multitude*, xiii sq.
23 Hardt and Negri, *Multitude*, 101.

be included in the biopolitical organization of social life, the life renovated by the multitude, only when every discipline of labor, affect, and power that makes gender difference into an index of hierarchy is destroyed." Multitude itself is deemed to become possible only "on the basis of such differences."[24]

Does this hope for a world of differences without negative consequences dislocate gender to Utopia in a sense of having no space or place for ever? Gender seems to have arrived in the middle of a society which beyond all even offers the opportunities of globalization. Answers to this complex question should consider the differing regimes of spatiotemporality implied in the underlying concepts of gender. Hardt and Negri locate gender difference without destructive features in a boundless, globalized space of future times and do not concentrate on contemporary concerns of gender. Critics, however, would maintain the need of a concept working in the actual world which means applying the gender category to Empire and/or the Counter-Empire[25] of the present. Furthermore, the concept of gender should be at hand for historical analysis e.g. of past empires. Thus the "*transformative potential for thinking through* the structures, processes and relationships", which the "conceptual tools developed within gender theory"[26] might develop continues to foster the applicability of the notion of gender before the multitude emerges more fully.

References

Boron, Atilio A. 'Empire' *and Imperialism. A Critical Reading of Michael Hardt and Antonio Negri*, translated by Jessica Casiro. London and New York: Zed Books, 2005.

Dean, Robert D. *Imperial Brotherhood. Gender and the Making of Cold War Foreign Policy.* Amherst: University of Massachusetts Press, 2001.

Hardt, Michael and Antonio Negri. *Empire.* Cambridge, Mass., London: Harvard University Press, 2000.

Hardt, Michael and Antonio Negri. *Multitude. War and Democracy in the Age of Empire.* New York: Penguin Books, 2005.

Said, Edward W. *Culture and Imperialism.* New York: Alfred A. Knopf, 1993.

Scott, Joan W. Gender. A Useful Category of Analysis. *American Historical Review* 91:5 (1986): 1053–1075.

24 Hardt and Negri, *Multitude*, 355.

25 Hardt and Negri, *Empire*, 205–218.

26 de Groot, Beyond Blindness, see below, 139.

Silvan Niedermeier

"If I were King" – Photographic artifacts and the construction of imperial masculinities in the Philippine-American War (1899–1902)

On September 6, 1898, Ashley R. Farless, a twenty-one-year-old sergeant of the 1st Regiment California Infantry, sent a letter from Manila to his uncle in Westchester County, New York:

> Dear Uncle Jim,– It has been my intention to write you ever since we sailed from home. To tell the truth, I have done very little writing even to Papa and Mamma. I thought you would like to have a photo of me as I am now. They were anxious at home to have a picture of me as soon as I got a chance. These I enclose were taken at a Philippino gallery. They are not very good at the art.
>
> I suppose the papers keep you posted with official and other (principally the latter) kinds of news about Dewey and Merritt before Manila. We are now maintaining Martial Law and waiting for (I suppose) the Paris conference and our 2 month pay.
>
> Do you remember the copy of the Manual of Arms you gave me that you carried in '61? I have it at home, and hope I will be able to put my own with it, as a relic. What a difference time has made! I do not pretend to say that this is a letter, but will try to write you later. Papa writes you once in a while does he not?
>
> Give the small picture to Mrs. Keeler with my love. My regards to Charlie.
>
> Hoping to see or hear from you before many months, I am,
>
> your affectionate nephew,
>
> Ashley[1]

Three and a half months before, on May 25, 1898, Farless and his regiment had boarded a transport ship to the Philippines, as part of a military expedition of 2,500 U.S. soldiers under the leadership of Major General Wesley Merritt. On May 1, six days after the United States' declaration of war against Spain, Admiral George Dewey had defeated the Spanish fleet in Manila Bay. Farless and his reg-

1 Letter of Ashley R. Farless to James Farless, Manila, September 8, 1898. In: U.S. Army Military History Institute, Carlisle, PA (USAMHI), Spanish American War Veterans Survey Collection (SAWVSC), Box 1, Folder 31, Farless, Ashley, Company K, 1st Regiment, California Infantry.

DOI 10.1515/9783110418750-007

iment arrived in the Philippines on June 30, 1898, with the mission to subdue the Spanish colonial troops in the Philippines. Three weeks before their arrival, General Emilio Aguinaldo declared the Philippines Islands independent from Spain and set out to establish a revolutionary government under his leadership.

After the victory against the Spanish troops in Cuba in July 1898, the U.S. government under President William McKinley decided to annex the Philippines. To save the honor of Spain and thwart the Philippine independence movement, Admiral Dewey and General Merritt negotiated a secret agreement with the Spanish Governor-General and staged a mock battle against the Spanish colonial troops in Manila on August 13, 1898. U.S. units took control over the city of Manila and barred the Philippine revolutionary troops from access to *Intramuros* (also called the "Walled City"), the fortified historic city center, and most of Manila's suburbs. In the meantime, the U.S. government prepared negotiations for the peace conference in Paris in November 1898, where Spain would agree to sell the Philippines to the United States for $20,000,000.

Farless's letter indicates that U.S. soldiers were well aware of the tense situation that arose out of the U.S. government's ambition to seize the Philippines Islands in spite of the Filipinos' strife for independence. Yet, the letter reveals another seemingly more pressing concern that occupied Farless's mind: The desire to retain a picture of himself and send it to his family in the United States. Unfortunately, no exemplar of Farless's portrait cards is preserved. Figure 1, however, shows the portrait of Robert W. Dodd, one of his fellow soldiers in Company K of the 1st California Regiment.

It shows a *carte de visite*[2] bearing the golden imprinting of the studio *Fotografia Pertierra* and its address in Manila. The size of the cardboard is 2 ½ x 4″. The gilt-edged card has further information on its owner and a medal of the Filipino Exposition in Madrid of 1887 on its back. It bears a 2 ¼ to 3 ½″ print of a photographic portrait of Robert W. Dodd. He sits on a small artificial stonewall in front of a studio wallpaper showing a lake scene with a Calibre .45 Springfield rifle on his knees. He wears a cartridge belt and a so-called "blue shirt" which were issued to U.S. volunteer soldiers in San Francisco in 1898 in lack of khaki uniforms. As Dodd and Farless served in the same company and knew each other, they probably went to the same studio.

Farless's apologetic and derogatory remark "They are not very good at the art," expresses a strong notion of cultural superiority towards the Philippine

---

2 The *carte de visite* was invented in France during the 1850s. First used by urban elites it soon became a popular medium of the mid and late nineteenth century, when people in Europe, the United States, and beyond traded their photographic portraits cards among visitors and friends.

Fig. 1: *Carte de visite* of Robert W. Dodd. In: U.S. Army Military History Institute (USAMHI), Spanish American War Veterans Survey Photograph Collection (SAWVSPC), Box 1.

photographer and the Filipinos in general whom he perceives as dilettante imitators of the Western "art" of photography. At the same time, does his letter express the special meaning those portraits had for Farless as they provided a material authentication of his military service abroad. As Susan Stewart tells us, souvenirs – such as photographs or photo albums but also other things kept as a reminder of experiences, persons, or places –, exemplify the "capacity of objects to serve as traces of authentic experience."[3] Dislocated from the time and space of its production, Stewart argues, the souvenir "transform(s) and collapse(s) distance into proximity to, or approximation with, the self. The souvenir (...) contracts the world in order to expand the personal."[4] She concludes that the ability of souvenirs to act as a form of self-expression and -expansion is secured by creating a narrative around them.[5]

---

3 Susan Stewart. *On Longing. Narratives of the Miniature, the Gigantic, the Souvenir, the Collection.* Durham: Duke University Press, 1993, xii.

4 Stewart, *On Longing*, 135.

5 Stewart, *On Longing*, 132–150.

Although Ashley Farless states that he took his photographic portrait to answer a wish of his family, there is evidently more at stake. Most importantly, his sending of his photographic portraits back "home" appears to be driven by a longing to authenticate and accentuate his personal involvement in the war abroad and document his maturing as a man through this experience. Most directly, this longing finds its expressions in Farless's remark, "I thought you would like to have a photo of me as I am now". The desire to situate oneself in the historical moment of America's imperial war overseas and gain male respectability also becomes evident in Farless's reference to his uncle's "manual of arms" of the American Civil War – an army instruction manual for using weapons such as rifles in formation – and his endeavor to bring back his own manual as a war "relic." Evidently, the photos Farless sent to his uncle and "Mrs. Keeler", not only answered his relatives' wish for a visual documentation of his well-being abroad, but also enacted a desire to picture oneself in the exotic scenery of war in the Philippines and gain recognition as a man and soldier.

My chapter follows these subtle interrelations of photographic artifacts and the construction of imperial masculinities in the Philippine-American War at the turn of the nineteenth to the twentieth centuries. It analyzes how U.S. soldiers used photographs of the Philippine-American War to create and retell a story of male imperial self-becoming that drew on spatiotemporal concepts of empire such as race, civilization, and modernity. Engaging in close readings of the materiality and imagery of soldiers' photo albums created in the course or aftermath of the war in the Philippines, I argue that photographic objects played a crucial role in the construction, authentication, and preservation of white male imperial subject positions at the time.

1 "↓ my blanket + towel": The two albums of Ashley R. Farless

My first example is two photo albums Ashley R. Farless created of his service in the Philippines from June 1898 to July 1899. They are today stored in the archival depositories of the U.S. Army History and Education Center in Carlisle, Pennsylvania, as part of the Spanish-American War Veterans Survey Collection. This collection grew out of an ambitions survey started in 1968, when the U.S. Military History Research Collection used pension records of the Veterans Administration to send letters to 8,000 identified veterans of the Spanish American War, the Philippine-American War and the Boxer War resulting in a vast collection of questionnaires, letters, diaries, certificates, and photographic records. The majority of these collections stem from participants of the Philippine-American War.[6]

2 Material Frames

The size of Farless's albums is 12 x 10″ (by way of comparison, 12 x 12″ is a much-used size of photo albums today). Both albums are composed of 25 pages made of thick but flexible grey-tone cardboard alternating with transparent acetate sheets to protect the photograph from gluing to each other. The albums are bound in thin light-brown nubuck leather. As figures 2 and 3 show, the covers have different designs. The first album bears a ribbon hovering above the inscription "Photos", written in ornamental glyphs. The second album, too, bears the

Figs. 2 and 3: Album Covers, Albums of Ashley R. Farless (USAMHI, SAWVSPC, Box 2).

inscription "Photos", this time written in slightly curved glyphs framed by leaf- and flowerlike decorative ornaments. The designs were most likely drawn with a heated pyrographic tool. As one looks closer at the leather's surface, one can observe depressions at the corner points of the glyphs and decorative elements indicating the spot where the heated needle of such a tool was placed before

6 The number of U.S. soldiers serving in the Philippines from 1898 on far outnumbered those in the other settings of the Spanish-American War and the Boxer War. By 1900, at the height of the Philippine American War, 70,000 U.S. volunteer and regular soldiers were stationed in the Philippines to secure American control of the islands. See Brian McAllister Linn. *The Philippine War, 1899–1902*. Lawrence: University Press of Kansas, 2000, ix.

the decorations were drawn. We do not know if Farless himself or a professional leather imprinter created the designs. In any case, the unique handcrafted designs of the covers add to the albums' individualistic character.

As one starts to page through the two albums, the photographs appear in varying formats and sizes, with different tinting and paper. Several of the images collected by Farless are of square 3½ to 3½" format. The majority is of rectangular shape, ranging from 2¼ x 3¼ ″ to 3½ x 5″[2] to as a large as 6 x 8″. While some pages hold up to five images of smaller size, other pages hold just a single large print. In several instances, prints of the same format are collected on consecutive pages indicating that they were most likely taken with the same camera.[7]

The different shapes and sizes of the images in the albums of Ashley R. Farless raise the question of who actually took these pictures. Although there is no written documentation on the production of the albums by Farless or his descendants, there are formal and material hints that provide some answers to this question.

Several larger prints included in Farless's albums have white words, glyphs, or numbers in the lower right or left corner. They stem from scratches made into the original glass plate negatives with a sharp stylus, a technique used by professional photographers to identify the negatives of the prints they sold to soldiers or other buyers in the Philippines or in the United States. Occasionally, the notes reveal the identity of the photographer by providing his – in all probability a man's[8] – name or initials. Most often, however, they merely give the number of the print within the photographer's collection of prints available for purchase. In Farless's albums, several pictures bear the imprint "J. Reyes 1898" or "J. Reyes. Manila", which indicates that they were taken and sold by the Filipino photographer Joaquin Reyes, who owned a well-known photo studio in Manila.

Most of the pictures in Farless's album, however, provide no hint of who actually made them. In many cases, however, their print size indicates that they were taken by amateur photographers who used a camera and roll film by Eastman

7 The first album contains 136 photographs, the second 154 plus several newspaper clippings pasted on its last pages, which relate to the Philippine-American and to the liberation of the Philippines from Japanese occupation by the U.S. Army in 1945.

8 To my knowledge there were no female professional photographers documenting the Philippine-American war. Laura Wexler's discussion of Frances Benjamin Johnston's portraits of Admiral Dewey on his battleship during the return from Manila to the United States shows, however, that female photographers, too, took an active role in the shaping of gendered imperial subject position at the time. See Laura Wexler. *Tender Violence. Domestic Visions in an Age of U.S. Imperialism*. Chapel Hill: University of North Carolina Press, 2000.

Kodak, which was already the largest and most popular producer of photographic goods in the United States at the time.[9]

When one looks closer at the format of the images collected in Farless's album and relates them to standard roll film sizes carried by the different Kodak camera models sold in the United States in the late nineteenth century, one is able to identify the possible camera models used for taking the images. For example, the square 3½ x 3½ format of the images in figure 5, showing scenes of the departure of Farless's regiment to the Philippines, indicates that they were shot with a camera holding a Kodak No. 101 or 106 roll film.[10] They were carried by Kodak's most popular box cameras of the late nineteenth century, such as the "Bullet No. 2" and "Bull's Eye", which were sold for $8 in 1898.[11]

The 2¼ x 3¼ prints pasted on the album page shown in figure 6 – with pictures from the Farless's regiment quarter in Tanjai, Negros – were in all probability taken by a Kodak No. 105 roll film. The most popular camera holding these films was the "Folding Pocket Kodak Camera". Sold from 1898 on for $10, the Folding Pocket was marketed as a specifically handy and flexible camera. In an advertisement of 1898, Kodak promoted its Folding Pocket Cameras as the "simple", "practical", "compact," and "mechanically perfect" alternative to previous glass plate cameras with their "heavy, fragile glass plates and cumbersome plate holders" and stressed their much easier portability: "They slip easily into an ordinary coat pocket and a half dozen cartridges of dozen exposures each can be carried in another pocket without annoyance."[12]

The images included in Farless's albums indicate that American soldiers made ample use of Kodak's portable roll film cameras during their stay in the Philippines. In fact, the Spanish-American War in Cuba, Puerto Rico, Guam and the Philippines and the following war between the United States and Philippine revolutionary army were among the first wars in history documented by its practitioners, preceding in many respects the widespread use of handheld roll film cameras during the First World War.[13]

At the same time, it needs to be noted that only a fraction of the U.S. soldiers took their own camera with them. Apparently, not every soldier was willing to invest one third or more of his prospective monthly salary of $15,60 in a roll film

9 Brian Coe and Paul Gates. *The Snapshot Photograph. The Rise of Popular Photography 1888–1939*. London: Ash & Grant, 1977.

10 See History of Kodak Camera, wwwuk.kodak.com/global/en/consumer/products/techInfo/aa13/aa13.pdf (November 4, 2015).

11 History of Kodak Camera.

12 Kodak Advertisment. *McLure's Magazine*, 12 (1898), 109.

13 Janina Struk. *Private Pictures. Soldiers' Inside View of War*. London: I. B. Tauris, 2001, 49–76.

camera. Also was the practice of photography around 1900 structured by categories of social difference such as race, class, and gender and coded as a pastime of white middle and upper class male and – to a lesser degree – female practitioners.[14]

This might explain why there are no self-produced photographs preserved in the archival records of the Spanish-American War Veterans Survey Collection, despite the fact that more than 7,000 black soldiers served in the Philippines and at least 31 veterans filled out and returned the questionnaires. All veterans answered that they do not have any photos of their service in the Philippines. In addition, many declared that they left their salary with the paymaster instead of spending it in the Philippines. The collection, however, contains some photographic portraits of African American soldiers taken in Army camps or photo studios before or after their return from the Philippines.[15]

Furthermore, the photograph collections of veterans of the 1st California Infantry Regiment indicate that soldiers exchanged the photographs taken by members of the regiment during the war. In fact, several of the photographs collected in Farless's albums re-appear in the photo collections of his fellow soldiers, including the above-mentioned Robert W. Dodd.[16] The apparent exchange of prints from negatives taken in the Philippines supports the assumption that individual soldiers and veterans made efforts to enlarge their collection of images of the war after their return. It also shows that these photographs circulated in the social realm of veterans' meetings long after the soldiers' return from the Philippines.[17] A close analysis thus reveals that Farless's albums must in fact be under-

---

14 Nancy Martha West. *Kodak and the Lens of Nostalgia*. Charlottesville: University Press of Virginia, 2000.

15 See questionnaires of African American soldiers. In: USAMHI, SAWVSC, Boxes 9, 31, 43, 58, 67, 74. I will discuss the absence of vernacular photographs in the collections of African American soldiers and the cultural work of studio portraits of African American soldiers and officers who participated in the Philippines American War in more detail in my book on the role of amateur photography in the Philippine-American War.

16 See Folders of Robert W. Dodd and Alfred Dole, In: USAMHI, Spanish American War Veterans Survey Photograph Collection (SAWVSPC), Box 1.

17 Sources on the afterlife of those photographs are scarce. There are, however, strong indications that soldiers not only showed these images to family members upon their return but also brought them to veterans' meetings. An article on the front page of the *San Francisco News*, of September 20, 1957, with a report on the above mentioned Robert W. Dodd, who met with members of the California Veterans Association of the Spanish American War to celebrate the 58th anniversary of their war service, includes a picture of the aged Dodd looking at a group portrait of the members of his company after their return from the Philippines. "Old Soldiers Will Relive 1898 Days", *San Francisco News*, September 20, 1957, 1.

stood as carefully crafted compositions of photographs, each of which carried its respective history of production, collection, and circulation.

3 Authentication and Narrative

As one opens the first of the two albums, two visual strategies catch the eye of the observer. First, Farless's unveiled impetus to put himself into the visual imagery of the war in the Philippines, and secondly, the ambition to create a theme-driven and self-aggrandizing narrative of his war service abroad.

Among the first images of Farless's album is a picture of a group of six volunteer soldiers with their field packs. As we can see, Farless marked himself as the person on the left with his initials A.R.F., while identifying his fellows with their full last names (fig. 4). The size of the print indicates that an army photographer or a commercial photographer took this image with a glass plate camera in the "Presidio", the military base in San Francisco Bay, before the regiment's departure to the Philippines.

Fig. 4: Ashley R. Farless, First Album (USAMHI, SAWVSPC, Box 2).

The albums, however, also include images of Ashley R. Farless and other regiment soldiers, which were taken by a fellow soldier who used a Kodak roll-film camera. One page in album No. 2, for example, shows four pictures of soldiers in the field during entrenchment work (fig. 5). In the picture to the lower right, Farless marked himself again with his initials. Apparently, the inclusion of these images in the album served to capture and authenticate his participation in the war abroad. The desire to put oneself into the photographic record of the war becomes most apparent in an image taken at Tanjai on the island of Negros, where Farless's regiment was stationed from March to July 1899 (fig. 6). The image includes the note "my blanket+towel" and an arrow directed to a window of the building.

Figs. 5 and 6: Ashley R. Farless, Second Album (USAMHI, SAWVSPC, Box 2).

The private use of the medium of photography thus enabled an individualized documentation of one's war experiences. Indeed, those personal "snap shots" or "snaps" – as they were called by U.S. soldiers –, opened up new and exclusive possibilities of authenticating and framing one's self-image as a man and soldier abroad.[18]

18 The concept of the "snap shot" or „instantaneous shot" developed parallel to the introduction of roll film cameras. The defining feature of the "snap shot" was that it was an exposure "made while the camera is held in the hand" with the shutter set for an instantaneous exposure (other than a time exposure, where the shutter was opened manually for specific amounts of seconds depending on light conditions). See Eastman Kodak Company. *Picture Taking and Picture Making*. Rochester, NY: Eastman Kodak Company, 1898, 13–15, 13.

4 Photo albums and male imperial self-formation

As Gail Bederman, Amy Greenberg, Kristin Hoganson, and others have shown, the male-centered discourse accompanying the United States' entry into the Spanish-American and the Philippine-American Wars gained its strength from the notion of a crisis in masculinity during the 1890s. Fostered by economic decline, rising immigration rates, and a burgeoning women's rights movement, a new cult of martial masculinity among white American middle-class men developed.[19] Its most prominent proponent was the later U.S. president Theodore Roosevelt. In order to transform his image of an effeminate "dude", the young politician Roosevelt successfully fashioned his male self-image by embracing the figure of the virile Western ranchman. In his speeches and widely read books, Roosevelt argued that white American men had built their manliness through the relentless fight against nature and the allegedly "savage" Native American. In this fight, the American frontiersman had at times to regress to brutal means. Roosevelt deemed this as necessary for the advancement of civilization and the Darwinist struggle for the maintenance of the virtues of the "American race".[20] During the 1890s, he and many of his contemporaries began to describe imperialism as a new outlet for the re-vitalization of white American manhood and as a "prophylactic means of avoiding (...) racial decadence".[21]

The broad appeal of their gender-saturated imperial rhetoric became evident in the enthusiastic support of the Spanish-American War in 1898, when thousands of young American men answered President McKinley's call for volunteer soldiers. Theodore Roosevelt himself fostered his masculine self-image by taking over the command of the 1st U.S. volunteer cavalry regiment, the so-called Rough Riders, to fight against the Spanish army in Cuba.[22] While the wars in Cuba and

19 Gail Bederman. *Manliness & Civilization. A Cultural History of Gender and Race in the United States, 1880–1917*. Chicago: University of Chicago Press, 1995; Amy Greenberg. Männlichkeiten, territoriale Expansion, und die amerikanische Frontier im 19. Jahrhundert. In *Väter, Soldaten, Liebhaber. Männer und Männlichkeiten in der nordamerikanischen Geschichte. Ein Reader*, Jürgen Martschukat and Olaf Stieglitz (eds.), 103–121. Bielefeld: Transcript, 2007; Kristin L. Hoganson. *Fighting for American Manhood. How Gender Politics Provoked the Spanish-American and Philippine-American Wars*. New Haven: Yale University Press, 1998.

20 See Theodore Roosevelt. *Hunting Trips of a Ranchman*. New York: G. P. Putnam's Son, 1885; Theodore Roosevelt. *The Winning of the West*, 4 volumes. New York: G. P. Putnam's Son, 1889–1896.

21 Bederman, *Manliness & Civilization*, 187.

22 Theodore Roosevelt. *The Rough Riders*. New York: C. Scribner's Sons, 1899; Hoganson, *Fighting*, 112–125.

the Philippines provided an outlet for the re-vitalization of American manhood, soldiers' photographs and photo albums provided means to document this undertaking a popular as they allowed U.S. soldiers to frame and construct their subject positions in relation to the dominant discursive notions of their time. Their photographs and photo albums enabled them to construct their war experiences as a racially coded project of masculinization and connect their biographies to the discourse of white Western superiority and imperial expansion.

As one opens the first of Ashley R. Farless's albums, a story of military preparation and departure unfolds. The first page shows two photographs. One portrays U.S. troops during maneuvers – the San Francisco Bay in the background – with spectators and carriages in the foreground. The second image, placed in the lower right corner, shows five U.S. volunteers during a break, three of them holding coffee mugs. Again, we see Farless marked by his initials "A. R. F." (fig. 7). Two pages onwards, four images appear that were taken from the railing of the U.S. steamship City of Peking that brought the soldiers to the Philippines. The pictures, bearing the handwritten captions "Farewell," "Salute," and "Goodbye," show U.S. soldiers during the farewell ceremonies in San Francisco Bay and waving spectators on other ships (fig. 8).

Fig. 7: Ashley R. Farless, First Album (USAMHI, SAWVSPC, Box 2).

Fig. 8: Ashley R. Farless, First Album (USAMHI, SAWVSPC, Box 2).

The focus of the album then moves to the Philippines with several photographs of Spanish ship wrecks, the result of Admiral Dewey's victorious sea battle against the Spanish fleet in Manila Bay. A widespread motive in American soldiers' photo collections, these images both emblematized Spain's decline and America's arrival as an imperial power as well as the U.S. soldiers' arrival in the Philippines as the U.S. Army transports usually passed the shipwrecks before anchoring in Manila Bay.

As one continues to page through the album, several motives begin to appear that center around the picturesque, curious, and touristic. The choice in motives illustrates in several regards the experiences of U.S. soldiers in the month after the Spanish capitulation in August 1898, when U.S. troops were stationed in military camps in and around Manila and engaged in leisurely activities outside of the camps such as visiting the "Walled City" and the rural areas surrounding

Manila. The following page, for example, bears a photo depicting a group of Filipino women, men, and children standing with working tools in front of a house (fig. 9), entitled "Back Country Native Philippinos (sic!)". Image and caption imitate the photographic practice and classification work of American anthropologists at the time, which became integral for the legitimization of U.S. colonial rule in the Philippines.[23] This photo also appeared in Volume 2 of José de Olivarez's highly popular illustrated books on the new American colonial possession in the Caribbean and the Pacific entitled "Our Islands and Their People. As seen with Camera and Pencil" published in 1899.[24] Farless probably bought a print of this picture in Manila or after his return to the United States in July 1899. Similar to its function in Olivarez's volume, the photograph and caption in Farless's album invite its viewers to gaze onto the alleged "otherness" of Philippine society. The photograph itself, however, is open to different readings as it depicts the Filipino women, men, and children in self-confident poses while looking earnestly and resolutely into the camera.

Fig. 9: Ashley R. Farless, First Album (USAMHI, SAWVSPC, Box 2).

23 See Mark Rice. *Dean Worcester's Fantasy Islands. Photography, Film, and the Colonial Philippines*. Ann Arbor: The University of Michigan Press, 2014.

24 See José Olivares. *Our Islands and Their People. As Seen with Camera and Pencil*. New York: N. D. Thompson Publishing Company, 1899, 552.

Another page in Farless's first album shows a photograph of an unidentified woman next to three smaller pictures that, according to their size, were taken by a Kodak roll film camera (fig. 10).

Fig. 10: Ashley R. Farless, First Album (USAMHI, SAWVSPC, Box 2).

The upper image of the three small images, shows a U.S. soldier posing in front of a cemetery wall. The photo in the middle depicts a skull fresco, probably a grave decoration the photographer found on the cemetery wall. The lower image depicts three soldiers striking their pose in a stonewall corner around what appears to be a pile of human bones. The soldier in the middle has his one hand resting upon a wooden stick and the other upon the shoulder of the soldier next to him while his right foot stands on the bones. Judging from the recurrent inclusion of such motives in soldiers' photo collections, the posing with skulls and bones in Manila's cemeteries was a common practice among U.S. soldiers.[25] In diary entries, soldiers described the sight of pits filled with human bones as a chilling and fascinating experience. In a letter to his fiancée Maude written on December 20, 1898, Edward F. Dunbar of the 1st California Regiment recounted his visit to one of the so called "bone yards" of Manila as follows:

> You have probably read of the cemeteries here. I visited one a few days ago and saw a sight that I will never forget. There is a tomb several hundred feet long, like a great wall, with three rows of niches, one above the other. The bodies are placed in these and sealed up, (...). As long as the relatives keep the rent paid, all is well, but as soon as it runs behind, the niche is opened and bones taken out and dumped in a pile at the rear. Here there is a pile of crumbling bones and dust and hair thirty feet long and five feet high.[26]

Dunbar's first sentence indicates that he envisioned his relatives in the United States to be informed about the open disposal of human remains in the Philippine cemeteries, which was in fact a frequent topic in American newspaper and magazine reports on the Philippines at the time.[27] The three photographs in Farless's album apparently responded to Americans' morbid fascination with Filipinos' burial practices at the time that was shaped by notions of racial, religious, and cultural difference and a hierarchical concept of civilization.

This might also apply to the image on the left side of the page, albeit in a somewhat different manner. It shows a woman leaning with her right arm on a stone-like object. Her right hand supports her head, while her eyes glance toward a point in the distance. Her long hair falls over her left shoulder, the rolled-up sleeve of her blouse reveals parts of her upper arm. It is very likely that American viewers perceived this as an erotically charged image, since loose hair and bare arms were common signifiers of erotic femininity on pornographic photographs and postcards at the time.[28]

---

25 The collection of and posing with skulls and bones of the enemy has been a central feature of modern wars. As Simon Harris has argued, the trophy hunting and presentation of bones by American or European soldiers in last two centuries rose parallel with the emergence of ideologies of race as soldiers used them were used propagate notions of racial difference. See Simon Harrison. *Dark Trophies: Hunting and the Enemy Body in Modern War*. New York; Oxford: Berghahn Books 2012.
26 Letter of Edward F. Dunbar to Maude (?), Manila, Dec. 20, 1898. In: USAMHI, SAWVSC, Box 1, Folder 30, Dunbar, Edward F., Company C, 1st Regiment, California Infantry.
27 See for example the picture captioned "Boneyard in Paco Cemtetery, Manila." In: Frank Tennyson Neely. *Fighting in the Philippines; Authentic Original Photographs*. London; Chicago: F. T. Neely, 1899. unp., 147.
28 On the display of loose hair in 19th century pornographic photography Europe see for example Isabel Richter. Der phantasierte Tod. Bilder und Vorstellungen vom Lebensende im 19. Jahrhun-

Pictures like this were included in many photo collections of U.S. soldiers, who often captioned them by designations such as "Philippine Belle" or "Philippine Beauty." They are indicative of an eroticized orientalist perception of upper-class Mestiza or Tagalog women by U.S. soldiers, that found its echoes both in popular accounts of the Philippines published in the late nineteenth and early twentieth-century United States as in soldiers' diaries.[29] At the same time, they may hint at soldiers' sexual encounters with and exploitations of women in the Philippine-American War. Parallel to the arrival of American troops in the Philippines in 1898, an elaborate system of commercial prostitution emerged which drew several thousand sex workers from Philippine rural areas and Japan to Manila and other Philippine cities.[30]

While the majority of the images collected in Farless's first album appears to be informed by the curious gaze of a touristic explorer, his second album shifts the attention to the war against the Philippine Revolutionary Army that started on February 4, 1899. During the first month of the war, the Philippine Army engaged in a trench war against its American adversaries, which led to heavy losses on its part due to the vast military superiority of the U.S. troops. In November 1899, General Emilio Aguinaldo dissolved the revolutionary army and ordered the establishment of decentralized guerilla units to continue the fight against the American occupiers.

The photographs collected in Farless's second album reflect his regiment's role in the war from February 1899 until its return to the United States in July 1899. After the outbreak of the war, the 1st California Infantry Regiment participated in the first battles between U.S. and Filipino troops southeast of Manila, including the so-called "Battle of Manila" at Santa Ana on February 5, 1899. On March 1, 1899, the regiment was sent to the island of Negros, where it stayed until its return to the United States.[31]

The first page of Farless's second album bears three pictures (fig. 11). Two of them are captioned. The picture in the left upper corner shows a volunteer soldier posing with his rifle in an army camp captioned "Pvt. Grayson, 1st Nebraska Vol.

dert. Frankfurt am Main: Campus, 2010, 191–92; Lisa Z. Sigel, *Governing Pleasures Pornography and Social Change in England, 1815–1914*. New Brunswick, NJ: Rutgers University Press, 2002.

29 See the descriptions and captioned photographs in Olivarez, *Our Islands and Their People*, 589–604.

30 Paul Kramer. The Darkness That Enters the Home. The Politics of Prostitution during the Philippine-American War. In *Haunted by Empire. Geographies of Intimacy in North American History, American Encounters/Global Interactions*, Ann Laura Stoler (ed.), 367–404. Durham: Duke University Press, 2006.

31 Linn, *Philippine War*, 50–55.

who fired the first shot from American end of Santa Mesa bridge, Febr. 4, 1899." The picture to the left is entitled "In Paco Cemetery" and shows the gravestones of several American volunteer soldiers who, according to the inscribed dates, had died in November 1898, most likely due to illness. Farless probably intended to juxtapose the portrait of Private Grayson, as a visual marker of the outbreak of the war, with an image that documented the sacrifices of his fellow soldiers in the commencing fights.

Fig. 11: Ashley R. Farless, Second Album (USAMHI, AWVSPC, Box 2).

Two further pages may serve as exemplary of the framing and representation of the war against the Philippine revolutionaries in Farless's second album. The following page shows six snapshots taken by a Kodak Folding Pocket camera. The images show scenes at the Pasig River, including a picture of an American gunboat. The three other images apparently show a deserted battlefield. A soldier of Farless's regiment most likely took them after the fight had ended. The image on the mid-left is captioned "Insurgent trench" while the picture to the right bears the note "Philippino trench filled in over the dead", indicating that these served to authenticate the soldiers' personal presence on the Philippine battlefields and their physical proximity to scenes of fighting and death (fig. 12).

Fig. 12: Ashley R. Farless, Second Album (USAMHI, SAWVSPC, Box 2).

While continuing to page through the album, one encounters several images depicting dead Philippine soldiers. The left picture on the following page (fig. 13) shows a trench filled with corpses entitled "36 Insurgent Dead in one trench in Santa Ana, Luzon, P.I." The image on the right is captioned "The Morque (sic!) at General Hospital, Manila." Similar to the picture of "Paco Cemetery" above, the image of the morgue at General Hospital in Manila seems to allude to the sacrifice of American soldiers in the Philippine-American War. The photo on the left, however, documents the heavy toll of the war against the militarily superior U.S. Army for the Philippine revolutionaries. The repeated inclusion of those images in soldiers' photo collections indicates that soldiers attached a specific meaning to them. This image, too, has a scratched note in the lower right corner indicating that is was made by a professional photographer who visited the battle scene after the fighting had ensued. Different versions of it appeared in several illustrated publications on the Philippine American War, including Frank Tennyson Neely's *Fighting in the Philippines. Authentic Original Photographs* published in 1899. In Neely's widely distributed picture book of the war in the Philippines, this image held the telling caption: "The American Artillery did wonderful execution in the battles with the insurgents. In a trench in Santa Ana the Tagal dead lay in piles. The group shown in the picture consisted of thirty-eight bodies."[32] Since Farless's caption also mentions an exact yet slightly different number of dead captured by the photo, one could even assume that he had seen a published version of the photograph with a similar caption before he created his album. Copies of it appear in many photo collection and albums of U.S. soldiers, indicating that it was a much-coveted souvenir of the war.

Fig. 13: Ashley R. Farless, Second Album (USAMHI, SAWVSPC, Box 2).

A second picture in Farless's album shows a group of four U.S. soldiers posing with their guns in front of four Philippine corpses. The caption says "Insurgent casualties". In the back one sees two further U.S. soldiers and several people in white clothing, presumably Philippine civilians with shovels in their hands to dig a pit. An abandoned hat lies on the ground, probably belonging to one of the four

32 Neely. *Fighting in the Philippines*, unp., 55.

dead or one of the digging Filipinos. The four soldiers on the right take relaxed postures, while leaning on their rifles. The one to the left has his foot resting on the fresh clumps of earth on the ground. The man to the right looks in direction of the corpses, the three other have their gazes turned to the camera/photographer. The structure of the picture reminds one directly of hunting pictures. In fact, hunters posing with their kill were a common photographic motif in the United States at the time. With the introduction of roll film cameras, many American hunters began to use their hand cameras on hunting trips and hunting pictures were frequently published in late 19th century hunting and fishing magazines.[33] In their diaries and letters U.S. soldiers' widely used the motif of hunting to describe warfare in the Philippines, often dubbing the killing of Filipino soldiers a "rabbit hunt" or "turkey shot".[34] The hunter poses of the U.S. soldiers on the photograph turns the dead Filipinos into trophies of conquest. At the same time, the photo itself can be interpreted as a trophy picture both for the soldiers depicted and for those soldiers such as Farless who bought and included it in their albums.

Deprived of their ability to influence their photographic self-presentation, the dead Filipinos on this image serve as a backdrop for the performance of white, virile-masculine and able-bodied selves. The Philippine workers in the background further reinforce this effect as they appear as those who have to cope with the deadly consequences of their compatriots' resistance. The motif of white men posing in front of dead non-whites takes up the visual tropes of lynching photographs of African American men killed by white mobs in the American South and the photographs of the Wounded Knee Massacre showing U.S. cavalry and artillery soldiers posing next to the corpses of Lakota men, women, and children and transfers them to the setting of a colonial war in the Pacific thus equalizing African Americans, Native Americans and Filipinos Revolutionaries as depraved and inferior non-white 'others.'[35]

---

33 See Frank Luther Mott. *A History of American Magazines, 1865–1885*. Cambridge, MA: Harvard University Press, 1938, 209–211. See also Jay Ruby. *The World of Francis Cooper: Nineteenth-Century Pennsylvania Photographer*. University Park, Pa.: Pennsylvania State University Press, 1999;

34 See for example Albert Sonnichsen. *Ten Months a Captive among Filipinos; Being a Narrative of Adventure and Observation during Imprisonment on the Island of Luzon*. New York: C. Scribner's Sons, 1901, 385. Frank Schumacher. "Marked Severities". The Debate over Torture during America's Conquest of the Philippines, 1899–1902. *Amerikastudien / American Studies* 51, no. 4 (2006): 475–498, 481; Paul A. Kramer, *The Blood of Government: Race, Empire, the United States, & the Philippines*. Chapel Hill: University of North Carolina Press, 2006, 144.

35 See Amy Louise Wood. Lynching Photography and the "Black Beast Rapist" in the Southern White Masculine Imagination. In *Masculinity Bodies, Movies, Culture*, edited by Peter Lehman, 193–212. New York [u.a.]: Routledge, 2001; Richard E. Jensen, R. Eli Paul, and John Carter. *Eye-*

Fig. 14: Ashley R. Farless, Second Album (USAMHI, SAWVSPC, Box 2)

The recurrent inclusion of these photographs in soldiers' war albums indicates that U.S. soldiers not only viewed those pictures as a much valued reminders and trophies of the war but that they also fulfilled a narrative function within the albums. Apparently, these images were sought to complement and complete a biographical narrative of their white male heroic self-becoming in the war abroad.

The intricate relationship of pictures of war violence and racialized notions of masculinity can be observed in closer detail in another photo album of the Philippine-American War.

---

witness at Wounded Knee. The Great Plains Photography Series. Lincoln: University of Nebraska Press, 2011.

5 "CAN the 33rd shoot?" The album of Colonel Luther R. Hare

The album of Colonel Luther Rector Hare is a most distinct example of U.S. soldiers' male imperial self-fashioning via photographs of violence in the course of the Philippine-American War. As in the case of Farless's albums, there is no information preserved about who actually created this album and to which occasion it was used and by whom. Its luxuriant design, however, suggests that it was probably a gift presented to Colonel Luther R. Hare by the officers and soldiers of the 33rd Regiment after its return from the Philippines.

The album has the size of 14¼ x 10¾ x 2¼″ and is bound by heavy wooden covers of 3/8″ diameter each, consisting of twenty-five gilt-edged cardboard leaves of 1/16″ diameter. The anterior cover shows a drawing of the American eagle sitting on the poles of two large crossed American flags, framed by palm leafs and blossoms on the left and right side. The letters below read "Souvenir Philippines 33rd Reg't".

On many album pages, the photographs and captions are accompanied by elaborate watercolor paintings. Both the artistic decorations and the artful layout indicate that professional artists and book designers participated in the creation of the album.

The rich features of Hare's album attest to the military rank and social status of its possessor. Luther R. Hare was born in Indiana in 1851 before his family moved to Texas. Following the example of his father, who had served as an officer in the Confederate Army, Hare entered West Point in 1870. Graduating four years later, he joined the 7th Cavalry Regiment. As a Lieutenant, he participated in the famous Battle of Little Bighorn in 1876 that led to the death of General George Armstrong Custer and the soldiers under his command. Between 1876 and 1899, Hare participated in further war expeditions against Native Americans, including the above mentioned Massacre at Wounded Knee in 1891.[36]

On July 5, 1899 Hare was appointed Colonel of the 33rd U.S. Volunteer Infantry Regiment. Upon receiving orders to enlist and train the new regiment, Hare recruited about 1,300 soldiers from Texas and New Mexico and the so-called "Indian Territory". In September 1899 the regiment left to the Philippines, where it participated in several battles with Philippine revolutionary troops in San

36 Ray Meketa. *Luther Rector Hare, a Texan with Custer A Biography of an American Hero*. Mattituck: J.M. Carroll, 1983, 21–47.

Jacinto and northern Luzon. Hare and his unit gained fame for their participation in the rescue of Navy Lieutenant James C. Gilmore along with twenty-five soldiers and sailors who had been held as captives by General Manuel Tinio y Bundoc.[37]

Hare's Contemporaries perceived Hare as an untiring and fearless military officer. One described him as "as our best type of cavalry soldier – an outdoor man, fine horseman, fine shot, alert, daring and practical, had a good eye for the country".[38]

The appreciation of Hare's military ability and leadership is also apparent in the images collected in his album. For the purpose of my argument, however, I will confine my observations to the representation of violence and soldiership on select pages of the album.

Similar to the first album of Ashley R. Farless, the album starts with snapshots of Luther Hare in his white leisure suite on board the transport ship Sheridan that brought the 33rd Regiment to the Philippines. They are followed by a colorized photograph of the wreck of the Spanish Cruiser *Castilla*, again a visual reminder of Admiral Dewey's successful sea battle against the Spanish fleet in Manila Bay. The following pages depict the 33rd Regiment's mobilization for the so-called Wheaton Expedition in November 1899, which involved the search for the fleeing General Emilio Aguinaldo in Northern Luzon and the above-mentioned search for Navy Lieutenant Gilmore and further Americans captives.

As one continues to turn the album sheets, one encounters a page that bears four photographs taken by roll-film cameras that are arranged in the pattern of a four-leaf clover (fig. 14). The 3½ x 3½″ size of the upper and lower picture again indicates that they were taken by a Kodak box camera holding a No. 101 or 106 film. The two pictures in the middle are of 2¼ x 3¼″ size which indicates that they were taken by a Folding Pocket Kodak camera, also mentioned earlier in the essay.

The image above shows four U.S. soldiers sitting in relaxed poses on the ground. In the lower picture, three soldiers take aim with rifles and pistols behind cover. The picture in the mid-left depicts an American soldier striking his pose behind an ox cart that appears to be loaded with at least one corpse, as one can see two legs and an arm hanging over its edge. The soldier stands with his legs apart and his arms akimbo, while he looks toward the cart in front of him. The picture to the right shows a row of corpses lined up along a stone balustrade.

---

37 Meketa, *Luther Rector Hare.*

38 Cited after Meketa, *Luther Rector Hare*, 8.

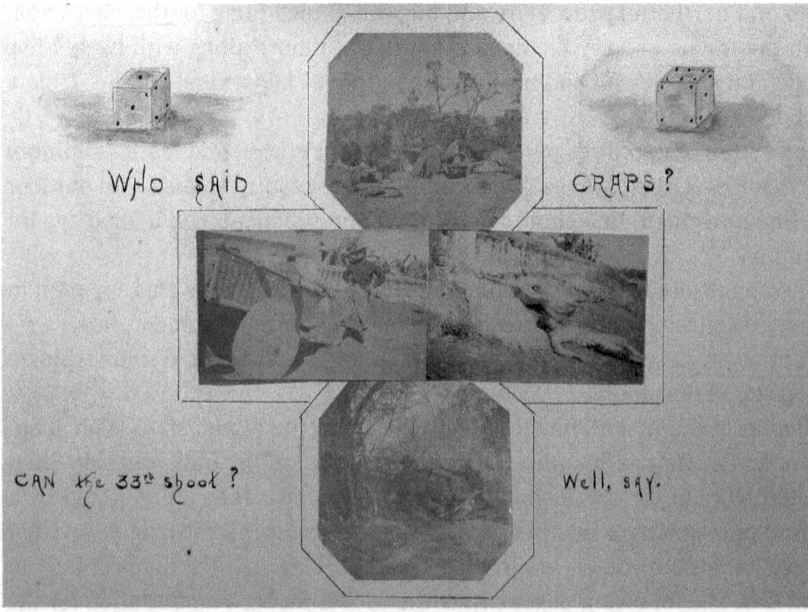

Fig. 15: Album of Luther R. Hare (USAMHI, Luther R. Hare Photograph Collection).

The images are accompanied by the caption "Who Said Craps? Can the 33rd shoot? Well, say". As also suggested by the two dice in the upper corners, the caption equates the shooting and killing of Filipino soldiers with the "shooting" of dice. Craps was a highly popular dice game among U.S. soldiers in the Philippines. In fact, the soldiers in Hare's regiment gained notoriety for playing this game. In his biography of Luther Hare, Ray Meketa recounts an anecdote that most likely served as an inspiration for the page's design:

> The 33rd was assigned to the forces under General Lloyd Wheaton and on November 7th landed in Lingayen Gulf, capturing the town of San Fabian. Three days later, in San Fabian, the General summoned General Hare to his headquarters and advised him that Colonel Freeman and the 13th had reconnoitered toward San Jacinto that day. The reports came back that Filipinos were entrenched along the river and were of such strength that the 13th had turned back without engaging them. Wheaton instructed Luther to take the 33rd, and such troops from the 13th as were necessary, and be in San Jacinto by sundown the following day. He asked Colonel Hare how many men he had ready for the mission. "By gad, General", he replied, "I have 1246 men ready. 1200 of them are sharpshooters and all of them are crapshooters … We will be in San Jacinto by sundown tomorrow night and we won't need the 13th![39]

39 Meketa, *Luther Rector Hare*, 54.

According to Mekata, Hare's use of the term "crapshooters" referred to "a disturbance" on board of the Sheridan on the passage to the Philippines – probably a fight arising out of crap gambling – "that was ended with the dice on board being collected and tossed overboard."[40]

The disturbing equation of "shooting" the dice and killing Filipino soldiers in battle on the album page hints again at the wide distribution of hunting and playing metaphors within the racial-exterminist rhetoric of American soldiers in the Philippine-American War.[41]

At the same time, it accounts for the ways in which soldiers used hand-held roll film cameras to frame their violent engagements on the battlefield as an experience of imperial masculinization. Several pictures in Hare's album underscore this observation. The following page shows eight portraits of officers in Hare's regiment, which were again taken by a Folding Pocket Kodak camera. The page is entitled "Hare's Willies" and decorated with palm trees on the sides, a piece of Pears soap and a can of tomatoes. The soldiers pose individually with their uniforms and pistol holsters. Seven soldiers pose in front of a house on a street, which indicates that the same photographer took the photographs in turns. Five of the eight soldiers take their pose with a spear, the traditional hunting and battle weapon of different ethnic groups in the Philippines. I suggest that the officers' seemingly casual and ironic appropriation of the spear can be read in at least two ways. First, as a performance of white racial and cultural superiority, and second, as a reference to the officers' subscription to a code of archaic masculinity in their fight against their Philippine adversaries. Another portrait on the following page seems to support this interpretation, although not all the visual references encoded in this image seem clear to me.

The portrait shows another officer of Hare's regiment, Lieutenant John Lipop. He stands in contrapposto, his left hand rests on his pistol holster, with a spear in his right hand. The caption adds "If I were KING" in quotation marks (fig. 16). Obviously, the officer's pose expresses a claim to power and one might guess to which "king" the caption alludes: Does the comment link Lipop's pose to classical European representations of kings and emperors, dating back to sculptures of Alexander the Great with the spear? Does it take up the figure of the enigmatic white colonizer who becomes the idolized leader of the natives, famously embodied by

---

40 Meketa, *Luther Rector Hare*, 61.

41 See Frank Schumacher. The Philippine-American War and the Creation of American Empire in Asia. In *The Routledge Handbook of American Military and Diplomatic History, 1865 to the Present*, Antonio S. Thompson and Christos G. Frentzos (eds.), 45–52. New York: Routledge, 2013, 49.

Fig. 16: Album of Luther R. Hare (USAMHI, Luther R. Hare Photograph Collection).

Fig. 17: Album of Luther R. Hare (USAMHI, Luther R. Hare Photograph Collection).

Mr. Kurtz in Joseph Conrad's novel Heart of Darkness[42] (published in the very same year of the 33rd Regiment's engagement in the Philippines)? Or, is Lipop just playfully mimicking what he perceives as a typical pose of a "native king"?

In any case, Lipop's triumphalist pose and the seemingly ironic comment capture the transformative potential of the colonial war experience for the

42 Joseph Conrad. *Heart of Darkness*. New York: Penguin Books, 2012.

Fig. 18: Album of Luther R. Hare (USAMHI, Luther R. Hare Photograph Collection).

participating soldiers and officers, as it provided them with the opportunity to act and present themselves as manly and virile representatives and agents of an allegedly superior race. At the same time, the pose and comment can be read as a crude and uninhibited expression of the spatiotemporal dimension that was the heart of American and European colonial policy in the age of high-imperialism: the self-righteous conquest and exploitation of colonial territories and their inhabitants under the maxim of social evolutionary progress and civilization.

Conclusion

As my discussion has shown, photographic artifacts in general and photo albums in particular constituted much-valued material devices for the creation and authentication of white male subject positions in the course and aftermath of the Philippines-American War. The blank pages and book format of photo albums provided the material setting for their use as a creative means and medium of soldiers' self-expression. The albums of Ashley R. Farless and Luther R. Hare document that these albums framed the experience of war as a heroic story of male imperial conquest, exotic encounter, and racial triumph. Pictures of dead Filipino soldiers in particular consolidated and enforced this narrative. Enclosed by scenes of departure and return, the visual arrangement of the photo albums linked this story of male imperial self-becoming to the concept and geography of the nation. In addition, it framed their war experiences in the narrative plot of a quest. Both albums end with pictures of their regiments' departure from the Philippines and return to the United States. Soldiers' use of personal hand cameras further accentuated this practice, as it allowed them to personalize their visual accounts of their service in the Philippine-American War. Through their material quality and their indexicality – that is, their functioning as a referent of past reality created by the radiation of the objects in front of the camera – photographs from the Philippines provided a powerful means of authentication of soldiers' war experiences.

Although we do not know much about the usage of these photographic artifacts, it is safe to assume that they served as important devices in the retelling of soldiers' war experiences in the realm of their families and veterans meetings. Their imagery and materiality allow us to grasp the multiple perceptions, desires, and expectation of the white men who fought in America's first colonial war overseas. They thus offer an important insight into the creation of masculine-imperial subject positions within American society at the turn of the nineteenth to the twentieth century.

Two aspects, however, need to be added to this conclusion. First, one needs to be aware of the visibilities created *by* and *through* photographic artifacts such as the albums of Ashley R. Farless and Luther R. Hare. Evidently, the visual encoding of masculinity in their albums is deeply immersed in a continuing process of hegemonic "whitening". While white soldiers used images of the war in the Philippines to enhance their manliness, they also participated in the creation of a Euro American image of white imperial masculinity in the turn-of-the-century United States that concealed the participation of African American or Native American soldiers in the war. Second, it is important to note that soldiers' photographic artifacts stand in a complex and strained relationship to their actual experiences in the war. Soldiers' diaries tell about a range of physical and emotional states and conditions that shaped their individual war experience in the Philippines, such as excitement, curiosity, pride, and euphoria but also illness, boredom, fear, pain, home-sickness, and desperation. The visual narrative of war created via the medium of the photo album is in many ways detached from this range of experiences, omitting what one might call the underside of war. In fact, one needs to perceive war albums just as photo albums in general, as more or less consciously created semi-public displays of one's self that settles upon the anecdotal, martial, exotic, and picturesque and reflects upon what is deemed as observable and memorable. At the same time, one might interpret those albums as elaborate cultural devices of coping and coming to terms with the experience of war, as they provided soldiers with a means to create an exciting and maybe even enjoyable narrative of this experience, both for themselves as for their intended viewers at home.

References

Archives and Newspapers/ Magazines

U.S. Army Military History Institute, Carlisle, PA (USAMHI), Spanish American War Veterans Survey Collection (SAWVSC)
U.S. Army Military History Institute, Carlisle, PA (USAMHI), Spanish American War Veterans Survey Photograph Collection (SAWVSPC)
U.S. Army Military History Institute, Carlisle, PA (USAMHI), Luther R. Hare Photograph Collection
Mc Lure's Magazine
San Francisco News

Literature

Bederman, Gail. *Manliness & Civilization. A Cultural History of Gender and Race in the United States, 1880–1917.* Chicago: University of Chicago Press, 1995.
Batchen, Geoffrey. *Each Wild Idea. Writing, Photography, History.* Cambridge: MIT Press, 2001.
Coe, Brian and Paul Gates. *The Snapshot Photograph. The Rise of Popular Photography 1888–1939.* London: Ash & Grant, 1977.
Conrad, Joseph. *Heart of Darkness.* New York: Penguin Books, 2012.
Eastman Kodak Company. *Picture Taking and Picture Making.* Rochester, NY: Eastman Kodak Company, 1898.
Edwards, Elizabeth. Material Beings. Objecthood and Ethnographic Photographs. *Visual Studies* 17:1 (2002): 67–75.
Elizabeth Edwards and Janice Hart. *Photographs Objects Histories: On the Materiality of Images, Material Cultures.* London; New York: Routledge, 2004.
Greenberg, Amy. Männlichkeiten, territoriale Expansion, und die amerikanische Frontier im 19. Jahrhundert. In *Väter, Soldaten, Liebhaber. Männer und Männlichkeiten in der nordamerikanischen Geschichte. Ein Reader*, Jürgen Martschukat and Olaf Stieglitz (eds.), 103–121. Bielefeld: Transcript, 2007.
History of Kodak Camera, wwwuk.kodak.com/global/en/consumer/products/techInfo/aa13/aa13.pdf (November 4, 2015).
Harrison, Simon. *Dark Trophies: Hunting and the Enemy Body in Modern War.* New York; Oxford: Berghahn Books 2012.

Hoganson, Kristin L. *Fighting for American Manhood. How Gender Politics Provoked the Spanish-American and Philippine-American Wars*. New Haven: Yale University Press, 1998.

Jensen, Richard E., R. Eli Paul, and John Carter. *Eyewitness at Wounded Knee*. The Great Plains Photography Series. Lincoln: University of Nebraska Press, 2011.

Kramer, Paul. The Darkness That Enters the Home. The Politics of Prostitution during the Philippine-American War. In *Haunted by Empire. Geographies of Intimacy in North American History, American Encounters/Global Interactions*, Ann Laura Stoler (ed.), 367–404. Durham: Duke University Press, 2006.

Kramer, Paul A., *The Blood of Government: Race, Empire, the United States, & the Philippines*. Chapel Hill: University of North Carolina Press, 2006.

Langford, Martha. *Suspended Conversations. The Afterlife of Memory in Photographic Albums*. Montreal: McGill-Queen's University Press, 2001.

Linn, Brian McAllister. *The Philippine War, 1899–1902*. Lawrence: University Press of Kansas, 2000.

Meketa, Ray. *Luther Rector Hare, a Texan with Custer A Biography of an American Hero*. Mattituck: J.M. Carroll, 1983.

Mott, Frank Luther. *A History of American Magazines, 1865–1885*. Cambridge, MA: Harvard University Press, 1938, 209–211.

Neely, Frank Tennyson. *Fighting in the Philippines; Authentic Original Photographs*. London; Chicago: F. T. Neely, 1899.

Olivares, José. *Our Islands and Their People. As Seen with Camera and Pencil*. New York: N. D. Thompson Publishing Company, 1899.

Rice, Mark. *Dean Worcester's Fantasy Islands. Photography, Film, and the Colonial Philippines*. Ann Arbor: The University of Michigan Press, 2014.

Richter, Isabel. Der phantasierte Tod. Bilder und Vorstellungen vom Lebensende im 19. Jahrhundert. Frankfurt am Main: Campus, 2010.

Roosevelt, Theodore. *The Rough Riders*. New York: C. Scribner's Sons, 1899.

Roosevelt, Theodore. *Hunting Trips of a Ranchman*. New York: G. P. Putnam's Son, 1885.

Roosevelt, Theodore. *The Winning of the West*, 4 volumes. New York: G. P. Putnam's Son, 1889–1896.

Ruby, Jay. *The World of Francis Cooper: Nineteenth-Century Pennsylvania Photographer*. University Park, Pa.: Pennsylvania State University Press, 1999.

Schumacher, Frank. "Marked Severities". The Debate over Torture during America's Conquest of the Philippines, 1899–1902. *Amerikastudien / American Studies* 51, no. 4 (2006): 475–498

Schumacher, Frank. The Philippine-American War and the Creation of American Empire in Asia. In *The Routledge Handbook of American Military and Diplomatic History, 1865 to the Present*, Antonio S. Thompson and Christos G. Frentzos (eds.), 45–52. New York: Routledge, 2013.

Stewart, Susan. *On Longing. Narratives of the Miniature, the Gigantic, the Souvenir, the Collection*. Durham: Duke University Press, 1993.

Sigel, Lisa Z., *Governing Pleasures Pornography and Social Change in England, 1815–1914*. New Brunswick, NJ: Rutgers University Press, 2002.

Sonnichsen, Albert. *Ten Months a Captive among Filipinos; Being a Narrative of Adventure and Observation during Imprisonment on the Island of Luzon*. New York: C. Scribner's Sons, 1901.

Struk, Janina. *Private Pictures. Soldiers' Inside View of War*. London: I. B. Tauris, 2001.

West, Nancy Martha. *Kodak and the Lens of Nostalgia*. Charlottesville: University Press of Virginia, 2000.

Wexler, Laura. *Tender Violence. Domestic Visions in an Age of U.S. Imperialism*. Chapel Hill: University of North Carolina Press, 2000.

Wood, Amy Louise. Lynching Photography and the "Black Beast Rapist" in the Southern White Masculine Imagination. In *Masculinity Bodies, Movies, Culture*, edited by Peter Lehman, 193–212. New York [u.a.]: Routledge, 2001.

Figures

Fig. 1: *Carte de visite* of Robert W. Dodd. In: U.S. Army Military History Institute (USAMHI), Spanish American War Veterans Survey Photograph Collection (SAWVSPC), Box 1.

Figs. 2, 3: Album Covers, Albums of Ashley R. Farless (USAMHI, SAWVSPC, Box 2).

Fig. 4: Ashley R. Farless, First Album (USAMHI, SAWVSPC, Box 2).

Figs. 5, 6: Ashley R. Farless, Second Album (USAMHI, SAWVSPC, Box 2).

Figs. 7, 8: Ashley R. Farless, First Album (USAMHI, SAWVSPC, Box 2).

Fig. 9: Ashley R. Farless, First Album (USAMHI, SAWVSPC, Box 2).

Fig. 10: Ashley R. Farless, First Album (USAMHI, SAWVSPC, Box 2).

Fig. 11: Ashley R. Farless, Second Album (USAMHI, AWVSPC, Box 2).

Fig. 12: Ashley R. Farless, Second Album (USAMHI, SAWVSPC, Box 2).

Fig. 13: Ashley R. Farless, Second Album (USAMHI, SAWVSPC, Box 2).

Fig. 14: Ashley R. Farless, Second Album (USAMHI, SAWVSPC, Box 2).

Fig. 15: Album of Luther R. Hare (USAMHI, Luther R. Hare Photograph Collection).

Fig. 16: Album of Luther R. Hare (USAMHI, Luther R. Hare Photograph Collection).

Fig. 17: Album of Luther R. Hare (USAMHI, Luther R. Hare Photograph Collection).

Fig. 18: Album of Luther R. Hare (USAMHI, Luther R. Hare Photograph Collection).

All photographs: Courtesy of the U.S. Army Military History Institute.

Joanna de Groot

Beyond Blindness, Bias, and Marginalisation

Gender and the Making/Unmaking of Colonial, Anti-Colonial, and Post-Colonial Analyses

The publication of Hardt and Negri's text *Empire* has stimulated varied responses, both enthusiastic and critical, among social theorists and commentators on the contemporary scene, but much less discussion among historians.[1] This may be understandable, since it is a work of grand theory offering an analysis of circumstances at the start of the twenty-first century, in which the rich complexities of the past are given less consideration than current political problems and opportunities. Indeed the level of abstraction and generalisation in their work often does significant epistemic violence to the past in the name of conceptual analysis. It must be stated at the outset that such violence seriously damages the very aims of the project, whether this argument is made from the dialectical perspective of historical materialism, from that of time/space narrative, or from a Foucauldian genealogical perspective. Nonetheless, the flawed relationship of past, present and future in the Hardt/Negri text should encourage historians to respond constructively to the poverty of its engagement with the past, as well as to its undoubted intellectual energy, as happened at the colloquium which gave rise to this volume. Clearly there can and should be systematic general critiques of the structural consequences of the problematic assumptions and inadequate treatment of the past for the text as a whole, but this piece centres on its gender bias and gender blindness. Stimulated by Quinby's perceptive, if predominantly conceptual and methodological, critique it focusses on the gender blind readings of 'empire' as a category, a historical phenomenon, and

1 Michael Hardt and Antonio Negri. *Empire*. Cambridge: Harvard UP, 2000; Paul Thompson. Foundation and Empire. A Critique of Hardt and Negri. *Capital and Class* 29:2 (2005): 73–98; Tarak Barkawi and MarkLaffey. Retrieving the Imperial. Empire and International Relations. *Millennium: A Journal of International Studies* 31:1 (2002): 109–127; Peter Green. 'The Passage from Imperialism to Empire'. A Commentary on *Empire* by Michael Hardt and Antonio Negri. *Historical Materialism* 10:1 (2002): 29–77; Philip Pomper. The History and Theory of Empires. *History and Theory* 44:4 (2005): 1–27; Alex Callinicos. The Actuality of Imperialism. *Millennium: A Journal of International Studies* 31:2 (2002): 319–326; Atilio Borón. *Empire and Imperialism. A Critical Reading of Hardt and Negri*. London: Zed Books, 2005.

DOI 10.1515/9783110418750-008

a social formation.[2] While the gender critique in this piece will draw attention to empirical absences and misunderstandings, its central argument is that such bias and blindness have major conceptual and theoretical implications for the whole Hardt/Negri project. The discussion will deploy the rich resources of feminist and gender theory and scholarship on empire and related topics, which in itself demonstrates the centrality and potential of a gendered approach. The length of some the footnotes which readers will encounter in this piece is a deliberate and material demonstration of the range and depth of scholarship which has been ignored in the investigations of time space and empire.

The question of how to position gender in the investigation and interpretation of the past poses itself in a number of ways. At the substantive level existing scholarly practice needs to be modified by the recuperation of the large quantity of ignored or marginalized ideas and information regarding gender for use in historical research and writing. The study of ancient slavery or modern labour movements, like accounts of religious practice or of monarchical governance are transformed once attention is paid to evidence about the distinctive roles and perceptions of women, men, families, and sexualities.[3] At the conceptual level such information requires a rethinking of terms regularly used by historians to depict and comment on past societies. Notions of 'work', 'politics', 'nation', 'family' and 'religion', to name just a few concepts regularly used in historical writing, acquire changed meanings once that evidence is taken into account.[4]

---

2 Lee Quinby. Taking the Millennial Pulse of *Empire*'s Multitude. A Genealogical Feminist Diagnosis. *Empire's New Clothes. Reading Hardt and Negri*, Paul Passavant and Jodi Dean (eds.), 231–251. New York: Routledge, 2004.

3 Elena Woodacre (ed.). *Queenship in the Mediterranean. Negotiating the Role of the Queen in the Medieval and Early Modern Eras*. New York: Palgrave Macmillan, 2013; Walter Scheidel. The most Silent Women of Greece and Rome. Rural Labour and Women's Life in the Ancient World. *Greece and Rome* 42:2 (1995): 202–217; Lex Heerma van Voss and Marcel van der Linden. *Class and Other Identities. Gender, Religion and Ethnicity in the Writing of European Labor History*. New York: Berghahn, 2002; Sonya Rose. Gender Antagonism and Class Conflict. Exclusionary Strategies of Male Trade Unionists in Nineteenth-century Britain. *Social History* 13:2 (1988): 191–208; Laura Levitt and Miriam Peskowitz (eds.). *Judaism since Gender*. London: Routledge, 1997; Fatima Mernissi. *Women's Rebellion & Islamic* Memory. London: Zed Books, 1996; Sue Morgan and Jaqueline de Vries (eds.).*Women, Gender and Religious Cultures in Britain, 1800–1940*. London: Routledge, 2010; Caroline Walker Bynum. *Jesus as Mother. Studies in the Spirituality of the High Middle Ages*. Berkeley: University of California Press, 1984.

4 See Sue Morgan. Rethinking Religion in Gender History. Theoretical & Methodological Reflections. In *Gender, Religion and Diversity. Cross-Cultural Perspectives*, Ursula King and Tina Beattie (eds.), 113–124. London: Continuum, 2005; Joanna de Groot and Sue Morgan. Beyond the 'religious turn'? Past, Present and Future Perspectives in Gender History. *Gender and History*, 25:3 (2013): 395–421; Ida Blom, Karen Hagemann and Catherine Hall (eds.). *Gendered Nations. Nationalisms*

At the level of general analysis the 'big stories' of, for example, Christianity in pre-modern western Europe or nineteenth century Africa, or of nineteenth and twentieth century 'industrialisations' and political changes, are also transformed once due attention is given to the gendered character of governance, production or culture.[5] Beyond this, the work of gender aware scholars in a range of disciplines has generated original concepts and theoretical insights with the potential to transform the whole field of historical practice. These range from analyses of patriarchies and sexualities as structures shaping human existence over time, and of men as gendered subjects, to the development of queer and postcolonial theory. Deconstructive strategies using material from various periods and places, alongside critiques of categories like 'experience' challenged many assumptions behind conventional social and political history.[6]

---

and Gender Order in the Long Nineteenth Century. Oxford: Berg, 2000; Himani Bannerji, Shahrzad Mojab and Judith Whitehead (eds.). *Of Property and Propriety. The Role of Gender and Class in Imperialism and Nationalism.* Toronto: University of Toronto Press, 2001; Anne McClintock, Aamir Mufti, Ella Shohat, Social Text Collective (eds.). *Dangerous Liaisons. Gender, Nation, and Postcolonial Perspectives*, Minneapolis: University of Minnesota Press, 1997; Anne Phillips and Barbara Taylor. Sex and Skill. Notes towards a Feminist Economics. *Feminist Review*, 6:1 (1980): 79–88; Sandy Bardsley. Women's Work Reconsidered. Gender and Wage Differentiation in Late Medieval England. *Past and Present* 165 (1999): 3–29; Deborah Valenze. *The First Industrial Woman*. Oxford: Oxford UP, 1995.

5 See Sonya O. Rose. Proto-Industry, Women's Work and the Household Economy in the Transition to Industrial Capitalism. *Journal of Family* History 13:2 (1988): 181–93; Katrina Honeyman and Jordan Goodman. Women's Work, Gender Conflict and Labour Markets in Europe 1500–1900. *Economic History Review* 44:4 (1991): 608–628; Janet Hunter. Gender, Economics and Industrialization. Approaches to the Economic History of Japanese Women, 1868–1945. In: *Japanese Women, Emerging From Subservience, 1868–1945*, Hiroko Tomida, and Gordon Daniels (eds), 119–144. Folkestone: Global Oriental, 2005; Jan De Vries. The Industrial Revolution and the Industrious Revolution. *Journal of Economic History* 54:2 (1994): 249–270; Beshara Doumani (ed.). *Family History in the Middle East. Household, Property, and Gender*. Albany: State University of New York Press, 2012; Merry Wiesner. Beyond Women and the Family. Towards a Gender Analysis of the Reformation. *Sixteenth Century Journal* 18:3 (1987): 311–321.

6 See Sylvia Walby. Theorising Patriarchy *Sociology* 23:2 (1989): 213–234; Aviar Brah. Difference, Diversity, Differentiation. *International Review of Sociology* 2:2 (1991): 53–71; Deniz Kandiyoti. Bargaining with Patriarchy. *Gender & Society* 2:3 (1988): 274–290; Siegfried Gruber and Mikołlaj Szołtysek. The Patriarchy Index. A Comparative Study of Power Relations across Historical Europe. *History of the Family* (2015), published online: http://dx.doi.org/10.1080/1081602X.2014.1001769, February 08, 2016; Vrushali Patil. From Patriarchy to Intersectionality. A Transnational Feminist Assessment of How Far We've Really Come. *Signs* 38:4 (2013): 847–867; Joan W. Scott. On Language, Gender, and Working-Class History. *International Labor and Working-Class History* 31 (1987): 1–13; and Joan W. Scott. Experience. In: *Feminists Theorize the Political*, Judith Butler and Joan W.Scott (eds.), 22–40. New York: Routledge, 1992; Elizabeth A. Clark.

Crucial within that work is the argument, originally made by the historian Joan Scott, for the central importance of gender as a *category of analysis* rather than just as a 'topic' for research and discussion.[7] It demonstrates how gendered concepts are core constituents of social and political meaning, and underpin the languages and images of power which shape social and political action, and consequently that no serious analysis of these phenomena can ignore gender as a category. While this argument contributed to the 'cultural turn' in historical studies, it is equally notable as a significant development in gender theory, and as a key intervention in historical practice grounded in social history and feminist politics as well as cultural theory. This becomes clear when Scott's piece is read in conjunction with her essays on nineteenth century French and English history published alongside it in *Gender and the politics of history*. Its challenge to established historical practice has stimulated a range of analyses of the gendered formation of social, religious, and political movements and discourses in different places and periods, and of various slave, peasant, and working class communities.[8] These pieces demonstrate the mutually constitutive role of gender,

---

The Lady Vanishes. Dilemmas of a Feminist Historian after the 'linguistic turn'. *Church History* 67:1 (1998): 1–31; Laura L. Frader. Dissent over Discourse. Labor, History, Gender, and the Linguistic Turn. *History and Theory* 34:3 (1995): 213–230; John Tosh. The History of Masculinity. An Outdated Concept? In *What is masculinity? Historical Dynamics from Antiquity to the Contemporary World*, John H. Arnold and Sean Brady (eds.), 17–34. Basingstoke: Palgrave Macmillan, 2011; Geoff Eley. *A Crooked Line. From Cultural History to the History of Society*. Ann Arbor: University of Michigan Press, 2005.

7 Joan W. Scott. Gender. A Useful Category of Analysis. *American Historical Review* 91:5 (1986): 1053–1075.

8 early examples are Deborah G. White. *'Ain't I a woman?' Female Slaves in the Plantation South*. New York: Norton, 1985; Barbara Bush. *Slave Women in Caribbean Society, 1650–1838*. Bloomington: Indiana UP, 1990; Gay L. Gullickson. *Spinners and Weavers of Auffay. Rural Industry and the Sexual Division of Labor in a French Village, 1750–1850*. Cambridge: Cambridge UP, 1986; Elinor Accampo. *Industrialization, Family Life, and Class Relations in St Chamond*. Berkeley: University of California Press, 1989; Judith M. Bennett. Medieval Women, Modern Women. Across the Great Divide. In *Culture and History 1350–1600. Essays on English Communities, Identities, and Writing*, David Aers (ed.), 47–176. Detroit: Wayne State UP, 1992; Ava Baron. *Work Engendered. Toward a New History of American Labor*. Ithaca: Cornell UP, 1991; Ruth L. Smith and Deborah Valenze. Mutuality and Marginality. Liberal Moral Theory and Working-Class Women in Nineteenth Century England. *Signs* 13:2 (1988): 277–298; Judith E. Tucker. *Women in Nineteenth Century Egypt*. Cambridge: Cambridge UP, 1985; Julia Clancy-Smith. A Woman without Her Distaff. Gender, Work, and Handicraft Production in Colonial North Africa. In *A Social History of Women and the Family in the Middle East*, Margaret L. Meriwether and Judith E.Tucker (eds.), 25–53, Boulder: Westview Press, 1999; Donald Quataert. Ottoman Handicrafts and Industry 1800–1914. *Review Fernand Braudel Center* 11:2 (1988): 169–178; Iris Berger and Claire Robertson (eds.). *Women and*

class, family, occupation, and community in the formation of nineteenth century urban discourses and activisms. They have opened up avenues of investigation and interpretation which will be referred to throughout this paper.

This conceptual argument for the role of gender as a crucial category for historical and cultural analysis, stimulated subsequent critiques and developments, responding in particular to its very Eurocentric perspective. Thus discussions by Sinha on constructions of colonial masculinity and the 'imperial social formation', and by Higginbotham on the 'meta-language of race', developed and adapted Scott's approach in transnational, colonial, and multi-cultural histories. Using insights from historical materialist, feminist, world systems, and Foucauldian theory, Sinha explores the intersectionality of empire, gender, and race in the formation and enactment of power relations within social and cultural formations. While her focus is on British ruled India, her analyses and proposal of the category 'imperial social formation' offer key insights for historical practice in general. In the case of Higginbotham the influence of the particular approach to race emerging from African American studies shapes a vigorous and nuanced analysis of the importance of understanding the interrelationship of race, gender and slavery as markers of difference and power.[9] A systematic critique of ethnocentric approaches to gender issues emerged from scholars whose work was grounded in analyses of gender, colonialism and contemporary issues of race and ethnicity. Exemplary texts which undertake this critique from multidisciplinary standpoint include those of Mohanty, Kandiyoti, and Najmabadi, and recent work

Class in Africa. New York: Africana Publishing, 1986; Martin Klein and Claire Robertson (eds.). *Women and Slavery in Africa*. Madison: University of Wisconsin Press, 1983; Joan W. Scott and Louise Tilly. *Women, Work, and Family*. New York: Holt, Rinehart & Winston, 1978; Verena Martinez-Alier. *Marriage, Class and Colour in Nineteenth Century Cuba. A Study of Racial Attitudes and Sexual Values in a Slave Society*. London: Cambridge UP, 1974.

9 Mrinalini Sinha. Mapping the Imperial Social Formation. A Modest Proposal for Feminist History. *Signs* 25:4 (2000): 1077–1082; Mrinalini Sinha. Teaching Imperialism as a Social Formation. *Radical History Review* 67 (1997): 175–186; Mrinalini Sinha. Gender in the Critiques of Colonialism and Nationalism: Locating the Indian Woman. In *Feminism and History*, Joan W. Scott (ed.), 499–537. Oxford: Oxford UP, 1996 and Mrinalini Sinha. *Colonial Masculinity. The 'manly Englishman' and the 'effeminate Bengali' in the Late Nineteenth Century*. Manchester: Manchester UP, 1995; Evelyn B. Higginbotham. African-American Women's History and the Meta-Language of Race. *Signs* 17:2 (1992), 251–274; Deborah Gaitskell. From 'Women and Imperialism' to Gendering Colonialism. *South African Historical Journal* 39:1 (1998), 176–193; Jean M. Allman and Antoinette Burton. Destination Globalization? Women, Gender and Comparative Colonial Histories in the New Millenium. *Journal of Colonialism and Colonial History* 4:1 (2003), published online: muse.jhu.edu, February 9, 2016.

on intersectionality.[10] Most recently historians have re-evaluated Scott's original argument for the use of gender as a category of analysis more fully. Deploying a cross cultural approach to suggest that gender needs to be posed as a question rather than assumed as a category, this work deepens the debate on gender while drawing attention to its fluidity, and to the challenges that poses for theoretical and for historical practice.[11]

It is significant that this rich and nuanced scholarship has been underpinned by the research and reflection on histories and concepts of gender, race, and empire undertaken by theorists of gender and feminist theorists. Historians of slavery and apartheid, and those who have developed the so-called 'new imperial history', have made path-breaking contributions to this field. This involves conceptualising 'empire' not as a dyadic relationship between distinct domains of 'colony' and 'metropole', but as a unified terrain (material, political, discursive) on which they are *mutually constitutive*. Pioneered in work by Hall, Sinha, Burton and Wilson analysing British imperialism, work along those lines has also been developed by Rogers and Thompson for France, Stoler, Thomas, Grever and Waldijk for the Netherlands, and Guy, Silverblatt and Kuznesof on Latin America. This is paralleled by the studies of Bozzoli, Marks, and Bradford on South Africa and Lake, Curthoys and Damousi on Australia.[12] All these historians combine

---

10 Chandra Mohanty. Introduction. Cartographies of Struggle. Third World Women and the Politics of Feminism. In T*hird World Women and the Politics of Feminism*, Chandra Mohanty, Ann Russo, Lourdes Torres, 1–50. Bloomington: Indiana UP, 1991; Jacqui Alexander and Chandra Mohanty (eds.). *Feminist Genealogies, Colonial Legacies, Democratic Futures*. New York: Routledge, 1997; Deniz Kandiyoti. Islam and Patriarchy. A Comparative Perspective. In *Women in Middle Eastern History. Shifting Boundaries in Sex and Gender*, Nikki Keddie and Beth Barons (eds.), 23–44. New Haven: Yale UP, 1991; Avtar Brah and Ann Phoenix. 'Ain't I a woman?' Revisiting Intersectionality. *Journal of International Women's Studies* 5:3 (2004), 75–86; Frances Gouda. *What's to be Done with Gender and Post-colonial Studies?* Amsterdam: Amsterdam UP, 2001.

11 Jeanne Boydston. Gender as a question of historical analysis. *Gender & History* 20:3 (2008): 558–583; Afsaneh Najmabadi. From Supplementarity to parasitism? *Journal of Women's History* 16:2 (2004): 30–35, and Beyond the Americas. Gender and Sexuality as Useful Categories of Analysis. *Journal of Women's History* 18:1 (2008): 11–21.

12 Catherine Hall. *Civilising Subjects. Metropole and Colony in the English Imagination, 1830–1867.* Oxford: Polity Press 2002; Antoinette Burton. *Burdens of History. British Feminists, Indian Women, and Imperial Culture, 1865–1915*. Chapel Hill: North Carolina UP, 1994; and Antoinette Burton (ed.). *Empire in Question. Reading, Writing, and Teaching British Imperialism*. Durham: Duke UP, 2011; Helen Bradford. Women, Gender and Colonialism. Rethinking the History of the British Cape Colony and Its Frontier Zones, c. 1806–70. *Journal of African History* 37:3 (1996): 351–370 and Peasants, Historians and Gender. A South African Case Study Revisited. *History and Theory* 39:4 (2000): 86–110; Iris Berger. African Women's History. Themes and Perspectives. *Journal of Colonialism and Colonial History* 4:1 (2003). Maria Grever and Berteke Waaldijk. *Transforming the Public*

engagement with histories of empire with a commitment to gender based narrative and analysis. While dealing with many places and periods and using a range of approaches, this varied scholarship shares three key positions. It takes a view of imperial and ethnic power relations as shaping *both* dominant and subaltern participants in those relations; it interprets those power relations as entwined with those of gender, class and sexuality; it places those relations within narratives and analyses of the formation of societies which were practitioners of colonial dominance as well as those subordinated to it. Scholarly and analytical work by historians like Hall on the Caribbean, Sinha on India, Silverblatt on Peru, Marks on South Africa, or Stoler on Indonesia exemplifies the creative potential of such approaches.[13]

By contrast very few among the general analysts of empire have engaged with gender and feminist theory to any significant degree. At the conceptual level the focus has been on the value, or otherwise of Marxian, development centred, or culturalist approaches, without serious discussion of feminist or gender theory, while the accounts of empire in most narrative treatments marginalize or ignore gender as a shaping influence or factor. From the comparative histories of Doyle, Osterhammel, and Burbank to Cooper's conceptual discussion, and the recent work of Cain, Hopkins and Eley, recent work of this type has not engaged much with gender as a central theme or conceptual tool.[14] Apart from a few feminist

Sphere. The Dutch National Exhibition of Women's Labor 1898. Durham: Duke UP, 2004; Maria Grever. The Gender of Patrimonial Pride. In *Travelling Heritages. New Perspectives on Collecting, Preserving, and Sharing Women's History*, Saskia Wieringa (ed.), 285–302. Chicago: University of Chicago Press, 2008.

13 Hall, *Civilising Subjects*; Sinha, *Colonial masculinity*; Sinha, Mrinalini. Introduction to Katherine Mayo: *Mother India. Selections from the Controversial 1927 Text* 1–62. Ann Arbor: University of Michigan Press, 2000; Shula Marks (ed.). *Not either an Experimental Doll. The Separate Worlds of three South African Women*. Bloomington: Indiana UP, 1987, and Changing History, Changing Histories. Separations and Connections in the Lives of South African Women. *Journal of African Cultural Studies* 13:1 (2000): 94–106; Julia Clancy-Smith. The Intimate, the Familial, and the Local in Transnational Histories of Gender. *Journal of Women's History* 18:2 (2006): 174–183; Mineke Bosch. Colonial Dimensions of Dutch Women's Suffrage. Aletta Jacobs's Travel Letters from Africa and Asia. *Journal of Women's History* 11:2 (1999): 8–34; Alice Conklin. Boundaries Unbound. Teaching French History as Colonial History and Colonial History as French History. *French Historical Studies* 23:2 (2000): 215–238; Louis Montrose. The Work of Gender in the Discourse of Discovery. *Representations* 33 (1991): 1–41.

14 Jürgen Osterhammel. *Colonialism. A Theoretical Overview*. Princeton: Wiener, 1997 and Jürgen Osterhammel and Niels Peterson. *Globalisation. A Short History*. Princeton: Princeton UP, 2005; Jane Burbank and Frederick Cooper. *Empires in World History. Power and the Politics of Difference*. Princeton: Princeton UP, 2010; Michael Doyle. *Empires*. Ithaca: Cornell UP, 1986; Frederick Cooper. *Colonialism in Question. Theory, Knowledge, History*. University of California Press, 2005;

scholars, the 'subaltern studies' school's critique of colonial and nationalist views of empire made similarly limited use of gendered approaches. The multi-volume *Oxford History of the British Empire* offers a volume on gender in its 'companion series', published well after the completion of the main series, sending a clear message as to how gender was seen either as concept or as theme by the series editors.[15] One issue worth pursuing further would be a systematic analysis of the basis for the general resistance within much social and cultural theory, as well as within historical practice, to gender and feminist thought.

In order to facilitate that discussion I shall now focus on the centrality of gender theory as a source of conceptual tools and insights which ought to inform investigations and analyses of empire. In its exploration of the structures and relations of difference and power, that body of theory has given substance to ideas which can be used to investigate and interpret past societies and which intersect with other conceptual markers of those structures and relations. It has delineated patriarchy as a set of structures shaping power and production, elaborated various versions of masculinity and femininity as discursive and normative concepts, and developed analyses of sexuality and domesticity as constitutive features of material, cultural and political life. This conceptual and theoretical work can be used to enhance the study of past 'empires', whether the "pre-modern" systems of the Romans, Chinese, and Ottomans, or the various European empires constructed from the sixteenth century onwards. The argument here is that the conceptual tools developed within gender theory *have transformative potential for thinking through* the structures, processes and relationships which have shaped empires, rather than just indicating a set of topics for study. The organisation of material productive and commercial activity and power relations within the imperial social formation, like the organisation of kin, family, reproductive and sexual activity in that setting, can only be fully understood and depicted through use of such tools.

The notion of patriarchy has been developed to think about the various power relations which have subordinated women to men. Initially proposed as a rather broad notion of male power, it has been refined to conceptualise the complex interactions of age, status and gender which have placed male heads of house-

Peter Cain and Anthony Hopkins. *British Imperialism 1688–1990*, 2 vols. London: Longman, 1993; Peter Cain and Mark Harrison (eds.). *Imperialism. Critical Concepts in Historical Studies*. London: Routledge, 2001; Geoff Eley. Imperial Imaginary, Colonial Effect. Writing the Colony and the Metropole Together. In *Race, Nation, and Empire. Making Histories from 1750 to the Present*, Catherine Hall and Keith McClelland (eds.), 217–237. Manchester: Manchester UP, 2010.

15 Philippa Levine (ed.). *Gender and Empire*. Oxford: Oxford UP, 2004.

hold or groups of privileged, propertied and elite men in positions of control and authority in past societies.[16] From Roman citizenship to racialised power systems in the post Reconstruction southern US, senior male control of women, younger men, employees, or slaves in households, communities, workplaces, governance, and cultural life can be understood as a constituent feature of their institutions and practices. On the one hand Weberian ideas of patrimonial forms of power and governance have been reconfigured around the understanding of gender, kinship and age relations as agents of control and/or co-operation within those settings, from European family firms to Mughal dynastic and clan power.[17] Such work has developed an understanding that the role of gender and familial relations has been central to the formation and maintenance of power relations and power structures, rather than just a superstructural addition. On the other hand studies of material life have been refigured around clearer analyses of gender, familial and generational divisions and relations within productive and reproductive labour as well as exchange and consumption in different places and periods.[18]

---

16 Sylvia Walby. *Theorising Patriarchy*. Chichester: Wiley-Blackwell, 1990; Kandiyoti, Bargaining with Patriarchy; Kandiyoti, Islam and Patriarchy; Heidi Gottfried. Beyond Patriarchy? Theorising Gender and Class. *Sociology* 32:3 (1998): 451–468; Rosemary Hennessy (ed.). *Materialist Feminism and the Politics of Discourse*. New York: Routledge, 2012; Judith Squires. *Gender in Political Theory*. Hoboken: Wiley, 2013; Anna Pollert. Gender and Class Revisited; Or, the Poverty of 'Patriarchy'. *Sociology* 30:4 (1996): 639–659; Phillips and Taylor, Sex and Skill; Sheila Rowbotham. The Trouble with Patriarchy. *People's History and Socialist Theory*, Raphael Samuels (ed.), 364–369. London: Routledge, 1981; Karen Sacks. Toward a Unified Theory of Class, Race, and Gender. *American Ethnologist* 16:3 (1989): 534–550; Julia Adams. The Rule of the Father. Patriarchy and Patrimonialism in Early Modern Europe. In *Max Weber's Economy and Society. A Critical Companion*, Charles Camic et al. (eds.), 237–266. Stanford: Stanford UP, 2005; Patil, From Patriarchy to Intersectionality; Mary Murray. *The Law of the Father? Patriarchy in the Transition from Feudalism to Capitalism*. Florence: Routledge, 2005.

17 Bonnie G. Smith. *Ladies of the Leisure Class. The Courgeoises of Northern France in the Nineteenth Century*. Princeton: Princeton UP, 1981; Ruby Lal. *Domesticity and Power in the Early Mughal World*. Cambridge: Cambridge UP, 2005.

18 Martha Howell. *Women, Production, and Patriarchy in Late Medieval Cities*. Chicago: University of Chicago Press, 1986; Jan De Vries. Between Purchasing Power and the World of Goods. Understanding the Household Economy in Early Modern Europe. In *Consumption and the World of Goods*, Brewer, John and Roy Porter (eds.), 85–132. London: Routledge, 1994; Jaqueline Jones. *Labor of Love, Labor of Sorrow. Black Women, Work, and Family from Slavery to the Present*. New York: Basic Books, 1995; Evelyn Glenn. From Servitude to Service Work. Historical Continuities in the Racial Division of Paid Reproductive Labor. *Signs* 18:1 (1992): 1–43; Leonardo Helfgott. *The Ties that Bind. A Social History of the Iranian Carpet*. Washington: Smithsonian Press, 1994; Lourdes Beneria and Ggita Sen. Accumulation, Reproduction, and 'Women's Role in Economic Development': Boserup Revisited. *Signs* 7:2 (1981): 279–298; Joan Acker. Class, Gender, and the Relations of Distribution. *Signs* 13:3 (1988): 473–497.

They have shown that gender interacting with other differences has been constitutive of material processes from nineteenth century European manufacturing to medieval household production, or Iranian craft exports and pastoral economies. Better understanding of the skill, effort, time and labour involved in family reproduction, and of the analytical significance of consumption in socio-economic history have transformed simplistic productionist approaches to that history.

The development of patriarchy as a category of analysis has distinctive implications for histories of empire. Awareness of the gender and familial dynamics of colonisation and plantation slavery, like the role of family in the commercial and governmental networks which shaped both formal and informal empire, entail at least three levels of rethinking of 'empire' as an object of enquiry and/or analysis. At the empirical level it requires appreciation of the specific functioning of those dynamics in particular settings. At the conceptual level it stimulates analyses of the gendered character of the imperial social formation from material, cultural, and political perspectives; at the historiographical level it encourages critiques of the gender blind and gender biased approaches of existing work on empire. This problematises conventional narratives of colonial power, of migration, and of economic activity in imperial and colonial contexts. Work like that of Adams' on the 'familial state', like that on global migration and indigenous contacts with colonists and imperial traders, develops Stoler's insights into the interactive relations and coloniser and colonised, and links gender relations to the material structures of empire. They are amplified by studies of patriarchal and familial aspects of global slave trade and colonial settlement and the sheer quantity and range of the material cited below is an indicator of the rich potential for analyses of empire grounded in an awareness of the complex, contested but illuminating category of patriarchy.[19] The 'imperial' dimensions of the world after decolonisa-

19 Ann Stoler. Rethinking Colonial Categories. European Communities and the Boundaries of Rule. *Comparative Studies in Society and History* 31:1 (1989): 134–161; Julia Adams. *The Familial State, Ruling Families and Merchant Capitalism in Early Modern Europe.* Ithaca: Cornell UP, 2005; Bosma, Ulbe and Remco Raben. *Being 'Dutch' in the Indies. A History of Creolisation and Empire, 1500–1920.* Singapore: NUS Press, 2008; Marlou Schrover (ed.). *Illegal Migration and Gender in a Global and Historical Perspective.* Amsterdam: Amsterdam UP, 2008; Anne-Meike Fechter. Gender, Empire, Global Capitalism. Colonial and Corporate Expatriate Wives. *Journal of Ethnic and Migration Studies* 36:8 (2010): 1279–1297; Pamela Sharpe (ed.). *Women, Gender and Labour Migration. Historical and Cultural Perspectives.* London: Routledge, 2001, chs. 1, 2, 5, 8, 10, 11; Mary Procida. *Married to the Empire. Gender, Politics and Imperialism in India, 1883–1947.* Manchester: Manchester UP, 2002; Durba Ghosh. *Sex and the Family in Colonial India. The Making of Empire.* Cambridge: Cambridge UP, 2006; Ondina González and Bianca Premo (eds.). *Raising an Empire. Children in Early Modern Iberia and Colonial Latin America.* Alberquerque: University of New Mexico Press, 2007; Michael Zuckerman. Endangered Deference, Imperiled Patriarchy.

tion, and theories of the 'post-colonial', require attention to the gendered dynamics of migration (male construction workers in the Gulf, female care workers in the Middle East and North America), or the trafficking of sex workers. One scholarly indicator of this need is the foundation of the journal *Gender and migration*. Links of work and familial responsibility structure the post-colonial world around the global power relations of exploitation and caring, entailing gendered conceptualisations of the neo-liberal world order.[20] Whether seen through the lens of global divisions and exchanges of labour or through patterns and experiences of marriage, family and household, the gender dynamic is constitutive of the processes and relationships which make and maintain that world order.

Central to the integration of gender as a category of analysis into scholarly practice is the deployment of notions of masculinity and femininity. While needing approaches which are attentive to cultural and historical specificities, and interrogate the relative importance of these notions in particular contexts, there is a weight of scholarship demonstrating their conceptual and analytical value. Productive and political relations have been shaped by ideas of mascu-

---

Tales from the Marchlands. *Early American Studies* 3:2 (2005): 232–252; Sarah Franklin. *Women and Slavery in Nineteenth Century Colonial Cuba*. Rochester: University of Rochester Press, 2012; Jennifer Morgan. *Laboring Women. Reproduction and Gender in New World Slavery*. Philadelphia: University of Pennsylvania Press, 2004; Pamela Scully. *Liberating the Family? Gender and British Slave Emancipation in the Rural Western Cape, South Africa, 1823–1853*. Portsmouth: Heinemann, 1997; Orlando Patterson. Slaver, Gender, and Work in the Pre-Modern World and Early Greece. A Cross-Cultural Analysis. In *Slave Systems. Ancient and Modern*, Enrico Dal Lago und Katsari Constantina (eds.), 32–69. Cambrige: Cambridge UP, 2008; Kevin Roberts. Yoruba Family, Gender, and Kinship Roles in New World Slavery. In *The Yoruba Diaspora in the Atlantic World*, Toyin Falola und Matt Childs (eds.), 248–259. Bloomington: Indiana UP, 2004; Julia Clancy-Smith. Women, Gender and Migration along a Mediterranean Frontier. Pre-Colonial Tunisia, c.1815–1870. *Gender & History* 17:1 (2005): 62–92; Raffaella Sarti. Fighting for Masculinity. Male Domestic Workers, Gender, and Migration in Italy from the Late Nineteenth Century to the Present. *Men and Masculinities* 13:1 (2010): 16–43; Stephen Blake. Returning the Household to the Patrimonial-Bureaucratic Empire. Gender, Succession, and Ritual in the Mughal, Safavid and Ottoman Empires. In *Tributary Empires in Global History*, Peter Fibiger Bang und C.A. Bayly (eds.), 214–226. New York, Palgrave Macmillan, 2011; Barbara Potthast-Jutkeit. The History of Family and Colonialism. *History of the Family* 2:2 (1997): 115–121.

20 Rajni Palriwala and Patricia Uberoi (eds.). *Marriage, Migration and Gender*. Los Angeles: Sage Publications, 2008; Suzanne Sinke. Gender and Migration. Historical Perspectives. *International Migration Review* 40:1 (2006): 82–103; Pierrette Hondagneu-Sotelo. Gender and Migration Scholarship. An Overview from a 21st Century Perspective. *Migraciones Internacionales* 6:1 (2011): 219–234; Janet Momsen [ed.] *Gender, Migration and Domestic Service*. London: Routledge, 2013; Sarti, Fighting for Masculinity; Catherine Scott. *Gender and Development. Rethinking Modernization and Dependency Theory*. Boulder: Lynne Rienner, 1995.

linity and femininity, whether the organisation of production in many past communities, in the principles and practices of politics, religion, and governance, or modern systems of industrial production or political representation.[21] Complex interfaces of gendered material processes with gendered discursive or cultural norms and conflicts has combined with other markers and practices of difference to underpin the formation, maintenance and transformation of households, workplaces, social, political and religious movements. Close research and conceptual innovation on topics as diverse as sixteenth century Ottoman artisans, nineteenth century British industry, medieval Russian religious organisation, or Masai pastoralism, have embedded masculinity and femininity as core elements.[22]

---

21 See Lal *Domesticity and Power*; Louise Tilly. Gender, Women's History, and Social History. *Social Science History* 13:4 (1989): 439–462; Anna Clark. *The Struggle for the Breeches. Gender and the Making of the British Working Class.* Berkeley: University of California Press, 1995; Stefan Dudink (eds.). *Representing Masculinity. Male Citizenship in Modern Western Culture.* Basingstoke: Palgrave Macmillan, 2012; Catherine Hall, Keith McClelland and Jane Rendall. *Defining the Victorian Nation. Class, Race and the British Reform Act of 1867.* Cambridge: Cambridge UP, 2000; Margaret Walsh (ed.). *Working out Gender. Perspectives from Labour History.* Aldershot: Ashgate, 1999; Deborah Valenze. Gender in the Formation of European Power, 1750–1914. In *A Companion to Gender History*, Teresa Meade and Merry Wiesner-Hanks (eds.), 459–476. Malden: Blackwell, 2004; Lynn Abrams und Elizabeth Harvey. *Gender Relations in German History. Power, Agency, and Experience from the Sixteenth to the Twentieth Century.* Durham: Duke UP, 1997; Petra Rostock and Sabine Berghahn. The Ambivalent Role of Gender in Redefining the German Nation. *Ethnicities* 8:3 (2008): 345-364; Gail Bederman. *Manliness and Civilization. A Cultural History of Gender and Race in the United States, 1880–1917.* Chicago: University of Chicago Press, 2008; Margot Badran. *Feminists, Islam, and Nation. Gender and the Making of Modern Egypt.* Princeton: Princeton UP, 1996; Lisa Pollard. *Nurturing the Nation. The Family Politics of Modernizing, Colonizing, and Liberating Egypt, 1805*–1923. Berkeley: University of California Press, 2005; Lahoucine Ouzgane and Robert Morrell (eds.). *African Masculinities. Men in Africa from the Late Nineteenth Century to the Present.* New York: Palgrave Macmillan, 2005.

22 Suraiya Faroqhi. *Artisans of Empire. Crafts and Craftsmen under the Ottomans.* London: IB Tauris, 2009; Basak Tug. Gender and Ottoman Social History. *International Journal of Middle East Studies* 46:2 (2014): 379–381; M. Erdem Kabadayi and Kate Creasey. Working in the Ottoman Empire and in Turkey: Ottoman and Turkish Labor History within a Global Perspective. *International Labor and Working-Class History* 82 (2012): 187–200; Nelly Hanna. *Artisan Entrepreneurs in Cairo and Early-Modern Capitalism 1600–1800.* Syracuse: Syracuse UP, 2011; Clark, *Struggle for the Breeches*; Sonya Rose. 'Gender at Work'. Sex, Class and Industrial Capitalism. *History Workshop Journal* 21:1 (1986): 113–132; Judy Lown. *Women and Industrialization. Gender at Work in Nineteenth-Century England.* Cambridge: Polity Press, 1990; Valenze, *The First Industrial Woman*; Honeyman and Goodman, Women's Work; Katrina Honeyman. Following Suit. Men, Masculinity and Gendered Practices in the Clothing Trade in Leeds, England, 1890–1940. *Gender & History* 14:3 (2002): 426–446; Isolde Thyrêt. *Women and the Orthodox Faith in Muscovite Russia. Spiritual Experience and Practice.* University Park: Pennsylvania State University, 2003; Dorothy Hodgson. *Once Intrepid Warriors. Gender, Ethnicity, and the Cultural Politics of Masai Development.* Bloom-

Such developments have been as much a feature of work on imperial governance and ideologies, on colonial settlement, and on cross cultural contacts, as in studies of culture, work and politics in other settings. From gender divisions and interdependence in the work of both dominant and subaltern groups to the construction of the assumptions, images and values which shaped imperial power and resistance, discourses and norms of masculine or feminine conduct, roles or qualities have been shown to play constitutive roles in imperial lives.[23] It is significant that the productionist veer of much imperial theorising obscures, or entirely ignores, the constitutive role of (gendered) consumption in the making and maintenance of the material fabric of empire. Similarly treatments of men's involvement in empire which fail to recognise that men are gendered as well as classed and raced subjects, marginalize and disregard the gendering of dominance, resistance, and ethnicity in imperial situations. Beyond that, work on the cultures and ideologies of empire and race which is not attentive to the tropes of masculinity, femininity and sexuality fails to analyse or interpret them adequately.

Thus the making of new relations of global/imperial consumption and production around tobacco, tea, sugar, cottons and ceramics in the eighteenth century was central to the formation of gendered imperial power networks linking Europe to India, China and the Americas.[24] This was characterised by the construction or

---

ington: Indiana UP, 2001 and Pastoralism, Patriarchy & History among Maasai in Tanganyika, 1890–1940. In *Rethinking Pastoralism in Africa. Gender, Culture & the Myth of the Patriarchal Pastoralist*, Dorothy Hodgson (ed.), 97–120. Oxford: James Currey, 2000.

23 Kathleen Wilson (ed.). *A New Imperial History. Culture, Identity and Modernity in Britain and the Empire, 1660–1840*. Cambridge: Cambridge UP, 2004; Catherine Hall and Keith McClelland. *Race, Nation and Empire. Making Histories, 1750 to the Present*. Manchester: Manchester UP, 2010; Wilma Dunaway. The Double Register of History. Situating the Forgotten Woman and her Household in Capitalist Commodity Chains. *Journal of World-Systems Research* 7:1 (2001): 1–29; Verene Shepherd et al. (eds.). *Engendering History. Caribbean Women in Historical Perspective*. London: James Currey, 1995; John Tosh. *Manliness and Masculinities in Nineteenth-Century Britain. Essays on Gender, Family, and Empire*. Harlow: Pearson, 2005.

24 Wilson, *A New Imperial History*; Joanna de Groot. Metropolitan Desires and Colonial Connections. Reflections on Consumption and Empire. In *At Home with the Empire. Metropolitan Culture and the Imperial World*, Catherine Hall and Sonya Rose (eds.), 166–190. Cambridge: Cambridge UP, 2006; Maxine Berg. In Pursuit of Luxury. Global History and British Consumer Goods in the Eighteenth Century. *Past and Present* 182 (2004): 85–142; Beverly Lemire and Giorgio Riello. East & West. Textiles and Fashion in Early Modern Europe. *Journal of Social History* 41:4 (2008): 887–916; Ghosh, *Sex and the Family in Colonial India*; Trevor Burnard. *Mastery, Tyranny, and Desire. Thomas Thistlewood and his Slaves in the Anglo-Jamaican World*. Chapel Hill: University of North Carolina Press, 2004; Brooke Newman. Gender, Sexuality and the Formation of Racial Identities in the Eighteenth Century Anglo-Caribbean World. *Gender & History* 22:3 (2010): 585–602;

reconfiguring of gendered plantation economies, of cross cultural reproductive intimacies, of new masculine/homosocial relations and organisations in colonial trade, finance, and government, and female networks of production, consumption and colonial involvement. It was reinforced by the subsequent growth of colonies of settlement, of global networks of investment, production and exchange, and of indigenous adaptations and resistances, in all of which male and female roles were developed and contested in conjunction with class structures and ethno-cultural encounters.[25] Similarly the processes and structures of decolonisation, of postcolonial state and society formation, and of neo-liberal 'empire', have involved the refiguring and management of masculinities and femininities, whether in revivalist religions, the trafficking of sex workers, or patterns of migrant domestic labour and sweatshop production. The construction of imperial exploration, surveillance, and imaginings as well as of work, war, and governance around powerful and contested notions of 'manly' or 'womanly' characteristics, aptitudes and roles is as striking as the making of anti-colonial thought and action around alternative and equally contested versions of such notions.[26] If

William Clarence-Smith and Steven Topik (eds.). *The Global Coffee Economy in Africa, Asia, and Latin America, 1500–1989*. Cambridge: Cambridge UP, 2003.

25 Maria Lugones. Heterosexualism and the Colonial/ Modern Gender System. *Hypatia* 22:1 (2007): 186–209; Durba Ghosh. Gender and Colonialism. Expansion or Marginalization? *Historical Journal* 47:3 (2004): 737–755; Adele Perry. *On the Edge of Empire. Gender, Race, and the Making of British Columbia, 1849–1871*. Toronto: University of Toronto Press, 2001; Scott Morgensen. Theorising Gender, Sexuality and Settler Colonialism. An Introduction. *Settler Colonial Studies* 2:2 (2012): 2–22; Darcy Leigh. Colonialism, Gender and the Family in North America. For a Gendered Analysis of Indigenous Struggles. *Studies in Ethnicity and Nationalism* 9:1 (2009): 70–88; Sheila McManus. Review: *Taking Land, Breaking Land. Women Colonizing the American West and Kenya, 1840–1940*, Glenda Riley. *Western Historical Quarterly 36:2*, 2005: 217; Susan Socolow. *The Women of Colonial Latin America*. New York: Cambridge UP, 2015; Antonia Castañeda. Engendering the History of Alta California, 1769–1848. Gender, Sexuality, and the Family. *California* History 76:2/3 (1997): 230–259; Michael Hardin. Altering Masculinities. The Spanish Conquest and the Evolution of the Latin American Machismo. *International Journal of Sexuality and Gender Studies* 7:1 (2002): 1–22; Robert McCaa. Marriageways in Mexico and Spain, 1500–1900. *Continuity and Change* 9:1 (1994): 11–43; Donna Gabaccia. Spatializing Gender and Migration. The Periodization of Atlantic Studies, 1500 to the Present. *Atlantic Studies* 11:1 (2014): 7–27; Jane Samson (ed.). *British Imperial Strategies in the Pacific, 1750–1900*. Aldershot: Ashgate, 2003; Ghosh, *Sex and the Family in Colonial India* and Who Counts as 'Native'? Gender, Race, and Subjectivity in Colonial India. *Journal of Colonialism and Colonial History* 6:3 (2005); Mrinalini Sinha. Reading *Mother India*. Empire, Nation, and the Female Voice. *Journal of Women's History* 6:2 (1994): 6–44.

26 Graham Dawson. *Soldier Heroes. British Adventure, Empire, and the Imagining of Masculinities*. London: Routledge, 1994; Kumari Jayawardena. *The White Woman's Other Burden. Western Women and South Asia during British Colonial* Rule. New York: Routledge, 1995; Richard Phillips. *Sex, Poli-*

historians and theorists of empire are to work effectively with concepts of colonial settlement and colonial rule, of colonial subalterns and colonial dominance, or of colonial resistance and indigenous response, they need to inflect them with due attention to ideas of masculinity and femininity.

As shown by the scale of what is merely an indicative set of references offered here, there is a dense and varied body of gender and feminist scholarship which makes both analytical and empirical challenges to gender blind scholarship on empire. In addition to reframing understandings of labour, politics, and family this scholarship has pioneered new analytical links between intimate aspects of past experience and those which are more publicly manifested. In conjunction with work in the history of medicine and cultural history, historical thought on gender has influenced the emergence of new scholarship on history of the body, which in turn has implications for the conceptualisation of empire, whose structures and practices have been founded on the movement and management of human bodies. Such embodied structures and practices range from global networks of commerce, migration, and trade in enslaved people to the technologies of punishment and incarceration practiced on the bodies of imperial subjects, and cross cultural bodily contacts and intimacies. These considerations are central to the dynamics of eighteenth century plantation slavery, the construction of nineteenth century racial thought around concepts of sexed gendered repro-

tics and Empire. A Postcolonial Geography. Manchester: Manchester UP, 2006 and *Mapping Men and Empire. Geographies of Adventure.* Hoboken: Taylor and Francis, 2013; Ben Shephard. Showbiz Imperialism. The Case of Peter Lobengula. In *Imperialism and Popular Culture*, John MacKenzie (ed.), 94–112. Manchester: Manchester UP, 1986; Esme Cleall. *Missionary Discourses of Difference. Negotiating Otherness in the British Empire, 1840–1900*. Basingstoke: Palgrave Macmillan, 2012; Kelly Boyd. *Manliness and the Boys' Story Paper in Britain. A Cultural History, 1855–1940.* Basingstoke: Palgrave Macmillan, 2003, 123–152; Sinha, *Colonial Masculinity*; Aviston Downes. Boys of the Empire. Elite Education and the Construction of Hegemonic Masculinity in Barbados, 1875–1920. In *Interrogating Caribbean Masculinities. Theoretical and Empirical Analyses*, Rhoda Reddock (ed.), 105–136. Mona: University of the West Indies Press, 2004; Julia Clancy-Smith and Frances Gouda (eds.). *Domesticating the Empire: Race, Gender, and Family Life in French and Dutch Colonialism*. Charlottesville: UP of Virginia, 1998; Vron Ware. *Beyond the Pale. White Women, Racism and History.* London: Verso, 1992; Kathleen Brown. *Good Wives, Nasty Wenches, and Anxious Patriarchs. Gender, Race, and Power in Colonial Virginia*. Chapel Hill: University of North Carolina Press, 1996; Ann Twinam. *Public Lives, Private Secrets. Gender, Honor, Sexuality and Illegitimacy in Colonial Spanish America*. Stanford: Stanford UP, 1999; Floya Anthias and Nira Yuval-Davis. *Racialized Boundaries. Race, Nation, Gender, Colour and Class and the Anti-Racist Struggle*. London: Routledge, 1995; Stephen Legg. Gendered Politics and Nationalised Homes. Women and the Anti-Colonial Struggle in Delhi 1930–47. *Gender, Place and Culture: A Journal of Feminist* Geography 10:1 (2003): 7–27; Tanika Sarkar. *Hindu Wife, Hindu Nation. Community, Religion and Cultural Nationalism*. Bloomington: Indiana UP, 2010.

ductive bodies, or the role of gendered body cultures in the formation of imperial rulers and subjects. We may note the significant role of the colonial state and medical 'expertise' in the management of diseases in the shaping of imperial relations and identities, whether through research into malaria, cholera quarantine, or missionary medical activity.[27] The inspection and control of colonial bodies has served to embed bio-power within gendered colonial power relations, while sustaining powerful imperial imaginings around health, climate and gendered bodies. Intrusions by colonial authorities into subaltern households, like the gendered tropes of health and disease used by both opponents and proponents of imperial power, were key elements in that process. Power relations were shaped by transfers and exclusions which established gendered and ethnicised hierarchies of gendered medical knowledge and expertise.[28]

One paradigm case illustrating the embedding of gender within the processes and relationships underpinning empire is that of colonial involvement with sexually transmitted conditions. Nineteenth century preoccupations with the trans-

27 David Arnold (ed.). *Imperial Medicine and Indigenous Societies.* Manchester: Manchester UP, 1988; Mark Harrison. *Medicine in an Age of Commerce and Empire. Britain and Its Tropical Colonies, 1660–1830.* Oxford: Oxford UP, 2010; Randi Davenport. Thomas Malthus and Maternal Bodies Politic. Gender, Race, and Empire. *Women's History Review* 4:4 (1995): 415–439; Kalpana Ram and Margaret Jolly. *Maternities and Modernities. Colonial and Postcolonial Experiences in Asia and the Pacific.* Cambridge: Cambridge UP, 1998; Alison Bashford. Medicine, Gender, and Empire. In *Gender and Empire*, Philippa Levine (ed.), 112–133. Oxford: Oxford UP, 2009; Philippa Levine. What Difference Did Empire Make? Sex, Gender and Sanitary Reform in the British Empire. In *National Healths. Gender, Sexuality and Health in a Cross-cultural Context*, Nana Wilson-Tagoe and Michael Worton (eds.), 71–82. London: UCL Press, 2004; Rod Edmond. *Leprosy and Empire. A Medical and Cultural History*, Cambridge: Cambridge UP, 2006; Judith Farquhar and Marta Hanson (eds.). *Positions: East Asia Cultures Critique* (Empire of Hygiene) 6:3 (1998); Alison Bashford 'Is White Australia Possible?' Race, Colonialism and Tropical Medicine. *Ethnic and Racial Studies* 23:2 (2000): 248–271; Biswamoy Pati and Mark Harrison (eds.). *The Social History of Health and Medicine in Colonial India.* London: Routledge, 2009.

28 Hibba Abugideiri. Egyptian Women and the Science Question. Gender in the Making of Colonized Medicine, 1893–1919. *Arab Studies Journal* 4:2 (1996): 46–78; Piya Chatterjee. An Empire of Drink. Gender, Labor and the Historical Economies of Alcohol. *Journal of Historical Sociology* 16:2 (2003): 183–208; Nigel Chancellor. A Picture of Health. The Dilemma of Gender and Status in the Iconography of Empire, India c. 1805. *Modern Asian Studies* 35:4 (2001): 769–782; Morag Bell. 'The Pestilence that Walketh in Darkness'. Imperial Health, Gender and Images of South Africa c. 1880–1910. *Transactions of the Institute of British Geographers* 18:3 (1993): 327–341; Sheryl Neste. (Ad)ministering Angels. Colonial Nursing and the Extension of Empire in Africa. *Journal of Medical Humanities* 19:4 (1998): 257–277; Narin Hassan. *Diagnosing Empire. Women, Medical Knowledge and Colonial Mobility.* Farnham: Ashgate, 2011; Waltraud Ernst. European Madness and Gender in Nineteenth-century British India. *Social History of Medicine* 9:3 (1996): 357–382.

mission of syphilis by sex workers in colonial settings, like global responses to HIV/AIDS issues in recent times, have produced and reinforced the gendered, raced and sexualised features of empire. This involved the reconfiguring of both the practices and the ideologies of interventionist colonial states, the foregrounding of racialised sex and gender discourses within the practices of medicine and reform, and the strengthening of the surveillance and regulation of bodies in imperial settings. The interactions between the growth of debate and intervention around sexually transmitted 'diseases' and their transmitters in European and American settings and in the wider world exemplify the mutually constitutive relationships between raced and gendered sexualities both in so-called colonies and so-called metropoles. The formation of new legal codes and police practices, medical interventions, and international movements around the "problem" of something called the "white slave trade" in Europe, North America and the wider world encapsulate just such relationships.[29]

From the physical invasion of individual bodies to global regulation, and from the practicalities of urban planning to medical thought and eroticised racial romance, the imperial story of sexually transmitted 'diseases' evokes a larger context in which 'empire' was shaped by and shaped sexualities and genders. Thus proper understanding of the imperial social formation benefits from an interactive and transnational analysis of both the sexualised dynamics of empire and imperial elements in histories of sexuality. There is now a considerable body of social and cultural theory which offers a basis for such an analysis.[30] While this

---

29 Philippa Levine. *Prostitution, Race, and Politics. Policing Venereal Disease in the British Empire*. New York: Routledge, 2003; Phillips, *Sex, Politics and Empire*; Benedict Carton. 'We Are Made Quiet by This Annihilation'. Historicizing Concepts of Bodily Pollution and Dangerous Sexuality in South Africa. *International Journal of African Historical Studies* 39:1 (2006): 85–106; Philip Howell. *Geographies of Regulation. Policing Prostitution in Nineteenth-Century Britain and the Empire*. Cambridge: Cambridge UP, 2009; Marc Epprecht. Sexuality, Africa, History. *American Historical Review* 114:5 (2009): 1258–1272; Daniel Walther. Sex, Public Health and Colonial Control. The Campaign against Venereal Diseases in Germany's Overseas Possessions, 1884–1914. *Social History of Medicine* 26:2 (2013): 182–203; Trevor Burnard and Richard Follett. Caribbean Slavery, British Anti-Slavery, and the Cultural Politics of Venereal Disease. *Historical Journal* 55:2 (2012): 427–451; Daniel Gorman. Empire, Internationalism, and the Campaign against the Traffic in Women and Children in the 1920s. *Twentieth Century British History* 19:2 (2008): 186–216.

30 Ann Stoler. *Race and the Education of Desire. Foucault's History of Sexuality and the Colonial Order of Things*. Durham: Duke UP, 1995; Joanna de Groot. 'Sex' and 'Race'. The Construction of Language and Image in the Nineteenth-Century. In *Sexuality and Subordination. Interdisciplinary Studies of Gender in the Nineteenth Century*, Susan Mendus und Jane Rendall (eds.), 89–128. London: Routledge, 1989; Tamara Loos. Transnational Histories of Sexualities in Asia. *American Historical Review* 114:5 (2009): 1309–1324; Richard Phillip. Histories of Sexuality and Imperialism.

work argues that imperial processes and power relations are partly constituted through processes and relations of sex and reproduction, it also shows that ideas and norms of sex and reproduction have themselves been partly constituted by imperial and colonial involvements. Weinbaum's subtle analysis of the intersections of empire, race, gender and sexuality in Darwinian, anti-racist, Marxian and feminist thought is paralleled by Brody and Burton's work on nineteenth century Britain and Frankenberg's discussion of 'whiteness', and by explorations of sexualised aspects of racial and nationalistic thought.[31] Scientific ideas and fantasies about 'mixed race' reproduction, or about the sexual conduct of particular ethnic or cultural groups provided cultural and intellectual scaffolding not only for western theories of race and nation, but also for views of sexual behaviour and desires within western cultures. Whether considering the flavouring of evolutionary and historical materialist thought with views of race or savagery drawn from western imperial involvements, or the making of western tropes of romance and pleasure through images of the non-western 'exotic', we are speaking of entanglements of sexuality and empire.

By combining conceptual insights with substantive research historians have shown how such elements have been articulated through varied forms of intimate connection between dominant and subaltern groups ranging from practical household arrangements to cultural and ideological imaginaries. Just as some of the everyday workings of power relations in colonial and metropolitan settings

What's the Use? *History Workshop Journal* 63:1 (2007): 136–153; Joanne Nagel. Ethnicity and Sexuality. *Annual Review of Sociology* 26 (2000): 107–133; Greg Thomas. Erotics of Aryanism/Histories of Empire. How 'White Supremacy' and 'Hellenomania' Construct 'Discourses of Sexuality'. *New Centennial Review* 3:3 (2003): 235–255; Donna Haraway. Monkeys, Aliens, and Women. Love, Science, and Politics at the Intersection of Feminist Theory and Colonial Discourse. *Women's Studies International Forum* 12:3 (1989): 295–312; Inderpal Grewal and Caren Kaplan. Global Identities. Theorizing Transnational Studies of Sexuality. *GLQ. A Journal of Lesbian and Gay Studies* 7:4 (2001): 663–679.

31 Alys Weinbaum. *Wayward Reproductions. Genealogies of Race and Nation in Transatlantic Modern Thought.* Durham: Duke UP, 2004; Jennifer Brody. *Impossible Purities. Blackness, Femininity and Victorian Culture.* Durham: Duke UP, 1998; Burton, *Burdens of History*; Antoinette Burton. *At the Heart of the Empire. Indians and the Colonial Encounter in Late-Victorian Britain.* New Delhi: Munshiram Manoharlal, 1998; Burton, *Empire in Question*; Ruth Frankenberg. *White Women, Race Matters. The Social Construction of Whiteness.* Minneapolis: University of Minnesota Press, 1993; Ann Stoler (ed.). *Haunted by Empire. Geographies of Intimacy in North American History.* Durham: Duke UP, 2006; Irvin Schick. *The Erotic Margin. Sexuality and Spatiality in Alteritist Discourse.* London: Verso, 1999; Grewal, Global Identities; Ann McClintock. *Imperial Leather. Race, Gender, and Sexuality in the Colonial Contest.* New York: Routledge, 1995; Revathi Krishnaswamy. *Effeminism. The Economy of Colonial Desire,* Ann Arbor: U of Michigan Press, 2001.

were constituted through sexual practices and relations, so the understandings and imaginings of the imperial social formation entwined the sexual, the racial and other markers of power and difference.[32] Such a combination of material with cultural elements, and of intimate with global structures, has been a powerful structuring feature of empire whether in its early, high colonial or post-colonial formations. The sexualised race-ing of US society, the association of non-Europeans with non 'normal' sexual desires or practices and *vice versa*, like anxieties around cross cultural sexual and reproductive activity have been powerful drivers of the thoughts and feelings which shaped imperial policy and social practice.[33] They exemplify the power of intimate relationships, and the intimacy which can colour power relations, and the significant connections between imperial governmentalities, sexual and gender politics, and bio-power. They also underpin analyses which encompass the complexity of imperial engagements in which distance, intimacy, violence, and negotiation, both divided and entangled the dominant and subaltern, while also enabling analyses of the mutually constitu-

32 Leonore Manderson, Colonial Desires. Sexuality, Race, and Gender in British Malaya. *Journal of the History of Sexuality* 7:3 (1997): 372–388; Daniel Walther. Sex, Race and Empire. White Male Sexuality and the "Other" in Germany's Colonies, 1894–1914. *German Studies Review* 33:1 (2010): 45–71; Newman, Gender, Sexuality and the Formation of Racial Identities; Ghosh, *Sex and the Family in Colonial India*; Julia Clancy-Smith. *La Femme Arabe*. Women and Sexuality in France's North African Empire. In *Women, the Family, and Divorce Laws in Islamic Society*, Amira El Azhary Sonbol (ed.), 52–63. New York: Syracuse UP, 1996; Owen White, *Children of the French Empire. Miscegenation and Colonial Society in French West Africa 1895–1960*. Oxford: Clarendon Press, 1999; Felicity Nussbaum. *Torrid Zones. Maternity, Sexuality, and Empire in Eighteenth-Century English Narratives*. Baltimore: Johns Hopkins UP, 1995; Helen Callaway. Purity and Exotica in Legitimating the Empire. Cultural Constructions of Gender, Sexuality and Race. In *Legitimacy and the State in Twentieth Century Africa*, Terence Ranger und Olufemi Vaughan (eds.), 31–88. Houndmills: Palgrave, 1993; Joy Damousi. *Depraved and Disorderly. Female Convicts, Sexuality and Gender in Colonial Australia*. Cambridge: Cambridge UP, 1997; Jennifer Spear. Colonial Intimacies. Legislating Sex in French Louisiana. *William and Mary Quarterly* 60:1 (2003): 75–98; David Alderson. *Mansex Fine. Religion, Manliness and Imperialism in Nineteenth-Century British Culture*. Manchester: Manchester UP, 1998.

33 Weinbaum, *Wayward Reproductions;* Robert Aldrich. *Colonialism and* Homosexuality. London: Routledge, 2002; Daniel Boyarin. *Unheroic Conduct. The Rise of Heterosexuality and the Invention of the Jewish Man*. Berkeley: University of California Press, 1997; Robert Young. *Colonial Desire. Hybridity in Theory, Culture and* Race. London: Routledge, 1996; Joanna de Groot. Hybridising Past, Present, and Future. Reflections on the 'Sexology' of R. F. Burton. In *Sex, Knowledge and Receptions of the Past*, Kate Fisher and Rebecca Langlands (eds.), 135–159. Oxford: Oxford UP, 2015; Siobhan Somerville. *Queering the Color Line. Race and the Invention of Homosexuality in American Culture*. Durham: Duke UP, 2000; Sander Gilman. *Difference and Pathology. Stereotypes of Sexuality, Race, and Madness*. Ithaca: Cornell UP, 1985, chapter: The Hottentot and the Prostitute. Toward an Iconography of Female Sexuality.

tive features of 'metropole' and 'colony'. While some critiques of the marginalisation of the material and political analysis consequent on the 'cultural turn' are well taken, use of the bodily and sexual as axes of enquiry and interpretation allows scholars to draw on those categories as well as exploring constructions of meaning, imagining and knowledge.

This brings us back to the Hardt/Negri text with its invocations of bio-power as a constitutive feature of empire and of body politics and desire as energising forces.[34] The over-homogenised depictions of power and resistance on these terrains ignore gender not just as an empirical complexity, but as an organising category, structural element, or pertinent question needed to underpin any serious analysis. As Quinby notes, this has a particularly limiting effect on the discussion of migration, a phenomenon structured by gender specifics, and by culturally diverse and politically charged familial and sexual practices.[35] The *constitutive role* of gender alongside ethnicity, age and class has been central to this phenomenon, whether in the Americas in the seventeenth century, or movements of Thai, east European and Filipina sex or sweatshop workers today. Patterns of the Atlantic trade in enslaved Africans, of assisted nineteenth century colonial migration, and of twenty first century trafficking need to be understood not only as empirical data but as analytical features of migration.[36] It is likewise worth noting that

34 Hardt and Negri, *Empire*, 27–30.

35 Quinby, Taking the Millennialist Pulse, 241, 248–50.

36 David Eltis and Stanley Engerman. Was the Slave Trade Dominated by Men? *Journal of Interdisciplinary History* 23:2 (1992): 237–257; Maggie Sale. *The Slumbering Volcano. American Slave Ship Revolts and the Production of Rebellious Masculinity*. Durham: Duke UP, 1997; Sandra Gunning et al. Gender, Sexuality, and African Diasporas *Gender and History* 15:3 (2003): 397–408; Edward Rugemer. The Development of Mastery and Race in the Comprehensive Slave Codes of the Greater Caribbean during the Seventeenth Century. *William & Mary Quarterly* 70:3 (2013): 429–458; Rita Kranidis. *The Victorian Spinster and Colonial Emigration. Contested Subjects*. Basingstoke: Macmillan 1999; Clancy-Smith, Women, Gender and Migration; Janet Myers. Performing the Voyage out. Victorian Female Emigration and the Class Dynamics of Displacement. *Victorian Literature and Culture* 29:1 (2001): 129–146; Lisa Chilton. A New Class of Women for the Colonies. The *Imperial Colonist* and the Construction of Empire. *Journal of Imperial and Commonwealth History* 31:2 (2003): 36–56; Kerry Ward. *Networks of Empire. Forced Migration in the Dutch East India Company*. Cambridge: Cambridge UP, 2009; Schrover, *Illegal Migration*; Phil Williams. *Illegal Immigration and Commercial Sex. The New Slave Trade*. London: Frank Cass, 1999; Jose Moya. Domestic Service in a Global Perspective. Gender, Migration, and Ethnic Niches. *Journal of Ethnic and Migration Studies* 33:4 (2007): 559–579; Elisabetta Zontini. *Transnational Families, Migration and Gender. Moroccan and Filipino Women in Bologna and Barcelona*. New York: Berghahn Books, 2010; Andrea Bertone. Sexual Trafficking in Women. International Political Economy and the Politics of Sex. *Gender Issues* 18:1 (1999): 4–22; Laura Shepherd (ed.). *Gender Matters in Global Politics. A Feminist Introduction to International* Relations. London: Routledge,

the section of the Hardt/Negri text entitled 'The production of life', observations about the non-somatic aspects of production are undercut by the lack of any attempt to theorise its gendered, reproductive, sexual and familial aspects. The use of an image of empire/global capitalism as a 'machine' throughout the text limits the effectiveness of their deployment of categories of bio/body politics and their discussion of power and sovereignty is likewise weakened by an equally limiting focus on policing and governance.[37] A similarly restrictive and old-fashioned treatment of early modern history and of modern formations of ethnic and racial power and division are both reductionist and gender blind. Beyond that the notion of 'desire', which plays such an interesting role in the analysis of potential resistance, is used without reference to the insights of gender and queer theory, remaining confined within the hetero-normative and androcentric paradigm of 1960s discourses of 'sexual liberation'.[38]

This critique of the highly abstract, monolithic, and dated construction of 'empire' could be pursued in greater detail, supplementing Quinby's challenges to their unhelpfully universalising notion of the multiple moments and spaces/places of empire.[39] There are similarly serious difficulties with their gender blind articulation of meanings for power and violence. However this is perhaps not the most productive approach for scholars exploring the historico-cultural dynamics of time and space to take. I have chosen instead to offer a more constructive demonstration that the analytical categories, explanatory frameworks and investigative questions created within gender thought and scholarship merit incorporation at the core of our theoretical, methodological and empirical practices.

2014; Rhacel Parreñas. *Servants of Globalization. Women, Migration and Domestic Work*. Stanford: Stanford UP, 2001.

37 Hardt and Negri, *Empire*, 13–20, 386–389.

38 Hardt and Negri, *Empire*, 8–19, 70–83: for fuller views of the Renaissance see Joan Kelly. Did Women have a Renaissance? In Joan Kelly. *Women, History and Theory*. Chicago: University of Chicago Press, 1986; Merry Wiesner. *Women and Gender in Early Modern* Europe. Cambridge: Cambridge UP, 2000; Marilyn Migiel and Juliana Schiesari (eds.). *Refiguring Woman. Perspectives on Gender and the Italian Renaissance*. Ithaca: Cornell UP, 1991; relevant material from queer theory includes Tamsin Spargo. *Foucault and Queer Theory*. Duxford: Icon Books, 1999; Andrea Smith. Queer Theory and Native Studies. The Heteronormativity of Settler Colonialism. *GLQ: A Journal of Lesbian and Gay Studies* 16:1–2 (2010): 41–68; John Hawley (ed.). *Postcolonial, Queer. Theoretical Intersections*. Albany: SUNY Press, 2001; Natalie Oswin. Critical Geographies and the Uses of Sexuality. Deconstructing Queer Space. *Progress in Human Geography* 32:1 (2008): 89–103; Nan Seuffert. Reproducing Empire in Same-Sex Relationship Recognition and Immigration Law Reform. In *Queer Theory. Law, Culture, Empire*, Robert Leckey and Kim Brooks (eds.), 173–190, London: Routledge, 2010.

39 Quinby, Taking the Millennialist Pulse.

The formation of developed concepts of patriarchy, masculinity, femininity and domesticity allows scholars to rethink conventional notions of 'labour', of 'politics', and of 'social structure', whether familial, local, 'national' or global. The categories of gender and sexuality enable richer more rigorous investigations and interpretations of the material, political and cultural terrains on which the activities, relationships and institutions have been constructed, and fuller understandings of their trajectories and contradictions, notably in the field of empire. There are fine exemplars of such work which work outwards from individual persons and situations, such as Hall's study of the Macaulays, Davis's treatment of three early modern women, or Painter's exploration of the life and image of Sojourner Truth. Their work situates particular lives in larger national, social, and global formations of empire and power.[40] Taking other approaches, work by Burton, Ware, Carby, Blunt and Wilson offers exemplars which bring broad gender perspectives and conceptual structures to the study and analysis of imperial, global, racial and transnational histories. This material demonstrates a dynamic interaction between grounded historical scholarship and the construction of conceptual frameworks and extended narratives making a serious contribution at the leading edge of both humanistic and social scientific disciplines.[41]

Beyond that there is a body of explicitly theoretical work engaged with the challenging project of connecting the insights of feminist, trans-national and postcolonial thought to generate useful concepts and categories with which to think gender, space, history and empire. Thinkers like Eisenstein, Mohanty, and Kaplan, among others, have produced texts which offer to those studying space, time, power, and empire powerful and stimulating insights frameworks and cri-

---

40 Catherine Hall. *Macaulay and Son. Architects of Imperial Britain*. New Haven: Yale UP, 2012; Natalie Davis. *Women on the Margins. Three Seventeenth-Century Lives*. Cambridge: Harvard UP, 1995; Nell Painter. *Sojourner Truth. A Life, a Symbol*. New York: Norton, 1996.

41 Burton, *Empire in Question*, part 1, and Tony Ballantyne and Antoinette Burton. *Bodies in Contact. Rethinking Colonial in World History*. Durham: Duke UP, 2005; Kathleen Wilson. Introduction. Histories, Empires, Modernities. In *A New Imperial History. Culture, Identity and Modernity in Britain and the Empire, 1660–1840*, Kathleen Wilson (ed.), 1–28. Cambridge: Cambridge UP, 2004; Kathleen Wilson. Rethinking the Colonial State. Family, Gender and Governmentality on Eighteenth-Century British Frontiers. *American Historical Review* 116:5 (2011): 1269–1293; Kathleen Wilson. The Nation Without. Practices of Sex and State in the Early Modern British Empire. In *Race, Nations and Empire. Making Histories, 1750 to the Present*, Catherine Hall and Keith McClelland (eds.), 177–198. Manchester: Manchester UP, 2010; Hazel Carby. *Cultures in Babylon. Black Britain and African America*. London: Verso, 1999; Alison Blunt and Gillian Rose (eds.). *Writing Women and Space. Colonial and Postcolonial Geographies*. New York: Guilford Press, 1994; Ware. *Beyond the Pale*; Vron Ware. Moments of Danger. Race, Gender and Memories of Empire. *History and Theory* 31:4 (1992): 116–137.

tiques.[42] Their foregrounding of gender provides a dimension manifestly lacking in other theoretical work on empire and temporality, converging with the other types of work described above. Each approach yields its own results, and the key point to grasp is that whatever their chosen point of departure, scholars studying time, space and empire will find that whichever point of departure they choose, their journeys will go better if they travel with the wonderfully useful and attractive luggage created within gender theory and research.

References

Abrams, Lynn und Elizabeth Harvey. *Gender Relations in German History. Power, Agency, and Experience from the Sixteenth to the Twentieth Century*. Durham: Duke UP, 1997.

Abugideiri, Hibba. Egyptian Women and the Science Question. Gender in the Making of Colonized Medicine, 1893–1919. *Arab Studies Journal* 4:2 (1996): 46–78.

Accampo, Elinor. *Industrialization, Family Life, and Class Relations in St Chamond*. Berkeley: University of California Press, 1989.

Acker, Joan. Class, Gender, and the Relations of Distribution. *Signs* 13:3 (1988): 473–497.

Adams, Julia. *The Familial State, Ruling Families and Merchant Capitalism in Early Modern Europe*. Ithaca: Cornell UP, 2005.

Adams, Julia. The Rule of the Father. Patriarchy and Patrimonialism in Early Modern Europe. In *Max Weber's Economy and Society. A Critical Companion*, Charles Camic et al. (eds.), 237–266. Stanford: Stanford UP, 2005.

Alderson, David. *Mansex Fine. Religion, Manliness and Imperialism in Nineteenth-Century British Culture*. Manchester: Manchester UP, 1998.

Aldrich, Robert. *Colonialism and* Homosexuality. London: Routledge, 2002.

Alexander, Jacqui and Chandra Mohanty (eds.). *Feminist Genealogies, Colonial Legacies, Democratic Futures*. New York: Routledge, 1997.

Allman, Jean M. and Antoinette Burton. Destination Globalization? Women, Gender and Comparative Colonial Histories in the New Millenium. *Journal of Colonialism and Colonial History* 4:1 (2003), published online: muse.jhu.edu, February 9, 2016.

42 Zillah Eisenstein. *Against Empire. Feminists, Islam and the West*. London: Zed Books, 2004; Chandra Mohanty. Cartographies of Struggle. Third World Women and the Politics of Feminism. In *Third World Women and the Politics of Feminism*, Chandra Mohanty (ed.), 1–50. Bloomington: Indiana UP, 1991; Chandra Mohanty. Under Western Eyes. Feminist Scholarship and Colonial Discourses. In *Third World Women and the Politics of Feminism*, Chandra Mohanty (ed.), 51–80. Bloomington: Indiana UP, 1991; Alexander and Mohanty, *Feminist Genealogies*; Chandra Mohanty. *Feminism without Borders. Decolonising Theory, Practicing* Solidarity. Durham: Duke UP, 2003; see also Raewyn Connell. Rethinking Gender from the South. *Feminist Studies* 40:3 (2014): 518–539; Inderpal Grewal and Caren Kaplan (eds.). *Scattered Hegemonies. Postmodernity and Transnational Feminist Practices*. Minneapolis: University of Minnesota Press, 1994.

Anthias, Floya and Nira Yuval-Davis. *Racialized Boundaries. Race, Nation, Gender, Colour and Class and the Anti-Racist Struggle*. London: Routledge, 1995.

Arnold, David (ed.). *Imperial Medicine and Indigenous Societies*. Manchester: Manchester UP, 1988.

Badran, Margot. *Feminists, Islam, and Nation. Gender and the Making of Modern Egypt*. Princeton: Princeton UP, 1996.

Ballantyne, Tony and Antoinette Burton. *Bodies in Contact. Rethinking Colonial in World History*. Durham: Duke UP, 2005.

Bannerji, Himani, Shahrzad Mojab and Judith Whitehead (eds.). *Of Property and Propriety. The Role of Gender and Class in Imperialism and Nationalism*. Toronto: University of Toronto Press, 2001.

Bardsley, Sandy. Women's Work Reconsidered. Gender and Wage Differentiation in Late Medieval England. *Past and Present* 165 (1999): 3–29.

Barkawi, Tarak and Laffey, Mark. Retrieving the Imperial. Empire and International Relations. *Millennium: A Journal of International Studies* 31:1 (2002): 109–27.

Baron, Ava. *Work Engendered. Toward a New History of American Labor*. Ithaca: Cornell UP, 1991.

Bashford, Alison 'Is White Australia Possible?' Race, Colonialism and Tropical Medicine. *Ethnic and Racial Studies* 23:2 (2000): 248–271.

Bashford, Alison. Medicine, Gender, and Empire. In *Gender and Empire*, Philippa Levine (ed.), 112–133. Oxford: Oxford UP, 2009.

Bederman, Gail, *Manliness and Civilization. A Cultural History of Gender and Race in the United States, 1880–1917*. Chicago: University of Chicago Press, 2008.

Bell, Morag. 'The Pestilence that Walketh in Darkness'. Imperial Health, Gender and Images of South Africa c. 1880–1910. *Transactions of the Institute of British Geographers* 18:3 (1993): 327–341.

Beneria, Lourdes and Sen, Gita. Accumulation, Reproduction, and 'Women's Role in Economic Development': Boserup Revisited. *Signs* 7:2 (1981): 279–298.

Bennett, Judith M. Medieval Women, Modern Women. Across the Great Divide. In *Culture and History 1350–1600. Essays on English Communities, Identities, and Writing,* David Aers (ed.), 147–176. Detroit: Wayne State UP, 1992.

Berg, Maxine. In Pursuit of Luxury. Global History and British Consumer Goods in the Eighteenth Century. *Past and Present* 182 (2004): 85–142.

Berger, Iris and Claire Robertson (eds.). *Women and Class in Africa*. New York: Africana Publishing, 1986.

Berger, Iris. African Women's History. Themes and Perspectives. *Journal of Colonialism and Colonial History* 4:1 (2003).

Bertone, Andrea. Sexual Trafficking in Women. International Political Economy and the Politics of Sex. *Gender Issues* 18:1 (1999): 4–22.

Blake, Stephen. Returning the Household to the Patrimonial-Bureaucratic Empire. Gender, Succession, and Ritual in the Mughal, Safavid and Ottoman Empires. In *Tributary Empires in Global History*, Peter Fibiger Bang und C.A. Bayly (eds.), 214–226. New York, Palgrave Macmillan, 2011.

Blom, Ida, Karen Hagemann and Catherine Hall (eds.). *Gendered Nations. Nationalisms and Gender Order in the Long Nineteenth Century*. Oxford: Berg, 2000.

Blunt, Alison and Gillian Rose (eds.). *Writing Women and Space. Colonial and Postcolonial Geographies*. New York: Guilford Press, 1994.

Borón, Atilio. *Empire and Imperialism. A Critical Reading of Hardt and Negri*. London: Zed Books, 2005.

Bosch, Mineke. Colonial Dimensions of Dutch Women's Suffrage. Aletta Jacobs's Travel Letters from Africa and Asia. *Journal of Women's History* 11:2 (1999): 8–34.

Bosma, Ulbe and Remco Raben. *Being 'Dutch' in the Indies. A History of Creolisation and Empire, 1500–1920*. Singapore: NUS Press, 2008.

Boyarin, Daniel. *Unheroic Conduct. The Rise of Heterosexuality and the Invention of the Jewish Man*. Berkeley: University of California Press, 1997.

Boydston, Jeanne. Gender as a question of historical analysis. *Gender & History* 20:3 (2008): 558–583.

Bradford, Helen. Peasants, Historians and Gender. A South African Case Study Revisited. *History and Theory* 39:4 (2000): 86–110.

Bradford, Helen. Women, Gender and Colonialism. Rethinking the History of the British Cape Colony and Its Frontier Zones, c. 1806–70. *Journal of African History* 37:3 (1996): 351–370.

Brah, Avtar. Difference, Diversity, Differentiation. *International Review of Sociology* 2:2 (1991): 53–71.

Brah, Avtar and Ann Phoenix. 'Ain't I a woman?' Revisiting Intersectionality. *Journal of International Women's Studies* 5:3 (2004), 75–86.

Brody, Jennifer. *Impossible Purities. Blackness, Femininity and Victorian Culture*. Durham: Duke UP, 1998.

Brown, Kathleen. *Good Wives, Nasty Wenches, and Anxious Patriarchs. Gender, Race, and Power in Colonial Virginia*. Chapel Hill: University of North Carolina Press, 1996.

Burbank, Jane and Frederick Cooper. *Empires in World History. Power and the Politics of Difference*. Princeton: Princeton UP, 2010.

Burnard, Trevor and Richard Follett. Caribbean Slavery, British Anti-Slavery, and the Cultural Politics of Venereal Disease. *Historical Journal* 55:2 (2012): 427–451.

Burnard, Trevor. *Mastery, Tyranny, and Desire. Thomas Thistlewood and his Slaves in the Anglo-Jamaican World*. Chapel Hill: University of North Carolina Press, 2004.

Burton, Antoinette (ed.). *Empire in Question. Reading, Writing, and Teaching British Imperialism*. Durham: Duke UP, 2011.

Burton, Antoinette. *At the Heart of the Empire. Indians and the Colonial Encoutner in Late-Victorian Britain*. New Delhi: Munshiram Manoharlal, 1998.

Burton, Antoinette. *Burdens of History. British Feminists, Indian Women, and Imperial Culture, 1865–1915*. Chapel Hill: North Carolina UP, 1994.

Bush, Barbara. *Slave Women in Caribbean Society, 1650–1838*. Bloomington: Indiana UP, 1990.

Cain, Peter and Anthony Hopkins. *British Imperialism 1688–1990*, 2 vols. London: Longman, 1993.

Cain, Peter and Mark Harrison (eds.). *Imperialism. Critical Concepts in Historical Studies*. London: Routledge, 2001.

Callaway, Helen. Purity and Exotica in Legitimating the Empire. Cultural Constructions of Gender, Sexuality and Race. In *Legitimacy and the State in Twentieth Century Africa*, Terence Ranger und Olufemi Vaughan (eds.), 31–88. Houndmills: Palgrave, 1993.

Callinicos, Alex. The Actuality of Imperialism. *Millennium: A Journal of International Studies* 31:2 (2002): 319–326.

Carby, Hazel. *Cultures in Babylon. Black Britain and African America*. London: Verso, 1999.

Carton, Benedict. 'We Are Made Quiet by This Annihilation'. Historicizing Concepts of Bodily Pollution and Dangerous Sexuality in South Africa. *International Journal of African Historical Studies* 39:1 (2006): 85–106.
Castañeda, Antonia. Engendering the History of Alta California, 1769–1848. Gender, Sexuality, and the Family. *California History* 76:2/3 (1997): 230–259.
Chancellor, Nigel. A Picture of Health. The Dilemma of Gender and Status in the Iconography of Empire, India c. 1805. *Modern Asian Studies* 35:4 (2001): 769–782.
Chatterjee, Piya. An Empire of Drink. Gender, Labor and the Historical Economies of Alcohol. *Journal of Historical Sociology* 16:2 (2003): 183–208.
Chilton, Lisa. A New Class of Women for the Colonies. The *Imperial Colonist* and the Construction of Empire. *Journal of Imperial and Commonwealth History* 31:2 (2003): 36–56.
Clancy-Smith, Julia and Frances Gouda (eds.). *Domesticating the Empire: Race, Gender, and Family Life in French and Dutch Colonialism*. Charlottesville: UP of Virginia, 1998.
Clancy-Smith, Julia. A Woman without Her Distaff. Gender, Work, and Handicraft Production in Colonial North Africa. In *A Social History of Women and the Family in the Middle East*, Margaret L. Meriwether and Judith E. Tucker (eds.), 25–53, Boulder: Westview Press, 1999.
Clancy-Smith, Julia. *La Femme Arabe*. Women and Sexuality in France's North African Empire. In *Women, the Family, and Divorce Laws in Islamic Society*, Amira El Azhary Sonbol (ed.), 52–63. New York: Syracuse UP, 1996.
Clancy-Smith, Julia. The Intimate, the Familial, and the Local in Transnational Histories of Gender. *Journal of Women's History* 18:2 (2006): 174–183.
Clancy-Smith, Julia. Women, Gender and Migration along a Mediterranean Frontier. Pre-Colonial Tunisia, c.1815–1870. *Gender & History* 17:1 (2005): 62–92.
Clarence-Smith, William and Steven Topik (eds.). *The Global Coffee Economy in Africa, Asia, and Latin America, 1500–1989*. Cambridge: Cambridge UP, 2003.
Clark, Anna. *The Struggle for the Breeches. Gender and the Making of the British Working Class*. Berkeley: University of California Press, 1995.
Clark, Elizabeth A. The Lady Vanishes. Dilemmas of a Feminist Historian after the 'linguistic turn'. *Church History* 67:1 (1998): 1–31.
Cleall, Esme. *Missionary Discourses of Difference. Negotiating Otherness in the British Empire, 1840–1900*. Basingstoke: Palgrave Macmillan, 2012.
Conklin, Alice. Boundaries Unbound. Teaching French History as Colonial History and Colonial History as French History. *French Historical Studies* 23:2 (2000): 215–238.
Connell, Raewyn. Rethinking Gender from the South. *Feminist Studies* 40:3 (2014): 518–539.
Cooper, Frederick. *Colonialism in Question. Theory, Knowledge, History*. University of California Press, 2005.
Damousi, Joy. *Depraved and Disorderly. Female Convicts, Sexuality and Gender in Colonial Australia*. Cambridge: Cambridge UP, 1997.
Davenport, Randi. Thomas Malthus and Maternal Bodies Politic. Gender, Race, and Empire. *Women's History Review* 4:4 (1995): 415–439.
Davis, Natalie. *Women on the Margins. Three Seventeenth-Century Lives*. Cambridge: Harvard UP, 1995.
Dawson, Graham. *Soldier Heroes. British Adventure, Empire, and the Imagining of Masculinities*. London: Routledge, 1994.
Doumani, Beshara (ed.). *Family History in the Middle East. Household, Property, and Gender*. Albany: SUNY Press, 2012.

Downes, Aviston. Boys of the Empire. Elite Education and the Construction of Hegemonic Masculinity in Barbados, 1875–1920. In *Interrogating Caribbean Masculinities. Theoretical and Empirical Analyses*, Rhoda Reddock (ed.), 105–136. Mona: University of the West Indies Press, 2004.

Doyle, Michael. *Empires.* Ithaca: Cornell UP, 1986.

Dudink, Stefan (eds.). *Representing Masculinity. Male Citizenship in Modern Western Culture.* Basingstoke: Palgrave Macmillan, 2012.

Dunaway, Wilma. The Double Register of History. Situating the Forgotten Woman and her Household in Capitalist Commodity Chains. *Journal of World-Systems Research* 7:1 (2001): 1–29.

Edmond, Rod. *Leprosy and Empire. A Medical and Cultural History*, Cambridge: Cambridge UP, 2006.

Eisenstein, Zillah. *Against Empire. Feminists, Islam and the West.* London: Zed Books, 2004.

Eley, Geoff. *A Crooked Line. From Cultural History to the History of Society.* Ann Arbor: University of Michigan Press, 2005.

Eley, Geoff. Imperial Imaginary, Colonial Effect. Writing the Colony and the Metropole Together. In *Race, Nation, and Empire. Making Histories from 1750 to the Present*, Catherine Hall and Keith McClelland (eds.), 217–237. Manchester: Manchester UP, 2010.

Eltis, David and Stanley Engerman. Was the Slave Trade Dominated by Men? *Journal of Interdisciplinary History* 23:2 (1992): 237–257.

Epprecht, Marc. Sexuality, Africa, History. *American Historical Review* 114:5 (2009): 1258–1272.

Ernst, Waltraud. European Madness and Gender in Nineteenth-century British India. *Social History of Medicine* 9:3 (1996): 357–382.

Faroqhi, Suraiya. *Artisans of Empire. Crafts and Craftsmen under the Ottomans.* London: IB Tauris, 2009.

Farquhar, Judith and Marta Hanson (eds.). *Positions: East Asia Cultures Critique (Empire* of Hygiene) 6:3 (1998).

Fechter, Anne-Meike. Gender, Empire, Global Capitalism. Colonial and Corporate Expatriate Wives. *Journal of Ethnic and Migration Studies* 36:8 (2010): 1279–1297.

Frader, Laura L. Dissent over Discourse. Labor, History, Gender, and the Linguistic Turn. *History and Theory* 34:3 (1995): 213–30.

Frankenberg, Ruth. *White Women, Race Matters.* The Social Construction of Whiteness. Minneapolis: University of Minnesota Press, 1993.

Franklin, Sarah. *Women and Slavery in Nineteenth Century Colonial Cuba.* Rochester: University of Rochester Press, 2012.

Gabaccia, Donna. Spatializing Gender and Migration. The Periodization of Atlantic Studies, 1500 to the Present. *Atlantic Studies* 11:1 (2014): 7–27.

Gaitskell, Deborah. From 'Women and Imperialism' to Gendering Colonialism. *South African Historical Journal* 39:1 (1998), 176–193.

Ghosh, Durba. Gender and Colonialism. Expansion or Marginalization? *Historical Journal* 47:3 (2004): 737–755.

Ghosh, Durba. *Sex and the Family in Colonial India. The Making of Empire.* Cambridge: Cambridge UP, 2006.

Ghosh, Durba. Who Counts as 'Native'? Gender, Race, and Subjectivity in Colonial India. *Journal of Colonialism and Colonial History* 6:3 (2005).

Gilman, Sander. *Difference and Pathology. Stereotypes of Sexuality, Race, and Madness.* Ithaca: Cornell UP, 1985.

Glenn, Evelyn. From Servitude to Service Work. Historical Continuities in the Racial Division of Paid Reproductive Labor. *Signs* 18:1 (1992): 1–43.

González, Ondina and Bianca Premo (eds.). *Raising an Empire. Children in Early Modern Iberia and Colonial Latin America.* Alberquerque: University of New Mexico Press, 2007.

Gorman, Daniel. Empire, Internationalism, and the Campaign against the Traffic in Women and Children in the 1920s. *Twentieth Century British History* 19:2 (2008): 186–216.

Gottfried, Heidi. Beyond Patriarchy? Theorising Gender and Class. *Sociology* 32:3 (1998): 451–468.

Gouda, Frances. *What's to be Done with Gender and Post-colonial Studies?* Amsterdam: Amsterdam UP, 2001.

Green, Peter. 'The Passage from Imperialism to Empire'. A Commentary on *Empire* by Michael Hardt and Antonio Negri. *Historical Materialism* 10:1 (2002): 29–77.

Grever, Maria and Berteke Waaldijk. *Transforming the Public Sphere. The Dutch National Exhibition of Women's Labor 1898.* Durham: Duke UP, 2004.

Grever, Maria. The Gender of Patrimonial Pride. In *Travelling Heritages. New Perspectives on Collecting, Preserving, and Sharing Women's History*, Saskia Wieringa (ed.), 285–302. Chicago: University of Chicago Press, 2008.

Grewal, Inderpal and Caren Kaplan (eds.). *Scattered Hegemonies. Postmodernity and Transnational Feminist Practices.* Minneapolis: University of Minnesota Press, 1994.

Grewal, Inderpal and Caren Kaplan. Global Identities. Theorizing Transnational Studies of Sexuality. *GLQ. A Journal of Lesbian and Gay Studies* 7:4 (2001): 663–679.

de Groot, Joanna and Sue Morgan. Beyond the 'religious turn'? Past, Present and Future Perspectives in Gender History. *Gender and History*, 25:3 (2013): 395–421.

de Groot, Joanna. Hybridising Past, Present, and Future. Reflections on the 'Sexology' of R. F. Burton. In *Sex, Knowledge and Receptions of the Past*, Kate Fisher and Rebecca Langlands (eds.), 135–159. Oxford: Oxford UP, 2015.

de Groot, Joanna. Metropolitan Desires and Colonial Connections. Reflections on Consumption and Empire. In *At Home with the Empire. Metropolitan Culture and the Imperial World*, Catherine Hall and Sonya Rose (eds.), 166–190. Cambridge: Cambridge UP, 2006.

de Groot, Joanna. 'Sex' and 'Race'. The Construction of Language and Image in the Nineteenth-Century. In *Sexuality and Subordination. Interdisciplinary Studies of Gender in the Nineteenth Century*, Susan Mendus und Jane Rendall (eds.), 89–128. London: Routledge, 1989.

Gruber, Siegfried and Mikołlaj Szołtysek. The Patriarchy Index. A Comparative Study of Power Relations across Historical Europe. *History of the Family* (2015), published online: http://dx.doi.org/10.1080/1081602X.2014.1001769, February 08, 2016.

Gullickson, Gay L. *Spinners and Weavers of Auffay. Rural Industry and the Sexual Division of Labor in a French Village, 1750–1850.* Cambridge: Cambridge UP, 1986.

Gunning, Sandra et al. Gender, Sexuality, and African Diasporas *Gender and History* 15:3 (2003): 397–408.

Hall, Catherine and Keith McClelland. *Race, Nation and Empire. Making Histories, 1750 to the Present.* Manchester: Manchester UP, 2010.

Hall, Catherine, Keith McClelland and Jane Rendall. *Defining the Victorian Nation. Class, Race and the British Reform Act of 1867.* Cambridge: Cambridge UP, 2000.

Hall, Catherine. *Civilising Subjects. Metropole and Colony in the English Imagination, 1830–1867.* Oxford: Polity Press 2002.

Hall, Catherine. *Macaulay and Son. Architects of Imperial Britain.* New Haven: Yale UP, 2012.

Hanna, Nelly. *Artisan Entrepreneurs in Cairo and Early-Modern Capitalism 1600–1800.* Syracuse: Syracuse UP, 2011.

Haraway, Donna. Monkeys, Aliens, and Women. Love, Science, and Politics at the Intersection of Feminist Theory and Colonial Discourse. *Women's Studies International Forum* 12:3 (1989): 295–312.

Hardin, Michael. Altering Masculinities. The Spanish Conquest and the Evolution of the Latin American Machismo. *International Journal of Sexuality and Gender Studies* 7:1 (2002): 1–22.

Hardt, Michael and Antonio Negri. *Empire*. Cambridge: Harvard UP, 2000.

Harrison, Mark. *Medicine in an Age of Commerce and Empire. Britain and Its Tropical Colonies, 1660–1830.* Oxford: Oxford UP, 2010.

Hassan, Narin. *Diagnosing Empire. Women, Medical Knowledge and Colonial Mobility.* Farnham: Ashgate, 2011.

Hawley, John (ed.). *Postcolonial, Queer. Theoretical Intersections.* Albany: SUNY Press, 2001.

Helfgott, Leonardo. *The Ties that Bind. A Social History of the Iranian Carpet.* Washington: Smithsonian Press, 1994.

Hennessy, Rosemary (ed.). *Materialist Feminism and the Politics of Discourse.* New York: Routledge, 2012.

Higginbotham, Evelyn B. African-American Women's History and the Meta-Language of Race. *Signs* 17:2 (1992), 251–274.

Hodgson, Dorothy. *Once Intrepid Warriors. Gender, Ethnicity, and the Cultural Politics of Masai Development.* Bloomington: Indiana UP, 2001.

Hodgson, Dorothy. Pastoralism, Patriarchy & History among Maasai in Tanganyika, 1890–1940. In *Rethinking Pastoralism in Africa. Gender, Culture & the Myth of the Patriarchal Pastoralist,* Dorothy Hodgson (ed.), 97–120. Oxford: James Currey, 2000.

Hondagneu-Sotelo, Pierrette. Gender and Migration Scholarship. An Overview from a 21st Century Perspective. *Migraciones Internacionales* 6:1 (2011): 219–234.

Honeyman, Katrina and Jordan Goodman. Women's Work, Gender Conflict, and Labour Markets in Europe 1500–1900. *Economic History Review* 44:4 (1991): 608–628.

Honeyman, Katrina. Following Suit. Men, Masculinity and Gendered Practices in the Clothing Trade in Leeds, England, 1890–1940. *Gender & History* 14:3 (2002): 426–446.

Howell, Martha. *Women, Production, and Patriarchy in Late Medieval Cities*. Chicago: University of Chicago Press, 1986.

Howell, Philip. *Geographies of Regulation. Policing Prostitution in Nineteenth-Century Britain and the Empire.* Cambridge: Cambridge UP, 2009.

Hunter, Janet. *Gender, Economics and Industrialization. Approaches to the Economic History of Japanese Women, 1868–1945.* In: *Japanese Women, Emerging From Subservience, 1868–1945,* Hiroko Tomida, and Gordon Daniels (eds), 119–144. Folkestone: Global Oriental, 2005.

Jayawardena, Kumari. *The White Woman's Other Burden. Western Women and South Asia during British Colonial* Rule. New York: Routledge, 1995.

Jones, Jaqueline. *Labor of Love, Labor of Sorrow. Black Women, Work, and Family from Slavery to the Present.* New York: Basic Books, 1995.

Kabadayi, M. Erdem and Kate Creasey. Working in the Ottoman Empire and in Turkey: Ottoman and Turkish Labor History within a Global Perspective. *International Labor and Working-Class History* 82 (2012): 187–200.

Kandiyoti, Deniz. Bargaining with Patriarchy. *Gender & Society* 2:3 (1988): 274–290.

Kandiyoti, Deniz. Islam and Patriarchy. A Comparative Perspective. In *Women in Middle Eastern History. Shifting Boundaries in Sex and Gender*, Nikki Keddie and Beth Barons (eds.), 23–44. New Haven: Yale UP, 1991.

Kelly Boyd. *Manliness and the Boys' Story Paper in Britain. A Cultural History, 1855–1940.* Basingstoke: Palgrave Macmillan, 2003.

Kelly, Joan. Did Women have a Renaissance? In Joan Kelly. *Women, History and Theory.* Chicago: University of Chicago Press, 1986.

Klein, Martin and Claire Robertson (eds.). *Women and Slavery in Africa.* Madison: University of Wisconsin Press, 1983.

Kranidis, Rita. *The Victorian Spinster and Colonial Emigration. Contested Subjects.* Basingstoke: Macmillan 1999.

Krishnaswamy, Revathi. *Effeminism. The Economy of Colonial Desire*, Ann Arbor: U of Michigan Press, 2001.

Lal, Ruby. *Domesticity and Power in the Early Mughal World.* Cambridge: Cambridge UP, 2005.

Legg, Stephen. Gendered Politics and Nationalised Homes. Women and the Anti-Colonial Struggle in Delhi 1930–47. *Gender, Place and Culture: A Journal of Feminist* Geography 10:1 (2003): 7–27.

Leigh, Darcy. Colonialism, Gender and the Family in North America. For a Gendered Analysis of Indigenous Struggles. *Studies in Ethnicity and Nationalism* 9:1 (2009): 70–88.

Lemire, Beverly and Giorgio Riello. East & West. Textiles and Fashion in Early Modern Europe. *Journal of Social History* 41:4 (2008): 887–916.

Levine, Philippa (ed.). *Gender and Empire.* Oxford: Oxford UP, 2004.

Levine, Philippa. *Prostitution, Race, and Politics. Policing Venereal Disease in the British Empire.* New York: Routledge, 2003.

Levine, Philippa. What Difference Did Empire Make? Sex, Gender and Sanitary Reform in the British Empire. In *National Healths. Gender, Sexuality and Health in a Cross-cultural Context*, Nana Wilson-Tagoe and Michael Worton (eds.), 71–82. London: UCL Press, 2004.

Levitt, Laura and Miriam Peskowitz (eds.). *Judaism since Gender.* London: Routledge, 1997.

Loos, Tamara. Transnational Histories of Sexualities in Asia. *American Historical Review* 114:5 (2009): 1309–1324.

Lown, Judy. *Women and Industrialization. Gender at Work in Nineteenth-Century England.* Cambridge: Polity Press, 1990.

Lugones, Maria. Heterosexualism and the Colonial/ Modern Gender System. *Hypatia* 22:1 (2007): 186–209.

Manderson, Leonore, Colonial Desires. Sexuality, Race, and Gender in British Malaya. *Journal of the History of Sexuality* 7:3 (1997): 372–388.

Marks, Shula. Changing History, Changing Histories. Separations and Connections in the Lives of South African Women. *Journal of African Cultural Studies* 13:1 (2000): 94–106.

Marks, Shula (ed.). *Not either an Experimental Doll. The Separate Worlds of three South African Women.* Bloomington: Indiana UP, 1987.

Martinez-Alier, Verena. *Marriage, Class and Colour in Nineteenth Century Cuba. A Study of Racial Attitudes and Sexual Values in a Slave Society.* London: Cambridge UP, 1974.

McCaa, Robert. Marriageways in Mexico and Spain, 1500–1900. *Continuity and Change* 9:1 (1994): 11–43.

McClintock, Ann. *Imperial Leather. Race, Gender, and Sexuality in the Colonial Contest.* New York: Routledge, 1995.

McClintock, Anne, Aamir Mufti, Ella Shohat, Social Text Collective (eds.). *Dangerous Liaisons. Gender, Nation, and Postcolonial Perspectives*, Minneapolis: University of Minnesota Press, 1997.

McManus, Sheila. Review: *Taking Land, Breaking Land. Women Colonizing the American West and Kenya, 1840–1940*, Glenda Riley. *Western Historical Quarterly 36:*2, 2005: 217.

Mernissi, Fatima. *Women's Rebellion & Islamic Memory*. London: Zed Books, 1996.

Migiel, Marilyn and Juliana Schiesari (eds.). *Refiguring Woman. Perspectives on Gender and the Italian Renaissance.* Ithaca: Cornell UP, 1991.

Mohanty, Chandra. Cartographies of Struggle. Third World Women and the Politics of Feminism. In *Third World Women and the Politics of Feminism*, Chandra Mohanty (ed.), 1–50. Bloomington: Indiana UP, 1991.

Mohanty, Chandra. *Feminism without Borders. Decolonising Theory, Practicing* Solidarity. Durham: Duke UP, 2003.

Mohanty, Chandra. Introduction. Cartographies of Struggle. Third World Women and the Politics of Feminism. In *Third World Women and the Politics of Feminism*, Chandra Mohanty, Ann Russo, Lourdes Torres, 1–50. Bloomington: Indiana UP, 1991.

Mohanty, Chandra. Under Western Eyes. Feminist Scholarship and Colonial Discourses. In *Third World Women and the Politics of Feminism*, Chandra Mohanty (ed.), 51–80. Bloomington: Indiana UP, 1991.

Momsen, Janet [ed.] *Gender, Migration and Domestic Service*. London: Routledge, 2013.

Montrose, Louis. The Work of Gender in the Discourse of Discovery. *Representations* 33 (1991): 1–41.

Morgan, Jennifer. *Laboring Women. Reproduction and Gender in New World Slavery.* Philadelphia: University of Pennsylvania Press, 2004.

Morgan, Sue and Jaqueline de Vries (eds.). *Women, Gender and Religious Cultures in Britain, 1800–1940*. London: Routledge, 2010.

Morgan, Sue. Rethinking Religion in Gender History. Theoretical & Methodological Reflections. In *Gender, Religion and Diversity. Cross-Cultural Perspectives*, Ursula King and Tina Beattie (eds.), 113–124. London: Continuum, 2005.

Morgensen, Scott. Theorising Gender, Sexuality and Settler Colonialism. An Introduction. *Settler Colonial Studies 2:*2 (2012): 2–22.

Moya, Jose. Domestic Service in a Global Perspective. Gender, Migration, and Ethnic Niches. *Journal of Ethnic and Migration Studies* 33:4 (2007): 559–579.

Murray, Mary. *The Law of the Father? Patriarchy in the Transition from Feudalism to Capitalism.* Florence: Routledge, 2005.

Myers, Janet. Performing the Voyage out. Victorian Female Emigration and the Class Dynamics of Displacement. *Victorian Literature and Culture* 29:1 (2001): 129–146.

Nagel, Joanne. Ethnicity and Sexuality. *Annual Review of Sociology* 26 (2000): 107–133.

Najmabadi, Afsaneh. Beyond the Americas. Gender and Sexuality as Useful Categories of Analysis. *Journal of Women's History* 18:1 (2008): 11–21.

Najmabadi, Afsaneh. From Supplementarity to parasitism? *Journal of Women's History* 16:2 (2004): 30–35.

Neste, Sheryl. (Ad)ministering Angels. Colonial Nursing and the Extension of Empire in Africa. *Journal of Medical Humanities* 19:4 (1998): 257–277.

Newman, Brooke. Gender, Sexuality and the Formation of Racial Identities in the Eighteenth Century Anglo-Caribbean World. *Gender & History* 22:3 (2010): 585–602.

Nussbaum, Felicity. *Torrid Zones. Maternity, Sexuality, and Empire in Eighteenth-Century English Narratives*. Baltimore: Johns Hopkins UP, 1995.
Osterhammel, Jürgen and Niels Peterson. *Globalisation. A Short History*. Princeton: Princeton UP, 2005.
Osterhammel, Jürgen. *Colonialism. A Theoretical Overview*. Princeton: Wiener, 1997.
Oswin, Natalie. Critical Geographies and the Uses of Sexuality. Deconstructing Queer Space. *Progress in Human Geography* 32:1 (2008): 89–103.
Ouzgane, Lahoucine and Robert Morrell (eds.). *African Masculinities. Men in Africa from the Late Nineteenth Century to the Present*. New York: Palgrave Macmillan, 2005.
Painter, Nell. *Sojourner Truth. A Life, a Symbol*. New York: Norton, 1996.
Palriwala, Rajni and Patricia Uberoi (eds.). *Marriage, Migration and Gender*. Los Angeles: Sage Publications, 2008.
Parreñas, Rhacel. *Servants of Globalization. Women, Migration and Domestic Work*. Stanford: Stanford UP, 2001.
Pati, Biswamoy and Mark Harrison (eds.). *The Social History of Health and Medicine in Colonial India*. London: Routledge, 2009.
Patil, Vrushali. From Patriarchy to Intersectionality. A Transnational Feminist Assessment of How Far We've Really Come. *Signs* 38:4 (2013): 847–867.
Patterson, Orlando. Slaver, Gender, and Work in the Pre-Modern World and Early Greece. A Cross-Cultural Analysis. In *Slave Systems. Ancient and Modern*, Enrico Dal Lago und Katsari Constantina (eds.), 32–69. Cambrige: Cambridge UP, 2008.
Perry, Adele. *On the Edge of Empire. Gender, Race, and the Making of British Columbia, 1849–1871*. Toronto: University of Toronto Press, 2001.
Phillip, Richard. Histories of Sexuality and Imperialism. What's the Use? *History Workshop Journal* 63:1 (2007): 136–153.
Phillips, Anne and Barbara Taylor. Sex and Skill. Notes towards a Feminist Economics. *Feminist Review* 6:1 (1980): 79–88.
Phillips, Richard. *Mapping Men and Empire. Geographies of Adventure*. Hoboken: Taylor and Francis, 2013.
Phillips, Richard. *Sex, Politics and Empire. A Postcolonial* Geography. Manchester: Manchester UP, 2006.
Pollard, Lisa. *Nurturing the Nation. The Family Politics of Modernizing, Colonizing, and Liberating Egypt, 1805–1923. Berkeley: University of California Press, 2005.*
Pollert, Anna. Gender and Class Revisited; Or, the Poverty of 'Patriarchy'. *Sociology* 30:4 (1996): 639–659.
Pomper, Philip. The History and Theory of Empires. *History and Theory* 44:4 (2005): 1–27.
Potthast-Jutkeit, Barbara. The History of Family and Colonialism. *History of the Family* 2:2 (1997): 115–121.
Procida, Mary. *Married to the Empire. Gender, Politics and Imperialism in India, 1883–1947*. Manchester: Manchester UP, 2002.
Quataert, Donald. Ottoman Handicrafts and Industry 1800–1914. *Review Fernand Braudel Center* 11:2 (1988): 169–178.
Quinby, Lee. Taking the Millennial Pulse of *Empire*'s Multitude. A Genealogical Feminist Diagnosis. *Empire's New Clothes. Reading Hardt and Negri*, Paul Passavant and Jodi Dean (eds.), 231–251. New York: Routledge, 2004.
Ram, Kalpana and Margaret Jolly. *Maternities and Modernities. Colonial and Postcolonial Experiences in Asia and the Pacific*. Cambridge: Cambridge UP, 1998.

Roberts, Kevin. Yoruba Family, Gender, and Kinship Roles in New World Slavery. In *The Yoruba Diaspora in the Atlantic World*, Toyin Falola und Matt Childs (eds.), 248–259. Bloomington: Indiana UP, 2004.
Rose, Sonya. Proto-Industry, Women's Work and the Household Economy in the Transition to Industrial Capitalism. *Journal of Family* History 13:2 (1988): 181–93.
Rose, Sonya. Gender Antagonism and Class Conflict. Exclusionary Strategies of Male Trade Unionists in Nineteenth-century Britain. *Social History* 13:2 (1988): 191–208.
Rose, Sonya. 'Gender at Work'. Sex, Class and Industrial Capitalism. *History Workshop Journal* 21:1 (1986): 113–132.
Rostock, Petra and Sabine Berghahn. The Ambivalent Role of Gender in Redefining the German Nation. *Ethnicities* 8:3 (2008): 345–364.
Rowbotham, Sheila. The Trouble with Patriarchy. *People's History and Socialist Theory*, Raphael Samuels (ed.), 364–369. London: Routledge, 1981.
Rugemer, Edward. The Development of Mastery and Race in the Comprehensive Slave Codes of the Greater Caribbean during the Seventeenth Century. *William & Mary Quarterly* 70:3 (2013): 429–458.
Sacks, Karen. Toward a Unified Theory of Class, Race, and Gender. *American Ethnologist* 16:3 (1989): 534–550.
Sale, Maggie. *The Slumbering Volcano. American Slave Ship Revolts and the Production of Rebellious Masculinity*. Durham: Duke UP, 1997.
Samson, Jane (ed.). *British Imperial Strategies in the Pacific, 1750–1900*. Aldershot: Ashgate, 2003.
Sarkar, Tanika. *Hindu Wife, Hindu Nation. Community, Religion and Cultural Nationalism*. Bloomington: Indiana UP, 2010.
Sarti, Raffaella. Fighting for Masculinity. Male Domestic Workers, Gender, and Migration in Italy from the Late Nineteenth Century to the Present. *Men and Masculinities* 13:1 (2010): 16–43.
Scheidel, Walter. The most Silent Women of Greece and Rome. Rural Labour and Women's Life in the Ancient World. *Greece and Rome* 42:2 (1995): 202–217.
Schick, Irvin. *The Erotic Margin. Sexuality and Spatiality in Alteritist Discourse*. London: Verso, 1999.
Schrover, Marlou (ed.). *Illegal Migration and Gender in a Global and Historical Perspective*. Amsterdam: Amsterdam UP, 2008.
Scott, Catherine. *Gender and Development. Rethinking Modernization and Dependency Theory*. Boulder: Lynne Rienner, 1995.
Scott, Joan W. and Louise Tilly. *Women, Work, and Family*. New York: Holt, Rinehart & Winston, 1978.
Scott, Joan W. Experience. In: *Feminists Theorize the Political*, Judith Butler and Joan W.Scott (eds.), 22–40. New York: Routledge, 1992.
Scott, Joan W. Gender. A Useful Category of Analysis. *American Historical Review* 91:5 (1986): 1053–1075.
Scott, Joan W. On Language, Gender, and Working-Class History. *International Labor and Working-Class History* 31 (1987): 1–13.
Scully, Pamela. *Liberating the Family? Gender and British Slave Emancipation in the Rural Western Cape, South Africa, 1823–1853*. Portsmouth: Heinemann, 1997.

Seuffert, Nan. Reproducing Empire in Same-Sex Relationship Recognition and Immigration Law Reform. In *Queer Theory. Law, Culture, Empire,* Robert Leckey and Kim Brooks (eds.), 173–190, London: Routledge, 2010.
Sharpe, Pamela (ed.). *Women, Gender and Labour Migration. Historical and Cultural Perspectives.* London: Routledge, 2001.
Shephard, Ben. Showbiz Imperialism. The Case of Peter Lobengula. In *Imperialism and Popular Culture*, John MacKenzie (ed.), 94–112. Manchester: Manchester UP, 1986.
Shepherd, Laura (ed.). *Gender Matters in Global Politics. A Feminist Introduction to International* Relations. London: Routledge, 2014.
Shepherd, Verene et al. (eds.). *Engendering History. Caribbean Women in Historical Perspective.* London: James Currey, 1995.
Sinha, Mrinalini. *Colonial Masculinity. The 'manly Englishman' and the 'effeminate Bengali' in the Late Nineteenth Century.* Manchester: Manchester UP, 1995.
Sinha, Mrinalini. Gender in the Critiques of Colonialism and Nationalism: Locating the Indian Woman. In *Feminism and History*, Joan W. Scott (ed.), 499–537. Oxford: Oxford UP, 1996.
Sinha, Mrinalini. Introduction to Katherine Mayo: *Mother India. Selections form the Controversial 1927 Text,* 1–62. Ann Arbor: University of Michigan Press, 2000.
Sinha, Mrinalini. Mapping the Imperial Social Formation. A Modest Proposal for Feminist History. *Signs* 25:4 (2000): 1077–1082.
Sinha, Mrinalini. Reading *Mother India.* Empire, Nation, and the Female Voice. *Journal of Women's History* 6:2 (1994): 6–44.
Sinha, Mrinalini. Teaching Imperialism as a Social Formation. *Radical History Review* 67 (1997): 175–186.
Sinke, Suzanne. Gender and Migration. Historical Perspectives. *International Migration Review* 40:1 (2006): 82–103.
Smith, Andrea. Queer Theory and Native Studies. The Heteronormativity of Settler Colonialism. *GLQ: A Journal of Lesbian and Gay Studies* 16:1–2 (2010): 41–68.
Smith, Bonnie G. *Ladies of the Leisure Class. The Courgeoises of Northern France in the Nineteenth Century.* Princeton: Princeton UP, 1981.
Smith, Ruth L. and Deborah Valenze. Mutuality and Marginality. Liberal Moral Theory and Working-Class Women in Nineteenth Century England. *Signs* 13:2 (1988): 277–298.
Socolow, Susan. *The Women of Colonial Latin America.* New York: Cambridge UP, 2015.
Somerville, Siobhan. *Queering the Color Line. Race and the Invention of Homosexuality in American Culture.* Durham: Duke UP, 2000.
Spargo, Tamsin. *Foucault and Queer Theory.* Duxford: Icon Books, 1999.
Spear, Jennifer. Colonial Intimacies. Legislating Sex in French Louisiana. *William and Mary Quarterly* 60:1 (2003): 75–98.
Squires, Judith. *Gender in Political Theory.* Hoboken: Wiley, 2013.
Stoler, Ann (ed.). *Haunted by Empire. Geographies of Intimacy in North American History.* Durham: Duke UP, 2006.
Stoler, Ann. *Race and the Education of Desire. Foucault's History of Sexuality and the Colonial Order of Things.* Durham: Duke UP, 1995.
Stoler, Ann. Rethinking Colonial Categories. European Communities and the Boundaries of Rule. *Comparative Studies in Society and History* 31:1 (1989): 134–161.
Thomas, Greg. Erotics of Aryanism/Histories of Empire. How 'White Supremacy' and 'Hellenomania' Construct 'Discourses of Sexuality'. *New Centennial Review* 3:3 (2003): 235–255.

Thompson, Paul. Foundation and Empire. A Critique of Hardt and Negri. *Capital and Class* 29:2 (2005): 73–98.
Thyrêt, Isolde. *Women and the Orthodox Faith in Muscovite Russia. Spiritual Experience and Practice.* University Park: Pennsylvania State University, 2003.
Tilly, Louise. Gender, Women's History, and Social History. *Social Science History* 13:4 (1989): 439–462.
Tosh, John. *Manliness and Masculinities in Nineteenth-Century Britain. Essays on Gender, Family, and Empire.* Harlow: Pearson, 2005.
Tosh, John. The History of Masculinity. An Outdated Concept? In *What is masculinity? Historical Dynamics from Antiquity to the Contemporary World*, John H. Arnold and Sean Brady (eds.), 17–34. Basingstoke: Palgrave Macmillan, 2011.
Tucker, Judith E. *Women in Nineteenth Century Egypt.* Cambridge: Cambridge UP, 1985.
Tug, Basak. Gender and Ottoman Social History. *International Journal of Middle East Studies* 46:2 (2014): 379–381.
Twinam, Ann. *Public Lives, Private Secrets. Gender, Honor, Sexuality and Illegitimacy in Colonial Spanish America.* Stanford: Stanford UP, 1999.
Valenze, Deborah. Gender in the Formation of European Power, 1750–1914. In *A Companion to Gender History*, Teresa Meade and Merry Wiesner-Hanks (eds.), 459–476. Malden: Blackwell, 2004.
Valenze, Deborah. *The First Industrial Woman.* Oxford: Oxford UP. 1995.
van Voss, Lex Heerma and Marcel van der Linden. *Class and Other Identities. Gender, Religion and Ethnicity in the Writing of European Labor History.* New York: Berghahn, 2002.
De Vries, Jan. Between Purchasing Power and the World of Goods. Understanding the Household Economy in Early Modern Europe. In *Consumption and the World of Goods*, Brewer, John and Roy Porter (eds.), 85–132. London: Routledge, 1994.
De Vries, Jan. The Industrial Revolution and the Industrious Revolution. *Journal of Economic History* 54:2 (1994): 249–270.
Walby, Sylvia. *Theorising Patriarchy.* Chichester: Wiley-Blackwell, 1990.
Walby, Sylvia. Theorising Patriarchy. *Sociology* 23:2 (1989): 213–234.
Walker Bynum, Caroline. *Jesus as Mother. Studies in the Spirituality of the High Middle Ages.* Berkeley: University of California Press, 1984.
Walsh, Margaret (ed.). *Working out Gender. Perspectives from Labour History.* Aldershot: Ashgate, 1999.
Walther, Daniel. Sex, Public Health and Colonial Control. The Campaign against Venereal Diseases in Germany's Overseas Possessions, 1884–1914. *Social History of Medicine* 26:2 (2013): 182–203.
Walther, Daniel. Sex, Race and Empire. White Male Sexuality and the "Other" in Germany's Colonies, 1894–1914. *German Studies Review* 33:1 (2010): 45–71.
Ward, Kerry. *Networks of Empire. Forced Migration in the Dutch East India Company.* Cambridge: Cambridge UP, 2009.
Ware, Vron. *Beyond the Pale. White Women, Racism and History.* London: Verso, 1992.
Ware, Vron. Moments of Danger. Race, Gender and Memories of Empire. *History and Theory* 31:4 (1992): 116–137.
Weinbaum, Alys. *Wayward Reproductions. Genealogies of Race and Nation in Transatlantic Modern Thought.* Durham: Duke UP, 2004.
White, Deborah G. *'Ain't I a woman?' Female Slaves in the Plantation South.* New York: Norton, 1985.

White, Owen, *Children of the French Empire. Miscegenation and Colonial Society in French West Africa 1895–1960*. Oxford: Clarendon Press, 1999.
Wiesner, Merry. Beyond Women and the Family. Towards a Gender Analysis of the Reformation. *Sixteenth Century Journal* 18:3 (1987): 311–321.
Wiesner, Merry. *Women and Gender in Early Modern* Europe. Cambridge: Cambridge UP, 2000.
Williams, Phil. *Illegal Immigration and Commercial Sex. The New Slave Trade*. London: Frank Cass, 1999.
Wilson, Kathleen (ed.). *A New Imperial History. Culture, Identity and Modernity in Britain and the Empire, 1660–1840*. Cambridge: Cambridge UP, 2004.
Wilson, Kathleen. Introduction. Histories, Empires, Modernities. In *A New Imperial History. Culture, Identity and Modernity in Britain and the Empire, 1660–1840*, Kathleen Wilson (ed.), 1–28. Cambridge: Cambridge UP, 2004.
Wilson, Kathleen. Rethinking the Colonial State. Family, Gender and Governmentality on Eighteenth-Century British Frontiers. *American Historical Review* 116:5 (2011): 1269–1293.
Wilson, Kathleen. The Nation Without. Practices of Sex and State in the Early Modern British Empire. In *Race, Nations and Empire. Making Histories, 1750 to the Present*, Catherine Hall and Keith McClelland (eds.), 177–198. Manchester: Manchester UP, 2010.
Woodacre, Elena (ed.). *Queenship in the Mediterranean. Negotiating the Role of the Queen in the Medieval and Early Modern Eras*. New York: Palgrave Macmillan, 2013.
Young, Robert. *Colonial Desire. Hybridity in Theory, Culture and* Race. London: Routledge, 1996.
Zontini, Elisabetta. *Transnational Families, Migration and Gender. Moroccan and Filipino Women in Bologna and Barcelona*. New York: Berghahn Books, 2010.
Zuckerman, Michael. Endangered Deference, Imperiled Patriarchy. Tales from the Marchlands. *Early American Studies* 3:2 (2005): 232–252.

Europe, Spatiotemporal Orientation, and the Imperial

Bärbel Frischmann

Introduction
Europe, Spatiotemporal Orientation, and the Imperial

Imperial aspirations are hallmarks not only of recent European history, but are also found spatially in other parts of the world and reach back temporally into the Antique. Given the conceptive impact of the processes of globalization on most aspects of contemporary life, various theoretical approaches present themselves for the comprehension of such processes as 'empire,' 'imperialism,' and the 'imperial.' These transnational developments have, as yet, been neither sufficiently examined and understood, nor have they been successfully regulated (and so rendered manageable) in regards to their political, economic, and legal repercussions. Such developments bear relevance for Europe particularly with regards to the development of the European Community, which on the one hand remains spatially and territorially grounded, but on the other hand must be understood through its expansion beyond such territorial limitations, constantly reckoning with further exclusion of individual states. Its self-conception as a community is thus constantly undermined through national interests and the fear of the loss of historical-cultural identities. Europe also remains an interesting subject of research due to its far-reaching imperial and colonial efforts since the inception of the Modern Era. Underlying questions remain concerning the development of empires, imperial claims, and their realization; as well as to the development and function of the narrative of 'Europe,' itself.

As Olaf Asbach argues in his article, the idea of 'Europe' as a wide-reaching political, economic, and cultural concept of self-definition was first developed and utilized as an institution of identity in the Early Modern Era. Different aspects have contributed to the narrative construction of 'Europe,' strengthening and determining one another in collocation. Several of these are worthy of mention, for example the successive erosion of *christianitas* as a unifying framework of orientation; external demarcations – in particular in opposition to the dangers from the East (the Ottoman Empire); the attempt at cultural-political self-positioning in relation to history, value systems, lifestyles, and those imperial ambitions and aggressive, global colonial interests emanating from a European nucleus. Asbach pursues the line of inquiry as to the extent to which imperial aspirations are iden-

---

Translated by Jon Cho-Polizzi.

DOI 10.1515/9783110418750-009

tifiable in the narrative of 'Europe' from the time of its modern constitution. How far-reaching were these ambitions for European influence and power over the entire world from the commencement of Europe's development onward? According to Asbach, the idea of a unique cultural and economic superiority was fundamental to the self-conception of the developing idea of Europe as a historically unified community, though it is by no means supported by the historical reality of the Early Modern. Nevertheless, this hegemonic self-fashioning developed such power that it sufficed to legitimize the colonial subjugation policies of the following centuries – though these did not transpire without opposition, but were constantly questioned from outside, and by anti-imperial images of Europe, ethics, and domestic political theories from inside.

Endeavors to explore the world and to collect knowledge in the most divergent of fields – geography, geology, botany, climatology, ethnology, etc. – can perhaps be understood as imperial gestures, in that these knowledge could then be made fruitful as an apparatus of domination. But this hardly means that it was imperial motivations alone which launched those individual researchers and explorers on their expeditions. Although the lives and works of the individual actors are embedded in the framework of the imperial, they do not necessarily arise from this same interpretational horizon. Elizabeth Millán, for example, has argued in the case of Alexander von Humboldt that the complexity of his scientific works and interests is difficult to locate within an imperial European discourse. Millán rejects an interpretation which would evaluate Humboldt's travels only from the perspective of an epistemological interest motivated by an "imperialist eye" (M.L. Pratt) which would then subjugate both people and landscapes to imperial or colonial interests. Humboldt's enormous appreciation for nature as an autarchic occurrence and as the source of life for all humanity stands in exemplary contrast to the imperial aspirations of the exercise of power over territories, peoples, and nature. In nature, Humboldt saw a separate realm of freedom, beyond human empowerment. The inexhaustible wealth of nature, its diversity, and its beauty enrich all of humanity. This is something universal, regardless of the social community to which a person belongs – an aesthetic evaluation of nature is known to all peoples and to all cultures. Through his descriptions, Humboldt brought environments and landscapes in other regions of the world closer to a European public, maintaining respect for the peoples who inhabited them. Not only in his travel narratives, but also in his political writings, Humboldt described and denounced the repression and the despotism of slavery and the colonial system.

A number of imperial perspectives are made evident through the example of encounters with the arctic regions, as well. Christian Holtorf's article concerns the function of the arctic in the establishment of a regulatory imperial knowledge of the world. Not only the arctic itself was to be explored, exploited, and commodified. With the exploration of the arctic, a cartographic net was drawn around the globe, eradicating the last 'blank spaces' on the map and creating imperial, spatiotemporal patterns of thought through measurement and structuring. In particular, the ways in which spatial fixations of imperial desire were condensed through various cartographic conceptions of the North Pole are made manifest, along with the manner in which cartography itself can be employed as a medium for hegemonic thought. Interestingly, the North Pole appears to have been so attractive for explorers and adventurers precisely because it – as an as yet indistinct and un-demarcated region – proved (and continues to prove) suitable as a site for the projection of the most diverse interests and fantasies. The polar ice is constantly in motion, and today, with the melting of this ice, the Arctic Ocean is seen increasingly as a shipping space with ever improving accessibility. Here economic and politically motivated imperial interests become visible which seem to be legitimized through maritime laws. Those interests are not simply imaginary constructs but produce tangible consequences: encroaching upon the habitation of the indigenous Inuit. Through this example, the manner in which the acquisition of knowledge, symbol systems, technological possibilities, interests, and consequent actions are interwoven becomes conspicuous. There is good reason to deliberate over the interdependences of the concept of 'Europe,' spatiotemporal orientation, and the imperial.

Olaf Asbach

Die neuzeitliche Narration „Europa" und ihr imperialer Anspruch

English abstract: "Europe" – The Modern Narration and Its Imperial Claim. In current political and scholarly debates, the European Union is taken more and more often to be a new empire. The contribution asks whether this "Empire Europe" is the continuation of the idea and practice of Europe since the early modern period, when the Euro-centric or Western-centric oriented world system was formed.

In order to address this question, the contribution starts by looking at what "Europe" and "the imperial" referred to in early modern Europe, keeping in mind that "Europe" came to be mainly a product of the disintegration of traditional empires, transnational institutions and conceptions of unity. It becomes apparent that political, economic and cultural structures, as well as logic and rationality came into being in *early modern Europe*, in turn, displayed traces of an active and permanent confrontation with the world and was characterized by material and intellectual appropriation and transformation. What is specific about this process is that it is dynamic and expansive – thus imperial – and is spurred by the interests of European players, but still never takes the form of a politically or militarily organized empire. The article illustrates this peculiarity of (early) modern Europe – being imperial without being an empire – by way of outlining three different ways in which the foundations and dynamics of the emerging modern world were conceptualized and organized: a system of free agents under international law, a system of free economic subjects mediated by markets, and a system of cooperation and competition which is regulated by transnational institutions.

It is thus not without irony that now the EU emerges as a politically organized empire: we witness the return to forms of political organization which were destroyed in the early modern period in order to be capable of acting in the modern world system which, being supposedly universalistic, has been dominated by European powers and actors. The informal global empire, initially shaped by European powers and actors, compels European powers and actors to construct a formal European empire.

DOI 10.1515/9783110418750-010

Einleitung

Der Begriff des Imperiums bzw. Empire hat in der politischen, aber auch in der wissenschaftlichen Debatte eine überraschende Renaissance erfahren. Sie erklärt sich vor allem aus den Unsicherheiten, die aus den als ‚Globalisierung' bezeichneten Entwicklungen resultierten, aus dem Ende des Ost-West-Antagonismus sowie aus der rapiden Zunahme substaatlicher, transnationaler und globaler Akteure und Institutionen. Diese Entwicklungen haben das Bewusstsein für die Notwendigkeit geschärft, die auf die einzelnen, vor allem national gefassten Gesellschaften und Staaten und das durch sie geprägte internationale System konzentrierte Perspektive zu erweitern. Viele Akteure und Analytiker erblicken in der Konzeption von Imperien und imperialen Strukturen die Möglichkeit, die Komplexität der globalen Ordnungs- und Herrschaftsverhältnisse zu konzeptualisieren, kritisch zu beleuchten oder politisch zu gestalten. Denn ‚Imperien' können allgemein als spezifische Formen der dauerhaften politischen Strukturierung und Beherrschung großer Räume und unterschiedlicher Völker und Herrschaftsverbände verstanden werden, die durch die Ausbildung von Zentrum-Peripherie-Beziehungen geprägt sind. Dabei übergreifen sie auf der einen Seite heterogene politische Einheiten, Grenzen und Regionen und sind in ihrer Intensität und ihrer räumlichen Ausdehnung im Zeitverlauf variabel. Auf der anderen Seite sind sie mit verschiedenen Formen von asymmetrischen Beziehungen und direkten und indirekten Formen politischer, ökonomischer, sozialer, rechtlicher und kultureller Einflussnahmen, Abhängigkeiten und Ungleichheit verkoppelt, die durch übergreifende Ordnungsideen gerechtfertigt und zusammengehalten werden.[1]

Die USA, immer öfter aber auch die Europäische Union werden als Prototypen bzw. als wesentliche Protagonisten einer solchen Restrukturierung der ‚Weltordnung' angesehen. Das legt nahe, hier Kontinuitäten in der Geschichte der imperialistischen Eroberung und Prägung der Welt zu vermuten, die im Europa der frühen Neuzeit ihren Ausgang nahmen. In der Tat hat die historische und ideengeschichtliche Forschung der letzten Jahre gezeigt, dass es fruchtbar ist, diese Geschichte unter der Perspektive des Imperialen neu zu beleuchten. Dies macht bewusst, dass solche transnationalen und globalen Verflechtungs- und Macht-

---

1 Zur Begriffsbestimmung von Empire/Imperium und Imperialismus vgl. etwa Stephen Howe. *Empire. A very short Introduction.* Oxford: Oxford University Press, 2002; Jürgen Osterhammel. Imperien. In *Transnationale Geschichte. Themen, Tendenzen und Theorien*, Gunilla Budde u.a. (Hrsg.), 56–67. Göttingen: Vandenhoeck & Ruprecht, 2006; Herfried Münkler. Imperialismus/Imperium. In *Politische Theorie und Politische Philosophie. Ein Handbuch,* Martin Hartmann, Claus Offe (Hrsg.), 98–103. München: C.H. Beck, 2011; Ulrich Leitner. *Imperium. Geschichte eines politischen Systems.* Frankfurt a. M. und New York: Campus Verlag, 2011.

prozesse keine neue Entwicklung, sondern für die politische Theorie und Praxis der von Europa bzw. dann dem Westen dominierten Ausbildung des modernen Weltsystems konstitutiv sind.[2]

Abb. 1: Franz Hogenberg. Europa auf dem Stier. In *De Europae virginis tauro insidentis, topographica atque historica descriptione liber*, Michael Eytzinger (Hrsg.). Köln 1588, 6.

2 Einen Überblick über die jüngst aufgeflammten Debatten über die Anwendungsbereiche und Fruchtbarkeit des Begriffs des *Empire* bietet Jennifer Pitts. Republicanism, Liberalism, and Empire in Postrevolutionary France. In *Empire and Modern Political Thought*, Sankar Muthu (Hrsg.), 351–387. Cambridge: Cambridge University Press, 2012.

Sind also die gegenwärtigen Tendenzen zur Ausbildung imperialer Strukturen nur eine neue Etappe der langen Geschichte von Versuchen europäischer Mächte, sich die außereuropäische Welt anzueignen und verfügbar zu machen? Ist dem modernen ‚Europa‘ also der Anspruch des Imperialen gleichsam eingeschrieben? – Stellt man aus dieser Perspektive des frühen 21. Jahrhunderts die Frage nach der Bedeutung des Imperialen in Bezug auf Europa und die Ideen, Vorstellungen und Narrationen zu ‚Europa‘ in und seit der frühen Neuzeit, scheint die Antwort unmittelbar klar zu sein. Dies lässt sich etwa an den ikonographischen Repräsentationen Europas, wie sie seit dem 16. Jahrhundert entstehen, ablesen und durchdeklinieren.

In dem 1588 publizierten Stich von Franz Hogenberg werden Geographie, Mythos und politisch-kulturelle Gestalt Europas für heutige Betrachter unmittelbar miteinander verkoppelt. ‚Europa‘ erscheint hier als gleichermaßen huldvoll-lächelnde wie machtvolle Königin, die die Territorien der verschiedenen europäischen Herrschaften umfasst und dynamisch zum Sprung über die Meere anzusetzen scheint, um die anderen Kontinente zu erobern.[3]

Um dieselbe Zeit, 1593, schreibt Cesare Ripa in einem vielfach übersetzten und immer wieder aufgelegten Werk, „dass Europa schon immer die Führerin und Königin der ganzen Welt gewesen ist“.[4] Von der zweiten Auflage im Jahre 1603 an illustriert er dies mit einer Allegorie, die die vier damals bekannten Kontinente darstellt. Lange Zeit diente sie als Vorlage zahlloser allegorischer Darstellungen und prägte die Repräsentationsweisen, Selbst- und Fremdbilder Europas. Während hier die anderen Kontinente als mehr oder weniger zivilisiert erscheinen – Asien mehr, Afrika und Amerika weniger –, ist Europa mit allen Insignien von Macht, Wohlstand, Zivilisation, Wissenschaft und Künsten ausgestattet.

In diesen beiden Abbildungen scheint also offenbar schon vereint, was heute in der Literatur als typische Merkmale des imperialen oder imperialistischen Europa der frühen Neuzeit und seines Blicks auf sich und die Welt gilt: Stärke, Einheit und Macht des modernen Europa verbinden sich mit militärischen und politischen Instrumenten, mit Wissen und Gelehrsamkeit, mit Religion und Kultur, mit technologischem Fortschritt, Handel und wirtschaftlichem Wohlstand.

3 Abgedruckt in: Michael Eytzinger (Hrsg.). *De Europa virginis tauro insidentis topographica*, Köln: Gottfried von Kempen, 1588. – Es geht hier, wohlgemerkt, nicht um die historische Angemessenheit einer solchen Interpretation, sondern um den heutigen Blick auf die Abbildung.

4 Cesare Ripa. Iconologia overo Descrittione Dell'imagini Universali cavate dall'Antichità et da altri luoghi; zit. nach John Hale. *Die Kultur der Renaissance in Europa*. München: Kindler, 1994, 25.

Abb. 2: Cesare Ripa. *Iconologia of Uytbeeldinghen des Verstants*. Amsterdam 1644, 602–605.

Solche im 16. und 17. Jahrhundert entstehende und in großer Zahl verbreitete Repräsentationsweisen ‚Europas‘ als in jeder Hinsicht fortgeschrittenster und mächtigster ‚Herrscherin‘ und ‚Führerin‘ der Welt sind jedoch mit Vorsicht zu interpretieren. Insbesondere wäre es ein Irrtum anzunehmen, sie gäben die real existierenden Verhältnisse wieder. Europa war im 16., 17. und selbst noch im 18. Jahrhundert faktisch weit davon entfernt, eine solch imperiale oder hegemoniale Position einnehmen zu können; es war insbesondere den asiatischen Mächten und Kulturen gegenüber in vielerlei Hinsicht unterlegen, sogar eher ein Nachzügler.[5] ‚Europa‘, so ließe sich ironisch anmerken, war offenbar von Anfang an durch ein reichlich verzerrtes Selbstbild und eine anmaßende und hegemoniale Anspruchshaltung gegenüber dem als nicht-europäisch bestimmten ‚Anderen‘ gekennzeichnet, die sich allerdings – und hier endet die Ironie – in gewaltsamen Formen der Aneignung, Unterwerfung und Ausbeutung materialisierte; dadurch erscheint die hier erst imaginierte ‚Imperatorin Europa‘ gleichsam als eine self-fulfilling prophecy, wie sie vor allem seit dem 19. Jahrhundert auf imperialistischem Wege realisiert wurde.

Doch *diese* Darstellung der Geschichte und des Narrativs des frühneuzeitlichen ‚Europa‘ als einer ‚imperialen Macht‘, die sich aufmacht, die Welt – die außerhalb befindliche wie die eigene und ihre Geschichte – zu dominieren, ist, wie die folgenden Ausführungen andeuten sollen, nicht ohne Probleme. Zum einen ist sie erkennbar unterkomplex, zum anderen und vor allem handelt es sich dabei um eine gewissermaßen zu ‚harmlose‘ Darstellung seiner imperialen Tendenzen. Sie tendiert nämlich dazu, das ‚Imperiale‘ an Erscheinungsformen festzumachen, deren – tatsächliche oder vermeintliche – Überwindung durch die gefällige Kritik euro- oder westernzentrischer Weltbilder und Praktiken noch nicht die Überwindung dieses ‚Imperialen‘ selbst bedeuten müssen.

Nach einer erkenntniskritischen Vorbemerkung zu der Frage, wovon wir eigentlich sprechen, wenn wir von ‚Europa‘ sprechen (II), soll kurz auf die Voraussetzungen der Entstehung Europas in der frühen Neuzeit eingegangen werden (III). Anschließend wird versucht, die Spezifik des ‚Imperialen‘ dieses Europa zu skizzieren (IV), bevor abschließend drei Konzeptionalisierungsweisen von staatlicher, sozioökonomischer und internationaler Ordnung skizziert werden (V), die im frühneuzeitlichen Europa ausgebildet wurden und imperiale Tendenzen aufweisen, die bis in die Gegenwart wirksam sind.[6]

5 Vgl. John Darwin. *After Tamerlan. The Global History of Empire since 1405*. London u.a.: Allen Lane, 2007.

6 Die folgenden Ausführungen bewahren den Charakter des Vortrags, der frühere Arbeiten mit Blick auf die Problematik des Imperialen reflektiert. Um angesichts der Komplexität der Sache den Rahmen nicht zu sprengen, bleiben Literaturhinweise auf das Notwendigste beschränkt.

1 ‚Europa' – Gibt es das? Erkenntniskritische Vorbemerkung

Bevor die Frage nach dem imperialen Anspruch der Narration ‚Europa' thematisiert werden kann, muss eine erkenntniskritische Reflexion vorgeschaltet werden.[7] Denn wovon spricht man eigentlich, wenn man von ‚Europa' oder von einer Narration, von Bildern, von Ideen oder gar Werten ‚Europas' spricht? Es ist für uns heute kaum möglich, Europa nicht unmittelbar als etwas ‚Wirkliches' und ‚Gegebenes' und als Darstellung von Vorstellungen von etwas Wirklichem oder Gegebenem zu fassen – ganz gleich, ob man sie bloß zu konstatieren glaubt, oder ob man sich positiv oder negativ auf sie bezieht. Um also nicht ungeklärten Vorannahmen, Projektionen, Vorurteilen oder Stereotypen zu unterliegen, wenn man nach dem ‚Imperialen' in Europa-Narrativen fragt, muss man sich den *konstruktivistischen Charakter* von Kategorien und Entitäten wie ‚Europa' bewusst machen und in die Reflexion einbeziehen. Analytisch umfasst der konstruktivistische Charakter Europas drei Aspekte:

1. ‚Europa' als ‚Wirkliches' gibt es nicht; es ist keine ‚an sich' existierende und greifbare Entität. Ganz gleich, was man darunter (politisch, geographisch, kulturell, normativ ...) versteht: ‚Europa' ist ein ‚Etwas', das nur Gegenstand sein und werden kann, wenn es als solches gedacht, bezeichnet und repräsentiert wird.
2. Ideen, Vorstellungen, Bilder oder Narrative ‚Europas' sind in diesem Sinne also gegenstandskonstitutiv: Man bezieht sich auf ‚etwas', das als solches erst dadurch ‚existiert', dass man sich in dieser Weise auf es bezieht.[8] – Dies bedeutet aber nicht die Auflösung der Wirklichkeit in rein sprachliche oder ‚subjektive', d. h. willkürliche oder zufällige Konstruktionen, denn auch solche Repräsentationsweisen sind keine letztbegründende Instanz.
3. *Dass* und *wie* ‚Europa' nämlich gedacht, repräsentiert, entworfen oder bewertet wird, ist abhängig von historischen und gesellschaftlichen, diskursiven und begrifflichen Voraussetzungen, Entwicklungen, Interessen und Ausein-

7 Die folgenden Ausführungen beruhen auf Kap. I aus Olaf Asbach. *Europa – Vom Mythos zur ‚Imagined Community'? Zur historischen Semantik ‚Europas' von der Antike bis ins 17. Jahrhundert.* Hannover: Wehrhahn, 2011; dort ausführlicher zu den hier aus Platzgründen knapp gehaltenen Hinweisen.

8 In welchen Medien, Formen und Gestalten dies geschieht, wie ‚Ideen', ‚Narrationen', ‚Bilder' usw. von sinnlich-übersinnlichen ‚Dingen' wie ‚Europa' sich im Einzelnen bilden und wirken, kann und muss an dieser Stelle nicht weiter diskutiert werden; vgl. hierzu Asbach, *Europa*, 16–17 (Anm. 6) und 24 ff.

andersetzungen und dabei je sich ausbildenden Positionen und Perspektiven der Sprechenden und Handelnden. Repräsentationsweisen ‚Europas‘ sind also stets Ausdruck und Elemente gesellschaftlicher Verhältnisse und Entwicklungen sowie von spezifischen Selbst- und Weltwahrnehmungen, und sie sind nicht zuletzt handlungsleitende Kategorien konkreter Akteure bzw. Sprecher in konkreten historischen Konstellationen.

Das heißt also: ‚Europa‘ ist nichts an sich Seiendes, immer schon Gegebenes, sondern als Idee und als Wirklichkeit etwas Gewordenes, Produziertes und Reproduziertes. Ideen und Wirklichkeit stehen in einem spezifischen Abhängigkeits- und Konstitutionsverhältnis zueinander. Die Erkenntnis von Europaideen und Europaplänen ist deshalb nur möglich, wenn man auch die historische Wirklichkeit in den Blick nimmt, auf die sie sich theoretisch und praktisch beziehen.

2 Die Genese ‚Europas‘ in der frühen Neuzeit als Zerfallsprozess universal-imperialer Strukturen

Folgt man dieser für die Erkenntnis notwendigen historisch-kritischen Verortung des Europabegriffs, so wird deutlich, dass die diesbezüglichen Ideen, Vorstellungen oder Narrationen in der frühen Neuzeit selbst ein ganz neuartiges Phänomen sind. Und wenn Begriff und Sache im Falle ‚Europas‘ so eng miteinander verkoppelt sind, wie gerade behauptet wurde, so gilt dies auch für die mit diesem Begriff bezeichnete Wirklichkeit. Dies bedeutet, wenn man es zugespitzt formulieren will: ‚Europa‘ wie auch Ideen, Konzepte oder Narrationen ‚Europas‘ sind etwas historisch Neuartiges, sie sind Produkte, ‚Erfindungen‘ oder Konstrukte der frühen Neuzeit, die es in der Antike und im Mittelalter nicht gab.

Das *Wort* ‚Europa‘ war natürlich seit der griechischen Antike im Umlauf, doch nur im Sinne des bekannten Mythos und als geographische Bezeichnung für einen der drei bekannten Erdteile. Demgegenüber finden sich weder in der Antike noch im Mittelalter jene Verwendungsweisen des Europabegriffs, wie sie heute selbstverständlich sind: als Kategorie der Identifikation eines geographisch und historisch verorteten Handlungs-, Lebens- und Erfahrungszusammenhangs, der mit bestimmten Akteuren, Institutionen, Praktiken, Normen und Eigenschaften verbunden sein soll und der eine identitätsstiftende Kategorie im Selbst- und Weltbild der Menschen als ‚Europäer‘ darstellen würde. Die antike wie die christlich-mittelalterliche Welt waren gänzlich anders gepolt und strukturiert; die politischen, kulturellen und mentalen Orientierungen, Grenzen und Handlungsformen verliefen in völlig anderer Weise.[9]

Dies ändert sich im Übergang zur frühen Neuzeit, v. a. im Laufe des 16. und 17. Jahrhunderts, die ich als ‚Inkubationsphase' von ‚Europa' und ‚Europaidee' bezeichnet habe.[10] Jetzt erst entstehen auf ‚objektiver' – materieller oder realgeschichtlicher – Ebene die Bedingungen der Herausbildung der seitdem als ‚Europa' bezeichneten Welt. Und erst jetzt lassen sich auch auf der ‚subjektiven' – mentalen oder repräsentierenden – Ebene die ersten Vorstellungen nachweisen, Teil eines solchen, spezifisch ‚europäischen' Zusammenhangs zu sein.

Dabei ist es mit Blick auf die hier interessierende Thematik zunächst einmal bemerkenswert, dass diese ‚Genese Europas', um es noch einmal zuzuspitzen, in gewisser Hinsicht geradezu als Prozess der *Auflösung* und *Negation* von imperial-universalen Strukturen, Institutionen, Konzepten und Orientierungen verstanden werden kann. Die Prägung und Ausbreitung von Begriff, Idee und Narrativen ‚Europas' können als Reaktion auf die säkularen Umbrüche zur modernen Welt interpretiert werden, die zur Entstehung gänzlich neuartiger Ordnungen von Gesellschaft, Wirtschaft und Politik wie auch der identitätsstiftenden und handlungsleitenden Normen und Weltbilder führen. Diese langfristigen und widerspruchsreichen Entwicklungstendenzen umfassen, knapp zusammengefasst, mindestens drei fundamentale Prozesse:

- *Erstens* der sukzessive Zerfall der einigenden Idee der *christianitas*, die mit ihrer transzendenten Orientierung und ihrem ideellen und institutionellen Universalismus einen ‚europäischen', d. h. auf Raum und Zeit gegründeten Bezugsrahmen in Denken und Handeln ausgeschlossen hatte.
- Dieser Zerfall steht *zweitens* im Zusammenhang mit der Ausbildung einer prinzipiell kompetitiven Struktur politischer, sozialer und rechtlicher Beziehungen. Dies zeigt sich an so heterogenen Phänomenen wie der Pluralisierung der religiösen Glaubens- und Normensysteme samt ihrer institutionel-

9 Wo es überhaupt übergreifende Ordnungsvorstellungen gab, bestanden diese, schematisch gesprochen, 1.) in Form einer Welt von griechischen *poleis*, die sich gar nicht als Teil der sie umgebenden Welt sahen – und schon gar nicht als Teil der Völker des geographischen Europa mit all seinen barbarischen Völkern wie Kelten oder Skythen –, 2.) in Form des römischen Imperiums, das um das Mittelmeer herum zentriert war und auf diese Weise alle bekannten Kontinente umfasste, oder aber 3.) in der Gestalt der in Spätantike und Mittelalter dominierenden Christenheit, die ihr heilsgeschichtliches Zentrum im vorderasiatischen Raum hatte, in ihrem Weltbild universalistisch war und auch praktisch von anderen räumlichen Bezügen als jener der Erdteilgeographie geprägt war; hierzu im Einzelnen und mit Hinweisen auf Quellen und Forschungsliteratur Asbach, *Europa*, Kap. II.

10 Asbach, *Europa*, 95.

len Verkörperungen, an der Auflösung ständischer Ordnungen durch das Entstehen neuer sozialer Schichten, Lebens- und Normordnungen, an der Verfestigung von Territorialherrschaften im Zuge der Staatenbildungsprozesse oder am Bedeutungszuwachs von Handel und Marktbeziehungen und ihren konkurrierenden Trägerschichten und Interessenlagen.

– All dies vollzieht sich *drittens* in einem spezifischen geographischen Zusammenhang. Der Vormarsch der Osmanen und das Ende des oströmischen Reiches (1453) führte zu neuen räumlichen und mentalen Grenzziehungen zwischen geographischem Europa und *Christenheit* einerseits und dem nichteuropäischen Asien andererseits; mental eingeprägt wird dies u. a. durch die sogenannten „Türken-Schriften", die durch die zeitgenössische Medienrevolution weite Verbreitung fanden und wohl bis heute bewusstseinsprägend nachwirken. Hinzu kommen mit der Ausweitung überseeischen Handels die Entdeckungen ‚neuer' Völker und Kulturen, die neue Selbst- und Weltbezüge generieren.

All diese Entwicklungen waren Teil einer langfristigen Säkularisierung der Handlungs-, Denk- und Orientierungsformen:[11] Die Genese prinzipiell kompetitiver Institutionen, Strukturen und Dynamiken in dem als ‚Europa' bezeichneten Raum und die Verschärfung der Gegensätze zu Akteuren und Orientierungen jenseits desselben erzeugt ein neues Selbst- und Weltbild. Möglich, aber auch praktisch notwendig wird jetzt eine genaue Verortung, ein Wissen um die räumlichen und die materiellen Bedingungen und Verhältnisse von ‚Selbst' und ‚Anderen', um sich in den europäischen und globalen Auseinandersetzungen behaupten zu können. Diese Verortung aber kann unter diesen Voraussetzungen nicht im Sinne eines ‚Imperium Europaeum' erfolgen, wie es die eingangs genannten Europa-Allegorien noch suggerierten. Entsprechend wurden in der Folge in Theorie und Praxis alle Vorstellungen eines einheitlichen, die Welt beherrschenden Reiches oder einer Universalmonarchie, wie es diese Allegorien noch vorzugeben scheinen, anachronistisch und verloren schnell ihre politische Relevanz.[12]

---

11 Diese ‚Säkularisierung' ist nicht mit der Abwendung von Religion und Kirchen gleichzusetzen, sondern erfolgte und erfolgt bis heute bekanntlich vielfach gerade in der Gestalt von religiösen Diskursen, Praktiken und Weltbildern.

12 So gibt es in der frühen Neuzeit nur wenige explizit ‚universalmonarchische' Konstruktionen ‚Europas' wie etwa, um 1600, die des auf eine ‚*spanische Monarchy*' setzenden Campanella. Seit dem 16. Jahrhundert ist sowohl für das allgemeine politische Denken wie auch für das Europadenken speziell vielmehr die polemische Wendung *gegen* ‚universalmonarchische' Aspirationen, wie sie etwa den Habsburgern oder, v. a. seit Ludwig XIV., Frankreich unterstellt werden, geradezu konstitutiv. – Zum Verschwinden des Konzepts des *Empire* in der staats- und

3 Imperial ohne Empire? ‚Europa' in der frühen Neuzeit

Auch wenn also die Ausbildung des ‚modernen Europa' gerade als radikale Negation von Vorstellungen eines christlichen oder weltlichen Imperiums erfolgt, ist doch weder an der Genese eines (wie auch immer bestimmten) ‚Europa' zu zweifeln noch daran, dass es mit (wie auch immer bestimmten) Dynamiken des Imperialen verbunden ist. Dass ein ‚Etwas' entsteht, das als ‚Europa' vorgestellt und wahrgenommen wird, lässt sich angesichts des Aufschwungs des Europabegriffs, der im Mittelalter lange Zeit kaum verbreitet war und nur selten Verwendung fand,[13] und seiner politisch-kulturellen Aufladung gewiss mit guten Gründen annehmen. Nach ersten Anfängen im 14./15. Jahrhundert und einem deutlichen Anstieg im 16. Jahrhundert explodiert die Verwendung des Europabegriffs im 17. und 18. Jahrhundert förmlich und findet sich in allen Bereichen der Welt- und Selbstwahrnehmung: in der Kartographie, in der Ikonographie, in den Reisebeschreibungen, in der anti-osmanischen Propaganda, in den Staats-Beschreibungen, in der Diplomatie und Politik. Und auch ohne politisch-institutionelle Einheit – oder vielleicht auch gerade deshalb – setzt hier eine imperiale und imperialistische Dynamik ein, die sowohl die als ‚Europa' identifizierten Regionen wie auch die Welt insgesamt transformieren und langfristig in spezifisch eurozentrischer Form restrukturieren sollte.

Wie passt dies zusammen? Inwiefern und in welcher Weise ist ‚Europa' als einheitliches und gemeinsam ‚Seiendes' und imperial ‚Wirkendes' möglich, ohne dass es ein einheitliches und gemeinsames ‚Sein' und imperiales ‚Wirken' in jenem Sinne gibt, wie es für das Imperium Romanum galt?

Es liegt nahe zu vermuten, dass dies an den spezifischen Formen, Institutionen, Praktiken und Triebkräften liegt, die das moderne Europa generiert und geprägt haben. Auf einer ganz allgemeinen Ebene kann man sagen, dass die angedeuteten Prozesse der Pluralisierung und Differenzierung, der Freisetzung und Dynamisierung partikularer Interessen und Kräfte *systemischen* Charakter

völkerrechtlichen Literatur vgl. Emmanuelle Jouannet. *The Disappearance of the Concept of Empire. Or, the Beginning of the End of Empires in Europe from the 18th Century.* [http://www.univ-paris1.fr/fileadmin/IREDIES/Contributions_en_ligne/E._JOUANNET/The_Disappearance_of_the_Concept_of_Empire_E._Jouannet-2.pdf, abgerufen: 17.8.2015].

13 Vgl. Asbach, *Europa*, 64–93. Einen Versuch, dieses Bild zu revidieren, ohne den gängigen anachronistischen Europa-Projektionen zu erliegen, unternimmt jetzt Oschema, ändert aber letztlich nichts am grundsätzlichen Befund; vgl. Klaus Oschema. *Bilder von Europa im Mittelalter.* Ostfildern: Thorbecke, 2013.

angenommen haben, indem Mannigfaltiges und Gegensätzliches auf spezifische Weise miteinander verbunden wird und interagiert. Genau diese systemische Verkopplung von Partikularem hat gleichsam strukturell imperial-expansive Dynamiken zur Folge, die stets mit Prozessen der Steigerung von Macht, Ungleichheit und Ausbeutung verbunden sind. Von besonderer Relevanz sind dabei vor allem drei – miteinander zusammenhängende – Bereiche:

1. *Die Ausbildung moderner Staaten* als neuartiger Form der Organisation politischer Herrschaft mit festen territorialen Grenzen und klaren Beziehungen der In- und Exklusion. Indem Staaten die souveräne Gewalt der politischen und rechtlichen Regulierung gesellschaftlicher Beziehungen im Inneren wie nach außen monopolisieren, werden sie zu zentralen Akteuren und Kampfplätzen politischer, ökonomischer und soziokultureller Auseinandersetzungen um die Organisation und die Macht innerhalb der gesellschaftlichen wie der internationalen Verhältnisse.
2. *Die Entfaltung von marktwirtschaftlichen Beziehungen, einer geld- und kreditgetriebenen Ökonomie und kapitalistischen Produktionsweise*, die von den entstehenden Staaten organisiert wird. Diese Entwicklung führte zur Umwälzung aller Lebensbereiche, Arbeits- und Selbstbeziehungen der Gesellschaftsmitglieder, insofern ihre Existenz, Interessen und Ansprüche nun immer stärker vom Kampf um Ressourcen und Chancen auf Märkten abhängen und so eine Dynamik unaufhörlicher Konkurrenz, Aneignung und Nutzung von Kapital, Arbeit und Ressourcen in Gang gesetzt wird.
3. *Die Entfaltung neuer Formen und Institutionen des Wissens, der Kultur und religiöser Glaubens- und Lebensweisen.* Die philosophischen, wissenschaftlichen und religiös-theologischen Entwicklungen der Neuzeit führen zur Unterminierung traditioneller Weltbilder und Sozialbeziehungen, angetrieben von den politischen, sozialen und wirtschaftlichen Umbrüchen, Konflikten und Konkurrenzbeziehungen und diese immer weiter befördernd. Hierzu trägt ihre unaufhörliche Steigerung von Kapazitäten der Herrschaft über Natur, Gesellschaft und Menschen bei, die durch Innovationen und Akkumulation von Wissen und Techniken, von immer effektiver und rationeller gestalteten Organisationsweisen und Einrichtungen erfolgt.

In der frühen Neuzeit entsteht so sukzessive eine Pluralität von Staaten und Interessengruppen, die von Logiken des Ringens um Territorien, Märkte und Handelswege, um die Akkumulation von Macht, Geld, Wissen und Ressourcen angetrieben werden. Sie beruhen auf und produzieren und reproduzieren Konkurrenz- und Ausschlussbeziehungen, die auf dem Kontinent und – worauf es in diesem Zusammenhang natürlich besonders ankommt – von Anfang an auch über den europäischen Kontinent hinaus wirksam sind und permanent Interessenkonflikte, Ungleichheiten, Gegensätze und Ausschlüsse und dabei neue Formen des Imperialen produzieren, die sich von traditionellen Formen imperialer Herrschaft spezifisch unterscheiden.

Und doch ist diese neue ‚Pluralität' nur eine Seite der Medaille. Denn historisch-praktisch wie systematisch sind solche ‚systemischen' Beziehungen von Konkurrenz, Konflikten und Kriegen, von Ausschluss und Ungleichheit eben zugleich auch – Beziehungen. Und als solche setzen sie soziale und geistige Zusammenhänge ebenso sehr voraus, stiften und reproduzieren sie, wie sie sie gefährden und immer wieder zerstören. Dies zeigt sich in jedem der eben genannten Bereiche:

1. Die Genese von Staaten generiert einerseits eine Pluralität solcher politischer Körperschaften, die, wenn sie bestehen wollen, einen modus vivendi finden müssen, um sich in Krieg und Frieden, d. h. politisch, militärisch, diplomatisch, wirtschaftlich usw., aufeinander beziehen zu können. Der moderne Staat entsteht in Europa insofern gleichursprünglich mit der Ausbildung eines Staatensystems; mit dem Territorialen und Nationalen entsteht das Inter- und Transnationale. Seinen Niederschlag findet dies in mannigfaltigen Formen: machtpolitisch im System von Allianzen, Bündnissen und Freund-Feind-Linien, wie sie als Mächtesystem, als ‚Balance of Power' oder ‚Europäisches Gleichgewichtssystem' vorgestellt und organisiert werden; hinzu kommt die Vielfalt von politischen, rechtlichen und intellektuellen Entwicklungen, die von politischen und diplomatischen Beziehungen bis zu den theoretischen und praktischen Entwicklungen hin zu einem Völker- und Kriegsrecht, das die freundlich-feindlichen Beziehungen regulieren soll.
2. Ebenso verhält es sich mit der ökonomischen Konkurrenz zwischen Staaten, Handelskompagnien, Produzenten und Kapitalbesitzern. Das Streben nach privater Aneignung, Gewinn, Akkumulation von Kapitel und effektiver Ausbeutung von Arbeit und Ressourcen sowie dem damit verbundenen steuer- und machtbasierten Reichtum von Fürsten und Staaten erzeugt ein immer dichter werdendes Netz von Beziehungen, Abhängigkeiten, Konkurrenz- und Kooperationsbeziehungen: Handelswege und Seerouten, Banken-, Kredit- und Finanzsysteme sowie Märkte, die Produktion und Konsum in unterschiedlichen Ländern und Regionen auf dem Kontinent und über ihn hinaus durch geldvermittelte Austauschprozesse verbinden und beeinflussen. Auch und gerade die Handel treibenden und zunehmend für überregionale und internationale Märkte produzierenden Akteure benötigen deshalb zudem ein umfassendes Wissen über politische, gesellschaftliche, kulturelle und natürliche Verhältnisse, Sitten, Bräuche und Bedürfnislagen, um gewinnbringende Produktions- und Austauschprozesse einschließlich möglichst effektiver Ausbeutung allererst möglich zu machen.

3. Dies verweist bereits auf die Bereiche von Wissenschaft, Kultur und Kommunikation. Sie entstehen in der frühen Neuzeit zwar in enger Beziehung mit den ‚dissoziierenden‘ Faktoren von Staatsbildung, der Steigerung politischer, ökonomischer, militärischer und weiterer, in den Macht- und Konkurrenzkämpfen nutzbarer Kenntnisse jeder Art, so dass sie insofern Medien und Faktoren im Prozess der Entzauberung und Auflösung traditioneller Ordnungen, Sozial- und Naturverhältnisse und der Durchsetzung neuer partikularer Mächte und Interessen sind. Zugleich erfolgt ihre Entwicklung jedoch intellektuell wie praktisch gleichsam immer auch in Form der Bildung immer neuer und dichterer Netzwerke, Lern- und Austauschprozesse. Dies beginnt seit dem Übergang zur frühen Neuzeit mit der Entstehung von Universitäten und Akademien, von konfessionellen und säkularen Vereinigungen und Gesellschaften, der Ausbildung einer Art von Gelehrtenrepublik, die durch Korrespondenzen und Reisen miteinander verflochten ist, durch einen immer dichter werdenden Markt von Büchern, Zeitschriften und Zeitungen, die Informationen und Wissen ebenso verbreiten wie Meinungen und Propaganda.

In der Summe entstehen auf dem europäischen Kontinent im Laufe der frühen Neuzeit also Ordnungen, Mächte und Strukturen, die einerseits durch die Entwicklung und Entfaltung *partikularer* Entitäten, Akteure und Interessen geprägt sind, die jede Bildung von übergreifenden Einheiten oder gar Imperien unterminieren. Zugleich aber konstituieren sie andererseits *allgemeine* Zusammenhänge und Verbindungen in Gestalt von Formen der Konkurrenz und Kooperation, des Aufeinanderprallens wie des wechselseitigen Nutzens bzw. Ausnutzens.[14]

Es ist genau dieser systemische, praktisch generierte, erfahrene und imaginierte, doch gerade nicht politisch-institutionell organisierte Zusammenhang, für den nun die Begriffe ‚Europa‘ und ‚europäisch‘ verwendet werden und der an die Stelle von Vorstellungen einer universellen *christianitas*, einer umfassenden Papstkirche oder imperialer Gebilde wie dem Römischen Reich tritt. Einheit

---

14 ‚Europäisierung‘ und ‚Nationalisierung‘ wie auch andere Formen von Verflechtung und Partikularisierung sind so keine Gegensätze, sondern verschiedene Seiten eines Prozesses. Das Europabewusstsein nimmt gerade in dem historischen Moment zu, „in dem die ‚Nationalisierung‘ der politischen Strukturen und der kulturellen Wahrnehmung mit Entschiedenheit vorangetrieben wurde. Kennzeichnend für die Situation seit dem 16. Jahrhundert ist demnach die Parallelität von europäischem und nationalem Diskurs, von Entgrenzung und Abschottung.“ (Silvia Serena Tschopp. Gegenwärtige Abwesenheit. Europa als Denkmodell im 17. Jahrhundert? In *Europa im 17. Jahrhundert. Ein politischer Mythos und seine Bilder*, Klaus Bußmann, Elke Anna Werner (Hrsg.), 25–36. Wiesbaden: Franz Steiner Verlag, 2004: 26–27).

und Substanz scheinen Konzepte von ‚Europa' für die Zeitgenossen dadurch zu gewinnen, dass es trotz und aufgrund seiner Gegensätze und Konflikte Potentiale und Formen von Macht, Kultur, Wissen und Reichtum produziert und akkumuliert, die als für diesen Kontinent spezifisch und eigentümlich vorgestellt werden.

Genau bei diesen politischen, ökonomischen und soziokulturellen Neuorientierungen und Praktiken also kann man ansetzen, wenn man die Logiken, Mechanismen und Ansprüche des ‚Imperialen' identifizieren will, die für das moderne ‚Europa' prägend werden und den Kontinent wie die Welt gleichermaßen umwälzen sollten. Denn wie wird der Europabegriff im 17. und 18. Jahrhundert politisch und kulturell repräsentiert und aufgeladen? Mit eben jenen oben genannten Strukturen, Praktiken und Rationalitäten der staatlichen und bürokratischen Organisation, mit der vernunftgemäßen Organisation und Selbstbestimmung von Staaten, Gesellschaften und (zumindest seit der Aufklärung zunehmend auch) Individuen, mit der Steigerung individuellen und gesellschaftlichen Reichtums durch Privateigentum und marktwirtschaftlich orientierte Produktion und Handel, mit der systematischen Entwicklung von Wissenschaft und Technik und ihrem praktischen Nutzbarmachen in allen Bereichen politischen, gesellschaftlichen und individuellen Lebens.

Entscheidend bei all diesen Entwicklungen ist nun, dass sie sich, wie angedeutet, von Anfang an mit Bewegungen und Zusammenhängen über den als Europa bezeichneten Kontinent hinaus verbinden und in spezifischer Abgrenzung zu diesem konstituieren. Dieses gerade entstehende ‚Europa' ist also als materiell-praktischer wie als imaginierter Zusammenhang immer schon ‚globalisiert', es ‚existiert' über den Kontinent hinaus und ist auf diese Weise seinerseits intrinsisch von Zusammenhängen und Entwicklungen beeinflusst, die außerhalb dieses Kontinents liegen.[15] Globale Zusammenhänge und Verflechtungen sind folglich keine den materiellen und intellektuellen Entwicklungen auf dem europäischen Kontinent äußerlichen Erscheinungen, sondern Teil seines Konstitutionsprozesses.

Nach den Anfängen der Spanier, Portugiesen und dann der Niederlande entstehen mit England und Frankreich die großen Staaten im europäischen System zugleich als imperiale Mächte, bei denen das globale Ausgreifen konstitutiver Bestandteil der Positionierung im Macht- und Wirtschaftskampf auf dem Kon-

15 Dies zeigt sich exemplarisch etwa am Siebenjährigen Krieg, dem eigentlich ersten Weltkrieg; vgl. Olaf Asbach. Die Globalisierung Europas und die Konflikte der Moderne. Dynamiken und Widersprüche in der Theorie und Praxis der internationalen Beziehungen in der frühen Neuzeit. In *Der Siebenjährige Krieg. Ein europäischer Weltkrieg im Zeitalter der Aufklärung*, Sven Externbrink (Hrsg.), 27–64. Berlin: Akademie Verlag, 2011.

tinent war. Wie die Stellung im europäischen System, so hing also auch umgekehrt die zunehmend imperialen Charakter annehmende Macht einzelner europäischer Staaten, Ökonomien und ihrer Zivilisierungsmissionen seit dem 15. und 16. Jahrhundert von überseeischen Handelswegen und -stützpunkten, von der Eroberung und Ausbeutung von Kolonien und Sklaven sowie von militärischen Kräfteverhältnissen auf den Meeren und auf anderen Kontinenten ab. All diese Prozesse prägten damit auch die materielle und Alltagskultur – von Textilien über Gewürze und Tee bis hin zu Luxusgütern und Moden – zunächst v. a. der Eliten und dann immer breiterer Schichten in Europa.[16]

Im Ergebnis führten diese kontinentalen, dabei stets überregional und global verflochtenen Prozesse zur Ausbildung eines Selbstbildes und Selbstverständnisses als ‚Europäer‘ – eines Selbstbildes, das offenbar mit der Vorstellung von Macht, Reichtum, Wissen und Dynamik assoziiert wurde, ohne sich durch Konflikte, Kriege, Gewalt und Ausbeutung innerhalb Europas wie nach außen irritieren zu lassen. Die praktische und intellektuelle Wahrnehmung der eigenen Partikularität als Besonderheit, der eigenen Macht und Eigenarten in den freundlich-feindlichen Beziehungen der politischen und ökonomischen Akteure innerhalb und außerhalb des Kontinents – auch wenn sie inhaltlich ganz unterschiedlich beschrieben und bewertet wurden – führt zur Konstruktion eines ‚Europa‘ oder ‚Europäischen‘, das etwas Partikulares ist und zugleich universale und imperiale Dimensionen und Implikationen hat.

Das Imperiale des modernen ‚Europa‘ entfaltet sich mithin nicht in Gestalt eines politisch-militärischen Imperiums, sondern als Genese eines neuen, von Europa dominierten Weltsystems, das imperiale Strukturen und Tendenzen besitzt. Diese bestehen in politischen, ökonomischen und kulturellen Strukturen, Rationalitäten und Logiken, die durch eine aktive und permanente materielle wie intellektuelle Auseinandersetzung mit der Welt in Zeit und Raum und ihre Aneignung und Transformation gekennzeichnet sind. Sie sind damit gleichsam intrinsisch dynamisch-expansiv und prägen beständig neue Formen von Herrschaft und Ungleichheit, der Ein- und Ausschlüsse. Dieses Weltsystem beginnt als europäisches, insofern seine zentralen Akteure und Institutionalisierungsweisen, Strukturen und Dynamiken historisch hier zuerst ausgebildet wurden und sowohl die Gesellschaften des Kontinents wie auch die der mit ihm verbundenen außereuropäischen Gebiete zuerst umfasst und inkorporiert haben.

---

16 Wendt macht diese Verflechtungen augenfällig, wenn er in seiner ‚europäischen Weltgeschichte‘ seit 1500 die Entwicklungen und Einflüsse aus den unterschiedlichen regionalen Perspektiven darstellt; vgl. Reinhard Wendt. *Vom Kolonialismus zur Globalisierung: Europa und die Welt seit 1500*. Paderborn: Schöningh, 2007.

Es ist aber, wie sich spätestens seit dem 20. Jahrhundert auch praktisch gezeigt hat, in seinen Strukturen und seinem Funktionieren von Europa unabhängig und mitsamt all seinen Dynamiken und Widersprüchen globalisiert.

4 Imperiale Dimensionen ‚europäischer' Diskurse und Konzepte

Die imperialen Ansprüche und Tendenzen, die im Europa der frühen Neuzeit erkennbar sind, wurden in einer Weise entwickelt und konzeptualisiert, dass sie auch in der Gegenwart noch wirksam sind, und zwar einschließlich ihrer negativen, Ungleichheit und Herrschaft, Widersprüche und Konflikte produzierenden Konsequenzen. Dies lässt sich an drei verschiedenen, genetisch wie systematisch jedoch miteinander verbundenen Formen zeigen, in denen in der frühen Neuzeit die Grundlagen und Dynamiken einer von heterogenen, miteinander konkurrierenden Akteuren und Interessen strukturierten Welt gedacht und organisiert wurden: als völkerrechtlich gestaltbares System freier Akteure (V.1), als über Märkte vermitteltes System freier Wirtschaftssubjekte (V.2) und als durch transnationale Institutionen geregeltes System friedlicher Kooperation und Konkurrenz (V.3). Diese Formen, die Beziehungen innerhalb und zwischen Gesellschaften zu verstehen und zu organisieren, sind besonders wirkmächtig, insofern sie von Anfang an in universelle und ‚rationale' Kategorien gefasst wurden. Dadurch perpetuieren sie zugleich Herrschaftsverhältnisse, Mächte und Kräfte, auch wenn und weil diese nicht mehr mit ‚europäischen' in Verbindung gebracht werden müssen.

4.1 Imperiale Tendenzen des modernen Völkerrechts

Die moderne Staatenwelt und das moderne Völkerrecht haben sich, wie schon bemerkt, in der Theorie wie in der Praxis im Europa der frühen Neuzeit entwickelt und stellen gleichzeitig ein wesentliches Konstitutionsmerkmal dieses ‚Europa' selbst dar. Ob es nun natur-, gewohnheits- oder positivrechtlich begründet wird, ist dieses Völkerrecht seiner Form nach universal und beruht auf der meist auch explizit formulierten Idee einer Gemeinschaft der Völker oder Staaten und eines sie verbindenden Rechts. Es dient der Bestimmung wechselseitiger Ansprüche, der Anerkennung als Rechtssubjekte, der Klärung von Eigentums- und territorialen Rechten, von Rechten im Handel und wirtschaftlichen Verkehr sowie der Explikation rechtlicher ‚Spielregeln' in Krieg und Frieden; Konflikte, Eroberungen und Kriege scheinen so noch zu rechtlich regulier- und berechenbaren Unternehmungen zu werden – bis dahin, dass niemand vernichtet werde, der dies nicht ‚rechtmäßig' verdient habe. Die Setzung, Bestimmung und Inhalte dieses Rechts, seiner Regeln und seines Geltungsbereichs sind jedoch ebenso wie die Frage, wer zu seiner Durchführung befugt ist, von Anfang an – und bis heute – notwendig von den je gegebenen Macht- und Kräfteverhältnissen abhängig. Die Attraktivität und der Erfolg des Völkerrechts ergibt sich dabei gerade nicht nur daraus, dass es internationale Beziehungen und Konflikte in spezifischer Weise organisieren und regulieren kann, sondern vor allem wohl auch daraus, dass es die mit der Setzung und Durchsetzung verbundenen asymmetrischen Machtbeziehungen und Ressourcen unsichtbar zu machen vermag.

Seit der Wende zum 16. Jahrhundert sind Genese und Geschichte des modernen Völkerrechts mit der Expansion europäischer Mächte verbunden, vor allem seit den Natur- und Völkerrechtlern der spanischen Spätscholastik, die sich in direkter Verbindung mit den Rechtsproblemen des entstehenden spanischen Weltreiches entwickelte.[17] Dies lässt sich an den Inhalten, den Problemstellungen und den dort entwickelten Konzepten und Kategorien erkennen. Diskutiert und bestimmt wird etwa, ob und welche Herrschaftsrechte in Übersee nun dem Kaiser, dem Papst oder Einzelstaaten zukommen, ob man die Indios oder andere Völker jenseits der Meere als Rechtssubjekte, als Eigentümer ihres Landes oder überhaupt als Menschen ansehen kann oder ob ein Recht auf freien Handel, die Bearbeitung wüsten Landes oder die Aneignung seiner Früchte durch Arbeit natur- und völkerrechtlich erlaubt ist oder nicht.[18] In solchen Fällen wird gar

17 Dieser Zusammenhang lässt sich bekanntlich auch für die ganze weitere Entwicklung beobachten: Grotius entwickelte das Völkerrecht der Spätscholastiker im Zuge der praktischen Probleme, die sich der neuen imperialen Handelsmacht der Niederlande stellten, weiter, bevor mit dem Aufstieg Englands und Frankreichs vor allem anglo- und frankophone Theoretiker und Praktiker bei der Völkerrechtsentwicklung führend wurden. In der Völkerrechtsgeschichte wird konsequenterweise auch von einer Abfolge der spanischen, französischen, englischen und amerikanischen Zeitalter des Völkerrechts gesprochen; vgl. etwa Wilhelm G. Grewe. *Epochen der Völkerrechtsgeschichte*. Baden-Baden: Nomos, 1984. Die heute diskutierte Ablösung dieses europäisch-westlichen Völkerrechts durch ein universelles oder kosmopolitisches bricht nicht mit dieser Beziehung von Macht und Recht, sondern macht lediglich weniger leicht erkennbar, wie sie jeweils aussieht und welche Interessen warum dominieren.

18 Ein besonders einschlägiges Beispiel ist angesichts des europäischen Interesses an der politischen und ökonomischen Expansion und Extraktion die von Vitoria und Grotius über Locke bis hin zu kolonialismuskritischen Autoren wie Raynal und Diderot diskutierte Frage, ob und unter welchen Bedingungen auf fremden Kontinenten Handel getrieben, Eigentum erworben und vermeintlich niemandem gehörendes Land durch Besetzung oder Arbeit angeeignet werden kann. – Zur Geschichte der Idee der *terra nullius* und ihrer Bedeutung für das moderne Völkerrecht vgl.

nicht erst in Frage gestellt, dass die von den europäischen Mächten behaupteten Ansprüche sowie ihre spezifischen rechtlichen Formen und Regelungsweisen die einzig legitimen und relevanten sind. Ebenso wenig scheint problematisch, dass insbesondere diese Mächte letztlich auch die faktisch entscheidenden Akteure sind, um das universal, für alle Völker der Welt geltende Recht zu konstituieren und durchzusetzen. Ihre global ausgreifende Macht schafft offenbar das globale Recht, in dem politische und rechtliche Institutionen der Durchsetzung ‚natürlicher Rechte' auf freie Verfolgung von individuellen und kollektiven Interessen und der damit verbundenen Eigentumsansprüche festgeschrieben werden.[19]

Die *Partikularität*, die Exklusionseffekte und Konfliktpotentiale, die dieses zunächst vor allem als *Jus Publicum Europaeum* entstandene *universelle* Völkerrecht in sich birgt, sind evident.[20] Und sie liegen auch noch der heute geltenden Völkerrechtsordnung zugrunde: Wer sich der so konstituierten, vermeintlich universellen Ordnung nicht fügt oder gar nicht erst einfügt, ist kein Teil der Rechtsgemeinschaft, kein Rechtssubjekt, sondern ein ‚Outlaw' oder ‚Schurkenstaat', dem man rechtmäßig mit Gewalt beikommen kann und muss, um die durch Macht und Recht gestiftete Welt globalen Völkerrechts zu sichern.[21] Die Geltung und die Mechanismen dieses ‚Imperiums' der europäischen Völkerrechtsordnung bestehen mitsamt ihrer Effekte der Exklusion, der Produktion von Ungleichheit, Konflikten und Gewalt auch heute noch fort, auch wenn sie nicht mehr allein von europäischen oder westlichen Mächten dominiert wird.

Lauren Benton und Benjamin Straumann. Acquiring Empire by Law. From Roman Doctrine to Early Modern European Practice. *Law and History Review* 28 (2010): 1–38.

19 Vgl. Martti Koskenniemi. International Law and the Emergence of Mercantile Capitalism: Grotius to Smith. In *The Roots of International Law / Les fondements du droit international*, Pierre-Marie Dupuy, Vincent Chetail (Hrsg.), 3–37. Leiden u.a.: Nijhoff, 2014: 18 ff.

20 In dieser Hinsicht hat Carl Schmitt mit Recht hervorgehoben, dass die frühneuzeitliche völkerrechtliche Ordnung auf impliziten historisch-konkreten Voraussetzungen beruht und eine „*europa-zentrische Raumordnung*" mit festen Grenzen begründet, die „von dem ‚freien', d. h. der europäischen Landnahme offenen Boden nicht-europäischer Fürsten und Völker unterschieden wird" (Carl Schmitt. *Der Nomos der Erde im Völkerrecht des Jus Publicum Europaeum*. Berlin: Duncker & Humblot, 1950, hier: 120).

21 Vor allem seit dem Ende des Ost-West-Dualismus ist dies bekanntlich die herrschende Legitimationsfigur, mit der je nach Interessenlage die herrschenden Mächte und Koalitionen innerhalb oder außerhalb der UNO – also mit Rückgriff auf positiv oder überpositiv begründete Rechtsansprüche – ‚gerechte Kriege', humanitäre Interventionen und sonstige ‚Strafmaßnahmen' rechtfertigen. – Vgl. zu dieser Kontinuität auch Antony Anghie. *Imperialism, Sovereignty and the Making of International Law*. Cambridge: Cambridge University Press, 2005.

4.2 ‚Imperiale‘ Tendenzen des globalen Systems freier Märkte

In derselben Weise verhält es sich mit den ‚imperialen‘ Tendenzen, die der kapitalistischen, profitgetriebenen Wirtschaftsordnung innewohnen. Dies ist keine zufällige Übereinstimmung, denn beide beruhen auf denselben historischen Grundlagen und Entwicklungen politischer und sozioökonomischer Theorie und Praxis.[22] Auch diese neue Wirtschaftsordnung ist jener Form nach, in der sie seit der Entwicklung der politischen Ökonomie und des liberalen politischen Denkens im 17. und 18. Jahrhundert beschrieben und gerechtfertigt wird, universal; sie beruht auf der Idee jener Gemeinschaft von Individuen, Völkern und anderen Akteuren, die zum Zweck der Produktion und Reproduktion ihrer Lebensgrundlagen miteinander in Verkehr treten. Das Fundament und die Vermittlungsform bestehen demnach im Tausch auf Märkten, auf denen sich Produzenten und Konsumenten, Käufer und Verkäufer von Arbeitskraft, Waren und Leistungen gegenübertreten, um aus dem jeweiligen Eigeninteresse heraus das Interesse der anderen zu verwirklichen.[23] Marktbeziehungen erscheinen so als Sphären vollendeter Freiheit und Gleichheit, reguliert durch den Mechanismus der Preise, die sich aus dem freien Spiel von Angebot und Nachfrage ergeben. Ungleichheit, Macht und Gewalt scheinen diesem Modell ökonomisch induzierter Sozialbeziehungen zufolge systemfremd und äußerlich zu sein, denn es kann als solches nur funktionieren, wenn das freie Spiel der Marktkräfte und der freien Interessenverfolgung der Marktteilnehmer ungehindert wirksam ist. Die Ausbildung des Weltmarktes und des Systems universell freien Handels stellt sich in dieser Perspektive als Gegenbild zu einer von Macht, Gewalt und Staatsraison geprägten Welt dar, als Produkt des *doux commerce*.[24]

---

22 Vgl. für die jüngere ideengeschichtliche Forschung zusammenfassend Koskenniemi. *International Law*.

23 Vgl. die klassische Formulierung, die Adam Smith dieser seit dem 17. Jahrhundert verbreiteten Idee gab: „Jeder, der einem anderen irgendeinen Tausch anbietet, schlägt vor: Gib mir, was ich wünsche, und du bekommst, was du benötigst. [...] Nicht vom Wohlwollen des Metzgers, Brauers und Bäckers erwarten wir das, was wir zum Essen brauchen, sondern davon, dass sie ihre eigenen Interessen wahrnehmen.“ (Adam Smith. *Der Wohlstand der Nationen. Eine Untersuchung seiner Natur und seiner Ursachen*, Horst Claus Recktenwald (Hrsg.), München: Dt. Taschenbuch-Verlag, 1978, 17).

24 Montesquieu etwa sieht eine „natürliche Neigung des Handels [...], zum Frieden geneigt zu machen“, denn „Völker, die miteinander Handel treiben, werden wechselseitig abhängig voneinander“ und durch das Interesse an dem für sie alle vorteilhaften Tausch miteinander verbunden (Charles Louis de Secondat Montesquieu. *Vom Geist der Gesetze*, Ernst Forsthoff (Hrsg.). *Tübingen*: Laupp, 1951, 2 [XX. 2]). Zu diesen Konzeptionen und Debatten vgl. Olaf Asbach. Politik, Handel und internationale Ordnung im Denken der Aufklärung. In *Der moderne Staat und*

Diese Konzeptionen marktwirtschaftlicher Idylle sind freilich in ihren Grundzügen im Zusammenhang mit Entwicklungen im frühneuzeitlichen Europa entstanden, die in der Retrospektive als Teil der Ausbildung dessen erkennbar sind, was man als kapitalistisches Weltsystem bezeichnen kann.[25] Dieses System aber ist auf mehrfache Weise imperial, imperialistisch und mit Gewalt, Herrschaft und Ungleichheit verbunden: historisch-genetisch und systematisch, theoretisch und praktisch, nach innen wie nach außen.

Die kapitalistische, auf die Verwertung von investiertem Kapital gerichtete, über freie Märkte vermittelte Organisation der Prozesse gesellschaftlicher Produktion und Verteilung beruht historisch-empirisch auf Voraussetzungen, die durch außerökonomische Faktoren und Prozesse geschaffen werden mussten und müssen. Dass dabei Macht, Gewalt und Ungleichheit der Kräfteverhältnisse, Formen der inneren und äußeren Landnahme und Expansion, der Umwälzung aller Formen der Produktion, der gesellschaftlichen Lebens- und Normordnungen bis hin zu den individuellen Bedürfnisstrukturen eine konstitutive Bedeutung zukam – und immer noch zukommt –, ist oft hervorgehoben worden.[26] Das kapitalistische Weltsystem entsteht schließlich als partikulare ‚Welt-Wirtschaft' im Sinne Braudels und Wallersteins, die einerseits dynamisch und expansiv ist, andererseits in Form, Ausdehnung und Intensität kontingent und wandelbar, je nach dem räumlichen und strukturellen Ausmaß, in dem Gesellschaften, Produktions- und Lebensweisen in diesen Zusammenhang integriert sind und von ihm bestimmt werden.[27]

Entstanden als *europäisches* und auf die Interessen der europäischen Mächte hin orientiertes Weltsystem, löst es sich im Zuge eben dieser Globalisierungsprozesse langfristig von diesen Grundlagen, ohne darum doch seine partikulare und

‚le doux commerce'. Politik, Ökonomie und internationale Beziehungen im politischen Denken der Aufklärung, Ders. (Hrsg.), 13–36. Baden-Baden: Nomos, 2014.

25 Die Entscheidung darüber, ob von der Etablierung eines von Europa ausgehenden Weltsystems nun ab dem 11. Jahrhundert (Fernand Braudel), ab dem 16. Jahrhundert (Immanuel Wallerstein) oder erst ab dem 19. Jahrhundert (Benno Teschke) gesprochen werden kann, hängt von der Interpretation der Kapitalismus- und Systembegriffe ab, die in diesem Zusammenhang nicht geleistet werden muss.

26 Vgl. hierzu schon Marx' Ausführungen über die „ursprüngliche Akkumulation" in: Karl Marx. *Das Kapital. Kritik der politischen Ökonomie. Erster Band (MEW Bd. 23).* Berlin/DDR: Dietz Verlag, 1983, 751 ff.

27 ‚Welt-Wirtschaft' setzt keine homogene ‚Welt' voraus, sondern Welt entsteht als sozialer – aber darum doch nicht homogener – Raum erst, indem „die wirtschaftlichen Beziehungen die Grenzen der sozialen Welt bestimmen" (Bernd Heiter. Immanuel Wallerstein: Unthinking Culture? In *Kultur. Theorien der Gegenwart*, 2. Aufl., Stephan Moebius, Dirk Quadflieg (Hrsg.), 708–721. Wiesbaden: VS Verlag für Sozialwissenschaften, 2011: 711).

imperiale Struktur zu verlieren. Denn auch als globales System ‚freier‘ Märkte und eines unbeschränkten, am Eigeninteresse der formal unabhängigen Akteure ausgerichteten Austauschs ist dieser Funktions- und Handlungszusammenhang auf doppelte Weise imperial: Imperial ist er einerseits gegenüber allen alternativen Formen gesellschaftlichen Lebens und Wirtschaftens, die dazu führen, dass am Ende die ganze Welt nichts als eine globalisierte Version des ‚modernen Europa‘ ist.[28] Imperial ist dieser ‚systemische Zusammenhang‘ andererseits, indem er die Produktion und Reproduktion von Verhältnissen von Ungleichheit, Macht und Abhängigkeit auf gesellschaftlicher wie auf internationaler Ebene und zugunsten je spezifischer Staaten, Regionen und Akteure befördert, die historisch langfristig wirksame Ungleichheiten erzeugen und perpetuieren.[29]

Dies lässt sich schon bei Adam Smith, dem einflussreichsten der ‚Väter‘ der liberalen politischen Ökonomie, erkennen, der sich bewusst ist, dass die Dynamik freier Märkte stets auch partikulare Interessen stärkt und deshalb Finanzspekulanten, kurzfristige Profitinteressen einzelner Produzenten und die Macht großer Handelskompanien kritisiert.[30] Anders sieht es aus, wenn er die Vor- und Nachteile von Imperien abwägt und zu dem Schluss kommt, dass

28 So liest man selbst bei Kritikern des europäischen Expansionsstrebens wie Raynal und Diderot: „Sanfter und schöner noch ist es aber vielleicht, ganz Europa mit arbeitsamen Nationen bevölkert zu sehen, die ohne Unterlass den Globus umfahren, um ihn nutzbar zu machen und den Menschen anzueignen, um mit dem belebenden Atem des Gewerbefleißes alle nutzbringenden Keime der Natur in Bewegung zu setzen, um den Abgründen des Ozeans und den Eingeweiden der Felsen neue Quellen der Nahrung oder des Vergnügens abzufordern [...]. Die Ursachen eines so glänzenden Wohlstandes werden bald gefunden sein, und die Nationen werden ihre alten Irrtümer und verderblichen Vorurteile ablegen und eilen, Grundsätze zu übernehmen, die so gute Ereignisse zeitigen. Die Umwälzung wird allgemein sein. Überall werden sich die Wolken zerstreuen. Ein heiterer Tag wird den ganzen Globus erleuchten. Die Natur wird wieder die Herrschaft über die Welt übernehmen. Dann oder nie wird jener allgemeine Frieden anbrechen.“ (Guillaume-Thomas Raynal. *Histoire philosophique et politique des Établissemens et du Commerce des Européens dans les Deux Indes*, 4 Bände, IV, 586 und 603 [Übers. O. A.]. Genf: J.L. Pellet, 1780.)

29 Auch wenn in diesem Zusammenhang vor allem auf die *strukturellen* Dimensionen dieses ‚Imperialen‘ eingegangen wird, darf nicht vergessen werden, dass es empirisch als koloniales wie post-koloniales System auf der geschichtlich generierten und akkumulierten Ungleichheit politischer, ökonomischer und soziokultureller Macht und Ressourcen der *europäischen* bzw. *westlichen* Mächte und Akteure beruht. Deren dauerhafte, trotz aller globalisierungsbedingten Anfechtungen immer noch bestehende Dominanz in konkreten Interessenlagen und Konflikten darf bei der Analyse des ‚Systemcharakters‘ des globalen Kapitalismus – wie auch des Völkerrechts oder internationaler Institutionen und Prozesse – nicht ausgeblendet werden.

30 Vgl. Sankar Muthu. Adam Smith's Critique of International Trading Companies: Theorizing ‚Globalization‘ in the Age of Enlightenment. *Political Theory* 36 (2014): 185–212.

informelle, auf freiem Handel basierende ‚Imperien‘ sehr viel effektiver funktionieren als förmliche, die auf direkter, politisch-rechtlich verfasster Herrschaft beruhen. In diesem Sinne empfahl er dem englischen Empire 1776, also inmitten des amerikanischen Unabhängigkeitskrieges, es sei sehr viel vorteilhafter, die amerikanischen Kolonien in die Unabhängigkeit zu entlassen und „mit ihnen einen Handelsvertrag zu schließen, der unbeschränkten Handel wirksam garantieren würde“. Dies nämlich werde die Zuneigung der Kolonisten zum britischen Empire und ihre „Bereitschaft stärken, nicht nur den bei der Loslösung mit uns geschlossenen Handelsvertrag auf Jahrhunderte hin zu respektieren, sondern auch im Krieg wie im Handel auf unserer Seite zu stehen und, anstatt zu unruhigen und aufrührerischen Untertanen, zu unseren treuesten, anhänglichsten und hilfreichsten Bundesgenossen zu werden“.[31] – Smith hat damit auf Aspekte der ‚imperialen‘ Strukturen und Tendenzen des kapitalistischen Weltsystems hingewiesen, die auch jenseits von der bei ihm noch unterstellten ‚natürlichen‘ Dominanz *europäischer* Mächte fortbestehen.

4.3 ‚Imperiale‘ Tendenzen transnationaler Institutionen

Ein dritter Fall, in dem sich die neuzeitliche Narration ‚Europa‘ von Anfang an auch mit imperialen Ansprüchen verbindet, die einerseits ‚europäisch‘ und deshalb ‚partikular‘, andererseits aber ‚universalistisch‘, doch darum nicht weniger ‚partikular‘ sind, sind die seit dem 17. und 18. Jahrhundert entstehenden Ideen, die endlosen Konflikte in den zeitgenössischen internationalen Beziehungen durch die Stiftung gemeinsamer transnationaler Institutionen zu überwinden und auf eine neue politische und rechtliche Grundlage zu stellen. Es wurden Projekte entworfen, die darauf abzielten, die entstehende Staatenwelt auf dem europäischen Kontinent zu einer politisch-rechtlichen Gemeinschaft zusammenzuschließen, um den inneren und äußeren Frieden zu sichern. Solche Bestrebungen waren in dieser Zeit zwar eher randständige Phänomene, doch wurden sie von Gelehrten und *philosophes*, aber auch von vielen Staatsmännern, Diplomaten und Praktikern in Wirtschaft und Verwaltung diskutiert. Dies zeigt einerseits, dass solche Konzeptionen auf grundlegende Probleme und Anforderungen des entstehenden Sozial- und Wirtschaftssystems reagierten. Andererseits wirft der Umstand, dass solche Institutionen in Europa, aber auch darüber hinaus in Theorie und Praxis immer wieder aufgegriffen wurden und seit der Mitte des 20. Jahrhunderts die internationale Ordnung immer stärker prägten, die Frage auf,

31 Smith, *Wohlstand*, 518–519.

inwiefern hier imperiale Tendenzen bestehen, die dafür (mit-)verantwortlich sind, dass dennoch ständig alte und neue Konflikte und Ungleichheiten zu konstatieren sind.

Eines der historisch und systematisch wichtigsten Projekte entwarf seit etwa 1710 der Abbé de Saint-Pierre, der den permanenten Konflikten und Kriegen zwischen den europäischen Staaten dadurch ein Ende bereiten wollte, dass er die Bildung eines Rechts- und Friedensbundes forderte, die er als *Union Européenne* bezeichnete.[32] Auch wenn dieses ‚EU-Projekt‘ insofern exklusiv ist, als es allein die europäischen Staaten umfasst, ist seine Begründung und Struktur zugleich universalistisch und inklusiv. Den Ausgangspunkt bildet nämlich die Erkenntnis, dass eine Welt freier und gleicher Akteure mit je eigenen Willen, Interessen und Ansprüchen prinzipiell instabil und konflikthaft ist, solange keine übergreifende Rechts- und Institutionenordnung existiert. Diese Konflikthaftigkeit besteht nicht, weil er ihnen spezifische Gründe oder Inhalte der Willensbestimmung unterstellen würde – dass sie etwa herrschsüchtig, gierig oder böse seien oder weil sie spezifischen Religionen, Ideologien oder Kulturen anhingen –, sondern aus der Widerspruchsstruktur freier Subjektivität selbst.

Gesicherte Stabilität und Garantie für Existenz, Freiheit und Besitz kann es Saint-Pierre zufolge nicht geben, solange man sich auf ein Gleichgewicht der Kräfte, auf (völker-)rechtliche Prinzipien oder ökonomischen Nutzen und Tauschinteressen verlässt. Denn jeder Akteur ist zu jeder Zeit frei, die eigenen Interessen, Ansprüche und Rechte als verletzt anzusehen und für ihre Durchsetzung oder Wiederherstellung zu Macht- und Gewaltmitteln zu greifen. Wie Individuen, um Rechtssicherheit zu erhalten, eine staatliche Ordnung bilden müssen, die über die Kompetenz und die Mittel zur Setzung und Durchsetzung von Recht und Frieden verfügt, müssen für Saint-Pierre auch Staaten vertraglich ein Gemeinwesen stiften. Gemäß der Theorie des Gesellschaftsvertrages, die er hier auf die internationale Ebene überträgt, soll so eine internationale politische Ordnung entstehen, die auf einem Grundlagenvertrag beruht, welcher als eine Art Verfassung die wichtigsten Prinzipien, Ziele und Einrichtungen festschreibt. Hierzu zählen Prinzipien wie die rechtliche Gleichheit der Mitgliedsstaaten, die Sicherung der Freiheit im Rahmen der vom Bund gegebenen Gesetze, die Öffent-

32 Ausführlich zu Saint-Pierre und seiner Stellung im politischen und völkerrechtlichen Denken vgl. Olaf Asbach. *Die Zähmung der Leviathane. Die Idee einer Rechtsordnung zwischen Staaten bei Abbé de Saint-Pierre und Jean-Jacques Rousseau.* Berlin: Akademie Verlag, 2002. Aus Platzgründen wird im Folgenden auf Belege aus Saint-Pierres Werken verzichtet; vgl. hierzu Olaf Asbach. Zur politischen Ökonomie des Friedens. Handel, Staat und internationale Ordnung beim Abbé de Saint-Pierre. In *Der moderne Staat*, Ders. (Hrsg.), 39–68. Baden-Baden: Nomos, 2014; die folgenden Passagen greifen hierauf zurück.

lichkeit und die Allgemeinheit von Gesetzen und Verträgen oder die Garantie von Eigentum und territorialer Integrität. Als zentrales Organ des Bundes fungiert ein Bundesrat, in dem die Mitgliedsstaaten Sitz und Stimme haben.

Zugleich zeigen schon diese wenigen Hinweise die historisch-gesellschaftliche Spezifik und Bedingtheit der so bestimmten Problemlage und Lösungskonzepte, denn sie thematisieren ganz offenbar in abstrakter Form eben jene Strukturen und Probleme, die das frühneuzeitliche Europa geprägt haben. Zum einen reagiert diese Konzeption auf die expansive Dynamik des frühneuzeitlichen Mächte- und Staatensystems: das Projekt der *Union européenne* soll die Strukturprobleme jener ‚Westfälischen Ordnung' souveräner Staaten lösen, die die internationale Ordnung Europas und zunehmend auch die der Welt wesentlich prägten. Zum anderen reagiert Saint-Pierres Konzeption offensichtlich auf die damit zusammenhängende Herausbildung der (früh-)kapitalistischen Wirtschaftsakteure und -systeme, die dynamisch und expansiv sind, bestimmt von der Konkurrenz um Märkte und Handelsrouten, Profite und Ressourcen, dabei Ungleichheit und Ausbeutung produzierend und dadurch prinzipiell konflikthaft.[33]

Die transnationale Rechts- und Friedensgemeinschaft Europas ist somit ein politisch-rechtliches Institutionensystem, das genau dieser historisch spezifischen Konstellation von politisch, sozial und ökonomisch prinzipiell kompetitiven Akteuren und Handlungsstrukturen entspringt und auf sie zugeschnitten ist. Zentral dabei ist: sie soll diese Konkurrenzbeziehungen nicht aufheben, sondern auf Dauer stellen. Sie zielt auf die Sicherung der Handlungsfreiheit von Individuen wie von Staaten im Prozess ihrer politischen und ökonomischen Interessenverfolgung; der Freiheits- und Souveränitätsverzicht der Individuen *im* Staat und der Staaten im internationalen System erfolgt zu eben diesem Zweck und unter der Bedingung, diese zu ermöglichen. Die Funktion politischer und rechtlicher Institutionen auf staatlicher wie europäischer Ebene ist die Sicherung von *life, liberty and estate*, d. h. der Freiheit von Eigentum, Produktion, Handel und der Möglichkeit der Akkumulation von allen Arten materieller, kultureller und ideeller Güter von Individuen, Unternehmen und Staaten.

Die *inhaltlichen* Aufgaben und Kompetenzen, die Saint-Pierre diesem Bund zuschreibt, zielen deshalb, sozusagen klassisch liberal, auf *marktschaffende* und *marktöffnende* Maßnahmen. Hierzu zählt als erstes das Erlassen von Handels-

33 Zu diesen politisch-ökonomischen Grundlagen und Zielen vgl. Asbach, *Zur politischen Ökonomie des Friedens* und Ders. Interests, Markets and the Modern Spirit of Institutions. The Totalization of the Principle of Competition in Early French Enlightenment. In *Les idées de l'abbé Castel de Saint-Pierre*, Simona Gregori, Patrizia Oppici (Hrsg.), 203–220. Macerata: eum, 2014.

gesetzen, die für alle gleich und verbindlich sind und den freien Warenverkehr festschreiben. Hierzu zählt zweitens die Etablierung eines gemeinsamen Marktes durch Aufhebung von Zöllen, Monopolen und Handelsbeschränkungen, so dass der Handelsverkehr unbeschränkt ablaufen und sich überall der günstigste Anbieter durchsetzen kann, was zum größten Nutzen aller sei. Und drittens zählt hierzu die Schaffung von Infrastrukturen für den ständig wachsenden Handel: das reicht von der Einführung gleicher Münzen, Gewichte und Maße über den Bau von Straßen und Kanälen für den internationalen Handelsverkehr bis zur Förderung einer einheitlichen europäischen Verkehrssprache.

Dieses ‚Europa‘ wird von Saint-Pierre dabei jedoch, wie schon gesagt, nicht partikular und exklusiv, sondern global und inklusiv verstanden, denn seine Prinzipien gelten ihm als universell gültige der modernen Welt. Die Etablierung transnationaler Institutionen und Räume des Friedens, der Rechtssicherheit, der Freiheit des Handels und generell des Strebens nach der Verwirklichung individueller Interessen setze eine Dynamik frei, die materiellen Wohlstand, individuellen Nutzen und dauerhafte Prozesse kulturellen und geistigen Fortschritts generiere. In Saint-Pierres Schriften lassen sich unterschiedliche Varianten erkennen, wo und wie diese politisch-institutionellen Rahmenbedingungen angesiedelt werden könnten: auf globaler Ebene als Schaffung einer globalen Rechts- und Friedensorganisation, auf regionaler Ebene durch die Schaffung einer Reihe von politisch verfassten Großräumen – neben der europäischen nennt er eine ‚Asiatische Union‘ und eine ‚Afrikanische Union‘ –, oder eben durch die Etablierung einer *Union européenne*, die mächtig genug ist, ihre Interessen nach innen und außen durch politische, diplomatische, militärische, rechtliche und ökonomische Maßnahmen zu verwirklichen.

Die eurozentrische und imperiale Ausrichtung dieser Konstruktion ist evident.[34] Sie wird jedoch stets universalistisch begründet, da die politischen,

34 Denn es ist gleichgültig, ob die europäischen Mächte ihre Interessen militärisch, diplomatisch oder durch rechtliche Kooperationen und Institutionen zur Geltung bringen, solange nur *ihre* Interessen gesichert werden. Man vergleiche dies mit den aufschlussreichen Bemerkungen eines frühen Verfechters der Wende der USA hin zu einer aktiv-imperialen Außenpolitik um 1900: In der Debatte über die Alternativen einer offen kolonialen oder einer informellen, auf militärische Stützpunkte, ökonomische, rechtliche und andere Einflusskanäle setzenden Strategie erklärte Charles A. Conant 1898, dies sei eine „Detailfrage“, da sie nur unterschiedliche Mittel zum Erreichen desselben Zieles betreffe, dass nämlich „die Vereinigten Staaten ihr Recht auf freien Handel mit all den alten Ländern geltend machen, die für die überschüssigen Ressourcen der kapitalistischen Staaten geöffnet werden und so in den Genuss der Vorzüge der modernen Zivilisation kommen.“ (Zit. n. James Tully. *Politische Philosophie als kritische Praxis*. Frankfurt a. M. und New York: Campus-Verlag, 2009, 155).

sozioökonomischen und kulturellen Strukturen der europäischen Staaten ganz selbstverständlich als Grundlage und Telos der globalen Ordnung verstanden werden. Dies wird augenfällig, wenn Saint-Pierre erklärt, die *Union européenne* müsse das betreiben, was man heute eine gemeinsame Außen- und Sicherheitspolitik nennt, um ihre politischen und wirtschaftlichen Interessen auch durch politische und rechtliche Vertretungen wie auch Truppen außerhalb Europas zu sichern. Eine solche globale Ordnungspolitik der europäischen Mächte nämlich sei aufgrund der Interdependenzen und Interessengleichheit aller Menschen und Völker schließlich zum Vorteil aller. Völker und Staaten anderer Kontinente würden die Aktivitäten und Präsenz europäischer Mächte bei ihnen deshalb auch nur begrüßen, denn

> sie werden sicher sein, dass der europäische Bund allein auf die Sicherheit seines Handels bedacht ist, dass dieser Handel für sie selbst äußerst vorteilhaft sein wird, dass der Bund niemals daran denkt, irgendwelche Eroberungen zu machen und niemals einen anderen als Feind betrachten wird als die Feinde des Friedens.[35]

Schlussbemerkung

Das ‚Imperiale' des frühneuzeitlichen ‚Europa' wird so als etwas erkennbar, was nicht einfach die Fortführung überkommener Imperien, sondern neuartig und erheblich komplexer ist. Das ‚Imperiale' der beschriebenen Entwicklungen und Konzeptionen des Völkerrechts, des entstehenden kapitalistischen Weltsystems wie auch einer europäischen oder selbst einer globalen Rechts- und Friedensordnung besteht nicht einfach – oder zumindest nicht nur – im Anspruch auf die Etablierung eines *europäischen* Imperiums.[36] Es zeigt sich insbesondere in der Tendenz, als Ausdruck von Prinzipien, Strukturen, Institutionen und Interessen

---

35 Charles-Irenée abbé de Saint-Pierre. *Projet pour rendre la paix perpétuelle en Europe*, Bd. I. Utrecht: Schouten, 1713: xx f. (Übers. O. A.).

36 Es kann hier nur darauf hingewiesen, aber nicht mehr diskutiert werden, dass genau 300 Jahre später aktuell erneut der Versuch gemacht wird, die Konzepte von EU, Friedensordnung und Empire als Garant regionaler und globaler Sicherheit sowie Wohlstandes und Zivilität generierenden Freihandels erneut miteinander zu verklammern und sogar offensiv in die Praxis umzusetzen; vgl. mit unterschiedlicher Ausrichtung Robert Cooper. *Order and Chaos in the Twenty-First Century*. London: Atlantic Books, 2003; Ulrich Beck und Edgar Grande. Empire Europa: Politische Herrschaft jenseits von Bundesstaat und Staatenbund. *Zeitschrift für Politik* 52 (2005): 397–420; Alan Posener. *Imperium der Zukunft. Warum Europa Weltmacht werden muss*. München: Pantheon Verlag, 2007.

aufzutreten, die keine spezifisch europäischen, sondern universale Ansprüche vertreten sollen und so die Verhältnisse auf dem Kontinent wie jenseits dessen in gleicher Weise betreffen und umwälzen. Hier dominiert noch die unmittelbare Überzeugung, dass die politischen und kulturellen Prinzipien der europäischen Welt und die dabei zentrale, auf der Freisetzung partikularer Interessen beruhende Dynamik von Produktion für freie Märkte auch tatsächlich universell gültig sind, keinesfalls aber, sofern nur angemessen realisiert, zu einer Perpetuierung von Ungleichheit, Herrschaft und Konflikten führen. Die Globalisierung der ‚europäischen' Lebens- und Denkformen erscheint daher insgesamt als ein universelles und kosmopolitisches Projekt, innerhalb dessen diejenigen, die sich ihm widersetzen, nur Feinde des Friedens, des Fortschritts und der Freiheit sein können und müssen – und die deshalb auch zu ihrem eigenen Besten gezwungen werden müssen, sich dieser Ordnung und den in ihr geltenden Prinzipien anzuschließen.[37]

Es ist sicher nicht falsch, diese Ideen und Ideologien als Ausdruck eines typisch eurozentrischen Denkens des Aufklärungszeitalters zu verbuchen. Es ist jedoch kaum zu übersehen, dass dies zugleich eine Argumentation und Denkfigur – und auch eine Praxis – ist, die gerade in der Gegenwart in immer stärkerem Maße wirksam ist. Sie findet geradezu ubiquitär Verwendung und wird regelmäßig zur Verteidigung von – wahlweise: europäischen, westlichen, universellen – Prinzipien, Werten, Rechten und den damit verbundenen Ansprüchen an politische, gesellschaftliche und kulturelle Institutionen vorgebracht – nicht zuletzt von sich aufgeklärt wähnenden Kritikern des modernen Eurozentrismus. Es ist also dringend notwendig, die imperialen Implikationen, die Partikularität und die konkreten Formen von Ungleichheit und Herrschaft bewusst und sichtbar zu machen, die den universalistischen Prinzipien und Institutionen innewohnen, die im Europa der frühen Neuzeit ihren Ausgang nahmen und sich bis heute reproduzieren.[38] Nur dann besteht die Aussicht, sie zum Gegenstand theoretischer und praktischer Auseinandersetzungen zu machen und die Hoffnung auf ihre Überwindung aufrechtzuerhalten.

---

37 Tully hält aktuellen Vertretern einer kosmopolitischen Weltordnung oder einer „von Europa vorangetriebene[n] globale[n] Verrechtlichung dieser asymmetrischen Institutionen der gegenwärtigen internationalen Ordnung und eine[r] Neuordnung der Weltregionen nach dem Bilde der EU" entgegen, sie seien am Ende „nicht weniger imperialistisch und antidemokratisch" als die Ordnung, gegen die sie sich wenden (Tully, *Politische Philosophie als kritische Praxis*, 245).

38 Denn dies vermeintlich ‚Universelle' wird stets in konkreten Konstellationen von konkreten Akteuren mit spezifischen Ressourcen und Interessen realisiert und produziert so immer auch Ausschlüsse, Verlierer, Widerstand und Gewalt.

Literaturverzeichnis

Anghie, Antony. *Imperialism, Sovereignty and the Making of International Law*. Cambridge: Cambridge University Press, 2005.

Asbach, Olaf. Die Globalisierung Europas und die Konflikte der Moderne. Dynamiken und Widersprüche in der Theorie und Praxis der internationalen Beziehungen in der frühen Neuzeit. In *Der Siebenjährige Krieg. Ein europäischer Weltkrieg im Zeitalter der Aufklärung*, Sven Externbrink (Hrsg.), 27–64. Berlin: Akademie Verlag, 2011.

Asbach, Olaf. *Die Zähmung der Leviathane. Die Idee einer Rechtsordnung zwischen Staaten bei Abbé de Saint-Pierre und Jean-Jacques Rousseau*. Berlin: Akademie Verlag, 2002.

Asbach, Olaf. *Europa – Vom Mythos zur ‚Imagined Community'? Zur historischen Semantik ‚Europas' von der Antike bis ins 17. Jahrhundert*. Hannover: Wehrhahn, 2011.

Asbach, Olaf. Interests, Markets and the Modern Spirit of Institutions. The Totalization of the Principle of Competition in Early French Enlightenment. In *Les idées de l'abbé Castel de Saint-Pierre*, Simona Gregori, Patrizia Oppici (Hrsg.), 203–220. Macerata: eum, 2014.

Asbach, Olaf. Politik, Handel und internationale Ordnung im Denken der Aufklärung. In *Der moderne Staat und ‚le doux commerce'. Politik, Ökonomie und internationale Beziehungen im politischen Denken der Aufklärung*, Ders. (Hrsg.), 13–36. Baden-Baden: Nomos, 2014.

Asbach, Olaf. Zur politischen Ökonomie des Friedens. Handel, Staat und internationale Ordnung beim Abbé de Saint-Pierre. In *Der moderne Staat und ‚le doux commerce'*, Ders. (Hrsg.), 39–68. Baden-Baden: Nomos, 2014.

Beck, Ulrich und Grande, Edgar. Empire Europa: Politische Herrschaft jenseits von Bundesstaat und Staatenbund. *Zeitschrift für Politik* 52 (2005): 397–420.

Benton, Lauren und Straumann, Benjamin. Acquiring Empire by Law. From Roman Doctrine to Early Modern European Practice. *Law and History Review* 28 (2010): 1–38.

Cooper, Robert. *Order and Chaos in the Twenty-First Century*. London: Atlantic Books, 2003.

Darwin, John. *After Tamerlan. The Global History of Empire since 1405*. London u.a.: Allen Lane, 2007.

Eytzinger, Michael. *De Europa virginis tauro insidentis topographica*, Köln: Gottfried von Kempen, 1588.

Grewe, Wilhelm G. *Epochen der Völkerrechtsgeschichte*. Baden-Baden: Nomos, 1984.

Hale, John. *Die Kultur der Renaissance in Europa*. München: Kindler, 1994.

Heiter, Bernd. Immanuel Wallerstein. Unthinking Culture? In *Kultur. Theorien der Gegenwart*, 2. Aufl., Stephan Moebius, Dirk Quadflieg (Hrsg.), 708–721. Wiesbaden: VS Verlag für Sozialwissenschaften, 2011.

Howe, Stephen. *Empire. A very short Introduction*. Oxford: Oxford University Press, 2002.

Jouannet, Emmanuelle. *The Disappearance of the Concept of Empire. Or, the Beginning of the End of Empires in Europe from the 18th Century*. [http://www.univparis1.fr/fileadmin/IREDIES/Contributions_en_ligne/E._JOUANNET/The_Disappearance_of_the_Concept_of_Empire_E._Jouannet-2.pdf, abgerufen: 17.9.2014].

Koskenniemi, Martti. International Law and the Emergence of Mercantile Capitalism: Grotius to Smith. In *The Roots of International Law / Les fondements du droit international*, Pierre-Marie Dupuy, Vincent Chetail (Hrsg.), 3–37. Leiden u.a.: Nijhoff, 2014.

Leitner, Ulrich. *Imperium. Geschichte eines politischen Systems*. Frankfurt a.M. und New York: Campus Verlag, 2011.

Marx, Karl. *Das Kapital. Kritik der politischen Ökonomie. Erster Band (MEW Bd. 23)*. Berlin/DDR: Dietz Verlag, 1983.

Montesquieu, Charles Louis de Secondat. *Vom Geist der Gesetze*, Ernst Forsthoff (Hrsg.). Tübingen: Laupp, 1951.

Münkler, Herfried. Imperialismus/Imperium. In *Politische Theorie und Politische Philosophie. Ein Handbuch*, Martin Hartmann, Claus Offe (Hrsg.), 98–103. München: C.H. Beck, 2011.

Muthu, Sankar. Adam Smith's Critique of International Trading Companies: Theorizing ‚Globalization' in the Age of Enlightenment. *Political Theory* 36 (2014): 185–212.

Oschema, Klaus. *Bilder von Europa im Mittelalter*. Ostfildern: Thorbecke, 2013.

Osterhammel, Jürgen. Imperien. In *Transnationale Geschichte. Themen, Tendenzen und Theorien*, Gunilla Budde u.a. (Hrsg.), 56–67. Göttingen: Vandenhoeck & Ruprecht, 2006.

Pitts, Jennifer. Republicanism, Liberalism, and Empire in Postrevolutionary France. In *Empire and Modern Political Thought*, Sankar Muthu (Hrsg.), 351–387. Cambridge: Cambridge University Press, 2012.

Posener, Alan. *Imperium der Zukunft. Warum Europa Weltmacht werden muss*. München: Pantheon Verlag, 2007.

Raynal, Guillaume-Thomas. *Histoire philosophique et politique des Établissemens et du Commerce des Européens dans les Deux Indes*, 4 Bände, Genf: J.L. Pellet, 1780.

Saint-Pierre, Charles-Irenée, abbé de. *Projet pour rendre la paix perpétuelle en Europe*, Bd. I. Utrecht: Schouten, 1713.

Schmitt, Carl. *Der Nomos der Erde im Völkerrecht des Jus Publicum Europaeum*. Berlin: Greven Verlag, 1950.

Smith, Adam. *Der Wohlstand der Nationen. Eine Untersuchung seiner Natur und seiner Ursachen*, Horst Claus Recktenwald (Hrsg.), München: Dt. Taschenbuch-Verlag, 1978.

Tschopp, Silvia Serena. Gegenwärtige Abwesenheit. Europa als Denkmodell im 17. Jahrhundert? In *Europa im 17. Jahrhundert. Ein politischer Mythos und seine Bilder*, Klaus Bußmann, Elke Anna Werner (Hrsg.), 25–36. Wiesbaden: Franz Steiner Verlag, 2004.

Tully, James. *Politische Philosophie als kritische Praxis*. Frankfurt a.M. und New York: Campus-Verlag, 2009.

Wendt, Reinhard. *Vom Kolonialismus zur Globalisierung: Europa und die Welt seit 1500*. Paderborn: Schöningh, 2007.

Abbildungsnachweise

Abb. 1: Franz Hogenberg. *Europa auf dem Stier*. In *De Europae virginis tauro insidentis, topographica atque historica descriptione liber*, Michael Eytzinger (Hrsg.). Köln 1588, 6. [http://digital.onb.ac.at/OnbViewer/viewer.faces?doc=ABO_%2BZ178291704 (12.07.2016)]

Abb. 2: Cesare Ripa. *Iconologia of Uytbeeldinghen des Verstants*. Amsterdam 1644, 602–605. [http://www.dbnl.org/tekst/pers001cesa01_01/pers001cesa01_01_1185.php (12.07.2016)]

Elizabeth Millán

Alexander von Humboldt's Interest in America: In the Service of Empire or of Humanity?[1]

To quickly reach the heart of darkness created by empire building, we may begin with the legacy of an empire that continues to exert its shadow over several regions today: I refer to the Catholic Empire of Spain and the colonization of its territories in America. The labor and subordination of non-European groups in Latin America created Spain's wealth during the colonial period, that is, from the time of the conquest in 1492, until the independence movements that liberated the colonial regions and led to the creation of sovereign states (beginning in the 1780s). In her study, *Imperial Eyes. Travel Writing and Transculturation*, Mary Louise Pratt presents a portrait of Alexander von Humboldt (1769–1859) that is most relevant to the themes of how empire shapes political work and our very understanding of and appreciation for other cultures. A particular chapter, "Alexander von Humboldt and the Reinvention of América," examines Humboldt's American writings "in their relation to prior paradigms of travel literature and to European ambitions in the region."[2] Pratt's focus is on "the relations between travel writing and processes of European economic expansion,"[3] that is, Pratt's focus is on empire building and Humboldt's role therein. As Pratt writes:

> The end of Spanish colonial rule entailed a full-scale renegotiation of relations between Spanish America and Northern Europe – relations of politics and economics, and with equal necessity, relations of representation and imagination. Europe had to reimagine América, and América, Europe. The reinvention of América, then, was a transatlantic process that engaged the energies and imaginations of intellectuals and broad reading publics in both hemispheres, but not necessarily in the same ways. For the elites of Northern Europe, the reinvention is bound up with prospects of vast systems of knowledge. The newly independent elites of Spanish America, on the other hand, faced the necessity for *self*-invention in relation both to Europe and the non-European masses they sought to govern. One can

1 With thanks to Bärbel Frischmann, Holt Meyer, Susanne Rau, and Katharina Waldner for their generous invitation to participate in the conference, *Space-Time of the Imperial*, at the University of Erfurt from October 8–October 11, 2014, an intellectual event that provided much productive discussion, collaboration, and pleasure.

2 Mary Louise Pratt. *Imperial Eyes. Travel Writing and Transculturation*. London: Routledge, 1992, 113.

3 Pratt, *Imperial*, 112.

DOI 10.1515/9783110418750-011

> only be fascinated, then, that the writings of Alexander von Humboldt provided founding visions to both groups.[4]

Several claims made above merit closer scrutiny: how do politics and economics merge (if at all) in Humboldt's work? Why is a concern with and representation of America necessarily bound to the political and economic? Pratt emphasizes the "elites" amongst the audience of Humboldt's writings – in general, her account of Humboldt overlooks the democratic goals he had in mind as he prepared his work for a broad audience. In the context of a discussion of Humboldt's role in the "invention" or "reinvention" (I prefer the more neutral term, "presentation") of America, we also need to look more carefully at what is meant with the reference to the "vast systems of knowledge" with which Pratt associates such an invention. Despite the many valuable insights to be found in her work on Humboldt and travel writing, especially within the context of postcolonial studies, Pratt scripts Humboldt's project too narrowly, and too ideologically. Before turning to the details of Humboldt's work and some of the problems with Pratt's reading of that work, some context is in order.

1 The Vision Guiding Humboldt's Voyage: Imperial Eyes or the Eyes of a Humanist?

Humboldt worked for years in the service of the Prussian government, reaching the position as inspector of mines (hence, we should not be surprised that Humboldt maintained a knack for uncovering subterranean connections in his work),[5] a position he carried out with the humanity that characterizes his work. Early on, he became appalled by the dangerous conditions in the mines and worked to educate the miners in such a way that their workplace would become safer

4 Pratt, *Imperial*, 112.

5 I make this connection inspired by Kutzinski and Ette's compelling claim, that, "[r]ather than providing closure,[the last endnote of Humboldt's *Views of the Cordilleras* text, in referring back to the very first plate] serves as an invitation for rereading what is now a familiar text to discover new associations. That many of these connections are subterranean, as it were, might be expected from this former mining inspector, who was eminently familiar with galleries below the surface." Vera Kutzinski and Ottmar Ette. The Art of Science: Alexander von Humboldt's Views of the Cultures of the World. In: Alexander von Humboldt. *Views of the Cordilleras and Monuments of the Indigenous Peoples of the Americas: A Critical Edition*, Vera Kutzinski and Ottmar Ette (eds.), xv–xxxv. Chicago: University of Chicago Press, 2012 [1810], xxxiv.

for them. In 1796, upon the death of his mother, a family inheritance enabled him to leave his government service and embark upon his voyage to America (1799–1804). With the French botantist, Aimé Bonpland, Humboldt left from La Coruña, Spain on June 4, 1799, spending a few days on the Canary Islands, and then leaving for South America, arriving first in Cumaná, Venezuela, exploring the Orinoco River, then travelling to Cuba, Colombia, Peru, Mexico, back again to Cuba and then for a short stay in the United States of America, where Humboldt met with Thomas Jefferson, urging him to do more to eradicate slavery. Humboldt and Bonpland returned to Europe on August 8, 1804, arriving to a warm reception in Bordeaux.

Humboldt's longing for an international voyage had been awakened in him during a trip with Georg Forster in 1790, during which they traveled up the Rhine to the Netherlands, France, and England. Forster had sailed with Captain Cook on Cook's second voyage, 1772–1775. Forster remained an influence upon Humboldt long after his (Forster's) early death in Paris in 1794 (Forster became a Jacobin and was exiled from his home city of Mainz). Humboldt gives Forster the credit for having inspired his work, *Essay on the Geography of Plants* (1807). In the preface to this volume he writes:

> I conceived of this book during my earliest years. I gave a first sketch of a *Geography of Plants* in 1790 to Cook's famous colleague, Mr. Georges Forster, with whom I had close ties of friendship and gratefulness. My later research in various areas of physics helped me reach a wider understanding of my initial ideas. My voyage in the tropics furnished me with precious materials for the physical history of the earth. I wrote the major part of this work in the very presence of the objects I was going to describe, at the foot of the Chimborazo, on the coasts of the South Sea.[6]

Although grateful for the inspiration behind his travel plans, Humboldt lamented the unlucky timing of those plans: The French Revolution and its aftermath created many hurdles for him as he attempted to get away from Europe to explore the rest of the world. First, Humboldt had plans for a voyage to Egypt, but Napoleon's invasion of Egypt thwarted those plans, and then a war and Franco-Prussian politics interfered with other travel plans. Finally, in 1799 Humboldt and Bonpland were stranded in Marseilles and set off to Madrid, hoping to get permission to travel to the Spanish colonies. As Pratt writes: "Perhaps Charles IV hoped Humboldt and Bonpland would help him regain control of his restless colonies. Certainly he was eager to make use of Humboldt's mining expertise, and

6 Alexander von Humboldt and Aimé Bonpland. *Essay on the Geography of Plants*, Stephen T. Jackson (ed.), Sylvie Romanowski (trans.). Chicago: University of Chicago Press, 2009 [1807], 61.

asked him to report back in particular on mineralogical findings."[7] Charles IV may have hoped to find in Humboldt a loyal servant, but permission to enter the Spanish territories in America was not tantamount to employment by the Spanish Crown: Humboldt maintained financial independence from Spain throughout his travels, the only thing granted to him by Spain was a passport for travel. There is no evidence in his writings (some of which were quite critical of Spain) or in what he did with his vast botanical collections (nothing was sent to Spain, but rather to France, Prussia, and even England), to suggest any desire on Humboldt's part to serve the Catholic Empire of Spain. Pratt reads Humboldt's work in the light of his "service" to Charles IV and the political-economic purposes of empire. I shall make a case for a different reading of Humboldt's American voyage, arguing that Humboldt's writings on Latin America did not serve the Spanish crown or the colonial structures of power, but rather served an aesthetic project that was politically liberating.

I have some company in this move to rescue Humboldt from Pratt's reading of him as "one more 'imperial eye.'"[8] As Laura Dassow Walls writes:

> Pratt [insists] that Humboldt invented America as "primal" nature, emptied of human history in order, in the classic imperial mode, to repopulate it with white European systems and goals. Here Humboldt becomes one more "imperial eye," handmaiden to colonial domination, blind to the Other and full of himself, omniscient and godlike lord of all he surveys. Her highly selective interpretation, driven by a need to privilege binary oppositions rather than pluralistic differentiations, has become canonical in postcolonial studies. The effect has been to silence Humboldt all over again. True, Humboldt cannot step very far outside the networks of colonial power; even his attempts to do so – if one is willing to grant they were more than mere self-delusion – ultimately made it stronger...However, to deny him the agency to recognize, protest, and on occasion even subvert those networks is to deny the moral reach of his arguments – worse, of anyone's arguments...There were, to be sure, plenty of "imperial eyes" stalking the New World, but Humboldt's project was different, and that difference matters. To begin with, he lacked imperial sponsorship...But perhaps most important, he felt (and that is precisely the right word) that nature without humanity lacked meaning.[9]

The typical imperial writer of the period would indeed present nature in the Eurocentric, anti-American mold, where all in Europe is bigger, better, more civilized. As Dassow Walls points out, Humboldt could not escape his time, that is, he could not step out of the networks of colonial power, but that does not entail that

7 Pratt. *Imperial*, 116.

8 Laura Dassow Walls. *The Passage to Cosmos. Alexander von Humboldt and the Shaping of America*. Chicago: University of Chicago Press, 2009, 19.

9 Dassow Walls, *Passage*, 19–20.

he worked in the service of the structures of colonial power. Indeed, as Vera Kutzinski and Ottmar Ette point out in, "The Art of Science: Alexander von Humboldt's Views of the Cultures of the World," while Humboldt "remained indebted to the idea of Europe as a cultural and political center ... he quite explicitly thematized his Eurocentrism, always turning it into an occasion for critical self-reflection."[10] Humboldt's critical self-reflection was key to his move away from "imperial eyes" – for with this critical self-reflection, Humboldt, as Kutzinski and Ottmar point out, finds a new way to order the cultures of the world. If we look more carefully at his presentation of the Latin American landscape, we see that it is indeed the case that for Humboldt "nature without humanity lacked meaning." In what follows, I shall attempt to uncover the reasons why for a thinker like Humboldt, nature is meaningless without humanity, and further, why Humboldt's presentation of the Latin American landscape was liberating precisely because it helped to wear away some typical structures of imperial power.

Humboldt's writings on Latin American culture and landscape are voluminous and I will not attempt to cover all of them here. I shall focus upon *Ansichten der Natur* (1808), *Relation Historique/Voyage to the Equinoctial Regions of the New Continent* (1805–1838) (which included the Political Essays on the Kingdom of New Spain (1811) and the Island of Cuba (1826)), the *Vues des Cordillères et Monumens des Peuples Indigènes de l'Amérique/Views of the Cordilleras and Monuments of the Indigenous Peoples of America* (1810–1813) and *Cosmos* (1845–1858 and 1862), for in these texts we find patterns that characterize his American writings.

Humboldt took an important stance on behalf of the "New World" in the "Dispute concerning the New World," which was largely an attempt by European thinkers to continue the pernicious barbarism/civilization dichotomy that had justified exploitation of Native Americans for centuries. According to the points established by this "Dispute," America and its inhabitants were inferior in all ways, and needed the aid of the more civilized Europeans to be rescued from their barbaric condition. The rescue plan was scripted by Juan Ginés de Sepúlveda (1490–1573), a leading ideologue of the *Conquista*, who claimed the natives were "by nature" slaves (invoking Aristotle to justify the enslavement of the natives). Perhaps the darkest chapter of Spain's imperialism is to be found in the deranged logic of thinkers such as Sepúlveda, a logic which fueled the extermination and enslavement of the Native Americans. If slavery is presented as the path to freedom, something is, as Shakespeare's Marcellus would say, rotten in the state of Denmark. Humboldt gave us a new lens through which to view America, a lens

10 Kutzinski and Ette, *Art of Science*, xix.

not darkened by the shadow of a distinction that trapped the inhabitants into the confines of Europe's barbaric "Other." As Kutzinski and Ette point out:

> Humboldt always called attention to the barbarity within civilization itself, notably to the barbaric aspects of so-called Western civilization, in full awareness that presumably gentle and peaceful Christianity got along rather well with colonialism and slavery in his day. He distanced himself from these implications of the civilizational process, adopting a critical attitude toward the myth of Western progress.[11]

In *Views of the Cordilleras and Monuments*, Humboldt himself observes that the people in "the holy city of Mexico" are far more civilized than some of the barbaric "systematic ideas" used by the Europeans to describe them would have us believe:

> This people, who based their festivals on the movements of the stars and who engraved their celebrations on a public monument, had likely reached a higher level of civilization than that accorded to them by DePauw, Raynal, and even Robertson, the most judicious of all the historians of the Americas. These authors regard as barbarous any state of humanity that diverges from the notion of culture that they have established based on their own systematic ideas. We simply cannot accept such sharp distinctions between barbarous and civilized peoples.[12]

In these lines condemning the civilization-barbarism dualism that trapped the indigenous within the confines of the barbaric, all by way of dubious "systematic ideas," we find an important glimpse of the promising new lens Humboldt used to investigate America: a lens free of the "vast systems of knowledge" referenced by Pratt and of the pernicious "systematic ideas" plaguing thinkers such as DePauw. To understand why Humboldt was able to escape the "systematic ideas" and "systems of knowledge" plaguing so many presentations of America, accounts afflicted by a bigotry that shrouded the American landscape and its inhabitants in a haze of barbarity, we have to get beyond the ways in which the "Dispute" shapes Humboldt's American writings. Humboldt's dedication to the cause of freedom fueled his battle against the European bigotry afflicting so many of his contemporaries. Humboldt's writings are characterized by a balancing between the empirical mastery of the diverse phenomena of nature with an aesthetic appreciation for the phenomena of nature. In order to address the question

11 Kutzinski and Ette, *Art of Science*, xxvi.

12 Alexander von Humboldt. *Views of the Cordilleras and Monuments of the Indigenous Peoples of the Americas: A Critical Edition*, Vera Kutzinski and Ottmar Ette (eds.), J. Ryan Poynter (trans.). Chicago: University of Chicago Press, 2012 [1810], 216.

of Humboldt and empire more accurately, we have to revisit his writings with new attention not only to the ends that are served in his presentation of the American landscape but also to the ideas that shape his views of nature and landscape.

2 Humboldt's Lens

Humboldt, in his approach to understanding the world around him (whether his object of study be Aztec culture or Andean mountain peaks), was never one of those scientists "in the dark when it [came] to anything beyond charts and graphs." What lies beyond the realm of the merely quantifiable? And why is it important for the scientist to go beyond this? For the merely empirical scientist, of course, nothing lies beyond the merely quantifiable, that is all we have. Humboldt was quite critical of the merely empirical approach to nature. In his final publication, *Cosmos*, Humboldt has this to say against vicious empiricism (*rohe Empirie*):

> It is the special object of [Cosmos] to combat those errors, which derive their source from a vicious empiricism (*rohe Empirie*) and from imperfect inductions. The higher enjoyments yielded by the study of nature depend upon the correctness and the depth of our views, and upon the extent of the subjects that may be comprehended in a single glance.[13]

The "higher enjoyments" are not yielded to the scientist who looks only for quantifiable facts in nature, that is, to the scientist who wants only to master nature. The scientist who can step out of mastery, who can find a way to join science and art, is the sort of scientist who is able to explore nature not for the sake of domination, that is, not in the service of a political empire or even in the service of a specific system of knowledge, but rather in the service of expanding our understanding and enriching our aesthetic experience. Humboldt's fusion of empirical mastery and the aesthetic appreciation of nature enabled him to approach nature (in all of its *living* manifestations) as a realm that could not be subsumed by any

13 Alexander von Humboldt, *Cosmos. A Sketch of the Physical Description of the Universe*, Nicolaas A. Rupke (ed.), Elise C. Otté (trans.). Baltimore: The Johns Hopkins University Press, 1997 [1858–1859], 38. "Es ist ein besonderer Zweck dieser Unterhaltungen über die Natur, einen Teil der Irrtümer, die aus roher und unvollständiger Empirie entsprungen sind und vorzugsweise in den höheren Volksklassen (oft neben einer ausgezeichneten literarischen Bildung) fortleben, zu berichtigen und so den Genuß der Natur durch tiefere Einsicht in ihr inneres Wesen zu vermehren." Alexander von Humboldt. *Kosmos. Entwurf einer physischen Weltbeschreibung*. Stuttgart: J.G. Cotta'scher Verlag, 1845, 18.

system of knowledge at all – grasping its meaning would have to be an infinite task, ever incomplete. His own description of his life's work provides good evidence of his scientific-aesthetic approach as part of an infinite task to get at the whole of nature:

> Although the outward relations of life, and an irresistible impulse toward knowledge of various kinds, have led me to occupy myself for many years – and apparently exclusively – with separate branches of science, as, for instance, with descriptive botany, geognosy, chemistry, astronomical determinations of position, and terrestrial magnetism, in order that I might the better prepare myself for the extensive travels in which I was desirous of engaging, the actual object of my studies has nevertheless been of a higher character. The principal impulse by which I was directed was the earnest endeavor to comprehend the phenomena of physical objects in their general connection, and to represent nature as one great whole, moved and animated by internal forces.[14]

Humboldt calls for a rational consideration of nature, that is, a nature submitted to the process of thought, finding a "unity in multiplicity, the connection of the manifold in form and composition (*Verbindung des Mannigfaltigen in Form und Mischung*), the essence of natural phenomena (*Naturdinge*) and forces of nature as a living whole.[15] Knowledge and reason ground Humboldt's lens to nature, but the lens is not exhausted by a desire for knowledge, but rather by something more. Humboldt's idea of nature, grounded in the view that nature is the realm of freedom (*Reich der Freiheit*), places the meaning of nature beyond human mastery. Freedom cannot be mastered or measured by any instrument. The awareness of the limits of human mastery over nature infused Humboldt's entire approach to America with an epistemological and cultural humility that was liberating. Imperial approaches are, of course, never humble. Eurocentrism is not a humble attitude, but rather a supremely arrogant attitude born of the view that one knows *with certainty* that European culture can and should set the standard by which all other cultures should be measured.

14 Humboldt, *Cosmos*, 7. "Wenn durch äußere Lebensverhältnisse und durch einen unwiderstehlichen Drang nach verschiedenartigen Wissen ich veranlaßt bin, mich mehrere Jahre und scheinbar ausschließlich mit einzelnen Disziplinen: mit bescreibender Botanik, mit Geognosie, Chemie, astronomischen Ortsbestimmungen und Erdmagnetismus als Vorbereitung zu einer großen Reiseexpedition zu beschäftigen, so war doch immer der eigentliche Zweck des Erlernens ein höherer. Was mir den Hauptantrieb gewährte, war das Bestreben, die Erscheinungen der körperlichen Dinge in ihrem allgemeinen Zusammenhang, die Natur als ein durch innere Kräfte bewegtes und belebtes Ganzes aufzufassen." Humboldt, *Kosmos*, v–vi.

15 My translation. Cf. Humboldt, *Cosmos*, 24. "Die Natur ist für die denkende Betrachtung Einheit in der Vielheit, Verbindung des Mannigfaltigen in Form und Mischung, Inbegriff der Naturdinge und Naturkräfte als ein lebendiges Ganzes." Humboldt, *Kosmos*, 5–6.

3 Nature and Culture in America[16]

In contrast to many of his contemporaries, Alexander von Humboldt certainly did not see America as any sort of wasteland – cultural or biological. He spent an important portion of his life in the New World collecting data and plant and mineral samples to send back to Europe for further investigation, and he was impressed with the wealth of biodiversity he encountered there. Yet it is critical to keep in mind that his conception of his work's importance was not limited to helping the cause of science understood as the mere amassing of data, but always also included the cause of *human* organization and progress. Unfortunately, Humboldt's dedication to the alleviation of conditions, which gave rise to social inequalities has been overlooked by some contemporary scholars. Consider, for example, Pratt's claim that: "Humboldt's eye depopulates and dehistoricizes the American landscape even as it celebrates its grandeur and variety."[17] One wonders, if such a claim were true, why Humboldt's *Essai politique sur le royaume de la Nouvelle-Espagne* (1811) would have been the cause of so much interest in Mexico, not only during the colonial period, but long into the 20th century, with various translations of the entire work from the French into Spanish. One wonders too, what would be left of a political essay without a group of people around which it were centered and the history that had given rise to the issues of freedom and justice that are addressed. In his work, *Ansichten der Natur* from 1807, a text upon which Pratt focusses, Humboldt is admittedly less concerned with presenting the people of Latin America, and more interested in presenting scenes of nature, but with one quite humanistic goal in mind: to reproduce the delights of nature, and hence to educate the public on the value that nature's splendor had for the enrichment of *human* life. Consider what he says in the prologue (*Vorrede*) to the first edition:

---

16 I have developed some of these ideas in two articles: Elizabeth Millán. Alexander von Humboldt's Role in the Decolonization of Spanish America. In *Alexander von Humboldt's American Journey*, Christine Knoop and Oliver Lubrich (eds.), 143–160. Bielefeld: Aisthesis Publishers, 2013; and Elizabeth Millán. Alexander von Humboldt's Scientific-Aesthetic Vision of America: A Move Away from Imperial Eyes. *Humanism and Revolution: The Continuing Impact of the European Enlightenment*. Uwe Steiner and Christian Emden (eds.), 209–226. Heidelberg: Winter Verlag, 2015.

17 Mary Louise Pratt. Humboldt and the Reinvention of America. In *Amerindian Images and the Legacy of Columbus*, René Jara and Nicholas Spadaccini (eds.), 585–606. Minneapolis: University of Minnesota Press 1992, 592.

> Overview of nature writ large, proof of the working together of forces, renewal of pleasure, which the unmediated view of the tropics grants to humans, are the goals towards which I strive.[18]

In the prologue to the second and third editions (1849), we find the following:

> The combination of a literary and purely scientific goal, the wish to, at one and the same time, engage the fantasy and through the increase of knowledge, to enrich life with ideas, make the order of the individual parts of the work and the demanded unity of the work most difficult to obtain.[19]

Humboldt, as he emphasizes in both prologues, is after a way to engage his audiences with the landscape of America, not for economic or political reasons, not to serve the empire of any European power, but rather for the sake of enlivening the mind of his readers, of enriching the mind with ideas (but not via the dubious "systematic ideas" used by figures such as DePauw and other anti-Americans). Humboldt combined scientific methods with aesthetic methods to create a harmonious view of nature and of America, one that would enable his readers to come to an understanding of America and to appreciate its natural splendors.

Thus, Humboldt did indeed celebrate the grandeur and variety of the American landscape, yet it is simply false to claim that his eye depopulated and dehistoricized that landscape.[20] In his hallmark work on America, *Voyages aux régions équinoxiales du Nouveau Continent*, Humboldt was primarily interested in providing an account of nature, yet never without concern for those who lived amidst the scenery described. In his *Essai sur la géographie des plantes*, Humboldt explores the issue of how the appearance of nature affects the customs and sensibility of the people of a given region. He was not, however, a geographical determinist, reminding us in his *Views of the Cordilleras*, that, "the more one examines the

---

18 „Überblick der Natur im großen, Beweis von dem Zusammenwirken der Kräfte, Erneuerung des Genusses, welchen die unmittelbare Ansicht der Tropenländer dem fühlenden Menschen gewährt, sind die Zwecke, nach denen ich strebe." Alexander von Humboldt. *Ansichten der Natur*. Hanno Beck (ed.). Darmstadt: Wissenschaftliche Buchgesellschaft, 1987 [1807], ix (my translation).

19 „Die Verbindung eines literarischen und eines rein szientifischen Zweckes, der Wunsch, gleichzeitig die Phantasie zu beschäftigen und durch Vermehrung des Wissens das Leben mit Ideen zu bereichern, machen die Anordnung der einzelnen Teile und das, was als Einheit der Komposition gefordert wird, schwer zu erreichen." Humboldt. *Ansichten*, xi (my translation).

20 For an excellent critique of Pratt's depiction of Humboldt, see Aaron Sachs. The Ultimate 'Other': Post-Colonialism and Alexander von Humboldt's Ecological Relationship with Nature. *History and Theory* 42 (2003), 111–135.

state of our species in various regions and becomes accustomed to comparing the physiognomy of countries with the people who have settled there, the less faith one has in the special theory that attributes exclusively to the climate the effects of a combination of numerous moral and physical circumstances."[21] Humboldt's texts on America provides abundant counterexamples to Pratt's claim that Humboldt's eye depopulates landscapes. Moreover, in his political essays, concern for human political structures and the inequalities suffered by the Americans under colonial rule are *the* central issues. In his *Essai politique sur le royaume de la Nouvelle-Espagne*, he described Mexico quite trenchantly (and not without risk of punishment from the Spanish Crown) as the land of inequality.[22] As Arthur Whitaker points out, "despite its modest title, [the *Essai politique sur le royaume de la Nouvelle-Espagne*] is much broader than "political" and much more substantial than an essay, in the usual sense of those terms. In fact, it is a study of the whole structure and functioning of Mexican society in the relation to the physical environment." [23] Humboldt was indeed a European acting as a cultural mediator in his role as travel writer, but because he is a travel writer engaged in a nuanced balancing act, an act involving a precise measuring of the phenomena he encounters along with a recognition that the full meaning of those phenomena cannot be mastered by his tools, but rather must be appreciated aesthetically, he is not guilty of an imperial view of the American landscape that strips the phenomena of freedom or of one that strips the landscape of the human inhabitants and of their agency in shaping the landscape. He is a self-critical European, and his Eurocentrism did not trap him or the people or places of Latin American in the dark place of barbarism that was typical of many accounts of his period. Indeed, Humboldt's presentation of nature grants a freedom to the objects and to

21 Humboldt. *Views of the Cordilleras*, Plate XXIX, 243.

22 Alexander von Humboldt. *Political Essay on the Kingdom of New Spain*. Mary Maples Dunn (ed., trans.). Norman, Oklahoma: University of Oklahoma Press, 1988 [1811], 64. "Le Mexique est le pays de l'inégalité." Alexander von Humboldt, *Essai politique sur le royaume de la Nouvelle-Espagne*. Paris: F. Schoell, 1811, 428.

23 Arthur Whitaker. Alexander von Humboldt and Spanish America. *Proceedings of the American Philosophical Society*, 104:3 (1960): 317–322, 320–21). Oliver Lubrich's article "In the Realm of Ambivalence: Alexander von Humboldt's Discourse on Cuba" provides an excellent account of the literary sophistication of another of Humboldt's political essays, *Essai politique sur l'ile de Cuba*. The reception of Humboldt is not hagiographical in Latin America: Lubrich points us to some of the critiques from Latin American scholars. Lubrich's account of Alejo Carpentier's reading of Humboldt in *La ciudad de las columnas* is especially illuminating. Oliver Lubrich. In the Realm of Ambivalence: Alexander von Humboldt's Discourse on Cuba. *German Studies Review* 26:1 (2003): 63–80.

the human subjects, who do not emerge as colonial subjects, but as free, human agents.

Contra Pratt's depiction of his work, there is no hint in any of Humboldt's work that he wants to "dehistoricize" anything. Quite the contrary, Humboldt incorporated history into his scientific approach in a comparative way that allowed him to free his observations of the cultural (and racial) bigotry that plagued the work of most of his contemporaries. In the *Essai politique*, rather than subsuming all his findings about the New World under what he already knows (for example, the indigenous 'barbaric' ways under the European 'civilized' ways), that is, rather than using the term 'European' as the universal standard by which to measure the degree of civilization that the American cultures possessed, Humboldt *compared* the American and European cultures, without appealing to European culture as the standard. He looked critically at *both* Europe and America, and then even looked critically at the two Americas he visited (i.e., U.S. and Spanish America). For example, he argues that in order to judge the worth of the indigenous cultures of New Spain (or Mexico, as the region came to be known after the colonial period) we must first make a proper comparison:

> How shall we judge, from these miserable remains of a powerful people, of the degree of cultivation to which it had risen from the twelfth to the sixteenth century and of the intellectual development of which it is susceptible? If all that remained of the French or German nation were a few poor agriculturists, could we read in their features that they belonged to nations which had produced a Descartes and Clariaut, a Kepler and a Leibniz?[24]

In the *Views of the Cordilleras*, Humboldt stresses the importance of pausing before we pass judgment on how developed a culture, reminding us that before one have sufficient data to support the judgment any such judgment will be rash. He writes:

> It is impossible to determine the accuracy of this theory [the theory concerning the accuracy of the Mexican calendar system] until a larger number of Mexica paintings in Europe and the Americas has been consulted; for – I cannot repeat this often enough – everything we have learned up to now about the former state of the peoples of the new continent will pale by comparison with the insights on this subject that will one day be reported once all the

24 Humboldt, *Political Essay*, 54. "Ou, comment juger, d'après ces restes misérables d'un peuple puissant, et du degré de culture auquel il s'étoit élevé depuis le douziéme jusqu'au seizième siècle, et du développement intellectuel dont il est susceptible? Si de la nation françoise ou allemande il ne restoit un jour que les pauvres agriculteurs, liroit-on dans leurs traits qu'ils appartenoient à des peuples qui ont produit les Descartes, les Clairaut, les Kepler et les Leibnitz." Humboldt, *Essai politique*, 401–402.

> materials that are scattered throughout the two worlds, and that have survived centuries of ignorance and barbarity, are successfully gathered in one place.[25]

Humboldt was well aware that the material necessary to critique Indigenous culture was simply not available due to the pillaging inflicted by the Spanish colonizers. He could comment on the dead remains, nothing more. The imposed European culture had stripped the American region of many of its indigenous roots. Humboldt knew that in order to understand the indigenous cultures and to judge their merits, presenting quantifiable data was not enough; rather what was needed was a proper sampling of evidence of their past achievements. Such a sampling to make a comparison with European intellectual culture would have to include the work of some of the leading intellectuals and scientists of the pre-colonial era, yet as most of the remains of the Indigenous culture had been destroyed, a reflection of European barbarism, it is hasty for the Europeans to assume that there were no intellectual figures or scientists of note. Furthermore, as Humboldt points out, the Indigenous people who survived the colonization have been oppressed, their character has been changed:

> As to the moral faculties of the Indians, it is difficult to appreciate them *with justice* if we only consider this long oppressed caste in their present state of degradation. ...All those who inhabited the *teocalli* or houses of God, who might be considered the depositories of the historical, mythological and astronomical knowledge of the country were exterminated. The monks burned the hieroglyphic paintings by which every kind of knowledge was transmitted from generation to generation. The people, deprived of these means of instruction, were plunged in an ignorance so much the deeper as the missionaries were unskilled in the Mexican languages and could substitute few new ideas in the place of the old.[26]

Humboldt was not willing simply to assume the superiority of the European culture based on a comparison with the scant historical evidence of the contributions of the Aztec civilization left in the wake of the devastation caused by the

25 Humboldt, *Views of the Cordilleras*, 211.

26 Humboldt, *Political Essay*, 53, (emphasis added). "Quant aux facultés morales des indigènes mexicains, il est difficile de les apprècier avec justesse si l'on ne considère cette caste souffrente sous uns longue tyrannie que dans son état actuel d'avilissement[...]tous ceux qui habitoient les teocalli ou les maissons de Dieu, et que l'on pourroit considerer comme dépositaires de connoissances historiques, mythologigues et astronomiques du pays, car c'etonient les prêtres que observoient l'ombre méridienne aux gnomons, et qui régloient les intercalations. Les moines firent bruler les peintures hiéroglyphiques par lesquelles des connoissances de tout genre se transmettoient de generation á generation. Privés de ces moyens d'instruction, le people retomba dans un ignorance d'autant plus profonde, que les missionaries, peu verses dans les langues mexicains substituouient peu d'idées nouvelles aux idées anciennes." Humboldt, *Essai politique*, 399–401.

Spanish *conquistadores*. He was, moreover, acutely aware of the problems left in the wake of uprooting a civilization from their traditions. The charts and graphs drawn by most scientists looking at the indigenous cultures were not prepared with sufficient attention to the historical factors, which may have accounted for the indigenous inhabitants *seeming* to be behind the Europeans in terms of intellectual contributions. The breath of life (*lebendiger Hauch*) had been snuffed out of those cultures, and so there was no hope of getting at the meaning of those cultures, of truly appreciating their worth.

Concluding Remarks

Humboldt's attention to the history of the indigenous cultures as a necessary condition for making a meaningful comparison of European and American cultures is a remarkable move, for it reveals an openness to acknowledging that the cultures of a continent that had been consistently labeled by Europeans as the dwelling place of beasts and barbarians, *deserved* more attention than they had received: the unfamiliar was not uncritically to be associated with the inferior. The respect Humboldt expressed for the Spanish-American region and its culture may very well have fueled the anti-colonial sentiment that was already building and which led to the independence movements of the early 1800s.[27] Humboldt saw a connection between scientific study and political change, and his approach throughout the *Essai politique sur le royaume de la Nouvelle-Espagne*, testifies to this connection. He connected social reforms to knowledge of the geography of a given region, guided by the belief that exact knowledge of the physical state of a given region was necessary before one could begin political change. This insight regarding the connection between political change and empirical knowledge of the territory where the change was desired was shared by local nation builders as well. In his *Carta de Jamaica*, Bolívar emphasizes the importance of such empirical knowledge (and laments that this knowledge is still "shrouded in mystery"). Humboldt linked the social inequality he observed in the region of New Spain to inequalities in land distribution, and he did believe (perhaps too optimistically), that a proper understanding of the territory in terms of its dimensions, and its richness, would enable political leaders to make more informed decisions on how to divide the territories of the region and how best to use the land to improve the

27 For more on this, see Michael Zeuske. ¿Padre de la Independencia? Humboldt y la transformación a la modernidad en la América Española. *Cuadernos Americanos* 78 (1999): 20–51.

lives of all the inhabitants. Here again, we see a fusion of both lines of Humboldt's thought: the natural scientific (dealing with empirical matters) and the aesthetic (dealing with values, in this case, justice). For Humboldt, issues of value (justice, freedom, goodness) fuse seamlessly into empirical matters of measurement. Humboldt saw great potential for social progress in the cultivation of both the natural sciences and art. His presentation of nature became also a presentation of America for the European audience, a presentation that fused elements of science and of art. Humboldt's presentation both of nature in general and of the Latin American landscape in particular was part of his commitment to justice: to creating a fair image of America to his European readers.

Humboldt's respect for freedom and change was widespread. It should, therefore, come as no surprise that he was one of the first Europeans to present Latin America to European readers with a marked respect for the people and places that are the subject of his voluminous writings on America. The very thinkers who introduced freedom and political change to the lands of the Spanish Empire, who were in effect decolonizing Spanish America and transforming it into an independent America, recognized in Humboldt a kindred spirit. In Humboldt, with his awareness of the limits of our mastery of nature, and in the freedom that flows in his presentation of nature, they found a fellow revolutionary. Humboldt's eyes are not imperial eyes, they are the eyes of a naturalist who has a deep appreciation for all of the beauty that he encounters and the proper indignation for the injustice he observes.

References

Dassow Walls, Laura. *The Passage to Cosmos. Alexander von Humboldt and the Shaping of America*. Chicago: University of Chicago Press, 2009.

Humboldt, Alexander von. *Ansichten der Natur*, Hanno Beck (ed.). Darmstadt: Wissenschaftliche Buchgesellschaft, 1987 [1807].

Humboldt, Alexander von. *Cosmos. A Sketch of the Physical Description of the Universe*, Nicolaas A. Rupke (ed.), Elise C. Otté (trans.). Baltimore: The Johns Hopkins University Press, 1997 [1858–59].

Humboldt, Alexander von. *Essai politique sur le royaume de la Nouvelle-Espagne*. Paris: F. Schoell, 1811.

Humboldt, Alexander von, and Aimé Bonpland. *Essay on the Geography of Plants*, Stephen T. Jackson (ed.), Sylvie Romanowski (trans.). Chicago: University of Chicago Press, 2009 [1807].

Humboldt, Alexander von. *Kosmos. Entwurf einer physischen Weltbeschreibung*. Stuttgart: J.G. Cotta'scher Verlag, 1845.

Humboldt, Alexander von. *Political Essay on the Kingdom of New Spain*. Mary Maples Dunn (ed., trans.). Norman, Oklahoma: University of Oklahoma Press, 1988 [1811].
Humboldt, Alexander von . *Views of the Cordilleras and Monuments of the Indigenous Peoples of the Americas: A Critical Edition*, Vera Kutzinski and Ottmar Ette (eds.), J. Ryan Poynter (trans.). Chicago: University of Chicago Press, 2012 [1810].
Kutzinski, Vera and Ottmar Ette. The Art of Science: Alexander von Humboldt's Views of the Cultures of the World. In: Alexander von Humboldt. *Views of the Cordilleras and Monuments of the Indigenous Peoples of the Americas: A Critical Edition*, Vera Kutzinski and Ottmar Ette (eds.), xv–xxxv. Chicago: University of Chicago Press, 2012.
Millán, Elizabeth. Alexander von Humboldt's Role in the Decolonization of Spanish America. In *Alexander von Humboldt's American Journey*, Christine Knoop, Oliver Lubrich, (eds.). Bielefeld: Aisthesis Publishers, 2013: 143–160.
Millán, Elizabeth. Alexander von Humboldt's Scientific-Aesthetic Vision of America: A Move Away from Imperial Eyes. *Humanism and Revolution: The Continuing Impact of the European Enlightenment*. Uwe Steiner and Christian Emden (eds.), 209–226. Heidelberg: Winter Verlag, 2015.
Pratt, Mary Louise. Humboldt and the Reinvention of America. In *Amerindian Images and the Legacy of Columbus*, René Jara and Nicholas Spadaccini (eds.), 585–606. Minneapolis: University of Minnesota Press 1992.
Pratt, Mary Louise. *Imperial Eyes. Travel Writing and Transculturation*. London: Routledge, 1992.
Lubrich, Oliver. In the Realm of Ambivalence: Alexander von Humboldt's Discourse on Cuba. *German Studies Review*, 26:1 (2003): 63–80.
Sachs, Aaron. The Ultimate 'Other': Post-Colonialism and Alexander von Humboldt's Ecological Relationship with Nature. *History and Theory* 42 (2003): 111–135.
Whitaker, Arthur. Alexander von Humboldt and Spanish America. *Proceedings of the American Philosophical Society*, 104:3 (1960): 317–322.
Zeuske, Michael. ¿Padre de la Independencia? Humboldt y la transformación a la modernidad en la América Española. *Cuadernos Americanos* 78 (1999): 20–51.

Christian Holtorf

Zum Pol

Positionen der Arktis in Raum und Zeit

English abstract: To the Pole. Positions of the Arctic in Space and Time. Cultural studies have always been skeptical about grand narratives. Often theories cannot keep the sweeping promises they make when it comes to historical details. The article sketches the spatiotemporal positioning of the Arctic on the basis of textual and visual discourses on Arctic expeditions of the 19th and 20th century. These discourses seem to contradict the theory of Hardt and Negri, for whom an ice landscape is a space without a horizon, without perspective, borders and definite distances. It is rather determined by heterogeneity and continuous gradual variation, by intensities, winds and sounds. For Hardt and Negri, ice is a classic example of "smooth space". They also call it "Eskimo space". However, Arctic space is not as smooth as it may appear from European cruise liners. The historicity of scientific methods, the multiplicity of historical imaginations, and the openness of the discourse on the social construction of space all speak against Hardt and Negri's view. Furthermore, representatives of the Inuits are resisting the conquest of the Arctic by "Empire". In the defense of their culture and their living space, they seem to be doing the very thing which Hardt and Negri view as the end of Empire: they ignore the determinations of territories and borders and live nomadically. They offer resistance in order to transform a transport region into a living region.

In einer Landschaft aus aufgetürmten, ineinander verkeilten Packeisschollen hat Robert E. Peary im Frühjahr 1909 den Nordpol erreicht. Europäische Geografen hatten den Pol seit der Antike als Ende der Erde betrachtet, ihn lebendig beschrieben und auf Karten verzeichnet. Über die Jahrhunderte hinweg war Expedition um Expedition in die Arktis aufgebrochen, doch die meisten von ihnen sind nur mit geringen Erkenntnissen oder gar nicht zurückgekehrt. Erst mit Hilfe moderner Kartografie und Navigationstechnik ist es zu Beginn des 20. Jahrhunderts gelungen, den Nordpol als einen der letzten „weißen Flecken" zu entdecken. Daraus haben sich freilich keine weiteren Folgen ergeben, so dass der Pol seine exponierte Bedeutung verlor. Der Wiener Schriftsteller Karl Kraus schloss daraus,

DOI 10.1515/9783110418750-012

dass "an dem Nordpol (...) nichts weiter wertvoll (war), als daß er nicht erreicht wurde."[1]

Verlief ungefähr so die Kulturgeschichte des Nordpols? Ist sie nichts anderes als Teil des europäischen Projekts, die Erde mit Hilfe geografischer Messpunkte möglichst vollständig erfassen, sich erfolgreich auf ihr orientieren und sie in Besitz nehmen zu können? Außer ihrer Entdeckungsgeschichte hätte die Arktis dann keine historische Entwicklung erlebt – abgesehen vom Wechselspiel zwischen wärmeren und kälteren Episoden der Erdgeschichte und variierenden Ausmaßen des „ewigen Eises". Der Abenteurer Arved Fuchs, der viele Expeditionen in die Polargebiete geleitet hat, hat die Vorstellung eines unberührten und geschichtslosen Nordens schon als Kind kennengelernt. Zur Arktis gehörten für ihn: „Schnee. Eis. Polare Nacht. Große Kälte. Packeis. Iglus. Eskimos. Schlittenhunde. Aber auch: Einsamkeit, die keine Fehler gestattet." Heute hat er Sorge, dass das Polarmeer aufgrund der steigenden Temperaturen regelmäßig für Frachtschiffe befahrbar wird. Er fürchtet: „Der Nordpol, den früher kaum einer erreichen konnte, wird zu einer Banalität."[2]

Hat der Polarabenteurer Recht? Verliert der Nordpol seine Bedeutung, wenn die Temperaturen in der Arktis ansteigen und der Schiffsverkehr zunimmt? Wenn mehr Menschen ihn erreichen? Was macht denn seine Bedeutung aus? Die beiden Sozialphilosophen Michael Hardt und Antonio Negri sehen den Nordpol vielleicht ganz ähnlich wie Arved Fuchs. In ihrer einflussreichen Studie „Empire. Die neue Weltordnung" von 2002 gehen sie zwar nicht auf die Arktis ein, beschreiben aber einen gegenwärtigen Prozess fortschreitender Globalisierung und Vernetzung, in den sich die Entdeckung des Nordpols, seine Kartierung und Vermessung leicht einfügen ließen. Sie behaupten, dass in der neuen Weltordnung des „Empire" nicht nur die territoriale Souveränität der Nationalstaaten, sondern die Festlegung jedweder Zentren und Grenzen ihre Bedeutung verloren hätte. In Anlehnung an Gilles Deleuze und Félix Guattari schreiben sie: „Der gekerbte Raum der Moderne schuf Orte, die beständig in einem dialektischen Spiel mit ihrem Außen standen und auf diesem Spiel gründeten. Der Raum imperialer Souveränität ist im Gegensatz dazu glatt."[3]

Deleuze und Guattari haben als Musterbeispiel für einen glatten Raums das Eis genannt und darin mit Steppe, Wüste und Meer verglichen. Diese Gebiete

---

1 Karl Kraus. Die Entdeckung des Nordpols. *Die Fackel* 287: IX (1909): 6.

2 Matthias Hannemann. Schmilzt die Arktis, Herr Fuchs? Gespräch mit einem Polar-Abenteurer. In *Frankfurter Allgemeine Zeitung* 4.2.2007.

3 Michael Hardt, Antonio Negri. *Empire. Die neue Weltordnung*. Frankfurt und New York: Campus, 2002, 202.

seien gekennzeichnet durch Heterogenität und kontinuierliche Variation, besetzt von Intensitäten, Winden und Geräuschen. Erst Positionsmessungen und Navigationskarten kerbten sie ein und ermöglichten die Orientierung. Die beiden Wissenschaftler fanden dafür den seltsamen Begriff „Eskimo-Raum“, der einen horizontlosen Raum ohne Perspektive, Grenze, Form und Abstände bezeichnen soll.[4] In gewisser Weise haben Hardt und Negri diesen Begriff nun verallgemeinert. „Die Kämpfe von heute“, schreiben sie, „winden sich lautlos über die glatten imperialen Landschaften. (...) Das Empire ist eine oberflächliche Welt.“[5] Die neue Macht des „Empire“ bestehe aus Strömen in einem vernetzten globalen Raum, die jederzeit fließen und keine spezifische Historizität besitzen.[6] Nicht neue „frontiers“ bildeten die Herausforderung, sondern die interne Koordination einer offenen Weltgemeinschaft. Natur werde darin nicht mehr entdeckt und erforscht, sondern graduell genutzt und je nach Kontext medial konstruiert.[7] Ob durch Klimawandel oder durch Eroberung – mit dem Nordpol wäre eine der letzten großen Kerben der Natur geglättet, und die Einsamkeit des Abenteurers in eine Mediengesellschaft überführt, die Extremerfahrungen für alle erlebbar macht.

Obwohl diese Interpretation zunächst plausibel klingt, besteht Anlass zur Diskussion. Gegenüber großen Erzählungen wie derjenigen vom „Empire“ hat sich in der Kulturwissenschaft Skepsis bewährt. Oft können die Theorien im historischen Detail nicht halten, was sie pauschal versprechen. Durch den Klimawandel schmilzt heute beispielsweise nicht nur das Packeis, sondern auch der Umfang der Arktis selbst: Hält man sich an eine der üblichen Definitionen, nach der die Arktis das Gebiet nördlich der 10°C-Juli-Isotherme umfasst, verringert sich mit den steigenden Temperaturen ihre geografische Fläche. Die Arktis ist also kulturell definiert und kein isolierter Naturraum. Dass der Nordpol jahrtausendelang imaginiert, immer wieder angesteuert und schließlich erreicht wurde, lässt sich durch Erwartungen erklären, die auf den Pol projiziert wurden, und durch die jeweiligen Möglichkeiten und Bedürfnisse der Medien. Dass wir wissen, dass sich das Ausmaß der Eisfläche seit einigen Jahren verringert, ist nicht nur eine Folge des Klimawandels, sondern auch das Ergebnis wissenschaftlicher Forschungen, die durch Weiterentwicklungen von Messinstrumenten und Abbildungsverfahren, aber auch durch Innovationen im Schiffsbau und in der Satellitentechnik möglich wurden. Durch Wissenschaften und Politik, durch Trends im Handel und neue Kommunikations-

4 Gilles Deleuze, Félix Guattari. *Tausend Plateaus. Kapitalismus und Schizophrenie*. Berlin: Merve, 1992, 664, 676, 683–684.

5 Hardt, Negri, *Empire*, 71.

6 Hardt, Negri, *Empire*, 11ff., vgl. 341.

7 Hardt, Negri, *Empire*, 194, 199.

weisen haben die Narrative und Bilder des Nordpols also keineswegs an Bedeutung verloren; sie sind durch ihren Wandel vielmehr zu einem wichtigen Element der Wissens- und Kulturgeschichte nicht nur der Meere geworden.

Hardt und Negri müssen sich genauer fragen lassen, wie sie ihre Begriffe definieren, worauf sie sich im Einzelnen stützen und ob es nicht auch Gegenbeispiele gibt, die zu falsifizieren wären. Wenn die Genealogie des Empire eurozentrisch ist, wie sie behaupten, muss ihre Arbeit dann nicht auch selbst in dieser Tradition verstanden werden?[8] Es besteht der Verdacht, dass ihre Darstellung global argumentiert, methodisch aber eurozentriert bleibt. Denn nicht nur sind Wissenschaften und Globalisierung von europäischen Werten und Normen ausgegangen, sondern auch Universalisierung und Pluralisierung sind Teil der Identität Europas. Worin aber könnte eine (nicht-europäische) Gegenposition bestehen, wenn die Einbeziehung von Widersprüchen, wie Hardt und Negri sagen, selbst zum (europäischen) Narrativ gehört? Im Folgenden werden raumzeitliche Positionierungen der Arktis skizziert, die zeigen, dass sich ihre Pluralität nicht leicht in die Entwicklung zunehmender Globalisierung und Vernetzung einordnen lässt, sondern im theoretischen Ansatz liegt, nicht im Befund. Gegen Hardt und Negri sprechen (1.) die Historizität der wissenschaftlichen Methoden, (2.) die Vielfalt von historischen Imaginationen und (3.) die Offenheit des Diskurses über die soziale Konstruktion des Raums.

Dieser Beitrag beruht auf einem Forschungsprojekt zur Wissensgeschichte des Nordpols, das textliche und bildliche Diskurse von Arktis-Expeditionen des 19. und frühen 20. Jahrhunderts analysiert. Es folgt kulturgeografischen Arbeiten, die Karten nicht als Abbilder verstehen, sondern als Gegenstand kultureller Konstruktionen in den Kontext der Wissens- und Kulturgeschichte einordnen.[9] Auf der Grundlage von nordamerikanischen und europäischen Quellen wird untersucht, was den Nordpol ausmacht, wie sich die Vorstellungen der Arktis historisch gewandelt haben und was es bedeuten kann, ihn zu erreichen. Dabei fließen auch öffentlich verbreitete Imaginationen und „mental maps" in die Produktion der Landkarten ein.[10] Gerade die im Justus Perthes Verlag in Gotha verlegten Karten, Atlanten und Zeitschriften, darunter am bekanntesten „Stielers Handatlas" und „Petermanns Geographische Mitteilungen", haben nicht nur das

8 Hardt, Negri, *Empire*, 14.

9 Zum Beispiel: Georg Glasze. Kritische Kartografie. *Geographische Zeitschrift* 4:97 (2009): 181–191; John B. Harley. *The New Nature of Maps. Essays in the History of Cartography*, Paul Laxton (Hrsg.). Baltimore und London: The Johns Hopkins University Press, 2001; Ute Schneider. *Die Macht der Karten. Eine Geschichte der Kartographie vom Mittelalter bis heute*. 2. Aufl. Darmstadt: Primus, 2006.

10 Vgl. Schneider, *Macht der Karten*, 81.

wissenschaftliche, sondern auch das populäre Bild der Erde geprägt. Aus ihrer Untersuchung ergibt sich zugleich ein Blick auf die Gesellschaften, von denen die Erforschung ausgegangen ist.[11]

1 Historizität

Die Vorstellung, dass es sich bei der Geschichte der Kartografie um einen Prozess handelt, der zur fortschreitenden Beseitigung von „weißen Flecken" dient, übersieht, dass sich Technik und Wissenschaft historisch verändern und Ausdruck ihrer sozialen und kulturellen Bedingungen sind. Es lässt sich kein neutraler wissenschaftlicher Standpunkt vorstellen. Auch der geografische Raum selbst ist keine Vorgabe der Natur, sondern kulturell konstruiert. Abseits von sichtbaren Landmarken und Grenzmarkierungen, die in der Arktis kaum zur Geltung kommen konnten, beruht er heute in der Kartografie meist auf der Anwendung von Geometrie. Räumliche Souveränität und Landbesitz sind dadurch zu einer Frage von wissenschaftlicher Argumentation, exakter Messung und kartografischer Darstellungsweise geworden.[12] „Weiße Flecken" sind nur dann entstanden, wenn Karten auf Angaben verzichtet haben, die nicht wissenschaftlich nachweisbar waren, und wenn weder Vermutungen aus Mythen und Sagen noch verdeckende Erläuterungstafeln und Kartuschen an ihre Stelle getreten sind – was allerdings den Nachteil hatte, dass bei in der Öffentlichkeit falsche Vorstellungen über „leere" Gebiete und ihre Zivilisierung hervorgerufen wurden.[13] Erst solche Grenzziehungen „mit Uhr und Kompass", schreibt Ulrike Jureit, haben eine Differenzierung zwischen Innen und Außen hergestellt, die Bewegungen durch erstarrte Linien ersetzt und einen „leeren Raum" erzeugt haben, wo der europäische Einfluss fehlte:[14] „In dieser Hinsicht ist der leere Raum keine Metapher, sondern demons-

---

11 Vgl. Michael Bravo und Sverker Sörlin (Hrsg.). *Narrating the Arctic. A Cultural History of Nordic Scientific Practices*. Canton: Watson Publishing International, 2002; Philipp Felsch. *Wie August Petermann den Nordpol erfand*. München: Luchterhand, 2010; Michael F. Robinson. *The Coldest Crucible. Arctic Exploration and American Culture*. Chicago und London: The University of Chicago Press, 2006; Christian Holtorf. Der Nordpol als Aufmerksamkeitsmagnet. Neue Monografien zur Konjunktur der Arktis im 19. und 20. Jahrhundert. *NTM Zeitschrift für Geschichte der Wissenschaften, Technik und Medizin* 4:18 (2010): 537–544.

12 Jeppe Strandsbjerg. *Cartography and Geopolitics in the Arctic Region. DIIS Working Paper*. Copenhagen: Danish Institute for International Studies, 2010, 12–13.

13 Schneider, *Macht der Karten*, 100, 112ff.

14 Ulrike Jureit. *Das Ordnen von Räumen. Territorium und Lebensraum im 19. und 20. Jahrhundert*. Hamburg: Hamburger Edition, 2012, 122ff.

triert die beanspruchte Überlegenheit europäischer Mächte sowie die beabsichtigte Installation des europäischen Territorialverständnisses“.[15]

Der Nordpol blieb für Kartografen und Naturforscher, für Walfänger und andere Abenteurer aus Europa und Amerika jahrhundertelang auch deswegen ein attraktives Reiseziel und Imaginationsobjekt, weil sich eine Reihe von imperialen Fantasien auf ihn projizieren ließen: Um den Pol schien sich die Erde zu drehen, während er selbst – wie ein Mittelpunkt oder Nabel – still stand; er galt als „Ende der Welt“. Man vermutete, dass er einem Felsen glich, aus „ewigem Eis“ bestand oder in anderer Weise substantialisiert war. Er könnte auch einen Zugang zu anderen Welten eröffnen: Bis zum Ende des 19. Jahrhundert waren Fantasien lebendig, die hier – da die Erddrehung an den Polen zu einer Abflachung der Erde führe – durch Vulkane oder Meeresstrudel einen Zugang ins Innere der Erde vermuteten. Andere hofften, durch einen Magnetberg oder ein physisches Heraustreten der Erdachse die kosmische Lage des Planeten vermessen zu können.[16]

Diese Überlegungen hat Shane McCorristine freilich mit dem Hinweis relativiert, dass Wissen in der Arktis stark von irritierenden sinnlichen Erfahrungen abhängt. Während Polarreisen lange als „körperlose Aktivität“ verstanden wurden, erkannte McCorristine die „Wichtigkeit von Körperlichkeit, Rhythmus und der Temporalität des Körpers für unser Denken über Bewegung durch den maritimen Raum“.[17] Dass Halluzinationen und Träume, aber auch Metaphern wie Labyrinth und Schleier häufig in Reisebeschreibungen auftreten, sei keine Verlegenheit, sondern eine körperliche Reaktion auf die ungewohnte Erfahrung eines amorphen heterogenen Raums. Die extreme sinnliche Verunsicherung, von der viele Polarreisende berichtet haben, habe das Bedürfnis nach einer Erklärung der Phänomene verstärkt und Anlass zu geografischen Spekulationen gegeben. McCorristine forderte daher dazu auf, „die Arktis als eine Mischwelt von Bewegung und Stillstand zu denken“.[18]

---

15 Jureit, *Ordnen*, 126.

16 Chet van Duzer. The Mythic Geography of the Northern Polar Regions: Inventio fortunata and Buddhist Cosmology. *Culturas Populares. Revista Electronica* 2 (2006); Christian Holtorf. Die Südsee im Norden. Technik und Transzendenz in Narrativen der Arktisforschung des 19. Jahrhunderts. In *Technik und Transzendenz. Zum Verhältnis von Technik, Religion und Gesellschaft*, Katharina Neumeister, Peggy Renger-Berka, Christian Schwarke (Hrsg.), 181–208. Stuttgart: Kohlhammer, 2012.

17 Shane McCorristine. Träume, Labyrinthe, Eislandschaften: Körper und Eis in Arktis-Expeditionen des 19. Jahrhunderts. In *Weltmeere. Wissen und Wahrnehmung im langen 19. Jahrhundert*, Alexander Kraus, Martina Winkler (Hrsg.), 103–126, hier 105–106. Göttingen: Vandenhoeck & Ruprecht, 2014.

18 McCorristine, Träume, 109.

Ein gutes Beispiel für dieses Bedürfnis nach klarer sinnlicher Orientierung ist in der Enttäuschung zu erkennen, die James C. Ross empfand, als er im Jahr 1831 den magnetischen Nordpol entdeckte und feststellen musste, dass die Stelle völlig „ungekerbt" war:

> We could have wished that a place so important had possessed more of mark or note. It was scarcely censurable to regret that there was not a mountain to indicate a spot to which so much interest must ever be attached; and I could even have pardoned any one among us who had been so romantic or absurd as to expect that the magnetic pole was an object as conspicuous and mysterious as the fabled mountain of Sinbad, that it was even a mountain of iron, or a magnet as large as Mont Blanc. But Nature had here erected no monument to denote the spot which she had chosen as the center of one of her great and dark powers; and where we could do little ourselves towards this end, it was our business to submit, and to be content in noting in mathematical numbers and signs, as with things of far more importance in the terrestrial system, what we could ill distinguish in any other manner.[19]

Der Nordpol ist nichts anderes als ein historisches Medienprodukt. Wer ihn nie gesehen hatte, konnte und musste auf kursierende Bilder zurückgreifen. Aber auch für diejenigen, die ihn erreichten, war es schwierig, ihn wahrzunehmen und zu beschreiben. Woran konnte man den Pol verlässlich erkennen, wenn nicht durch eine Anzeige der geografischen Position? Wie ließ er sich darstellen, wenn nicht durch eine angebrachte Markierung? Was bedeutete es, ihn zu erreichen, wenn nicht den Moment des Erreichens als Bericht oder Foto zu dokumentieren? Auch als der Journalist Matthias Hannemann im Jahr 2007 einen Sonderrundflug von Düsseldorf zum Nordpol und zurück begleitete, war von der Polarlandschaft nur ein „weißes Nichts" zu erkennen. Als die Maschine den Nordpol tatsächlich erreichte, schauten die meisten Passagiere nicht aus dem Fenster, sondern konzentrierten sich auf das Display über ihren Sitzen, denn nur dort wurde die genaue Position angezeigt und konnte abfotografiert werden – als Bild eines Bildes.[20]

Heutige Reise-Anbieter sind in derselben Verlegenheit wie frühere Expeditionsleiter: sie müssen einen in der Natur nicht erkennbaren Punkt zu einem Ereignis machen. Meist errichten sie inmitten der instabilen Packeislandschaft einen mitgebrachten Fahnenmast und arrangieren die Menschen darum herum. Währenddessen wird allerdings das gesamte Setting bereits zügig von der starken

---

19 John Ross. *Narrative of a second voyage in search of a North-west passage: and of a residence in the Arctic regions during the years 1829, 1830, 1831, 1832, 1833: including the reports of Commander, now Captain, James Clark Ross, R.N., F.R.S., F.L.S., &c. and the discovery of the Northern Magnetic Pole*. London: A.W.Webster, 1835, 555–556.

20 Matthias Hannemann. *Der neue Norden. Die Arktis und der Traum vom Aufbruch*. Bad Vilbel: Scoventa, 2010, 203–207.

Strömung mitgetragen, befindet sich also gar nicht mehr am Pol. Mit dem Hissen einer Fahne wird Land traditionell in Besitz genommen. Fahne und Mast, Foto und Erinnerung sind, wenn man so will, imperiale Gesten in der unendlichen Eiswüste, der Versuch, eine metrische Position mitten auf einem großen Ozean zu markieren. Aber sie sind auch das Ergebnis spezifischer Reisemöglichkeiten und historisch entstandener Nordpol-Inszenierungen, bei denen die Realität oft hinter den Erwartungen zurückblieb, wie wiederum Karl Kraus bemerkte: „Einmal erreicht, ist er eine Stange, an der eine Fahne flattert, also etwas, das ärmer ist als das Nichts, eine Krücke der Erfüllung und eine Schranke der Vorstellung."[21]

An Polarfahrer wurden nicht nur bestimmte soziale Erwartungen gerichtet, sie übertrugen auch verbreitete Bildstereotype einer erhabenen Natur auf die Arktis.[22] Die Romantik des nördlichsten Punktes der Erde etablierte einen arktischen Exotismus, der Wissenschaftler interessieren, Abenteurer motivieren und die Öffentlichkeit faszinieren konnte.[23] Nicht zuletzt erleichterte er auch die Finanzierung der Expeditionen. Dabei konnten interessanterweise sogar stark umstrittene Vorhaben die Unterstützung von wissenschaftlichen Gesellschaften finden, weil Einwände eine zusätzliche Motivation waren, den Dingen auf den Grund zu gehen – auch bevor andere Staaten dies taten. Manche Wissenschaftler waren sogar bereit, für die Gewinnung physikalischer Daten aus unzugänglichen Regionen erkennbare Fantastereien in Kauf zu nehmen.[24] Wissenschaft und Öffentlichkeit beeinflussten sich dabei gegenseitig, denn schon der Einsatz unterschiedlicher Medien wie Tageszeitungen, Tagebücher, Handskizzen und Fotografie hat die Möglichkeit der Erkenntnisse und ihrer Rezeption stark beeinflusst.[25]

---

21 Kraus, Nordpol, 6.

22 Chauncey C. Loomis. The Arctic Sublime. In *Nature and the Victorian Imagination*, U.C. Knoepflmacher, G.B. Tennyson (Hrsg.), 95–112. Berkeley und Los Angeles: University of California Press, 1977; Dorit Müller. Transformationen populären Wissens im Medienwandel am Beispiel der Polarforschung. In *Populäres Wissen im medialen Wandel seit 1850*, Petra Boden, Dorit Müller (Hrsg.), 35–79. Berlin: Kadmos, 2009; Robinson, *Coldest Crucible*.

23 Holtorf, Südsee; Friedhelm Marx. „Die Paradiese des Südpols". Phantastische Expeditionen ans Ende der Welt. In *Phantastik – Kult oder Kultur? Aspekte eines Phänomens in Kunst, Literatur und Film*, Christine Ivanovic, Jürgen Lehmann, Markus May (Hrsg.), 197–212. Stuttgart und Weimar: Metzler, 2003.

24 Christian Holtorf. Das offene Polarmeer. Ein Bilddiskurs im 19. Jahrhundert. In *Bilder in historischen Diskursen*, Franz K. Eder, Oliver Kühschelm, Christina Linsboth (Hrsg.), 145–172. Wiesbaden: VS Verlag für Sozialwissenschaften, 2014.

25 Beau Riffenburgh. *The Myth of the Explorer. The Press, Sensationalism, and Geographical Discovery*. Oxford: Oxford University Press, 1994; Janice Cavell. *Tracing the Connected Narrative. Arctic Exploration in British Print Culture, 1818–1860*. Toronto: University of Toronto Press, 2008;

Karten sind also keine neutralen Abbilder der Wirklichkeit, sondern sie organisieren und reproduzieren „soziale Ordnungen, festigen sie und tragen zu ihrer Naturalisierung bei."[26] Karten machen Herrschaft verhandelbar und ermöglichen die Vorbereitung von Handlungen aus der Distanz. Sie bildeten damit selbst eine historische Voraussetzung für internationalen Handel und territoriale Expansion. Dass sich ihre kulturell ganz unterschiedlichen Entwürfe aber alle auf einen einheitlichen natürlichen Raum beziehen, hat Bruno Latour die einflussreichste Idee des Westens genannt.[27] Wissenschaft sei schon deswegen nichts anderes als Politik mit anderen Mitteln.[28] Der Wissensvorsprung, den Karten für weitere Fahrten vermittelten, so Latour, wurde in ein europäisches Netzwerk des Wissens eingespeist und dadurch imperial nutzbar gemacht: „Die Kartografie hat den sicheren Pfad einer Wissenschaft betreten; ein Zentrum (Europa) war gebildet worden, um das sich der Rest der Welt zu drehen beginnt."[29]

2 Vielfalt

Wie nun wurde die Arktis in Karten konstruiert? Welches Wissen transportierten sie? Welche Bilder standen Modell und nach welchen Narrativen wurden erzählt? In der historischen Kartografie lässt sich eine Vielzahl unterschiedlicher Repräsentationsformen und Darstellungen des Nordpolargebiets finden. Seit dem Mittelalter wurde der Pol als imaginärer Fluchtpunkt von Raum und Zeit konstruiert, von dem aus sich das Netz der Längengrade über die Erde zog. Doch mindestens seit dem 19. Jahrhundert setzten die Wissenschaften zusammen mit neuen Navigations- und Medientechniken auch einen Prozess in Gang, der die historische Wahrnehmung der Arktis gewandelt hat: Neben die durch den Pol markierte,

---

Russell A. Potter. *Arctic Spectacles. The Frozen North in Visual Culture 1818–1875.* Seattle: University of Washington Press, 2007; Dorit Müller. Fotografie und Südpolforschung um 1900. In *Frosch und Frankenstein. Bilder als Medium der Popularisierung von Wissenschaft*, Bernd Hüppauf, Peter Weingart (Hrsg.), 233–253. Bielefeld: Transcript, 2009; Dorit Müller. Kartierungen polarer Räume. In *Karten-Wissen. Territoriale Räume zwischen Bild und Diagramm*, Stephan Günzel, Lars Nowak (Hrsg.), 377–395. Wiesbaden: Reichert Verlag, 2012.

26 Glasze, Kritische Kartographie, 184.

27 Zit. n. Strandsbjerg, *Cartography*, 20.

28 Bruno Latour. Give me a laboratory and I will rise the world. In *Science Observed*, Karin Knorr-Cetina, Michael Mulkay (Hrsg.), 141–169, hier 168. London, Beverly Hills und New Delhi: Sage, 1983.

29 Bruno Latour. Die Logistik der immutable mobiles. In *Mediengeographie. Theorie – Analyse – Diskussion*, Jörg Döring, Tristan Thielmann (Hrsg.), 111–144, hier 126. Bielefeld: Transcript, 2009.

vereiste Arktis trat ein dynamischer Ozean mit Tiefseegräben und Untiefen, mit Winden und Strömungen, mit Eisbewegungen und offenen Wasserstraßen.

Was bedeutet diese Alternative zweier Darstellungsmöglichkeiten?[30] Zunächst stellten sich die Polarforscher die Aufgabe, den Nordpol möglichst genau zu erforschen und zu kartieren. Expeditionsrouten nach diesem Muster waren möglichst geradlinig ausgerichtet, auch wenn sie aufgrund der Drift im Packeis in Schlangenlinien verliefen oder ihr Ziel aus anderen Gründen nicht erreicht haben. Allein das Vorhaben, den Nordpol zu erreichen und dies zu dokumentieren, schien den hohen Aufwand und die körperlichen Strapazen der Expeditionen zu rechtfertigen. 1865 erklärte der Gothaer Geograf August Petermann: „Bei der Erreichung der Pole kommt zunächst in Betracht, die Bestimmung der Hauptcontouren unseres Planeten zu vollenden, die Grenzen von Land und Meer festzustellen. (...) die Scheidung des Festen und Flüssigen gehört zur Basis und ersten Grundlage der Erdkunde".[31] Man fand dabei übrigens nicht nur einen geografischen Pol, sondern mehrere Pole, die sich in der Erdgeschichte auch noch verändert haben: einen Magnetpol, mehrere Kältepole und dazu noch sogenannte Unzugänglichkeitspole, die am weitesten von allen menschlichen Siedlungen entfernt liegen.

Das Denken in Achsen und Polen, Mittelpunkten und Horizonten verweist offensichtlich auf raumzeitliche Imperialität. Seit dem Mittelalter hatten die Polargebiete nicht nur geografische und politisch-ökonomische Bedeutung, sondern waren auch ein Medium kosmologischer Wissensordnungen. Ihre Erforschung stieß auf hohe wissenschaftliche und öffentliche Aufmerksamkeit. Paradoxerweise führte aber die Fixierung auf einen substantiellen Pol auch zur Imagination eines eisfreien Polarmeers.[32] Schon die antiken Autoren Herodot und Diodor hatten vage Vermutungen dokumentiert. Seit dem 14. Jahrhundert stellten sich europäische Kartografen den Nordpol ähnlich dem biblischen Paradies als

---

30 Zur Vielzahl der Narrative vgl. Christian Holtorf. Der Nordpol – eine Erzählung. In: *Erzählung und Geltung. Wissenschaft zwischen Autorschaft und Autorität*, Safia Azzouni, Stefan Böschen, Carsten Reinhardt (Hrsg.), 133–155. Weilerswist: Velbrück Wissenschaft, 2015.

31 August Petermann. Der Nordpol und Südpol, die Wichtigkeit ihrer Erforschung in geographischer und kulturhistorischer Beziehung. *Mittheilungen aus Justus Perthes' Geographischer Anstalt* 11 (1865): 146–160, hier 149.

32 van Duzer, Mythic Geography; Fergus Fleming. *Neunzig Grad Nord. Der Traum vom Pol*, Hamburg: Rogner & Bernhard, 2003; Holtorf, Südsee; Russell A. Porter. Open Polar Sea. In *Encyclopedia of the Arctic*, Vol. 3, Marc Nuttall (Hrsg.), 1578–1580. New York und London: Fitzroy Dearborn Publ., 2005; Michael F. Robinson. Reconsidering the Theory of the Open Polar Sea. In *Extremes. Oceanography's Adventures at the Poles*, Keith R. Benson, Helen M. Rozwadowski (Hrsg.), 15–29. Sagamore Beach: Science History Publications, 2007; Erki Tammiksaar, Natal'ja G. Suchova, I.R. Stone. Hypothesis Versus Fact: August Petermann and Polar Research. *Arctic* 3:52 (1999): 237–244; John K. Wright. The Open Polar Sea. *The Geographical Review* 3:43 (1953): 338–365.

Landfläche mit vier Flüssen vor, die aus einem offenen Gewässer im Zentrum entsprangen bzw. in es mündeten. Auf den ersten Globen von Martin Behaim und Gerhart Mercator im 15. und 16. Jahrhundert ist am Nordpol ein eisfreies Polarmeer abgebildet, das von einem Ringkontinent umgeben ist und in dessen Mitte ein fabulöser „Polus Arcticus“ aus dem Wasser ragt (Abb. 1). Begründet durch seine exponierte Lage im höchsten Norden und ausgestattet mit verschiedenen transzendenten und imperialen Bezügen, wurde seine Existenz über die Jahrhunderte immer wieder von Seefahrern bestätigt.

Abb. 1: Polarkarte von Gerhart Mercator (1569).

Bis zum Ende des 17. Jahrhundert waren europäische Walfänger in immer höhere Breitengrade vorgedrungen sind und hatten in der Grönland-See, im Smith-

Sund und in der Bering-Straße die Ränder des Arktischen Ozeans erreicht.[33] Die Berichte des englischen Walfängers William Scoresby Jr. über einen signifikanten Rückgang des Packeises zwischen 1814 und 1817 dienten der britischen Marine als Anlass, weitere Expeditionen in die Arktis zu schicken. Scoresby selbst blieb allerdings skeptisch gegenüber den weitreichenden Schlussfolgerungen bezüglich der Erreichbarkeit des Nordpols.[34] Am bekanntesten wurde die Expedition von John Franklin – nicht nur, weil sie im Eis verschollen war, sondern vor allem deswegen, weil sie das öffentliche Interesse jahrelang wach gehalten und zahlreiche Polarfahrten initiiert hat. Die beiden amerikanischen Arktisforscher Elisha K. Kane und Isaac I. Hayes behaupteten in den Jahren 1854 und 1861, dass ihre Expeditionen jeweils das Ufer des offenen Polarmeeres erreicht hätten. Zum Beweis brachten sie Zeichnungen und Schilderungen des Ausblicks mit nach Hause.[35] In der Öffentlichkeit wurde der Mythos so populär, dass er sich auch in Romanen, Science-Fiction-Fantasien und Jugendbücher, sogar in politischen Ideologien und Esoterik niederschlug.[36]

Nachdem jedoch die Polarexpeditionen zu wenig konkreten Ergebnissen, dafür aber vielen Todesfällen geführt hatten, war in der Öffentlichkeit zur Mitte des 19. Jahrhunderts der Eindruck entstanden, „that any further efforts toward the North Pole must be fruitless".[37] Die Expeditionsleiter brauchten für ihre Projekte daher die zusätzliche Unterstützung von wissenschaftlichen Gesellschaften. Kane und Hayes gewannen die American Geographical & Statistical Society; auf einer Sondersitzung der Gesellschaft am 22 März 1860 in New York stellte Hayes seinen Plan anhand einer Karte vor, die vermutlich derjenigen entsprach, die der Dokumentation der Sitzung beigefügt war (Abb. 2).[38] Sie soll hier exemplarisch vorgestellt werden.

---

33 John R. Bockstoce. *Whales, Ice, and Men: The History of Whaling in the Western Arctic.* Seattle und London: University of Washington Press, 1986.

34 C. Ian Jackson. William Scoresby the Scientist. *The Journal of the Hakluyt Society* (online), January 2010; Constance Martin. William Scoresby, Jr. (1789–1857) and the Open Polar Sea – Myth and Reality. *Arctic* 1:41 (1988): 39–47. Michael Bravo. Geographies of exploration and improvement: William Scoresby and Arctic whaling, 1782–1822. *Journal of Historical Geography* 32 (2006): 512–538.

35 Holtorf, Das offene Polarmeer; Douglas W. Wamsley. *Polar Hayes. The Life and Contributions of Isaac I. Hayes, M.D.* Philadelphia: American Philosophical Society, 2009.

36 Joscelyn Godwin. *Arktos. Der polare Mythos zwischen NS-Okkultismus und moderner Esoterik.* Graz: Ares, 1996.

37 Isaac I. Hayes. *The Open Polar Sea: A Narrative of a Voyage of Discovery Towards the North Pole, in the Schooner "United States"*. New York: R. Worthington, 1867, 3.

38 American Geographical & Statistical Society. *The polar exploring expedition. A special meeting of the American Geographical & Statistical Society*, held March 22, 1860. New York: American Geographical & Statistical Society, Anhang.

Abb. 2: Berghaus & Petermann: A Chart of the Arctic Regions (1860).

Die zirkumpolare Karte „Chart of the Arctic Regions“ von Berghaus & Petermann bildet den Nordpol und die angrenzenden Meere und Kontinente bis zum 50. Nördlichen Breitengrad ab. Unter Verwendung von Material des British Hydrographic Office bezieht sie sich auf Ergebnisse der Reise von Kane und die beabsichtigte Expedition von Hayes. Fokus der Karte ist der Nordpol, der genau in der Mitte emblemartig mit einem umrahmenden Schriftzug in Kapitalen eingezeichnet ist. Sehr klein sind die „ungefähren“ Lagen des amerikanischen und des asiatischen Kältepols verzeichnet, der Magnetpol ist nicht angegeben. Rings um den geografischen Nordpol fehlt bis zum 80. Breitengrad das Netz der Längen- und Breitengrade, so dass sich im Zentrum der Karte der Eindruck einer großen weißen Fläche ergibt, die zugleich mit Ausnahme von Spitzbergen und Smith-Sund auch dem bis dahin unerforschten Raum entspricht.

Die Karte setzt die Küstenlinien von Kontinenten und Inseln mit Schraffuren ab. Auf den Landflächen sind die wichtigsten geografischen Bezeichnungen, einige Flüssen und wenige Städte verzeichnet. Die nördlichen Flussmündungen Asiens und Amerikas werden als Zuflüsse in das Polarmeer bezeichnet. Auch die arktische Wasserscheide ist eingezeichnet, denn das Hauptaugenmerk der Karte liegt auf den Meeren. Dort werden besonders viele Angaben gemacht: Knapp über dem Nordpol verläuft quer der Schriftzug „Arctic Ocean". Auch die anderen großen Wasserflächen sind mit Kapitalbuchstaben bezeichnet und eine Fülle von Inseln, Kaps und Meeresstraßen wird namentlich benannt. Mit kleinen Pfeilen sind der Golfstrom und der Polarstrom von der Ostküste Grönlands bis nach Neufundland eingezeichnet. In diesem Gebiet sind auch Oberflächentemperaturen des Meeres angegeben; westlich von Spitzbergen gibt es wenige Angaben von Tiefenmessungen. Meerestiefen sind an keiner Stelle genannt. Nur in der Lancaster Street wird sehr klein auf den Walfang hingewiesen, obwohl von ihm immerhin ein Großteil des damaligen Wissens über das Nordpolarmeer stammte.

Der Arktische Ozean steckt auf der Karte voller Relationen und Vermutungen: die wesentlichen Erläuterungen sind mit den Zusätzen „approximate", „probably" und „probable", „farthest" und „proposed" versehen. An mehreren Stellen wird auf ein „offenes Polarmeer" hingewiesen: Nicht nur die Stelle, von der Kane das eisfreie Meer gesehen haben will, ist angegeben. Auf einer Einsatzkarte, die sonst nur wenig neue Informationen bietet, ist sogar das „Open water seen from C. Independence June 24 1854" verzeichnet. Zusätzlich steht über dem Pol die vage Erläuterung: „This Sea is probably never completely closed". Eine gestrichelte Linie markiert die Eisgrenze nach Norden hin, woraus sich der Umfang des eisfreien Polarmeeres in einer unregelmäßigen Elipse um den Nordpol ergibt. Die Angabe gilt nur für den Sommer, wie Legende und Erläuterung auf der Karte angeben. Augenfällig ist die Häufung von Grenzen: Eingezeichnet sind neben Längen- und Breitengraden auch der Polarkreis, das offene Polarmeer, ungefähre Begrenzungslinien von Packeis und driftenden Eisbergen, Regionen der Schiffbarkeit sowie die Wasserscheide (auf Land). Die Linien verweisen auf Wissens- und Erwartungshorizonte: sie stellen einen leeren Raum in der Arktis her, in dessen Zentrum der Pol wie Nabel und Achse der europäisch-amerikanischen Zivilisation in einem erscheint.

Die Karte vermittelt dadurch zwei wesentliche Aussagen. Erstens: es gibt im Arktischen Ozean ein Zentrum, das noch nicht bekannt ist. Zweitens: die Polarforschung ist auf dem Weg, es zu erreichen. Allerdings gaben Karten und Aufzeichnungen, die Polarexpeditionen mit nach Hause brachten, nicht nur wieder, was sie in der Ferne gesehen, sondern auch, was sie vorher erwartet hatten. In seinem Reisebericht vermischt Hayes den Blick auf das Polarmeer mit seinen Vorstellungen:

> It possessed a fascination for me, and it was with no ordinary sensations that I contemplated my situation, with one solitary companion, in that hitherto untrodden desert; while my nearness to the earth's axis, the consciousness of standing upon land far beyond the limits of previous observation, the reflections which crossed my mind respecting the vast ocean which lay spread out before me, the thought that these ice-girdled waters might lash the shores of distant islands where dwell human beings of an unknown race, were circumstances calculated to invest the very air with mystery, to deepen the curiosity, and to strengthen the resolution to persevere in my determination to sail upon this sea and to explore its furthest limits.[39]

Hayes entwarf das vermeintliche Ufer des offenen Polarmeeres hier als eine Außengrenze der bekannten Welt. Die vor ihm liegende Weite begriff er als Tor zu fernen Ländern. Halluziniert von der polaren Eiswüste, unter Verwendung exotistischer Motive und mit Hilfe einer lebendigen Fantasie imaginierte Hayes ein unbekanntes Volk, das er der europäischen Zivilisation zugänglich machen wollte. Das Polarmeer, seine möglichen Bewohner und schließlich der Pol selbst sollten in den eigenen Horizont integriert werden. Die Karte von Berghaus & Petermann dokumentiert zugleich, wie viele Anstrengungen dafür bereits unternommen worden waren. Denn nicht nur „Dr Hayes' proposed Line of approach" ist mit einer fast geraden Linie zum Pol eingezeichnet, sondern auch die jeweils nördlichsten Punkte, die über zwanzig Expeditionen seit dem 17. Jahrhundert erreicht hatten. Bei einigen sind Hunde-Schlitten als Transportmittel vermerkt – vielleicht weil sie damit naturgemäß an der vermeintlichen Küste des offenen Polarmeers scheitern mussten.

Das Projekt von Isaac I. Hayes fand die nahezu ungeteilte Unterstützung der American Geographical & Statistical Society. Der Reiseschriftsteller Bayard Taylor behauptete auf der Sitzung, dass die Existenz des offenen Polarmeeres „to my mind, does not admit of a doubt".[40] Der Astronom Benjamin A. Gould erklärte, "the almost universal faith of scientists in the existence of such a sea (...) is an inference from wellknown and undoubted physical facts."[41] Auch eine Reihe von prominenten Mitgliedern, die nicht zur Sitzung erschienen waren, schickten zustimmende Schreiben, darunter Alexander D. Bache, Direktor der US-Küstenvermessung, Joseph Henry, Leiter der einflussreichen Smithsonian

---

39 Hayes, *The Open Polar Sea*, 354.

40 American Geographical & Statistical Society, *The polar exploring expedition*, 13.

41 American Geographical & Statistical Society, *The polar exploring expedition*, 11.

Institution, und Matthew F. Maury, Begründer der amerikanischen Ozeanografie. Während Henry die Expeditionen anfangs kritisch sah, verband er mit ihnen später das Interesse, aus unzugänglichen Regionen wenigstens einige wissenschaftliche Daten zu erhalten. Henry hatte deshalb dafür gesorgt, dass schon der Expeditionsauftrag für Kane eigens um wissenschaftliche Fragen ergänzt wurde. Kane erhielt Messinstrumente, Einladungen zu Vorträgen und Empfehlungsschreiben also nicht weil, sondern obwohl er den Mythos des offenen Polarmeeres wach gerufen hatte.[42]

Die Vision, den Nordpol zu erreichen und nutzbar zu machen, lebte lange über das 19. Jahrhundert hinaus. Im Jahr 1949 griff beispielsweise Ernst Herrmann die Idee wieder auf. Die „unfruchtbarste, siedlungsfeindlichste Gegend der Erde“[43] sollte zum „Mittelmeer von morgen“ werden, weil die moderne Technik es ermögliche, die Arktis als Verkehrsraum zu nutzen: „Der Nordpol ist erreicht! Gut! Aber was nun? Der mathematische Punkt „Nordpol“ ist völlig wertlos. (…) Der Nordpol hat heutigentags nur eine einzige Bedeutung: man kann über ihn hinweg!“[44] Herrmann sprach zwar von Verkehr und Überquerung, sah den Pol aber weiterhin als Einzelpunkt. Seitdem hat das abschmelzende Meereseis den Zugang wesentlich erleichtert. Eine Kommission des US-amerikanischen National Research Council nahm dies im Jahr 2014 zum Ausgangspunkt der Schrift „The Arctic in the Anthropocene“:

> The Arctic has long been hidden from most of Earth's inhabitants. Physical access to key archives has been limited by sea ice cover, terrestrial ice cover, lack of research icebreakers, lack of terrestrial infrastructure, limited access, and the sporadic nature of international research campaigns. Much of what was previously concealed by logistical challenges is becoming increasingly accessible, aided by reduced sea ice, greatly improved remote sensing, and advances in instrumentation, analytical tools, and observational platforms. This means we can now discover what has long been unseeable.[45]

Für das National Research Council bedeutete das nichts anderes als ein spektakulärer Aufbruch in die wirtschaftliche Erschließung des Nordens. Die Inuit sollten darin eingeschlossen sein: „the region's indigenous peoples are now exercising greater political power: The Arctic is at the forefront of evolving gover-

42 Holtorf, Das offene Polarmeer.

43 Ernst Herrmann. *Das Nordpolarmeer – das Mittelmeer von morgen*. Berlin: Safari-Verlag, 1949, 7.

44 Herrmann, *Nordpolarmeer*, 264–265.

45 National Research Council, Committee on Emerging Research Questions in the Arctic. *The Arctic in the Anthropocene: Emerging Research Questions*. Washington D.C.: The National Academies Press, 2014, 56.

nance systems and cultural innovations compelled by rapid environmental and social changes."[46] Solche Visionen scheinen die Thesen von Hardt und Negri zu bestätigen: Aus der unzugänglichen, irritierenden und lebensfeindlichen Polarlandschaft wird eine in jeder Hinsicht geglättete Verkehrsfläche. An die Stelle des nationalstaatlichen Imperialismus treten in der Studie des National Research Council Kommunikation und Flexibilität, Netzwerke und Beratungsprozesse:

> Although national and regional governments remain powerful agents of policy making, global markets, intergovernmental forums, and nongovernmental organizations play an increasing role in determining the attractiveness and viability of economic development in the Arctic. Perhaps more important, though, is the evolving role of Arctic communities and institutions. In particular, the role of indigenous and other local communities, in an era where knowledge networks and consultative processes can play a prominent role in policy formation, is plausibly much greater than ever before. New and emerging research priorities need to focus on the ways that contemporary Arctic communities navigate and shape their evolving circumstances, drawing on a tradition of flexibility, resilience, and adaptive capacity in an environment of high natural variability.[47]

3 Offenheit

Doch haben sich nicht nur neue Chancen für Globalisierung und Marktwirtschaft eröffnet, sondern auch der Lebensraum Arktis hat sich durch das Abschmelzen der polaren Eiskappe verändert. Das National Research Council bemerkt zu Recht: „Never before has an ocean opened up before our eyes".[48] Auch Ernst Herrmann hatte schon bemerkt: „Dieses Nordpolarmeer ist an Flächeninhalt so groß wie ganz Europa. Es ist ganz oder in Teilen mehrfach überflogen worden. Trotzdem kennt man es nur wenig."[49] Der Blick richtet sich nun nicht mehr ausschließlich auf einen geografischen Punkt im Eis, sondern das Nordpolargebiet wird als Ozean bezeichnet, der von Strömungszirkeln und Gezeiten, Meerestiefen, Klimaschwankungen und periodisch veränderten Eisbewegungen bestimmt wird. Die naturräumliche Veränderung als „Glättung" zu interpretieren und in die Theorien von Hardt und Negri einzufügen, scheint naheliegend. Interessant ist jedoch erstens, dass sich der Blick auf das Nordpolarmeer als arktischem Ozean schon mindestens seit dem 19. Jahrhundert etabliert hat, und zweitens,

---

46 National Research Council, *The Arctic*, 1.
47 National Research Council. *The Arctic*, 41.
48 National Research Council, *The Arctic*, VII.
49 Herrmann, *Nordpolarmeer*, 291.

dass es heute Stimmen der Inuit gibt, die sich gegen ihre Inbesitznahme durch das „Empire“ zur Wehr setzen. Der arktische Raum und seine Geschichte sind also längst nicht so glatt, wie sie heute vielleicht von Bord europäischer Kreuzfahrtschiffe aussehen.

Dass es schon früh eine kartografische Alternative gab, zeigte ein gutes Jahrzehnt nach Berghaus & Petermann der österreichische General und Kartograf Franz Ritter von Hauslab. In einem Beitrag im Journal of the Royal Geographical Society befasste er sich 1875 mit Zugangsmöglichkeiten zum Arktischen Ozean; er bezog sich dabei auf einen 1871 aus Anlass der österreichischen Arktisexpedition in Wien gehaltenen Vortrag. Hauslab suchte weder Land noch Pol noch offenes Polarmeer, sondern analysierte das Verhalten des Meeres. Die Erdrotation bot für ihn nicht Anlass, am Pol nach der Achse zu suchen, sondern die Möglichkeit, Rückschlüsse auf ein globales ozeanisches Strömungssystem zu ziehen. Das Ziel von Expeditionen bestand demgemäß nicht im Erreichen eines einzelnen Punktes, sondern in der sicheren Überquerung der riesigen vereisten Wasserfläche. Starke Strömungen und Winde führten dazu, wie Hauslab erkannte, dass sich das arktische Packeis stark bewegt und ein fester „Nordpol“ auf der Oberfläche des Eises gar nicht existieren kann. Die Strecken der Expeditionsrouten und die jeweils nördlichsten von Menschen erreichten Punkte schufen bei dieser Betrachtung keinen Entwicklungsreihe, sondern die Möglichkeit, Einzelinformationen zu vernetzen und dadurch über einen längeren Zeitraum zu neuen Erkenntnissen zu gelangen.

Auf Hauslabs Karte „Map of the North Polar Region“ (Abb. 3) befand sich der Pol weder im Zentrum der Kartenprojektion noch der Strömungsbewegungen des Arktischen Meeres. Es war für Hauslab durchaus „possible, nay probable, that the centre of the great Polar basin is periodically free from ice“.[50] Allerdings verband er damit keine imperialen Fantasien, sondern glaubte, das Nordpolargebiet „will probably turn out to be a gulf rounded off at the end“. Zur Erläuterung des Strömungsverhaltens zog er Analogien aus der Beobachtung eines Wasserglases und empfahl: „explorers should in every case follow the ocean-currents“.[51]

Auch auf anderen Karten erschien das Meereis in Bewegung und hatte veränderliche Eigenschaften. Es gab Karten, die ihre Darstellung nach Jahren und Jahreszeiten differenzierten, permanentes, temporäres und treibendes Eis voneinander unterschieden, arktische Meeresströmungen abbildeten und nach Gezeiten

---

50 General Franz v. Hauslab. Inference Applied to Geography, with Special Reference to Ocean Currents and the Arctic Regions. *The Journal of the Royal Geographical Society* 45 (1875): 34–45, hier 44.

51 Hauslab, Inference, 44.

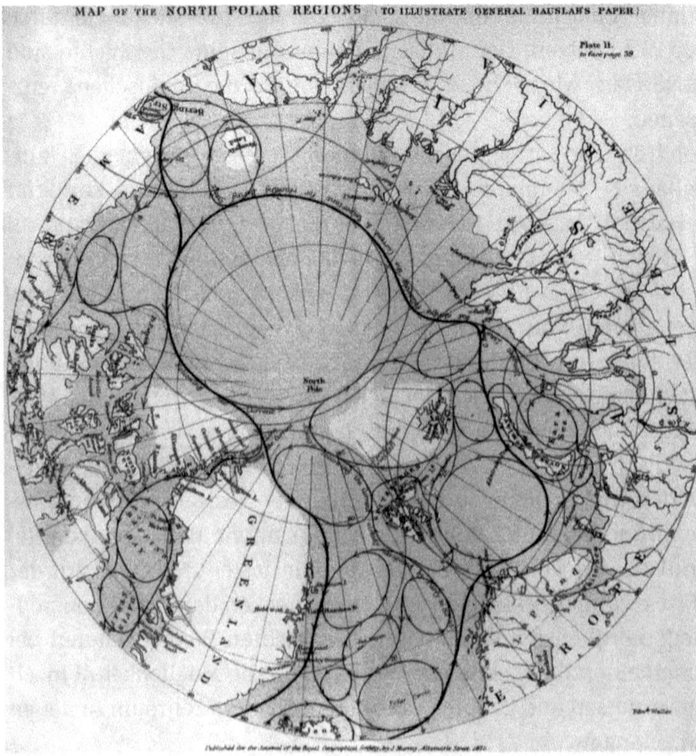

Abb. 3: General Franz von Hauslab: Map of the North Polar Region (1875).

und Erdrotation, Salzgehalt und Temperatur, Windrichtungen und Wolken differenzierten. Auf solchen Karten sind Richtungspfeile und Tiefenangaben, Wasserstraßen und Eisbewegungen zu erkennen; oft stand der geografische Punkt des Nordpols nicht im Zentrum oder war nicht einmal eingezeichnet. In den 1950er Jahren schrieb der Fotograf Ragnar Thorén über den Sinn von fotografischen Luftaufnahmen der Arktis, dass sie „give the interpreter a clear view of the conditions there and the ice drift. The photos are presenting him information about the ice not only concerning its concentration, size, topography, age and snow-coverage etc. but regarding temporary open or frozen cracks, leads, pools and polynias as well, and of the varying position of the ice edge."[52]

52 Ragnar Thorén. Fotografisk rekognoscering och folkring ar havsisen i Arktis. *Ymer* 4 (1958): 263–279, hier 277.

Der Nordpol unterliegt heute dem Seerecht. Es ermöglicht den westlichen Staaten, weite Teile der Arktis für die wirtschaftliche Nutzung zu beanspruchen, und bedroht dadurch das Überleben der jahrhundertealten Kultur der Inuit. Das National Research Council schreibt: „Internationally, the opening of the Arctic has raised issues of sovereignty and preparedness and spurred political realignment. Recently, the European Command identified the Arctic as a security concern. (...) With an abundance of fossil fuel deposits, minerals, and possible new fisheries, the Arctic attracts attention from industries and nations eager for new frontiers and opportunities for their economies and peoples."[53] Während der hier beschriebene Konflikt vor allem unter den angrenzenden Nationalstaaten geführt wird, sollen die Gefahren für die Inuit nur einerseits im Missverhältnis zwischen globalen Marktpreisen und lokaler Fischversorgung und andererseits in der zunehmenden Abwanderung und Alterung der Fischer bestehen.[54]

Doch der Konflikt mit den Inuit reicht tiefer, denn ihr Selbstbestimmungsrecht ist hier in Frage gestellt. Der dänische Politologe Jeppe Strandsbjerg behauptet, dass die eigentliche Auseinandersetzung zwischen unterschiedlichen Konstruktionen des Raums ausgetragen werde.[55] Auf der einen Seite stehe das Seerechtsübereinkommen der Vereinten Nationen, das davon ausgeht, dass den Kontinenten ein Festlandsockel von in der Regel 200 Seemeilen vorgelagert ist, der nicht zum Staatsgebiet des Küstenstaates gehört, aber dessen Ressourcen von diesem bis zu einer Meerestiefe von 2.500 Meter genutzt werden dürfen. Dieser Festlandsockel macht einen Großteil der Arktis aus. Strandsbjerg weist darauf hin, dass damit, wie von Latour behauptet, ein einheitlicher Naturraum vorausgesetzt werde, der nur mit Hilfe der in Europa entwickelten wissenschaftlichen Methoden der Geografie bestimmbar ist.[56] Der verwendete Begriff des Raums bleibt dabei abstrakt und nimmt keine Rücksicht auf Menschen, die darin leben, ihn nutzen und vielleicht ganz anders auffassen als die UN.[57]

Auf der anderen Seite steht das Raumverständnis der Inuit. Das Inuit Circumpolar Council Canada hat im Jahr 2008 Souveränitätsansprüche erhoben, die unterschiedliche Konstruktionen von Raum wieder zu einem Thema der Politik gemacht haben. Zunächst stellen die Inuit den messbaren Raum der Naturwissenschaften in Frage, indem sie erklären: „Inuit view the Arctic as the places

53 National Research Council, *The Arctic*, VII, 1.

54 National Research Council, *The Arctic*, 42.

55 Strandsbjerg, *Cartography*, 7.

56 Strandsbjerg, *Cartography*, 11.

57 Strandsbjerg, *Cartography*, 18.

where Inuit have traditionally lived."[58] Dabei unterscheiden sie nicht zwischen Festland und Eisfläche, denn „Land is anywhere our feet, dog teams, or snowmobiles can take us."[59] Die nomadische Lebensweise beruhe darauf, dass sie sich frei bewegen könnten, auch über das Meer, denn es verbinde ihre Gemeinschaften:[60] „The sea, for Inuit, is their highway. In wintertime, their highway is sea ice. In summertime, it is the open sea. The sea is integral to the Inuit way of life."[61]

Das Inuit Circumpolar Council Canada verweist daher auf die internationalen Vereinbarungen zum Selbstbestimmungsrecht der Völker. Es wehrt sich gegen eine Vereinheitlichung des Raums, die Grundlage der wirtschaftlichen Ausbeutung der Arktis sein würde. Diese Auseinandersetzung stellt auch die Thesen von Hardt und Negri in Frage, denn sie öffnet die Debatte über globale Interessenskonflikte in neuer Weise. Der Eindruck, dass das Schmelzen des Packeises einen bis dahin unzugänglichen „weißen Fleck" für die Förderung von Bodenressourcen erreichbar gemacht hat, habe, so Jeppe Strandsbjerg, „fantasies of exploration and expansion into terra nullius, and a new (black) gold rush" hervorgerufen.[62] Inwiefern solche „weiße Flecken" jedoch selbst nur eine Folge bestimmter Aufzeichnungsmethoden sind, die über diese Gebiete einen falschen Eindruck vermitteln, haben wir oben bereits gesehen.

Hardt und Negri räumen zwar ein, dass soziale Ungleichheiten und Spaltungen im Raum des „Empire" zunehmen können.[63] Sie geben auch zu, dass der „Raum imperialer Souveränität (...) kreuz und quer von so vielen Verwerfungen durchzogen" sei, dass von einer „Omni-Krise" gesprochen werden könne.[64] Sie erwähnen sogar das Vorhandensein von Alternativen und Gegenmächten des Empire.[65] Doch sie prognostizieren, dass „eine Situation permanenter sozialer Gefahren" lediglich „nach den mächtigen Apparaten der Kontrollgesellschaft verlangt, um die Demarkationslinien zu sichern und die neue Ordnung des gesellschaftlichen Raums zu garantieren."[66] Die „lokalen Differenzen" zu verteidigen sei „schädlich", doch ein universeller Rechtsbegriff ist für Hardt und Negri „das

58 Inuit Circumpolar Council. *The Sea Ice is Our Highway: An Inuit Perspective on Transportation in the Arctic. A Contribution to the Arctic Marine Shipping Assessment*, March 2008, 1.
59 Inuit Circumpolar Council, *The sea ice*, 2.
60 Inuit Circumpolar Council, *The sea ice*, i, vgl. 3.
61 Inuit Circumpolar Council, *The sea ice*, ii, vgl. 27.
62 Strandsbjerg, *Cartography*, 7.
63 Hardt, Negri, *Empire*, 345.
64 Hardt, Negri, *Empire*, 202.
65 Hardt, Negri, *Empire*, 14.
66 Hardt, Negri, *Empire*, 345.

Herzstück des Empire".[67] Die Frage ist also, ob der Protest der Inuit so tiefgreifend ist, dass er den Geltungsraum der „neuen Weltordnung", nämlich den von den empirischen Wissenschaften konstruierten, von internationalen Normen reglementierten und vom globalen Markt genutzten Naturraum selbst bestreitet.

Das Beharren der Inuit verschafft der europäischen Zivilisation eine neue Grenze. Es rückt den Erfahrungs- und Wissensraum der in der Arktis lebenden Menschen, aber auch der Reisenden, die der fremden Umgebung irritiert gegenüber stehen, in den Mittelpunkt. Die Konstruktion des Raums und des Globus steht dadurch in Frage. Als die amerikanischen Ureinwohner sich schon einmal nicht, wie Hardt und Negri schildern, „in die Logik der expansiven Grenzverschiebung integrieren" lassen wollten, wurden sie „vom Territorium ausgeschlossen". Denn „hätte man sie anerkannt, hätte es auf dem Kontinent keine wirliche frontier und keine offenen Räume, die man besetzen konnte, gegeben."[68] Wäre so etwas heute mit den Inuit wieder möglich?

Und was geschieht eigentlich, wenn der Nordpol als Punkt auf dem Eis verschwunden sein wird? Passiert überhaupt etwas? Möglicherweise richten sich die Ängste vor dem Klimawandel auch deswegen so stark auf die Arktis, weil das Bild eines maritimen Polarmeeres so ungewohnt ist und Irritationen über die Orientierung auslöst. Die Auseinandersetzung mit den Bildern der Arktis hat unter dem Eindruck der globalen Klima-Erwärmung eine hohe wissenschafts-, wirtschafts- und gesellschaftspolitische Aktualität gewonnen. In der Geschichte der Kartografie spielt das Nordpolargebiet zwar nur eine kleinere Rolle, doch wissensgeschichtlich hat die Idee des offenen Polarmeeres eine große Bedeutung. War es seit dem Mittelalter eine hoffnungsvolle Vermutung und am Ende des 19. Jahrhunderts ein widerlegter Irrglaube, ist es im dritten Jahrtausend – zumindest während der Sommermonate – zu einer Realität geworden, die das natürliche Gleichgewicht zu bedrohen scheint. In der Verteidigung ihrer Kultur und ihres Lebensraums scheinen die Inuit nun zu verfolgen, was Hardt und Negri selbst für das Ende des Empire gehalten haben: sie orientieren sich nicht mehr an der Bestimmung von Territorien und Grenzen, sondern leben nomadisch und leisten einen „Widerstand, der zu Liebe und Gemeinschaft wird".[69] Das Verständnis von Raum, dass die Inuit dem Empire entgegen halten, entspricht der Vision, die Hardt und Negri mit dessen Untergang verbinden: „Ein Raum, der lediglich durchquert werden kann, muss zu einem Lebens-Raum werden".[70]

---

67 Hardt, Negri, *Empire*, 59, 210.
68 Hardt, Negri, *Empire*, 182.
69 Hardt, Negri, *Empire*, 369.
70 Hardt, Negri, *Empire*, 369.

Literaturverzeichnis

American Geographical & Statistical Society. *The polar exploring expedition. A special meeting of the American Geographical & Statistical Society*, held March 22, 1860. New York: American Geographical & Statistical Society, 1860.

Bockstoce, John R. Whales, Ice, and Men. *The History of Whaling in the Western Arctic*. Seattle und London: University of Washington Press, 1986.

Bravo, Michael. *Geographies of exploration and improvement: William Scoresby and Arctic whaling, 1782–1822*. Journal of Historical Geography 32 (2006): 512–538.

Bravo, Michael und Sörlin Sverker (Hrsg.). *Narrating the Arctic. A Cultural History of Nordic Scientific Practices*. Canton: Watson Publishing International, 2002.

Cavell, Janice. *Tracing the Connected Narrative. Arctic Exploration in British Print Culture, 1818–1860*. Toronto: University of Toronto Press, 2008.

Deleuze, Gilles und Felix Guattari. *Tausend Plateaus. Kapitalismus und Schizophrenie*. Berlin: Merve, 1992.

Duzer, Chet van. The Mythic Geography of the Northern Polar Regions: Inventio fortunata and Buddhist Cosmology. *Culturas Populares. Revista Electronica* 2 (2006), 16 Seiten. http://www.culturaspopulares.org/textos2/articulos/duzer.pdf (Abruf am 4. Juni 2016).

Felsch, Philipp. *Wie August Petermann den Nordpol erfand*. München: Luchterhand, 2010.

Fleming, Fergus. *Neunzig Grad Nord. Der Traum vom Pol*. Hamburg: Rogner & Bernhard, 2003.

Glasze, Georg. Kritische Kartografie. *Geographische Zeitschrift* 4:97 (2009): 181–191.

Godwin, Joscelyn. Arktos. *Der polare Mythos zwischen NS-Okkultismus und moderner Esoterik*. Graz: Ares, 1996.

Hannemann, Matthias. *Der neue Norden. Die Arktis und der Traum vom Aufbruch*. Bad Vilbel: Scoventa, 2010.

Hannemann Matthias. Schmilzt die Arktis, Herr Fuchs? Gespräch mit einem Polar-Abenteurer. *Frankfurter Allgemeine Zeitung*, 4.2.2007.

Hardt, Michael und Antonio Negri. *Empire. Die neue Weltordnung*. Frankfurt und New York: Campus, 2002.

Harley, John B. *The New Nature of Maps. Essays in the History of Cartography*, Paul Laxton (Hrsg.). Baltimore und London: The Johns Hopkins University Press, 2001.

Hauslab, General Franz v. Inference Applied to Geography, with Special Reference to Ocean Currents and the Arctic Regions. *The Journal of the Royal Geographical Society* 45 (1875): 34–45.

Hayes, Isaac I. *The Open Polar Sea: A Narrative of a Voyage of Discovery Towards the North Pole, in the Schooner "United States"*. New York: R. Worthington, 1867.

Herrmann, Ernst. *Das Nordpolarmeer – Das Mittelmeer von morgen*. Berlin: Safari-Verlag, 1949.

Holtorf, Christian. Das offene Polarmeer. Ein Bilddiskurs im 19. Jahrhundert. In *Bilder in historischen Diskursen*, Franz K. Eder, Oliver Kühschelm, Christina Linsboth (Hrsg.), 145–172. Wiesbaden: VS Verlag für Sozialwissenschaften, 2014.

Holtorf, Christian. Der Nordpol als Aufmerksamkeitsmagnet. Neue Monografien zur Konjunktur der Arktis im 19. und 20. Jahrhundert. *NTM Zeitschrift für Geschichte der Wissenschaften, Technik und Medizin* 4:18 (2010): 537–544.

Holtorf, Christian. Der Nordpol – eine Erzählung. In *Erzählung und Geltung. Wissenschaft zwischen Autorschaft und Autorität*, Safia Azzouni, Stefan Böschen, Carsten Reinhardt (Hrsg.), 133–155. Weilerswist: Velbrück Wissenschaft, 2015.

Holtorf, Christian. Die Südsee im Norden. Technik und Transzendenz in Narrativen der Arktisforschung des 19. Jahrhunderts. In *Technik und Transzendenz. Zum Verhältnis von Technik, Religion und Gesellschaft*, Katharina Neumeister, Peggy Renger-Berka, Christian Schwarke (Hrsg.), 181–208. Stuttgart: Kohlhammer, 2012.

Inuit Circumpolar Council. *The Sea Ice is Our Highway: An Inuit Perspective on Transportation in the Arctic. A Contribution to the Arctic Marine Shipping Assessment*, March 2008. http://www.inuitcircumpolar.com/uploads/3/0/5/4/30542564/20080423_iccamsa_finalpdfprint.pdf (Abruf am 4. Juni 2016).

Jackson, C. Ian. William Scoresby the Scientist. *The Journal of the Hakluyt Society*, January 2010.http://www.hakluyt.com/PDF/Scoresby.pdf (Abruf 4. Juni 2016).

Jureit, Ulrike. *Das Ordnen von Räumen. Territorium und Lebensraum im 19. und 20. Jahrhundert.* Hamburg: Hamburger Edition, 2012.

Kraus, Karl. Die Entdeckung des Nordpols. *Die Fackel* 287, IX (1909).

Latour, Bruno. Die Logistik der immutable mobiles. In *Mediengeographie. Theorie – Analyse – Diskussion*, Jörg Döring, Tristan Thielmann (Hrsg.), 111–144. Bielefeld: Transcript, 2009.

Latour, Bruno. Give me a laboratory and I will rise the world. In *Science Observed*, Karin Knorr-Cetina, Michael Mulkay (Hrsg.), 141–169. London, Beverly Hills und New Delhi: Sage, 1983.

Loomis, Chauncey C. The Arctic Sublime. In *Nature and the Victorian Imagination*, U.C. Knoepflmacher, G.B. Tennyson (Hrsg.), 95–112. Berkeley und Los Angeles: University of California Press, 1977.

Martin, Constance. William Scoresby, Jr. (1789–1857) and the Open Polar Sea – Myth and Reality. *Arctic* 1:41 (1988): 39–47.

Marx, Friedhelm. „Die Paradiese des Südpols“. Phantastische Expeditionen ans Ende der Welt. In *Phantastik – Kult oder Kultur? Aspekte eines Phänomens in Kunst, Literatur und Film*, Christine Ivanovic, Jürgen Lehmann, Markus May (Hrsg.), 197–212. Stuttgart und Weimar: Metzler, 2003.

McCorristine, Shane. Träume, Labyrinthe, Eislandschaften: Körper und Eis in Arktis-Expeditionen des 19. Jahrhunderts. In *Weltmeere. Wissen und Wahrnehmung im langen 19. Jahrhundert*, Alexander Kraus, Martina Winkler (Hrsg.), 103–126. Göttingen: Vandenhoeck & Ruprecht, 2014.

Müller, Dorit. Fotografie und Südpolforschung um 1900. In *Frosch und Frankenstein. Bilder als Medium der Popularisierung von Wissenschaft*, Bernd Hüppauf, Peter Weingart (Hrsg.), 233–253. Bielefeld: Transcript, 2009.

Müller, Dorit. Kartierungen polarer Räume. In *Karten-Wissen. Territoriale Räume zwischen Bild und Diagramm*, Stephan Günzel, Lars Nowak (Hrsg.), 377–395. Wiesbaden: Reichert Verlag, 2012.

Müller, Dorit. Transformationen populären Wissens im Medienwandel am Beispiel der Polarforschung. In *Populäres Wissen im medialen Wandel seit 1850*, Petra Boden, Dorit Müller (Hrsg.), 35–79. Berlin: Kadmos, 2009.

National Research Council, Committee on Emerging Research Questions in the Arctic. *The Arctic in the Anthropocene: Emerging Research Questions*. Washington DC: National Academies Press, 2014.

Petermann, August. Der Nordpol und Südpol, die Wichtigkeit ihrer Erforschung in geographischer und kulturhistorischer Beziehung. *Mittheilungen aus Justus Perthes' Geographischer Anstalt* 11 (1865): 146–160.

Potter, Russell A. *Arctic Spectacles. The Frozen North in Visual Culture 1818–1875*. Seattle: University of Washington Press, 2007.

Porter, Russell A. Open Polar Sea. In *Encyclopedia of the Arctic*, Vol. 3, Marc Nuttall (Hrsg.), 1578–1580. New York und London: Fitzroy Dearborn Publ., 2005.

Riffenburgh, Beau. *The Myth of the Explorer. The Press, Sensationalism, and Geographical Discovery*. Oxford: Oxford University Press, 1993.

Robinson, Michael F. Reconsidering the Theory of the Open Polar Sea. In *Extremes. Oceanography's Adventures at the Poles*, Keith R. Benson, Helen M. Rozwadowski (Hrsg.), 15–29. Sagamore Beach: Science History Publications, 2007.

Robinson, Michael F. *The Coldest Crucible. Arctic Exploration and American Culture*. Chicago und London: The University of Chicago Press, 2006.

Ross, John. *Narrative of a second voyage in search of a North-west passage: and of a residence in the Arctic regions during the years 1829, 1830, 1831, 1832, 1833: including the reports of Commander, now Captain, James Clark Ross, R.N., F.R.S., F.L.S., &c. and the discovery of the Northern Magnetic Pole*. London: A.W.Webster, 1835.

Schneider, Ute. *Die Macht der Karten. Eine Geschichte der Kartographie vom Mittelalter bis heute*. 2. Aufl. Darmstadt: Primus, 2006.

Strandsbjerg, Jeppe. *Cartography and Geopolitics in the Arctic Region. DIIS Working Paper*. Copenhagen: Danish Institute for International Studies, 2010.

Tammiksaar, Erki, Natal'ja G. Suchova und I.R. Stone. Hypothesis Versus Fact: August Petermann and Polar Research. *Arctic* 3:52 (1999): 237–244.

Thorén, Ragnar. Fotografisk rekognoscering och folkring ar havsisen i Arktis. *Ymer* 4 (1958): 263–279.

Wamsley, Douglas W. *Polar Hayes. The Life and Contributions of Isaac I. Hayes, M.D.* Philadelphia: American Philosophical Society, 2009.

Wright, John K. The Open Polar Sea. *The Geographical Review* 3:43 (1953): 338–365.

Abbildungsnachweise

Abb. 1: Polarkarte von Gerhart Mercator (1569). (Gerardus Mercator. *Atlas sive Cosmographicae Meditationes de Fabrica Mundi et Fabricati Figura*. 1595. Wikimedia: https://commons.wikimedia.org/wiki/File:Mercator_north_pole_1595.jpg, 7.8.2015)

Abb. 2: Berghaus & Petermann: A Chart of the Arctic Regions (1860). (*American Geographical & Statistical Society. The polar exploring expedition. A special meeting of the American Geographical & Statistical Society*, held March 22, 1860. New York: American Geographical & Statistical Society 1860, Anhang. Library of Congress/Map Division).

Abb. 3: General Franz von Hauslab: Map of the North Polar Region (1875). (General Franz v. Hauslab. Inference Applied to Geography, with Special Reference to Ocean Currents and the Arctic Regions. In *The Journal of the Royal Geographical Society* 45 (1875), 34-45, Karte nach 38.) (Courtesy of JSTOR).

God(s) in the Empire: Mapping Imperial Religion

Katharina Waldner

Introduction
Empire and Religion

Empire and Religion, as well as empires and religions, are and were entangled in multifarious and complex ways through politics, philosophy, and theology as well as by religious practices. When I was confronted as a historian of ancient Mediterranean religion with Hardt and Negri's *Empire* and at the same time with the question of spatiotemporality, two fields of interest immediately caught my attention. First, spatiotemporality turned out to be a new perspective on central questions about the relation between power, religious practice, narratives and the body in ancient cultures, especially in the context of religion in the Roman Empire. Two of the contributors to this chapter, Alfred Schmid (Ancient History) and the Harry O. Maier (New Testament Studies) thus concentrate on religion in ancient Hellenistic empires and the Roman Empire itself. Second, the copious and somehow ambiguous use of the trope of the Roman Empire as well as of religion and especially of Christianity in Hardt and Negri's *Empire* is striking. Jens Kugele explores partly this second aspect in the third contribution of this chapter asking from the perspective of religious studies how Hardt and Negri use "religion" to reconstruct and analyse the interrelationship between power, souvereignty, individuals, normativity, and collective imaginations in their narrative of European national history. He then confronts their approach with his own clearly located study of this interrelationship analysing the spatiotemporality of a contemporary ceremony of American civil religion,[1] the memorial service for Neil Armstrong on September 13, 2012 at the National Cathedral in Washington D.C. In my following remarks I will try to show how these three case studies, two from the ancient Mediterranean and one from the contemporary United States, use a spatiotemporal perspective in different and highly enlightening ways. They thus help to clarify the way in which Hardt and Negri use "religion" and Christianity in their *Empire*. In the end, their discourse unintentionally and ironically mirrors not only Christian millenarianism but also fits new trends in the field of religious studies that turn from "religion and media" to "religion as media"[2].

If one tries to use the index at the end of Hardt and Negri's *Empire* to find passages on religion, one gets utterly disappointed: No entries for "religion" or "God" –

1 For the notion of American civil religion cf. Robert N. Bellah. Civil Religion in America. *Journal of the American Academy of Arts and Sciences* 96 (1967): 1–12.
2 Cf. Jens Kugele, see below, 322, esp. note 58.

DOI 10.1515/9783110418750-013

instead of that one for "superstition".[3] Of course, this is no surprise in a book written by Neo-Marxist intellectuals. On the other hand, what comes as a surprise is the fact that while reading the book, one recognizes that religion plays an important and sometimes ambiguous role on different levels of their argument. One could say the same about the Roman Empire,[4] which oscillates in comparable way between being pointed to as a classical example for cruel imperialism and being invoked as positive analogy to a new peaceful world order (Empire?) after the end of the Cold War.[5] Chris Fox aptly sums up the treatment of religion by Hardt and Negri by structuring his reading in the following chronological scheme:

> In the case of the past, this would be religion as the expression of transcendence. The present is the hybrid formed by combining the earlier legitimation of transcendence with some functions that belong to the structures of immanent rule. Finally, the future of religion appears to be its collapse as a distinct institution in favour of its uptake into a generalized theory of the multitude's constituent power. [6]

Beginning with the past, one can observe that Hardt and Negri's treatment of religion oscillates strongly. From the Renaissance onwards, religion is on the one hand on the side of the "second mode of modernity, constructed to wage war against the new forces", but otherwise it forms also part of modernity, being reformed "by a new life"[7]. Third, it is seen in a functionalist way, in its form as polytheistic civic religion as a kind of indispensible glue ("civic passion") for the archetypal Empire of the past overthrown by Christianity. Hardt and Negri thus follow Machiavelli's explanation of the decline of the Roman Empire:

> When Machiavelli discusses the fall of the Roman Empire, he focuses first and foremost on the crisis of civil religion, or really on the decline of the social relation that had unified the different ideological social forces and allowed them to participate together in open interac-

3 Michael Hardt and Antonio Negri, *Empire*. Cambridge, MA: Harvard University Press, 2001, 473–478.

4 For the many ways in which ancient Rome was used by European and American imagination cf. Richard Hingley, *Images of Rome. Perceptions of ancient Rome in Europe and United States in the modern age*. Ann Arbor, Michigan: Cushing-Malloy, 2001. For an overview on the discourse in Classics (esp. archaeology), cf. Richard Hingley, *Globalizing Roman Culture. Unity, Diversity, and Empire*. London: Routledge 2005, 14–48.

5 Hardt and Negri, Empire, 181; cf. also 167, where they prize „Roman peace" as described by Virgil.

6 Chris Fox. From Representation to Constituent Power: Religion, or something like it, in Hardt and Negri's *Empire*. *Journal for Cultural and Religious Theory* 9.2 (2008): 30–42, quote: 31.

7 Hardt and Negri, Empire, 74; Kugele, see below, 305.

> tion of counterpowers. Christian religion is what destroyed the Roman Empire by destroying the civic passion that pagan society had sustained, the conflictual but loyal participation of the citizens in the continuous perfecting of the constitution and the process of freedom.[8]

However, Alfred Schmid shows in his chapter that the pagan Roman Empire was not so much constituted by "civic passion" but rather by a thoroughly created image of the world that was based on the imagination of the world as a perfect globe. The possibility to combine power with the globe (in forging the concept of a "global" power) was provided by Greek metaphysics, when the world was newly conceived as a providential unity, symbolized by the globe. The highly efficient and complex "science" of astrology made it possible to combine space and time. The two were inseparably bound to the political order of the Empire conceived thus as "fatal", and especially to the person of the emperor with his individual horoscope. In this sense the Roman Empire was a kind of round, perfect "smooth space", but only for the elites who firmly believed in astrology.[9] In contrast to this, Harry O. Maier emphasizes in his chapter, that in the daily life of the Roman Empire "there were no 'smooth spaces', but rather a continuing contest over identity and time-space in innumerable social spheres, domains, and places."[10] He concentrates on some narrative texts, so called martyrdom acts, produced by the very first members of the Christian Church in the second and third century CE, which succeeded according the narrative by Machiavelli in bringing the Roman Empire to fall. Maier refers to Hardt and Negri's book *Commonwealth* of 2009, comparing the creation of these narrative texts to "events of resistance", pitted against the "smooth space". According to him, they create "a rupture and reconfiguration of hegemonic spaces".[11] The martyrdom acts recount the condemnation and/or the execution of Christians by the public authorities of the Roman Empire. In doing so they re-present place and time of the juridical procedure in a sophisticated way in order to transform the humiliating events into a triumph of Christian identity by means of narrative devices. To understand this kind of resistance Maier uses the concept of "thirdspace" first put forth by Edward W. Soja. Among other things, he can show that the authors implement the narra-

8 Hardt and Negri, *Empire*, 372–371.

9 This was often combined with a predilection for Stoic philosophy. For an account of astrology (subversively) used by minority groups in the Roman and Hellenistic empire(s) cf. Kocku von Stuckrad, *Das Ringen um die Astrologie. Jüdische und christliche Beiträge zum antiken Zeitverständnis*. Berlin: Walter de Gruyter 2000.

10 Harry O. Maier, see below, 254, note 3.

11 Maier, see below, 254–255, note 3, with reference to Micael Hardt and Antonio Negri. *Commonwealth*. Cambridge, MA: Belnap, 2009, 61.

tological device of focalization in order to create a kind of "cosmic" position for the reader/hearer from which he or she can observe the hostile observers of the cruel imperial spectacles and the martyrs. At the same time, it also gains insight into the victims' experience. This "cosmic position" precisely mirrors the "global" perspective of the imperial elite reconstructed by Alfred Schmid and will later turn into the figure of Christ as Kosmokrator. In this sense one might as well argue that the religious system did not abruptly change from pagan civic polytheism to the Christian monotheism which overturned the Roman Empire, but that the relationship between religion and politics as a whole was slowly transformed into a new shape.[12]

When dealing with contemporary issues, Hardt and Negri are even more vague about "religion", as Chris Fox has shown.[13] Fox especially draws our intention to the fact that although they were well aware of the position of religion as counterpart of the modern secular state and thus of its potential to be a moral corrective as well as in danger of being misused by the state, they put religion aside when describing "the Non-Governmental Organizations (NGO's) as the real vanguard of moral intervention", although they already before introduced religion and new media as tools of such interventions.[14] Elsewhere, Hardt and Negri state that "[r]eligious institutions are an even more long-standing sector of non-governmental institutions that represent the People", and immediately afterwards mention Islamic and Christian fundamentalism as such a form of resistance of the "People" against the state,[15] whereas they also call Islamic fundamentalism "a paradoxical kind of postmodernist theory – postmodern only because it chronologically follows and opposes Islamic modernism."[16] In sum, Hardt and Negri conceive in their construction of historian and contemporary issues religion from a mainly functional perspective. This dramatically changes in the very last chapter entitled "The Multitude Against Empire", dealing with the future and in which they describe the strength and political potential of the multitude in religious terms. Although they forcefully argue that there is "a world that knows no outside" that "knows only an inside" and refuse again and again any form of

12 For a historical perspective on religion in ancient Empires in general as well as in the Roman Empire cf. Greg Woolf. World Religion and World Empire in the Ancient Mediterranean. In *Die Religion des Imperium Romanum. Koine und Konfrontationen*, Jörg Rüpke and Hubert Cancik (eds.), 19–35. Tübingen: Mohr Siebeck, 2009.

13 Fox, Religion, 33–36.

14 Fox, Religion, 34 refering to Hardt and Negri, *Empire*, 36.

15 Fox, Religion, 35 quoting Hardt and Negri, *Empire* 311.

16 Hardt and Negri, *Empire*, 149, cf. the introduction to this volume, see above, 7 and Kugele, see below, 309.

transcendence,[17] they use religious tropes and motives from Christian religious tradition to describe this world to come. They start with an motto drawn from the *Das älteste Systemprogramm des deutschen Idealismus* which postulates a "material religion of the senses" (eine sinnliche Religion) for the great masses as well as for the philosophers.[18] The chapter gets a millenaristic touch by the repeated reference to the two cities of the *City of God* by St. Augustine (5th cent. CE). At first they are said to be abolished both,[19] whereas later it is insisted that there will be a "new city" [20] or that the "earthly city" has to be radically distinct from "the divine city" [21]. The radical immanence of the "earthly city" as well as the refusal of any forms and ideas of transcendence are a kind of leitmotiv in this very last chapter. But by the quote from the *Systemprogramm* this immanence is framed by "religion". Fox succeeds in interpreting this as consistent with the earlier passages in the book were religion is put together with mass media as a kind of communication mode:

> Religion here is redetermined through the immanence of materiality, and away from any form of transcendence. It therefore is the strength of the multitude that has found itself, and that has become able to dispel the parasitic order of sovereign power. Put somewhat differently, religion is the communicative matrix that compromises the interhuman world itself, and is identified with this self-related, immanent omniverse.[22]

In other words, on the map of the multitude, religion is a kind of blank space, because it only exists as a communicative function, and maybe also as the "civic passion" used in the description of the civic religion of the pagan Roman Empire by Machiavelli.[23] As if they wanted this blank spot to be inhabited at least by one person, Hardt and Negri introduce at the very end of the book the legend of St. Francis of Assisi who is imagined as having been able to live a religious life outside or beyond any institutional discipline.[24] But even if we might accept that religion in the future has to disappear in this way, should it not be possible to

---

17 Hardt and Negri, *Empire* 411.
18 Hardt and Negri, *Empire*, 393; cf. Fox, Religion, 40 and Kugele, see below, 310–311.
19 Hardt and Negri, *Empire*, 393.
20 Hardt and Negri, *Empire*, 395.
21 Hardt and Negri, *Empire*, 411.
22 Fox, Religion, 40.
23 It is remarkable, however, that they go on further in this direction in *Multitude*, by stating that "Christianity and Judaism (...) both conceive love as a political act that constructs the multitude": Michael Hardt and Antonio Negri, *Multitude. War and democracy in the age of Empire*. New York: The Penguin Press, 2004, 351.
24 Cf. Kugele, see below, 311.

map it in the present (empires)? That is exactly what happens in the second part of Jens Kugeles' chapter when he introduces a spatiotemporal perspective. Kugele does not only refer to Soja's Thirdspace but also to the anthropologist of religion Birgit Meyer who developed the notion of "aesthetic formation"[25] as valuable tool to analyze religious events and their production of so called "sacred space". Meyer insists on the materiality of human imaginations especially when "imagined communities" (Anderson) were produced by rituals; she calls them thus no longer "communities" but "aesthetic formations". Kugele can show by his analysis of the filmic documentary representation (DVD) of the memorial service for Neil Armstrong on September 13th , 2012 at the National Cathedral in Washington how this contingent "aesthetic formation" of American civil religion works precisely to perform a national memory by architectural and liturgical devices as well as to communicate a certain version of US imperialism which, of course, resembles Hardt and Negri's Empire. The film leaves out the passages of a speech by Kennedy in which he hinted on the military aspects of the Apollo 11 mission; instead it ends with his invocation of "God's blessing on the most hazardous and dangerous and greatest adventure on which man has ever embarked."[26] In contemporary cultures this kind of mass media performance of religious events and messages is not restricted to huge social units like the American state. Comparable to Hardt and Negri, Birgit Meyer's approach moves from the "religion and media" paradigm to the view of "religion as media". But in contrast to Hardt and Negri, for Meyer the "appeal of religious, mass mediated communities" is not a desirable future but a contemporary fact with challenging political consequences.[27] Mapping religion in a very concrete spatiotemporal sense may be useful for understanding this kind of challenge.

References

Fox, Chris. Form Representation to Constituent Power. Religion, or something like it, in Hardt and Negri's *Empire*. *Journal for Cultural and Religious Theory* 9.2 (2008): 30–42.

Hardt, Michael and Antonio Negri, *Multitude. War and democracy in the age of Empire*. New York: The Penguin Press, 2004.

Hardt, Michael and Antonio Negri, *Empire*. Cambridge, MA: Harvard University Press, 2001.

Hardt, Michael and Antonio Negri. *Commonwealth*. Cambridge, MA: Belnap, 2009.

25 Birgit, Meyer. *Aesthetic Formations: Media, Religion, and the Senses*. New York: Palgrave Macmillan. 2009.

26 See Kugele, below, 326.

27 See Kugele, below, 320, with reference to Meyer, *Aesthetic Formations*, 4.

Hingley, Richard. *Images of Rome. Perceptions of ancient Rome in Europe and United States in the modern age*. Ann Arbor, Michigan: Cushing-Malloy, 2001.

Hingley, Richard. *Globalizing Roman Culture. Unity, Diversity, and Empire*. London: Routledge 2005.

Meyer, Birgit. *Aesthetic Formations: Media, Religion, and the Senses*. New York: Palgrave Macmillan. 2009.

Stuckrad, Kocku von. *Das Ringen um die Astrologie. Jüdische und christliche Beiträge zum antiken Zeitverständnis*. Berlin: Walter de Gruyter 2000.

Woolf, Greg. World Religion and World Empire in the Ancient Mediterranean. In *Die Religion des Imperium Romanum. Koine und Konfrontationen*, Jörg Rüpke and Hubert Cancik (eds.), 19–35. Tübingen: Mohr Siebeck, 2009.

Harry O. Maier

Early Christian Martyrology, Imperial Thirdspace and Mimicry

1 Taking the Spatial Turn to the Arena

This discussion seeks a social-spatial analysis of early Christian martyrology. "Spatiality" is a term that describes "how space and social relations are made through each other; that is, how space is made through social relations, and how social relations are shaped by the space in which they occur."[1] Spatial theory contests a view that sees space as a kind of empty container, or an inert point on a map.[2] It rather conceives space as dynamic, living phenomena that people produce, imagine, and live in their interactions with one another and with the places they inhabit. As I hope to show, the dynamic features of space-time the martyr accounts open up belong to a world where there is a contest over place and identity and where imperial ideas are at once repudiated but also embraced to form new modalities. In this regard, these accounts conform to Hardt and Negri's bio-political conceptualization of space-time, adopted from Michel Foucault and Gilles Deleuze. "Events of resistance" they argue, "have the power not only to escape control but also to create a new world."[3] Placed against what they call the "smooth space" of the modern Empire that capital and its flow produces, "the productivity of the biopolitical event" produces a rupture and reconfigura-

1 Phil Hubbard and Rob Kitchin (eds.). *Key Thinkers of Space and Place*. 2n ed. Los Angeles: Sage, 2011, 499.

2 For this older conception of space, its philosophical and socio-political setting, and the ways in which spatial theory contests that view, David Harvey. *The Condition of Postmodernity. An Enquiry into the Origins of Cultural Change*. Cambridge, MA: Blackwell, 1990, 240–259; Edward S. Casey. *Getting Back into Place. Toward a Renewed Understanding of the Place-World*. Bloomington, IN: Indiana University Press, 1993, 1–40.

3 Michael Hardt and Antonio Negri. *Commonwealth*. Cambridge, MA: Belnap, 2009, 61; By Empire as "smooth space" they describe a place in which "subjectivities glide without substantial resistance or conflict". (Michael Hardt and Antonio Negri, *Empire*. Cambridge, MA: Harvard University Press, 2001, 98.) Their treatment in *Commonwealth* addresses key objections to their earlier formulation of "smooth space" as furnishing little socio-graphical room for local resistance and socio-geographical space-time practices. For this critique, David Featherstone. *Resistance, Space and Political Identities. The Making of Counter-Global Networks*. Oxford: Wiley, 2008, 124–127. Hardt and Negri's description of modern Empire contrasts sharply with the ancient Roman one, where there were no "smooth spaces," but rather a continuing contest over identity and time-space in innumerable social spheres, domains, and places. In that context the biopolitical account is more precise.

DOI 10.1515/9783110418750-014

tion of hegemonic space. A space-time consideration of martyrdom in the Roman Empire is an exploration of a biopolitical event from another period and time.

To take up this line of investigation is to bring spatial study into the discussion of early Christian texts, a discipline that has proven fruitful in the study of the Old Testament/Hebrew Bible and is proving its value in New Testament/Early Christian Studies.[4] So far as I am aware, with the exception of some discussion by Judith Perkins, no sustained social geographical analysis of the texts considered here has been undertaken, so this treatment is exploratory and invites further work.[5] Indeed, few topics present themselves so readily for social geographical considerations of emergent Christianity and the Roman Empire as the early accounts of the martyrs. Here Empire and Church are cast in settings charged with both temporal and spatial significance: household, courtroom, bridal chamber, birthing room, and of course the arena. The accounts place Christ followers in liminal spaces where they challenge traditional roles and expectations and reconfigure them for purposes of individual and communal self-definition.

The space of the arena, I hope to show, makes social relations, even as social relations marked out by the martyr accounts also make space – new spaces that engender new identities often in paradoxical and unpredictable ways. We see this in the two martyr reports this essay considers – the one that recounts the deaths of Christ followers at Vienne and Lugdunum ca. 177 C.E., recorded in a letter quoted by Eusebius of Caesarea in his Church History (5.1.1–5.3.4; hereafter *MVL*), and a second one recorded in the *Passion of Saints Perpetua and Felicitas* (hereafter *PPF*) in Carthage, traditionally dated to 7 March 203 CE – the date of the emperor Geta's birthday, but the discussion will also take us to other sites of emergent Christian spatial imagination of the first two centuries.[6] In particular I

4 For an overview with respect to biblical literature, Eric C. Stewart. New Testament Space/ Spatiality. *Biblical Theology Bulletin* 42 (2012): 139–150.

5 Judith Perkins. Social Geography in the Apocryphal Acts of the Apostles. In *Space in the Ancient* Novel, Michael Paschalis and Stavros A. Frangoulidis (eds.), 118–131. Groningen: Barkuis, 2002. In *Roman Imperial Identities in the Early Christian Era* she expands her discussion of the Acts to include martyrdom as a place of resistance and formulation of alternative civic identity. (Judith Perkins. *Roman Imperial Identities in the Early Christian Era*. London: Routledge, 2008, 107–126.) An earlier essay, "Space, Place, Voice in the Acts of the Martyrs and the Greek Romance" offers a similar spatial gesture. In these essays she deploys chiefly social geographical treatments of resistance. (Judith Perkins. Space, Place, Voice in the Acts of the Martyrs and the Greek Romance. In *Mimesis and Intertexuality in Antiquity and Christianity*, Dennis R. MacDonald (ed.), 117–137. Harrisburg: Trinity Press International, 2001.) My discussion should be read as complimentary to the excellent start she has made in her own approach.

6 For the backdrop and scholarly discussions of authorship, date, and background of the Lugdunum report, Candida R. Moss. *Ancient Christian Martyrdom. Diverse Practices, Theologies, and*

will analyse these accounts with the help of spatial theory outlined Edward Soja in his conceptualization of Thirdspace – a concept that proves especially fruitful for the kind of analysis undertaken here.[7]

Further, I will argue that the accounts of death in the arena created spectacles for early audiences to observe. By turning reading/listening audiences into observers these stories both describe and create space – the physical space of the amphitheatre and the viewers'/readers' space looking on. They conspire to create a form of theatricality, that is where observers see themselves observing and being observed, and are invited to adjust their reactions accordingly.[8] The vividly told stories of martyrs helped to create what Averil Cameron has described as an emergent Christian visuality that belongs more broadly to the spectator culture of the second century, and that was already inaugurated with the Gospel narratives and that in due course helped form what she describes as a new rhetoric of Empire.[9] The accounts of martyrs in the arena relish in the visual and devote a great deal of attention to detailed description of both martyrs and their tormentors.[10] The ancient amphitheatre was a place not only to see but also to be seen as differing social groups marked their identity through seating, viewing

---

Traditions. New Haven: Yale University Press, 2012, 100–116. For the Passion of Perpetua and Felicitas, Jan N. Bremmer. Perpetua and Her Diary. Authenticity, Family and Visions. In *Märtyrer und Märtyrerakten*, Walter Ameling (ed.), 77–120. Stuttgart: Franz Steiner, 2002; Jan N. Bremmer. The Vision of Saturus in the *Passio Perpetuae*. In *Jerusalem, Alexandria, Rome. Festschrift Ton Hilhorst*, F. Garcia Martinez and G. Luttikhuizen (eds.), 55–73. Leiden: Brill. 2003; Jan N. Bremmer and Marco Formisano (eds.). *Perpetua's Passions. Multidisciplinary Approaches to the Passio Perpetuae et Felicitatis*. Oxford: Oxford University Press, 2012.

In what follows I cite the text and translation of Herbert Musurillo (ed.). *The Acts of the Christian Martyrs*. Oxford: Clarendon, 1979, except for the *Martyrdom of Perpetua and Felicitas*, where I follow that of Bremmer and Formisano. I take for granted that the accounts of the martyrs functioned ideologically in a variety of different ways, that they are not all working in the same way, and that questions of audience and purpose are no less important than in the study of other forms of early Christian literature (Moss, *Martyrdom*, 164–166).

7 Edward W. Soja. *Thirdspace. Journeys to Los Angeles and Other Real-and-Imagined Places*. Oxford: Blackwell, 1996.

8 For theorization of theatricality and its application to the Roman Empire, Shadi Bartsch. *Actors in the Audience. Theatricality and Doublespeak from Nero to Hadrian*. Cambridge, Mass.: Harvard University Press, 1994.

9 Averil Cameron. *Christianity and the Rhetoric of Empire. The Development of Christian Discourse*. Berkeley: University of California Press, 1991, 47–88.

10 For an overview, Elizabeth A. Castelli. Persecution and Spectacle: Cultural Appropriation in the Christian Commemoration of Martyrdom. *Archiv für Religionsgeschichte* 7 (2005): 102–136 (a reworking of the chapter "Martyrdom and the Spectacle of Suffering" in her book: *Martyrdom and Memory. Early Christian Culture Making*. New York: Columbia University Press, 2004, 104–133).

one another, and witnessing the events that unfolded below them. In the martyrdom accounts readers/listeners have privileged ringside seats at the games; they become elite viewers, even as they are brought narratively into the arena and view the hostile crowd observing the events that unfold there. In the case of Perpetua, whose martyrdom in part takes the form of first person narrative, the account gets up close and personal, so that the reader's eye becomes the martyr's eyes and her ears become the audience's.[11] The textual amphitheatre of the Christian martyrology similarly places observers in place and marks their identity, in a cosmic sense that they are construed as being observed as fighters in a grand cosmic arena where God is in the audience, and in a more narratological sense, as audiences take on the point of view of the suffering protagonists.[12] Vivid spatiality served an important function in shaping emergent Christian identity. The authors who recounted the martyrdoms at Lyons and Carthage intended that their reports be read publicly to worshiping communities (*MVL* 5.1.1; *PPF* 8.6). As such they played a critical role in helping to form another spatiality associated with liturgical assembly and religious practice. There the martyrs were to be celebrated as models of endurance and true belief. To achieve that goal they deploy devices found in vivid Greek and Roman accounts of biographical reports of heroic virtue. Such accounts, as David Reis has shown, were designed to place a picture of virtuous accounts in the minds of readers. They formed a kind of mirror so that through meditation upon their example they might "play a catatropic role in self-improvement."[13] In the case of these martyrdom accounts, however, the role is not individual self-cultivation, but the formation of communal identity and self-definition. The accounts are not there to fashion a self but a communal spatial imagination, and, as we will see, an emerging religious identity in the Roman Empire.

---

11 This is what the narrative becomes as it is integrated into a third person narrative; Bremmer, Perpetua, 77–9, describes a graded history of composition through which a first person diary was adopted and included in a third person report.

12 In this aspect they are not dissimilar to the NT Apocalypse, where seeing and being seen functions as a critical hortatory device; see Harry O. Maier. Staging the Gaze. Early Christian Apocalypses as Narrative Self-Representation. *Harvard Theological Review* 92 (1997): 131–154; more recently, Christopher A. Frilingos. *Spectacles of Empire. Monsters, Martyrs, and the Book of Revelation*. Philadelphia: University of Pennsylvania Press, 2004.

13 David M. Reis. Spec(tac)ular Sights. Mirroring in/of Acts. In *Engaging Early Christian History. Reading Acts in the Second Century*, Rubén R. Dupertius and Todd Penner (eds.), 59–78. New York: Routledge, 2014, at 59–65. I am grateful to Professor Reis for making his essay available to me; also, Shadi Bartsch. *The Mirror of the Self. Sexuality, Self-Knowledge and the Gaze in the Early Roman Empire*. Chicago: University of Chicago Press, 2006, 15–56.

Finally, spatial analysis helps to lay bare modes of entanglement of the martyr stories with larger spatial and social realities. With the help of post-colonial analysis I will demonstrate that the martyrdom accounts discussed here mimic imperial claims and also revise them so that martyrology communicates a new set of claims that destabilize imperial ideas even as they echo them. We can see that the textual arena manifests hybridity and entanglement as imperial attitudes and values are reconfigured for new social identities and spatial practices. In this aspect, the narratives are both tokens and applications of lived ancient religion at once located in and removed from received practices and civic identities.

2 Remapping Roman Spaces

At the most general level, early Christian martyr stories remap the places where Rome expressed its claim to a universal imperium stretching across a vast geographical area, affecting perhaps some fifty million people: the courtroom and the arena, law and punishment. Martyrology configures an alternative Empire, a competing space, and another time – a universality on different terms. Two accounts offer striking examples, the first from the arena and the second from the courtroom.

The Martyrdom of Polycarp (hereafter *MP*) narrates the death of the second century bishop of Smyrna.[14] It offers a picture of a re-spatialized arena. The narrative gives the readers/listeners a front row seat at the stadion of Smyrna where the governor's interrogation of Polycarp unfolds before a bloodthirsty crowd. The story places the bishop in a *cognitione extra ordinem* – a public investigation of his crimes before a magistrate the function of which, amongst other things, was to make visible his social power.[15] The encounter between Polycarp and the governor, therefore, constitutes a reconfiguration of social power as the bishop trumps his interrogator. When the governor asks him to renounce his faith and swear by the genius of the emperor and curse Christ, Polycarp captures in a sentence a reformed spatiality: "For eighty six years I have served as his slave [ἔχω δουλεύων αὐτῷ] and he has done me no wrong. How can I blaspheme against my emperor and savior [τὸν βασιλέα μου τὸν σώσαντά με]" (*MP* 9.3). Slave, emperor, saviour: in a phrase Polycarp reconfigures the emperor's language and his claims to obedience centred in his salvific works and expresses an alternative

14 In what follows I cite translation of Musurillo, *Acts*.

15 Chris Epplett. Spectacular Executions in the Roman World. In *A Companion to Sport and Spectacle in Greek and Roman Antiquity*, Paul Christesen and Donald G. Kyle (eds.), 520–532. Oxford: Wiley-Blackwell, 2013, 527.

imperium, delivered before the tokens of emperor's presence – his governor, the arena, the *damnatio ad bestias*, and pagan imperial subjects. Polycarp's response is dramatic and memorable and conforms to a general pattern of martyrological speeches found across the reports. The language of martyrs is typically apophthegmatic and lapidary. It takes the form of pronouncements designed in an economy of perfectly timed and well-placed words to deliver a powerful punch. As such it conforms to the ideals of ancient rhetoricians, that less is more: the proper rhetor is to train himself in virtue and self-control of speech to know how to accomplish much in few words.[16] Indeed, like Jesus in the Synoptic Gospels, the events are sometimes arranged, as they surely are in this statement from Polycarp, for maximum impact. "If you delude yourself into thinking that I will swear by the emperor's Genius, as you say, and if you pretend not to know who I am. Listen and I will tell you boldly [παρρήσιας]: I am a Christian [Χριστανός εἰμι]" (*MP* 10.1). *Christianos eimi; Christianus sum*: the two words that mark, as Daniel Boyarin argues, the Christian martyr story, are here, as in the other accounts, voiced like thunder.[17] *Parrhēsia* – bold frank speech – was championed in the agonistic oral culture of Antiquity especially in situations of opposition and debate, and here Polycarp's biographer uses the term in a way exemplary of its larger cultural association.[18]

Similarly rhetorically and spatially conceived are the other forms of martyr stories, where it is not the arena but the courtroom that is the site for spatial practices. In these accounts another spatial configuration opens through legally styled interrogations or *colloquia*, presented as trial transcripts complete with witnesses and the paraphernalia of Roman law. In contrast to the focus on the passion of the martyr in the arena, the narratives create a different kind of spectacle amidst a cooler set of narrative devices.[19] The zinger responses Christians offer their interlocutors serve to turn the tables on the examining imperial authority so that it is no longer the Christian on trial but the one who holds judicial power.

The late third century *Acts of Maximilian* is typical in this regard. Maximilian, summoned to the forum of Tebessa in North African to be recruited for the army,

---

16 Harry O. Maier. The Politics of the Silent Bishop. Silence and Persuasion in Ignatius of Antioch. *The Journal of Theological Studies* 55 (2004): 503–519.

17 Daniel Boyarin. *Dying for God. Martyrdom and the Making of Christianity and Judaism*. Stanford: Stanford University Press, 1999, 93–126.

18 For bold speech, Rubén R. Dupertius. Bold Speech, Opposition, and Philosophical Imagery in Acts. In *Engaging Early Christian History. Reading Acts in the Second Century*, Rubén R. Dupertius and Todd Penner (eds.), 153–168. New York: Routledge, 2014.

19 For discussion of the accounts as akin to philosopher martyr stories, Moss, *Martyrdom*, 89–96, 125–29. As Moss suggests (129), the dialogues characteristic of these reports reflects the adoption rather than historical record of legal proceedings against Christians.

refuses enlistment. The martyrdom records the ensuing examination by proconsul Dion:

> Dion said to Maximilian: "Agree to serve [*milita*] and receive the military seal [*signaculum*]."
>
> "I will not accept the seal [*signaculum*]", he replied. "I already have the seal [*signum*] of Christ, who is my God."
>
> Dion said, "I shall send you to Christ directly."
>
> "I only wish you would, " he replied. "This would be my glory."
>
> Dion addressed his staff: "Let him be given the seal [*signetur*]."
>
> Maximilian resisted and said: "I will not accept the seal of this world [*signaculum saeculi*]; and, if you give it to me, I shall break it, for it is worthless. I am a Christian [*ego Christianus sum*]. I cannot wear a piece of lead around my neck after I have received the saving sign [*signum salutare*] of Jesus Christ my Lord, the son of the living God. You do not know him; yet he suffered for our salvation; God delivered him up for our sins. He is the one whom all we Christian serve [*servimus*]: we follow him as the prince of life and the author of salvation. [*vitae principem, salutis auctorem*]"
>
> "You must serve [*milita*]", said Dion, "and accept the seal [*signaculum*]– otherwise you will die miserably."
>
> "I shall not perish," said Maximilian. "My name is already before my Lord. I may not serve." (2.4–7; Musurillo, 246–247).

This is idealized Christian apology clearly and rhetorically designed to make a profound impression on the implied audience.[20] The transcript of Maximilian's trial overturns Roman claims to power by playing on the terms "seal," and "serve [*milita/servimus*]," as well as "*princeps*," "*auctoritas*," and "*salus*." Now the cosmos is courtroom, God is emperor/imperial commander, judge and ultimately

---

20 Robin Darling Young. *In Procession Before the World. Martyrdom as Public Liturgy in Early Christianity*. Milwaukee: Marquette University Press, 2001, 9–15 and Nicole Kelly. Philosophy as Training for Death. Reading the Ancient Christian Martyrs Acts as Spiritual Exercises. *Church History* 75 (2006): 723–747, read them as training for apologetics or preparation for martyrdom. These arguments both level the diverse nature of the reports by assigning them a single purpose and exceed the bounds of the evidence. Whatever else these reports are, they show the martyrs either engaging in spirited verbal combat with imperial opponents or using the spaces of punishment for communal self-definition, and as such contribute to a new formulation of the Roman Empire with the help of literary devices and rhetorical appropriation.

executioner or saviour, and guilt and innocence are referred to the conditions and achievements of Christ's atoning sacrifice. Space makes a new social relationship through inversion, even as Maximilian's and Deon's social relations are shaped by the articulation of space the transcript creates.

Transcripts such as this one parallel and are contemporary with the kinds of accounts found in the *Acta Alexandrinorum*, where we can read accounts of the so-called pagan martyrs.[21] Andrew Harker has argued that one should read these pagan acts as entertaining anti-Roman parodies of courtroom trials, composed for pro-Hellenic audiences who were convinced of the superiority of their Greek culture.[22] In the accounts the defendant typically bests Roman overlords in imagined examination and rebuttal. The courtroom provides a *mise-en-scène* for the reader/viewer to relish in dissent from Roman legal authority and imperial power. The Christian trial accounts, which were arguably built up from existing trial transcripts, function analogously, but place a strong emphasis on Christian endurance and courage. Through the form of a trial transcript they create a space also for challenge and response and to demonstrate the power of Christians to hold their own as witnesses to the true law and lawgiver.

3 Spectacle and Empire

Where the arena is the setting, ancient Christian martyrdom stories place their characters against a backdrop of a typical day of games. The accounts unfold in the judicial context of the *damnatio ad bestias*, the noontime event of a day of games, when criminals without status and guilty of capital crimes (*noxii/ damnati*) were executed. Sometimes, however, they portray them in gladiatorial contests, which were the afternoon entertainment. Gladiators were not always available in sufficient numbers and were expensive so sometimes they were matched against criminals, who scarcely could defend themselves.[23] This is the case in a vision of the North African martyr Perpetua who sees herself transformed as a man pitted

21 Herbert Musurillo (ed.). *Acta Alexandrinorum. De mortibus Alexandriae nobilium fragmenta papyracea Graeca*. Leipzig: Teubner, 1961.

22 Andrew Harker. *Loyalty and Dissidence in Roman Egypt. The Case of the Acts Alexandrinorum*. Cambridge: Cambridge University Press, 2011. Harker convincingly shows how unlikely it is that the transcripts are genuine, and argues instead that they are built from readily available public documents, which furnished a template for composing proceedings.

23 For an overview, Donald G. Kyle. *Sport and Spectacle in the Ancient World*. Oxford: Blackwell, 2007, 312–329.

against an Egyptian gladiator (*PPF* 10.1–15). Egyptians were celebrated for their prowess in the arena, and the narrative makes this one fiercer by associating him with the devil (10.14).[24] He is no match for Perpetua, however. Her vision is both vivid and entertaining and reveals knowledge of gladiatorial and gymnastic contests.[25] In the account she discovers herself transformed into a male, and, improbably, she is victorious. As such she conforms to the male gaze of the arena with its seats segregated according to gender and status and its top-down organization of observers that replicates an ideology of masculine self-mastery and regulation. "Then I was floating in mid-air and started hitting him without really touching the ground. But when I saw that there was a lull, I put my hands together, interlaced my fingers, and grabbed his head. He fell down flat on his face and I stepped on his head" (10.11; Bremmer and Formisano, 16). The account depicts the battle, seconds and assistants (*adiutores et fautores*), and even an *agônothetês/lanista/editor* – the host and referee who has the final decision whether the defeated gladiator should be executed by the victor.[26] The *lanista* is a man who is "huge, towering over the top of the amphitheater, with a loose robe on – a purple tunic framed by two stripes in the middle of the chest – and elaborate sandals made of gold and silver. He was holding a staff like a gladiatorial trainer, as well as a green branch which had golden apples on it" (10.7; Bremmer and Formisano, 19). This is exemplary *ekphrasis* and communicates accurately the costume of the official. Its vivid language conforms to the goal of ekphrasistic speech outlined in the *Progymnasmata* – elementary handbooks of the first three centuries C.E. for rhetorical training – to turn listeners into spectators.[27]

---

24 For discussion of the diabolic features of the Egyptian, that unfortunately sometimes moves far beyond the limits of the evidence, Peter Habermehl. *Perpetua und der Ägypter oder Bilder des Bösen im frühen afrikanischen Christentum. Ein Versuch zur Passio Sanctarum Perpetua et Felicitatis*. Berlin: De Gruyter, 2004, 145–177.

25 See Bremmer, Perpetua, 77–120.

26 Louis Robert. Une vision de Perpetue martyre a Carthage en 203. *Comptes rendus des séances de l'Académie des Inscriptions et Belles-Lettres* 126:2 (1982): 228–276, at 258–262 with a visual representation at 244, fig. 8; for a related figure and host of games, the *editor*, see Shelby Brown. Death as Decoration. Scenes from the Arena on Roman Domestic Mosaics. In *Pornography and Representation in Greece & Rome*, Amy Richlin (ed.), 180–211. Oxford: Oxford University Press, 1992, which also discusses the presence of iconography of the games in the *triclinia* of the homes of the games' patrons as entertainment.

27 For example, Theon, *Progymnasmata* 7.118: ekphrasis is "descriptive language, bringing what is portrayed clearly [ἐναργῶς] before the sight"; George A. Kennedy (ed., trans.). *Progymnasmata. Greek Textbooks of Prose Composition and Rhetoric*. Atlanta: Society of Biblical Literature, 2003, 45. For discussion of *ekphrasis* in the Roman period and its psychology, Ruth Webb. *Ekphrasis*

Vivid speech in ancient rhetorical practice was designed to awaken emotion and hence participation. Such vividness may also have functioned with a view to entertainment. Across the Empire people surrounded themselves with pictures of cruelty and violence; the material experience of violence in the arena was "freeze-framed" in representations of *damnatio ad bestias* in media ranging from mosaics in villas to terra sigilatta, to oil lamps manufactured for mass consumption.[28] However incomprehensible to a modern mindset, it is an inescapable that apart from serving the obvious function of reminding people of the consequences for breaking the law, everyday viewers found this imagery entertaining.[29]

This is in fact confirmed by Tertullian's treatise *On the Games* where he describes Christian attendance at the games and the theatre. Tertullian invites his audience to share in Schadenfreude where he uses *ekphrasis* to urge Christians to forsake the entertainment of the pagan games and instead imagine (and enjoy) another spectacle, the final one where God will liquefy his enemies with fire in the Last Judgment.[30] He uses vivid speech to awaken participation in the event and the emotions that typically went along with it. This alerts us to a possible role

---

Imagination, and Persuasion in Ancient Rhetorical Theory and Practice. Surrey: Ashgate Publishing, 2009, 87–130.

28 For the ubiquity of representation of cruelty and suffering in the visual world of the Roman Empire, Donald G. Kyle. *Spectacles of Death in Ancient Rome*. London: Routledge, 1998, 91–94; Melissa Barden Dowling. *Clemency and Cruelty in the Roman World*. Ann Arbor: University of Michigan Press, 2006, 226, 230–232. For the various contexts, also with examples, in which art of combat was displayed, especially baths and arenas, Steven L. Tuck. Representations of Spectacle and Sport in Roman Art. In *A Companion to Sport and Spectacle in Greek and Roman Antiquity*, Paul Christesen and Donald G. Kyle (eds.), 422–477. Oxford: Wiley Blackwell, 2013; for oil lamps and other media, David Potter. Martyrdom as Spectacle. In *Theatre and Society in the Classical World*, Ruth Scodel (ed.), 53–88. Ann Arbor: University of Michigan Press, 1993, at 68–70. Christine Kondoleon. Timing Spectacles. Roman Domestic Art and Performance. In *The Art of Ancient Spectacle*, Bettina Bergmann and Christine Kondoleon (eds.), 321–341. New Haven: Yale University Press, 1999, argues that the representation of the games in domestic mosaics also contains critical elements related to time and the imperial calendar. The mosaics show, she argues, a domestic ideology of mastery of time and space.

29 For a discussion John R. Clarke. *Art in the Lives of Ordinary Romans. Visual Representation and Non-Elite Viewers in Italy, 100 B.C.–A.D. 315*. Berkeley: University of California Press, 2003, 130–159. For the social psychology of cruelty as entertainment in the Roman world, Garrett G. Fagan. *The Lure of the Arena. Social Psychology and the Crowd at the Roman Games*. Cambridge: Cambridge University Press, 2011.

30 Tertullian, *On Spectacles*, 30; for imitation of the ideology Tertullian rejects, Simon Goldhill. The Erotic Eye. Visual Stimulation and Cultural Conflict. In *Being Greek Under Rome. Cultural Identity, the Second Sophistic and the Development of Empire*, Simon Goldhill and Robin Osborne (eds.), 154–194. Cambridge: Cambridge University, at 181–184.

of these martyrdom stories that a purely pious reading overlooks. The colourful account of Perpetua's battle with the Egyptian assumes that the audience hearing the episode knows what a typical combat of the type described looked like, and that its author could assume a set of emotions that would be triggered through its account. People came to the arena expecting to see certain things, but, as Garrett Fagan has argued, they were also socialized by it to respond to what they saw both behaviorally and affectively.[31] In his *ekphrasis* of fiery damnation, as Simon Goldhill aptly observes, "Tertullian is a wildly excited spectator of the judgment day...."[32] The *Martyrdom of Polycarp* describes the execution of the bishop by cremation – namely burning at the stake. The flames, however, do not kill Polycarp and so a *confector* plunges a dagger into his body, whereupon a gush of blood extinguishes the flames (*MP* 16). The scene is theological, of course. The entire narrative of Polycarp's arrest, trial, and execution is mimetic of Jesus' Passion; here Polycarp's blood teaches something about the salvific effects of a sacrificial atonement. But this is not only theological. Surviving iconography, depicting both the *damnatio ad bestias* and gladiatorial combat, oozes with blood. These are accounts for public consumption, not for discussion in university seminars.

4 Amphitheatres as Spaces of Imperial Power and Social Relations

Scholars have sometimes trivialized the arena as pure entertainment designed to tranquilize populations with bread and circuses.[33] Such accounts are too crude and ignore the critical ideological and social function of amphitheatres and the events hosted there.[34] The construction of amphitheatres went hand in hand

31 Fagan, *Lure*, 123–139; also David Potter, Martyrdom, 53–54.

32 Goldhill, Eye, 183.

33 For example, Paul Veyne. *Bread and Circuses. Historical Sociology and Political Pluralism*, Brian Pearce (trans.). London: Penguin, 1980; Paul Veyne. The Roman Empire. In *A History of Private Life*, Philippe Ariès and George Duby (eds.), vol. 1, *From Pagan Rome to Byzantium*, Paul Veyne (ed.), Arthur Goldhammer (trans.), 5–234. Cambridge, MA: Belnap, 1992, 200–202, where he lists the games as exemplary of the "pleasures and excesses" of private imperial life.

34 Erik Gunderson. The Ideology of the Arena. *Classical Antiquity* 15 (1996): 113–151; Erik Gunderson. The Flavian Amphitheatre. All the World as Stage. In *Flavian Rome. Culture, Image, Text*, A.J. Boyle and W.J. Dominik (eds.), 637–658. Leiden: Brill, 2003.

with expansion of and continuing renewal of Roman power.[35] As Jörg Rüpke has argued they were part of the communicative action of the Roman Empire – a means of making the Empire present and visible and assimilating dispersed populations into a shared identity.[36] Romans paid close attention to space – where to build cities, how to mark their boundaries, how to lay the grid of streets, and where to place temples and buildings. And they also attended to how to spatialize and thus create a Roman landscape. This is the case, for example, in the Lyonnais countryside in Gaul were some 30 arenas dotted the landscape, ranging in size from structures that could hold 7,000 to 20,000 spectators, each roughly two days journey from one another. This distribution has been explained as a means of creating a network of structures for the administration of Roman justice and as a strategy for creating Roman space out of territory populated by conquered Gallic tribes.[37]

The amphitheatre belonged also to the colonization of the countryside. "The spread of Roman forms was the spread of Roman Imperium."[38] In Gaul, as elsewhere, arenas took various forms: full amphitheatres like the Coliseum, semi-amphitheatres, and mixed edifices or theatre-amphitheatres with both an elliptical arena and a stage. These spaces were far more than mortar and bricks; they were signs of Roman hegemony. Army barracks were often connected with full amphitheatres, making them a ready venue for military parades and the public display of Roman military prowess. In Lyons, where the report of martyrs of Vienne and Lugdunum unfolds, the amphitheatre was structurally integrated within a sacred compound created through the presence of an imperial altar and temple, the first one to be erected in Roman Gaul, and which inaugurated the cult of the emperor in the Western Empire.[39] Carthage, where the martyrdoms of Perpetua and Felicitas most probably took place, was also a centre of the imperial cult. There, as in Lugdunum, entertainment facilities were conceived spatially to form an integrated Roman architectural complex. The amphitheatre and circus were erected

---

35 Alison Futrell. *Blood in the Arena. The Spectacle of Roman Power*. Austin: University of Texas Press, 2001, 53–76.

36 Jörg Rüpke. Das Imperium Romanum als religionsgeschichtlicher Raum. Eine Skizze. In *Ein Pluriverses Universum. Zivilisationen und Religionen im antiken Mittelmeerraum*, Richard Faber and Achim Lichtenberger (eds.), 333–351. Paderborn: Fink, 2015.

37 For the distribution and their function, Futrell, *Blood*, 66–76.

38 Futrell, *Blood*, 54. See also, Greg Woolf. *Becoming Roman. The Origins of Provincial Civilization in Gaul*. Cambridge: Cambridge University Press, 2000, 106–141, for the building of Roman monuments and colonization.

39 See Futrell, *Blood*, 79 for the complex at Lugdunum and 79–92 for the intimate link between the imperial cult and the arena in Roman Gaul.

perpendicularly along the city's two main arteries, the *decumanus* and *cardo* respectively, 750 metres from the forum. To date the imperial temple has not been found; archaeologists hypothesize it was either adjacent to the amphitheatre or in the nearby forum, proximate enough for processions to the arena.[40]

We remember that spatiality is a study of how space shapes social relations and social relations shape space. The amphitheatre is an excellent case study for spatial investigation. The games played an important role in marking identity and status in the Roman world. As Monique Clavel-Lévêque argues, the celebrations of the games worked both to instantiate and renew the view that the gods appointed the Romans to rule the world and to make visible and reinforce social hierarchies.[41] Their spatial form worked to mark and reinforce social relations.[42] People went to the games both to see and to be seen. From the time of Augustus, the seating of the *cavea* was regulated according to status and gender.[43] Those of highest status – governing elites – sat in the privileged seats closest to the arena; men sat in a second tier and women and children in the third most distant section. In a social world where visuality counted for much, this spatial configuration allowed those of lower status to see elites watching the entertainments unfolding below them. Those with less status were hierarchically organized further back. The lowest point of the amphitheatre, the sand-covered area where executions and combats were staged, was where the audience witnessed the erasure of status. Criminal executions were designed to strip the condemned of all social identity. Crowds came to the arena expecting to see prisoners dressed up in costumes to reenact what K. Coleman calls "fatal charades": the brutal deaths narrated in Greek and Roman myth.[44] Or they looked forward to seeing the condemned reveal their servile status by being compelled to fight unarmed with ferocious animals, or, watching them battle with gladiators, professional

---

40 For the location, construction, history of the arena at Carthage, D. L. Bomgardner. *The Story of the Roman Amphitheatre*. New York: Routledge, 2002, 128–146; for its relationship to the imperial cult, Duncan Fishwick. *The Imperial Cult in the Latin West. Studies in the Ruler Cult of the Western Provinces of the Roman Empire*. Leiden: Brill, 1987, 178–179.

41 Monique Clavel-Lévêque. L'espace des jeux dans le monde romain. Hégémonie, symbolique et pratique sociale. In *Aufstieg und Niedergang der römischen Welt*, 16.2.3, Hildegaard Temporini and Wolfgang Haase (eds.), 2405–2563. Berlin: De Gruyter, 1986, at 2513–2563.

42 Gunderson, Ideology, 123–126.

43 For widespread archaeological evidence of this from western arena, Fagan, *Lure*, 96–120.

44 Kathleen M. Coleman. Fatal Charades. Roman Executions Staged as Mythological Enactments. *Journal of Roman Studies* 80 (1990): 44–73, for erasure of identity, Potter, Martyrdom, 65–66.

fighters who slaughtered them.[45] The bodies of the dead were dragged from the arena by hooks and disposed of in ways that did not permit a proper burial and constituted a pagan form of damnation.[46] The corpses of the Lugdunum martyrs are ripped apart by dogs and left exposed or are incinerated and the ashes swept into the Rhone (*MVL* 5.1.59, 63). This was with a view to damnation: "And they did this as though they could overcome God and deprive the martyrs of the restoration, in order, as they themselves said, 'that they might have no hope in the resurrection in which they put their trust when they introduce this strange new cult among us…'." Ignatius, the early second century bishop of Antioch turns this practice into a reward when he uses ekphrastic speech to describe his anticipated execution as a *noxius* or *damnatus* in Rome: "Let me be food for the wild beasts…. [C]oax the wild beasts, so that they may become my tomb and leave nothing of my body behind, lest I become a burden to anyone once I have fallen asleep. Then I will truly be a disciple of Jesus Christ, when the world will no longer see my body" (*Ign. Rom.* 4.1–2). Oil lamps have been found all across the Empire depicting *damnatio ad bestias, venatores* (where animals fought to the death), and gladiator contests. They are tokens of the widespread popularity of the games, even as they show the internalization and penetration of the imperial spatial order across a vast geographical area.

These events were like everything else in the Roman world highly ritualized performances, and, as in the case of the games at Lugdunum and Carthage where the arenas were attached to imperial temples, explicitly laden with political meaning. Evidence from Lugdunum and Carthage, as well as other literary accounts of games, allows us to reconstruct the kinds of rituals involved to inaugurate these occasions. The event Eusebius reports probably took place at a chief imperial celebration, the annual summer meeting of the Gallic nations.[47] The event would have drawn great crowds for the celebratory games paid for by the national representative chosen to be the annual chief priest.[48] It began with a religious procession or *pompa* of magistrates and gladiators that started at the imperial temple.[49] The imperial arenas of the western Empire contained statues

45 Rose MacLean. People on the Margins of Roman Spectacle. In *A Companion to Sport and Spectacle in Greek and Roman Antiquity*, Paul Christesten and Donald G. Kyle (eds.), 569–578. Oxford: Wiley Blackwell, 2013.

46 Kyle, *Spectacles*, 131–133.

47 Duncan Fishwick. The Federal Cult of the Three Gauls. In *Les martyrs de Lyons*, J. Rougé and R. Turcan (eds.), 33–45. Paris: CNRS, 1978.

48 John F. Drinkwater. *Roman Gaul. The Three Provinces, 58 BC–AD 260*. New York: Routledge, 2014, 78.

49 For an outline of the ceremony, drawn from Roman practice, Clavel-Lévêque, L'espace, 2436–2446.

of gods, predictably like Mars or Hercules. Nemesis, however, was the chief god associated with the games, in this case not as vindictive judge and jury, but the power of changing fortune.[50] The sacral nature of amphitheatre space was marked by *nemesea* or shrines dedicated to Nemesis. Where found they always occupy conspicuous places, usually built beneath the short axis, with access to the presider's box or the main entrance, or at the extreme end of the long axis, either opposite the main entrance or part of it. Futrell explains how Nemesis furnished the arena with both sacred and political meanings. She gave poor spectators victim to often brutal forces outside their control a brief moment of respite to watch others battling with their destiny. Further, she was associated in Roman ideology with Fortuna and Pax and communicated the divinely appointed destiny of imperial rule. We might say that audiences at the games thus lived a double spatiality, one their own private destiny inhabited, and another mythic one of divine fortune appointing the world as space of Rome and its emperor. The arena as a mythic and imperial space formed a hierarchical set of relations of those without status, elites, emperors, and gods, even as the people who most benefited from that order created spaces to consolidate their power and legitimate it by imbuing it with a sacred character.

5 Martyrs and Thirdspace

It is against this spatial imperial backdrop that we should understand early Christian martyrology. The martyrdom accounts as narratives of execution in the time and space of the arena offer excellent material for the kind of spatial analysis outlined by the American social geographer Edward Soja. Soja draws on the socio-geographical theorization of Henri Lefebvre who distinguishes between perceived, conceived, and lived space. Soja renames these, first, second and thirdspace.[51] In Soja's appropriation of Lefebvre he designates perceived/

50 Futrell, *Blood*, 110–120.

51 Soja, *Thirdspace*, 53–83; Henri Lefebvre. *The Production of Space*, Donald Nicholson-Smith (trans.). Oxford: Blackwell, 1991, 31–33. Soja's use of Lefebvre has been criticized for drawing firm distinctions where the latter has not made them and for over simplifying Lefebvre's conceptualizations of the production of space; see the discussion of Christian Schmid, "Henri Lefebvre's Theory of the Production of Space: Towards a Three-Dimensional Dialectic." In *Space, Difference, Everyday life: Reading Henri Lefebvre*, ed. by Kanishka Goonewardena, Stefan Kipfer Richard Milgrom and Christian Schmid (London: Routlegde, 2008), 7–45, at 41–43; also Christian Schmid, *Stadt, Raum und Gesellschaft. Henri Lefebvre und die Theorie der Produktion des Raumes* (Stuttgart: Steiner, 2003), 308–313.

firstspace to describe the physical empirically defined world; conceived/secondspace represents the way the physical world is designed by people so that it functions according to particular social and cultural blueprints; lived/thirdspace refers to individual and social appropriations of space as practice. Soja's interest in thirdspace is in the surprising and unanticipated practices that arise out of conceived/second space blueprints for social organization and behaviour.

We might conceptualize this as follows: the Gallic countryside with its geological and topographical features constitutes firstspace. Roman imperial builders conceptualized this firstspace as a secondspace means of organization. For example, they built roads, founded cities, transformed existing towns through monumental structures, and erected amphitheatres for the distribution and exercise of power and the execution of justice. Thirdspace however arises when the practices of justice are appropriated: the arena conceived as a way to inscribe imperial power on the gathered viewers by spatial organization and on the bodies of criminals by violent entertainment becomes the occasion for a provocative set of new practices, including reading practices of commemoration and devotion. Soja considers thirdspaces as sites for creative and often ironic reversals of the perceived-conceived union.

The martyrdom accounts of Lugdunum and Carthage are nothing if not reversals. Each report of those who are allegedly stripped of all identity, in order to suffer the cruel punishment of the Empire and a bloodthirsty crowd, opens up a new kind of spectacle and creates a new mode of spatiality that resides in a narrative account and serves the liturgical functions of the martyr reports read in sacred assemblies. In the case of the Carthaginian martyrs, for example, the men are compelled to put on the robes of the priests of Saturn, and Perpetua and Felicitas those of the priestesses of Ceres (18.4). Perpetua resists this and then says, astonishingly, "The point of deliberately going this far was not to have our freedom taken away; the point of staking our lives was to avoid doing this kind of thing. That was our agreement with you" (18.5; Bremmer and Formisano, 22). Perpetua talks back to the Empire not as an erased subject but one with status and authority; she is not compelled to die, she chooses to do so and thus suffers a noble death rather than an ignoble one. Further the night before their execution, Perpetua and her companions hold a love feast in obvious imitation of the celebratory supper of gladiators the evening preceding the games (17.1–3).[52] The narrator consciously plays on imperial space here: "On the day before the event

52 Jan N. Bremmer. The Motivations of Martyrs: Perpetua and the Palestinians. In *Religion im kulturellen Diskurs. Festschrift für Hans Kippenberg*, Brigitte Luchesi and Kocku von Stuckrad (eds.), 535–554. Berlin: de Gruyter, 2004, at 536–537.

as well, when in place of that last supper that they call the 'dinner of the free [*liberam*],' they celebrated for all they were worth not a dinner of the free but a banquet of divine love [*non cenam liberam sed agapem*]" (17.1, Bremmer and Formisano, 21).

A similarly imperial space opens up in Perpetua's first vision, where she sees a narrow ladder with imperial weapons on its sides. Jan Bremmer convincingly argues that we should understand the ladder and its weapons as a reference to the *gradus* that appears explicitly in 6.2, the scaffold where the Roman judge was seated for interrogation and execution, also equipped with weapons and instruments of torture. If so, the ladder/*gradus* that in imperial spatial terms functions as an avenue of execution is the means for Perpetua's ascent to heaven and the trampling of dragon that would destroy her – a reference to Genesis 3.15, but also perhaps to the beast of Revelation, where it is closely associated with the Roman Empire (Revelation 12.13,18; 13.4,11). This is a dramatic reconceptualization of Roman torture and punishment. Finally, on the day of the games, it is the martyrs who perform a liturgical *pompa*, not the imperial dignitaries: "The day of their victory [*victoriae*] dawned and they processed [*processerunt*] from the prison into the amphitheater cheerfully, as if to heaven, with composed expressions, except perhaps for an excitement caused by joy and not by fear" (18.1, Bremmer and Formisano, 21). These are exemplary thirdspace formulations that reprise imperial configurations and improvise upon them to perform space in ways the builders of the arena could scarcely have conceived.

6 SpaceTime

The martyrdom stories we have considered in this essay are told with vivid speech. They are designed to place readers there where the action is happening. The narratives focalize on those who suffer rather than take on the point of view of the hostile audience. They reverse the flow of power the imperial arena seeks to inscribe, and turn observers into the observed. Or, more boldly, they adopt first person narration so that events unfold in real time with play-by-play reporting. The result is that the reader occupies a twofold space – both on the ground with the martyr, and somewhere above the arena, with a kind of bird's eye view.

In the case of the martyrs in Gaul the report comes in the form of a retrospective report of the persecution sent in a letter form, which allows the narrator both to number himself amongst those who suffered but to live to tell the tale. This permits a dynamic temporal and spatial perspective. In the case of the African martyrs, space is similarly created through report, but this time with first person narration of events first by Felicitas and then Saturus. The verb *video* appears often in the Latin MSS to inaugurate Perpetua's vision narratives (4.3; 7.4; 8.1; 10.1) and Saturus's visions regularly unfold in the first person plural, *videmus* (11.4; 12.3; 13.1). English translations of the first person accounts of Felicitas and Saturus as for example the one by Musurillo regularly mistranslate these as simple past tenses, I/we saw.[53] This occludes the Latin historic present that communicates a dramatic and even oracular quality to narration.[54] The result is that audiences are transported out of one time and into another, and hence from one space to another. I will return to this below. Here it is important to observe how effectively first person present tense narration contributes to an alternative creation of space because it both places the audience in the real time of the martyrs and it brings them into the visionary spaces that mark this martyrdom account as so unique.

Finally, the move to a third person narration at the end of the account gives an arena perspective, but heavily controlled by the narrator's interpretations. This is immediately evident in the narrator's report of Perpetua's battle with a heifer. From the perspective of the Roman crowd the battle with the heifer dehumanizes her in a typical combat of unequally matched opponents even as it associates her perhaps with the "fatal charade" of Dirce and the Bull and perhaps Pasiphaë. Both stories exploit gender – Dirce with the obvious reference to a woman gored by a bull and Pasiphaë through intercourse with one. For the audience, matching Perpetua with a heifer is a burlesque. The narrator, however, develops the scene to emphasize a different set of social realities, no less gender specific. S/he emphasizes Perpetua's modesty when she pulls down the tunic the heifer has torn, to cover her legs, and her comportment and matronly virtue when she asks for a pin to fasten her hair after it was mussed up by the attack. The narrator explains, "for it was inappropriate to suffer martyrdom with her hair in that state, lest she seem to be mourning in her hour of glory" (*PPF* 20.5; Bremmer and Formisano, 22). Correctly coifed hair was of course a means of signifying female virtue; imperial iconography relishes in depictions of the conquered, grief-stricken and terrorized female with unbound and messy hair to depict the lack of civilization and vice of those vanquished. The narrator also controls the point of view of the watching audience in the arena. Improbably, when it sees that Felicitas is a "beautiful girl [*puellam delicatam*]" and that Perpetua is "fresh from giving birth and with dripping breasts" the crowd is "aghast [*horruit]*" (20.2, Bremmer and Formisano, 22).

53 Musurillo, *Acts*, 107–131.

54 Erin Ronnse. Rhetoric of Martyrs. Listening to Saints Perpetua and Felicitas. *Journal of Early Christian Studies* 14:3 (2006): 283–327, at 318–323.

In *MVL*, Blandina is pitted against a bull. Here there is perhaps again an oblique reference to the "fatal charade" of mythical reenactment of the story of Pasiphaë mated with a bull, a sufficiently frequent motif at these executions to inspire capturing images of them on oil lamps. The account begins with Blandina entering the arena as a "noble mother [μήτηρ εὐγενής]" who encourages her children. The narrator continues, "After being tossed a good deal by the animal, she no longer perceived what was happening because of the hope and possession of all she believed in and because of her intimacy [ὁμιλία with Christ" (5.1.56; Musurillo, 79–81). The term ὁμιλία is here possibly a play on the word since it can mean both communion and sexual intimacy. In this case, the narrator controls the interpretation even as he gives a play-by-play description of the disturbing event. We will return to gender and space directly, but it is sufficient to notice here how the narrative transforms space into a bridal setting in the most unlikely of places, an event designed to conjure images of bestiality.

In the case of the *Martyrdom of Lyons* the arena forms a provocative anti-imperial space time. The whole document is framed by an *inclusio* portraying the events as a spectacle of a cosmic athletic contest.[55] It begins by stating that Satan has gone to all lengths "to train and prepare (προγυμνάζων) his minions against God's servants" (5.1.3) and ends with a summary of the martyrdom as "the greatest of all contests [μέγιστος...ὁ πόλεμος]... waged against the Demon" (5.2.5). As the narrative unfolds it deploys vivid speech in order to turn the entire event into a kind of Christian cosmic spectacle: the martyrdom's protagonists Maturus, Sanctus, Blandina and Attalus are led into the amphitheatre "to be exposed to the beasts and to make a public spectacle of the inhumanity of the people [εἰς κοινὸν τῶν ἐθνῶν τῆς ἀπανθρωπίας θέαμα]" (5.1.37). Blandina's torture is a "spectacle to the world [θέαμα γενόμενοι τῷ κόσμῳ]" (5.1.40). Thus the event works as a mirror that reflects a series of spatial realities. With this spectacle come surprising reversals.

The events the *inclusio* brackets upend the status of those whose identity is erased as *noxii*. It is not in the privileged seats of the stands where spectators see noble status displayed. It is amongst those stripped of identity. Blandina is a "noble athlete [γενναῖος ἀθλητής]" (5.1.19). Attalus enters the arena "as a warrior [ἀγωνιστής] well prepared for the contest by reason of his good conscience. For he had been nobly trained in the Christian discipline [γνησίως ἐν τῇ Χριστιανῇ συντάξει γεγυμνασμένος] and had always been a witness among us

55 For discussion of the way the *MVL* uses athletic body performance, especially with reference to Blandina where it bends culturally constructed gender expectations, see Brent D. Shaw. Body/Power/Identity. The Passion of the Martyrs. *Journal of Early Christian Studies* 4:3 (1996): 269–312, at 307–312.

to the truth" (5.1.43). "Surely it behooved these noble athletes, after sustaining a brilliant contest and a glorious contest, to win the great crown of immortality [τοὺς γενναίους ἀθλητὰς ποικίλον ὑπομείναντας ἀγῶνα καὶ μεγάλως νικήσαντας ἀπολαβεῖν τὸν μέγαν τῆς ἀφθαρσίας στέφανον]," comments the narrator (5.1.36; also 5.1.37). The martyrdoms offer a new grammar of seeing via a narrative spectacle. The narrator describes how Blandina when hung on a post as bait for wild animals "seemed to hang there in the form of a cross." Her companion martyrs "in their torment with the physical eyes saw in the person of their sister him who was crucified for them" (5.1.41). Here she takes on a masculine imperial cultural identity. They see that Blandina "has put on Christ, that mighty and invincible athlete, and had overcome the Adversary in many contests and through her conflict had won the crown of immortality [ἀθλητὴν Χριστὸν ἐνδεδυμένη, διὰ πολλῶν κλήρων ἐκβιάσασα τὸν ἀντικείμενονκαὶ δι' ἀγῶνος τὸν τῆς ἀφθαρσίας στεψαμένη στέφανον]" (5.1.42). The narrator focalizes the narration so that it is through the martyr's eyes that we see the spectacle Blandina performs and from the space occupied by the tortured that an Empire is turned on its head.

Whereas contemporary artistic representation of the executed typically represents the condemned as downcast and humiliated – the visual expression of the erasure of identity – the martyrs of Lyons advance into the arena "joyously, with majesty and great beauty mingled on their countenances, so that even their chains were worn on them like some lovely ornament." The narrative magnifies this posture when it contrasts the faces of the murderers and criminals who appear alongside them as "dejected, downcast, ill-favored, and devoid of all comeliness" (5.1.35). The contrast in facial expressions serves to magnify the innocence of the martyrs, but more importantly it helps to mark their conquest of the imperial space that would inscribe Roman victory upon their bodies. It relishes in the anti-imperial performance of Blandina the slave: she is someone whom "people think cheap, ugly, and contemptuous," whom Christ has nevertheless chosen to enjoy glory before God (5.1.17). Unlike her master who is afraid that her female weakness will make her less bold in Christian confession, Blandina "was filled with such power that even those who were taking turns to torture her in every way from dawn until dusk were weary and exhausted" (5.1.18). In all of this the arena becomes a thirdspace. The spatial configuration of the arena as a place for punishment becomes the locale for the confession of an alternative sociality and set of practices.

The Passion of Perpetua and Felicitas is a far more complex narrative from both a spatial and temporal point of view. Like the Lyons account the story places the martyrs in the arena with an explicit reference to *damnatio ad bestias* (*PPF* 6.6), Perpetua's battle with the Egyptian and the heifer, and the other martyrs' various battles with animals, as well as language of victory (18.1), the crown (19.2), and the final dispatch by the sword (8.7–9). But it also opens up a much more variegated topography and spatial imagination than the one found in Eusebius's *Church History*. The narrative makes imperial space and time explicit by stating the martyrdom occurred in military games held to celebrate the birthday of the emperor Geta (7.9). This creates a space-time context for another set of births. Significantly, Perpetua is a new mother and Felicitas is eight months pregnant when the passion story begins. The story turns repeatedly to details dedicated to giving labour, weaning, and proper care of the martyrs' newborns (2.2; 3.8,9; 6.7–8; 15.1–7; 18.3). When Felicitas appears in the arena the narrator describes her as "fresh from giving birth and with dripping breasts" (20.2). Felicitas' martyrdom is itself a birth: "rejoicing to have survived childbirth so that she might battle against beasts, going from one blood bath to another, from midwife to gladiator, to bathe after childbirth in her second baptism" (18.3; Bremmer and Formisano. 21). At the same time, even as she looks forward to being reborn, she remains pregnant with Jesus. As she goes into labour to give birth to her daughter guards tell her this does not compare with the pain she is about to endure. "Now I am the one suffering what I suffer," she replies, "but then another inside me [*alius erit in me*] will suffer for me, since I will also be suffering for him" (15.6; Bremmer and Formisano, 21).

The narrative inscribes domestic space when it emphasizes Perpetua's identity as "a newly married woman of good family and upbringing [*honeste nata, liberaliter instituta, matronaliter nupta*]" (*PPF* 2.1), as well as a nursing mother (2.2) and through her first person account of the appeal of her father to renounce Christ (3.1–5; 5.2–6; 6.2,5). Her good upbringing and identity as true matron are reinforced by the references to anxiety for the care of her baby (3.8; 6.7), and, of course, her care for her tunic and her hair in the battle with the Egyptian (20.5). The narrator describes her as a "bride of Christ [*matrona Christi*]" (18.2) and makes sure that we recognize her as a good one. Placed against the backdrop of games celebrated in honour of the emperor's birthday, we see a complex engagement with imperial time and space, as well as domestic rituals and identities. In one way, the narrative focalizes the women as ideal inhabitants of male spaces, even as they are seen from a male point of view. At the same time, it shifts traditional expectations by focusing on how these women turn away from the traditional obligations of daughters to fathers and mothers to infants, to faithful religious confession and thus practice their female identity in new ways.[56] The narrative

56 For further discussion, Hanne Sigismund-Nielsen. Vivia Perpetua. An Indecent Woman. In *Perpetua's Passions. Multidisciplinary Approaches to the Passio Perpetua et Felicitias*, Jan N. Bremmer and Marco Formisano (eds.), 103–117. Oxford: Oxford University Press, 2012.

views these women as enacting a culturally constructed masculinity even as they deflect the narrator's male gaze in favour of provocative gender performances.[57] The result is a new space-time imaginary and set of practices.

Further, the vision sequences open up space-time in dramatic ways. Five visions transport readers/listeners outside the prison cells where the martyrs await their death to other spaces: a garden (4.3–10); a pool of water (7.1–8 and 8.1–4); a gladiatorial combat in the amphitheatre (10.1–10); and a garden with a road that leads to a place with walls constructed of light (11.1–13.8). Each is a first person report and each is presented with a mixture present and past tense narration. Perpetua "sees" [*video*] a great ladder leading up to the garden (4.3), even as she sees her brother Dinocrates (7.4), and the Egyptian (10.1,7). Saturus and his companions "are carried [*gestamur*]" to heaven (11.5) and "come [*venimus*]' to the walls of light and "see [*vidimus*]" an aged man with white hair (12.3). They "stand [*stetimus*]" before a throne (12.4). Later the "go out [*exivimus*]" and "see [*vidimus*]" Optatus and Aspasius (13.1); they "speak [*loquimur*]" with them; they "begin to recognize [*ceopimus...cognoscere*] martyred brothers (13.6). These visions move the reader not only between spaces, but between times: Perpetua sees her coming battle; Optamus and Aspasius are dead as the martyred brethren are, and so on. Modern translations render these verbs in the past tense, but their present indicative form is significant. In the first place, it serves the apologetic purposes of the martyrdom as a whole, which is outlined in its preface and reiterated at its conclusion, that visions are not a thing of the past but continue into the present (1.3–5; 11). And secondly, the present tense conveys this dramatically: it underscores the active, present nature of the visions. As Erin Ronsse argues, translations of these active verbs into the past tense tend to turn the reports into a kind of dream or trance journey that is later recollected. Present tense verbs make the accounts active. Spatially they help to make the places they describe both vivid and contemporary.

Zalda Maldonado-Pérez, although not using the tools of spatial geography, captures the new sense of space-time that the narrative creates. She notes that the "dynamic ability of the martyr to exist between present-future horizons...defied all boundaries of time and space, life and death, powers and principalities. The

57 For the male point of view, L. Stephanie Cobb. *Dying to be Men. Gender and Language in Early Christian Martyr Texts*. New York: Columbia University Press, 2008, 60–123; for masculine culture constructions and improvisations, Rebecca Solevåg. *Birthing Salvation. Gender and Class in Early Christian Childbearing Discourse*. Leiden: Brill, 2013, 215–240; Elizabeth A. Castelli. 'I Will Make Mary Male'. Pieties of the Body and Gender Transformation of Christian Women in Late Antiquity. In *Body Guards. The Cultural Politics of Gender Ambiguity*, Julia Epstein and Kirstina Staub (eds.), 29–49. London: Routledge, 1991, at 33–43.

result is a subaltern existence that is uniquely subversive."[58] Here imperial time and space – the emperor's time of his birthday and his military arena and Empire – are destabilized and re-inscribed through spatial imagination. In Soja's terms, these are thirdspaces; the place created for the emperor, conceived for execution, has become the space for a whole new set of practices and relationships to emerge.

The narrative performance of space leaves nothing as it was – gender, domestic responsibilities, the normal succession of time, incarceration, torture, execution. Further the narrative turns the pagan observers into the witless observed. From a literary point of view, through first person accounts we the audience are privy to a set of events they can scarcely imagine. They are unaware that they have no control over the space they claim to govern. They cannot even kill Perpetua. "[S]he redirected the aimless hand of the inexperienced gladiator right at her neck. Perhaps such a great woman [*tanta femina*], who was feared even by an unclean spirit, could not have been killed unless she herself had wished it" (10.9; Bremmer and Formisano, 23). Perpetua dies a noble death in defiance of the space that exists to humiliate her. And once again, as in the case of the martyrs of Lyons, execution is victory and the arena is a spectacle for triumph: "The day of their victory dawned and they marched from the prison to the amphitheatre joyfully as though they were going to heaven, with calm faces, trembling, if at all, with joy rather than fear, Perpetua followed them with a steady gait, like a bride of Christ or the favorite of God, parrying the gaze of all with the strength of her own" (18.1–2; Bremmer and Formisano, 21). The vivid description plays in the contact zone of imperial ideology communicated through picture language. Readers/listeners shaped by ubiquitous imperial iconography depicting terrorized and downcast barbarians would have immediately understood what kind of counter-imperial description this is. From an imperial point of view, these executions fail to reach their intended outcome and imperial ideology is overturned. The arena does not strip the executed of identity, it rather is the place where it is won.

58 Zaida Maldonado Perez. *The Subversive Role of Visions in Early Christian Martyrs*. Lexington: Emeth, 2011, 103–104.

7 Mimicry and Entangled History

As we have seen, Soja describes thirdspace as "Other spaces." It resides in space and time and has as its precondition the ideologies and meanings of the places it contests and reverses. We have further seen that the Roman amphitheatre had a critical role in the communicative action of the Roman Empire as a locale through which social hierarchies could be both seen and practiced, and where it could inscribe its power on the natural world and criminals alike. In their studies of the role of the games in the capital, Egon Flaig and Keith Hopkins have shown how the games were highly performative and socially ritualized spaces where people and emperor (or the local governors) met, power was displayed, and authority claimed and conferred. Or not; it was also a space where the people could and did express their dissent to the emperor.[59] They have shown that it was not only in the arena that a contest for power unfolded; it could also occur in the stands when crowds expressed their dissatisfaction with the emperor and his policies. Another avenue for voicing resistance was the martyrdom report. Martyrdom accounts re-politicized the executed by turning them into models of virtue and piety and reworked humiliation narratives so that the spaces of power and social hierarchy were inverted. The thirdspace the martyrs' accounts create express a contest over imperial power and a paradoxical instantiation of it.

Thirdspaces are entangled spaces. They exist within society as reproductions of social orders and as improvisations of them. Such appropriation Michael Werner and Bénédicte Zimmermann entitle *histoire croiseé*.[60] As the name implies, *historie croisée* attends to historical entanglement, that is the way ideas, practices, and vocabularies intersect in history and their modes of transportation from one domain to another.[61] Histoire croisée "points toward an analysis of

59 Egon Flaig. Den Kaiser herausfordern. *Historische Zeitschrift* 253 (1991): 371–384, at 373–374. I am grateful to Katharina Waldner for bringing this article to my attention; Keith Hopkins. *Death and Renewal*. Cambridge: Cambridge University Press, 1983, 14–20. Increasingly representation of the games included happy spectators, as sign of successful rule by their emperor, see Richard Lim. 'In the Temple of Laughter'. Visual and Literary Representations of Spectators at Roman Games. In *The Art of Ancient Spectacle*, Bettina Bergmann and Christine Kondoleon (eds.), 343–365. New Haven: Yale University Press, 1999.

60 Michael M. Werner and Bénédicte Zimmermann. Beyond Comparison. *Histoire Croisée* and the Challenge of Reflexivity. *History and Theory* 45 (2006): 30–50.

61 The German term, *Verflechtung*, is a synonym for *histoire croisée*. For the relation between the terms, Michael M. Werner and Bénédicte Zimmermann. Vergleich, Transfer, Verflechtung. Der Ansatz der *Histoire Croisée* und die Herausforderung des Transnationalen. G*eschichte und Gesellschaft* 28 (2002): 607–636. The term also has parallels with the English phrase "entangled history," coined by Sidney W. Mintz. *Sweetness and Power. The Place of Sugar in Modern History*.

resistances, inertias, modifications – in trajectory, form, and content – and new combinations that can both result from and develop themselves in the process of crossing."[62] Homi Bhabha offers an analysis of such communication and transportation of ideas in his post-colonial treatment of mimicry.[63] For Bhabha colonial mimicry describes a process by which the colonized imitate the colonizer, but inflect the performances expected by the colonizer in unanticipated and unpredictable ways. The space that emerges between colonizer and colonized is an inchoate space that requires both imitation and its failure since the colonizer must always retain a superior position over the colonized. Since imitation is by definition not the thing it imitates, there is always more than can be done. From the perspective of entangled history, the space between colonizer and colonized in a crisscrossed space

As thirdspace articulation the martyrdom accounts treated in this discussion relish in both mimicry and entanglement. They have as their backdrop the physical and ideological spaces of Roman colonization. The letter from Gaul is in Greek to a community of believers in Phrygia. It is impossible to know how these believers came to Gallia Lugdunensis, but one possibility is through trade networks.[64] In a deeply romanized province they write in the language of the colonized immigrants who, like other eastern populations Gaul, have brought their religion with them. Mimicry of arena language marks then a twofold otherness – as migrant provincials and as Christ followers. An audience in Gaul acquainted with the site of the amphitheatre the letter describes would have had a picture in their mind of an arena structurally integrated into a temple of the imperial cult, and would have known that the time of the games in Lyons coincided with the annual summer meeting of Gallic nations, when consciousness of provincial identity would have been in the forefront of celebration. The story of Perpetua and her companions unfolds during games to mark the emperor's birthday, in a military amphitheatre. These are the spaces the Romans have allocated to inscribe the marks of

Harmondsworth: Penguin Books, 1986, which has come to refer in postcolonial studies to the contact, interaction and transfer between the cultures of colonizer and colonized. *Entangled Religions: Interdisciplinary Journal for the Study of Religious Contact and Transfer* (Ceres, Ruhr Universität Bochum), founded in 2014 and edited by V. Krech, attends to the history of contact and transfer in religion in past and present times.

62 Werner and Zimmermann, Beyond Comparison, 38.

63 Homi K. Bhabha. *The Location of Culture*. London: Routledge, 1994, specifically, "Of Mimicry and Man: The Ambivalence of Colonial Discourse," 121–131.

64 For the Greek speakers and Asian Christ followers in Gaul, John Behr. Gaul. In *The Cambridge History of Christianity, Volume 1: Origins to Constantine*, Margaret M. Mitchell and Frances M. Young (eds.), 366–379. Cambridge: Cambridge University Press, 2006, 368–370.

colonial power and domination on criminals and thus reinforce the social order through punishment and execution. The language of battle and athletics in the case of the Lyons account and that of the fight with the Egyptian gladiator in the report of Perpetua re-perform these ideological constructions. This space/time "emplotment" brings significant reversals as we have seen. It imitates imperial ideals even as it disavows them.

Further, the visionary flights out of the prison cells of Perpetua and Saturus require at least in part for their dramatic effect the idea that although they are incarcerated, they are free, in the Spirit, to come and go. The reversal of expectations in reports of martyrs not humiliated by death but enhanced in glory and dignity by it further requires a backdrop of imperial signification and expectations. The elites who win honour through hosting the games and the empire that would inscribe power on criminal bodies are seen from the bottom up, from the sand of the arena as it were, as a sham. It is they who are humiliated as the reader/audience takes on the point of view of characters dying in the arena below them. The crowd is more or less stripped of identity even as the narratives furnish the executed with noble deaths.

The narratives create new chronologies as part of their time-space productions. The games unfolded on important dates of the Roman calendar. This helped to inscribe a sense of mastery of both time and space. Such mastery was instantiated in the houses of the wealthy, who commissioned mosaics to represent games with clear markers linking them to annual festivals. As Christine Kondoleon points out, the person walking through the household was given the impression of Roman control over time and space through the careful selection of calendrically associated motifs.[65] But as texts read in liturgical celebrations of Christ followers, a new claim over time and space arises, even as the events recounted in the arena mark out a different kind of calendar. These events take place within a larger apocalyptic time frame marked by the narrative's preface where the narrator locates the visionary experiences of the martyrdom's protagonists as evidence of the advent of latter days (*PPF* 1.4). A contest for time may be seen in the jumble of time-space events in the arena in the case of Perpetua's and Saturus's visionary journeys. It is also present in a more apocalyptic form in the representation of Perpetua conquering the devil. The narratives thus challenge Roman control over time by inscriptions of their own temporality and as a consequence result in a kind of anti-chronology. [66]

65 Kondoleon, Timing, 336.
66 I am indebted to Katharina Waldner for this insight.

These martyrologies thus accomplish an imperial mimesis. They imitate events of the arena and associated meanings, adopt the arena's vocabulary, report the viewing audiences' responses, but all with a kind of burlesque effect. They talk back to imperial power. The accounts relish in the fact that the Empire's victims are not humiliated; on the contrary, there is, as we have seen, an emphasis on their greeting their deaths with heads held high and with eyes looking expectantly at the prize. As a consequence, the imperial strategy to use the arena to terrorize and make its prisoners cower fails, and with it, a colonizing discourse of power.[67] The martyrs gain athletic, civic, and military prizes for their performances. Through their deaths they beat imperial power at its own game.

References

Barton, Carlin. Savage Miracles. The Redemption of Lost Honor in Roman Society and the Sacrament of the Gladiator and the Martyr. *Representations* 45 (1994): 41–71.

Bartsch, Shadi. *Actors in the Audience. Theatricality and Doublespeak from Nero to Hadrian.* Cambridge, MASS: Harvard University Press, 1994.

Bartsch, Shadi. *The Mirror of the Self. Sexuality, Self-Knowledge and the Gaze in the Early Roman Empire.* Chicago: University of Chicago Press, 2006.

Behr, John. Gaul. In *The Cambridge History of Christianity, Volume 1: Origins to Constantine*, Margaret M. Mitchell and Frances M. Young (eds.), 366–379. Cambridge: Cambridge University Press, 2006.

Bhabha, Homi K. *The Location of Culture.* London: Routledge, 1994.

Bomgardner, D. L. *The Story of the Roman Amphitheatre.* New York: Routledge, 2002.

Boyarin, Daniel. *Dying for God. Martyrdom and the Making of Christianity and Judaism.* Stanford: Stanford University Press, 1999.

Bremmer, Jan N. Perpetua and Her Diary. Authenticity, Family and Visions. In *Märtyrer und Märtyrerakten*, Walter Ameling (ed.), 77–120. Stuttgart: Franz Steiner, 2002.

Bremmer, Jan N. The Motivations of Martyrs: Perpetua and the Palestinians. In *Religion im kulturellen Diskurs. Festschrift für Hans Kippenberg*, Brigitte Luchesi and Kocku von Stuckrad (eds.), 535–554. Berlin: de Gruyter, 2004.

Bremmer, Jan N. The Vision of Saturus in the *Passio Perpetuae*. In *Jerusalem, Alexandria, Rome. Festschrift Ton Hilhorst*, F. Garcia Martinez and G. Luttikhuizen (eds.), 55–73. Leiden: Brill. 2003.

67 Similarly, Judith Perkins. *The Suffering Self. Pain and Narrative Representation in the Early Christian Era.* London: Routledge, 1995, 117–118. For the importance of these references to direct eyesight in marking power and social identity in the arena, Carlin Barton. Savage Miracles. The Redemption of Lost Honor in Roman Society and the Sacrament of the Gladiator and the Martyr. *Representations* 45 (1994): 41–71, at 48–49.

Bremmer, Jan N. and Marco Formisano (eds.). *Perpetua's Passions. Multidisciplinary Approaches to the Passio Perpetuae et Felicitatis*. Oxford: University Press, 2012.

Brown, Shelby. Death as Decoration. Scenes from the Arena on Roman Domestic Mosaics. In *Pornography and Representation in Greece & Rome*, Amy Richlin (ed.), 180–211. Oxford: Oxford University Press, 1992.

Cameron, Averil. *Christianity and the Rhetoric of Empire. The Development of Christian Discourse*. Berkeley: University of California Press, 1991.

Casey, Edward S. *Getting Back into Place. Toward a Renewed Understanding of the Place-World*. Bloomington, IN: Indiana University Press, 1993.

Castelli, Elizabeth A. 'I Will Make Mary Male'. Pieties of the Body and Gender Transformation of Christian Women in Late Antiquity. In *Body Guards. The Cultural Politics of Gender Ambiguity*, Julia Epstein and Kirstina Staub (eds.), 29–49. London: Routledge, 1991.

Castelli, Elizabeth A. *Martyrdom and Memory. Early Christian Culture Making*. New York: Columbia University Press, 2004.

Castelli, Elizabeth A. Persecution and Spectacle: Cultural Appropriation in the Christian Commemoration of Martyrdom. *Archiv für Religionsgeschichte* 7 (2005): 102–136.

Clarke, John R. *Art in the Lives of Ordinary Romans. Visual Representation and Non-Elite Viewers in Italy, 100 B.C.–A.D. 315*. Berkeley: University of California Press, 2003.

Clavel-Lévêque, Monique. L'espace des jeux dans le monde romaín. Hégémonie, symbolique et pratique sociale. In *Aufstieg und Niedergang der römischen Welt*, 16.2.3, Hildegaard Temporini and Wolfgang Haase (eds.), 2405–2563. Berlin: De Gruyter, 1986.

Cobb, L. Stephanie. *Dying to be Men. Gender and Language in Early Christian Martyr Texts*. New York: Columbia University Press, 2008.

Coleman, Kathleen M. Fatal Charades. Roman Executions Staged as Mythological Enactments. *Journal of Roman Studies* 80 (1990): 44–73.

Young, Robin Darling. *In Procession Before the World. Martyrdom as Public Liturgy in Early Christianity*. Milwaukee: Marquette University Press, 2001.

Dowling, Melissa Barden. *Clemency and Cruelty in the Roman World*. Ann Arbor: University of Michigan Press, 2006.

Drinkwater, John F. *Roman Gaul. The Three Provinces, 58 BC–AD 260*. New York: Routledge, 2014.

Dupertius, Rubén R. Bold Speech, Opposition, and Philosophical Imagery in Acts. In *Engaging Early Christian History. Reading Acts in the Second Century*, Rubén R. Dupertius and Todd Penner (eds.), 153–168. New York: Routledge, 2014.

Epplett, Chris. Spectacular Executions in the Roman World. In *A Companion to Sport and Spectacle in Greek and Roman Antiquity*, Paul Christesen and Donald G. Kyle (eds.), 520–532. Oxford: Wiley-Blackwell, 2013.

Fagan, Garrett G. *The Lure of the Arena. Social Psychology and the Crowd at the Roman Games*. Cambridge: Cambridge University Press, 2011.

Featherstone, David. *Resistance, Space and Political Identities. The Making of Counter-Global Networks*. Oxford: Wiley, 2008.

Fishwick, Duncan. The Federal Cult of the Three Gauls. In *Les martyrs de Lyons*, J. Rougé and R. Turcan (eds.), 33–45. Paris: CNRS, 1978.

Fishwick, Duncan. *The Imperial Cult in the Latin West. Studies in the Ruler Cult of the Western Provinces of the Roman Empire*. Leiden: Brill, 1987.

Flaig, Egon. Den Kaiser herausfordern. *Historische Zeitschrift* 253 (1991): 371–384.

Frilingos, Christopher A. *Spectacles of Empire. Monsters, Martyrs, and the Book of Revelation*. Philadelphia: University of Pennsylvania Press, 2004.
Futrell, Alison. *Blood in the Arena. The Spectacle of Roman Power*. Austin: University of Texas Press, 2001.
Goldhill, Simon. The Erotic Eye. Visual Stimulation and Cultural Conflict. In *Being Greek Under Rome. Cultural Identity, the Second Sophistic and the Development of Empire*, Simon Goldhill and Robin Osborne (eds.), 154–194. Cambridge: Cambridge University.
Gunderson, Erik. The Flavian Amphitheatre. All the World as Stage. In *Flavian Rome. Culture, Image, Text*, A.J. Boyle and W.J. Dominik (eds.), 637–658. Leiden: Brill, 2003.
Gunderson, Erik. The Ideology of the Arena. *Classical Antiquity* 15 (1996): 113–151.
Habermehl, Peter. *Perpetua und der Ägypter oder Bilder des Bösen im frühen afrikanischen Christentum. Ein Versuch zur Passio Sanctarum Perpetua et Felicitatis*. Berlin: De Gruyter, 2004.
Harker, Andrew. *Loyalty and Dissidence in Roman Egypt. The Case of the Acts Alexandrinorum*. Cambridge: Cambridge University Press, 2011.
Hardt, Michael and Antonio Negri. *Commonwealth*. Cambridge, MA: Belnap, 2009.
Hardt, Michael and Antonio Negri, *Empire*. Cambridge, MA: Harvard University Press, 2001.
Harvey, David. *The Condition of Postmodernity. An Enquiry into the Origins of Cultural Change*. Cambridge, MA: Blackwell, 1990.
Hopkins, Keith. *Death and Renewal*. Cambridge: Cambridge University Press, 1983.
Hubbard, Phil and Rob Kitchin (eds.). *Key Thinkers of Space and Place*. 2n ed. Los Angeles: Sage, 2011.
Kelly, Nicole. Philosophy as Training for Death. Reading the Ancient Christian Martyrs Acts as Spiritual Exercises. *Church History* 75 (2006): 723–747.
Kennedy, George A. (ed., trans.). *Progymnasmata. Greek Textbooks of Prose Composition and Rhetoric*. Atlanta: Society of Biblical Literature, 2003.
Kondoleon, Christine. Timing Spectacles. Roman Domestic Art and Performance. In *The Art of Ancient Spectacle*, Bettina Bergmann and Christine Kondoleon (eds.), 321–341. New Haven: Yale University Press, 1999.
Kyle, Donald G. *Spectacles of Death in Ancient Rome*. London: Routledge, 1998.
Kyle Donald G. *Sport and Spectacle in the Ancient World*. Oxford: Blackwell, 2007.
Lefebvre, Henri. *The Production of Space*, Donald Nicholson-Smith (trans.). Oxford: Blackwell, 1991.
Lim, Richard. 'In the Temple of Laughter'. Visual and Literary Representations of Spectators at Roman Games. In *The Art of Ancient Spectacle*, Bettina Bergmann and Christine Kondoleon (eds.), 343–365. New Haven: Yale University Press, 1999.
MacLean, Rose. People on the Margins of Roman Spectacle. In *A Companion to Sport and Spectacle in Greek and Roman Antiquity*, Paul Christesen and Donald G. Kyle (eds.), 569–578. Oxford: Wiley Blackwell, 2013.
Maier, Harry O. Staging the Gaze. Early Christian Apocalypses as Narrative Self-Representation. *Harvard Theological Review* 92 (1997): 131–154.
Maier, Harry O. The Politics of the Silent Bishop. Silence and Persuasion in Ignatius of Antioch. *The Journal of Theological Studies* 55 (2004): 503–519.
Maldonado Perez, Zaida. *The Subversive Role of Visions in Early Christian Martyrs*. Lexington: Emeth, 2011.
Mintz, Sidney W. *Sweetness and Power. The Place of Sugar in Modern History*. Harmondsworth: Penguin Books, 1986.

Moss, Candida R. *Ancient Christian Martyrdom. Diverse Practices, Theologies, and Traditions.* New Haven: Yale University Press, 2012.

Musurillo, Herbert (ed.). *Acta Alexandrinorum. De mortibus Alexandriae nobilium fragmenta papyracea Graeca.* Leipzig: Teubner, 1961.

Musurillo, Herbert (ed.). *The Acts of the Christian Martyrs.* Oxford: Clarendon, 1979.

Perkins, Judith. *Roman Imperial Identities in the Early Christian Era.* London: Routledge, 2008.

Perkins, Judith. Social Geography in the Apocryphal Acts of the Apostles. In *Space in the Ancient* Novel, Michael Paschalis and Stavros A. Frangoulidis (eds.), 118–131. Groningen: Barkuis, 2002.

Perkins, Judith. Space, Place, Voice in the Acts of the Martyrs and the Greek Romance. In *Mimesis and Intertexuality in Antiquity and Christianity,* Dennis R. MacDonald (ed.), 117–137. Harrisburg: Trinity Press International, 2001.

Perkins, Judith. *The Suffering Self. Pain and Narrative Representation in the Early Christian Era.* London: Routledge, 1995.

Potter, David. Martyrdom as Spectacle. In *Theatre and Society in the Classical World*, Ruth Scodel (ed.), 53–88. Ann Arbor: University of Michigan Press, 1993.

Reis, David M. Spec(tac)ular Sights. Mirroring in/of Acts. In *Engaging Early Christian History. Reading Acts in the Second Century*, Rubén R. Dupertius and Todd Penner (eds.), 59–78. New York: Routledge, 2014.

Robert, Louis. Une vision de Perpetue martyre a Carthage en 203. *Comptes rendus des séances de l'Académie des Inscriptions et Belles-Lettres* 126:2 (1982): 228–276.

Ronnse, Erin. Rhetoric of Martyrs. Listening to Saints Perpetua and Felicitas. *Journal of Early Christian Studies* 14:3 (2006): 283–327.

Rüpke, Jörg. Das Imperium Romanum als religionsgeschichtlicher Raum. Eine Skizze. In *Ein Pluriverses Universum. Zivilisationen und Religionen im antiken Mittelmeerraum*, Richard Faber and Achim Lichtenberger (eds), 333–351. Paderborn: Fink, 2015.

Shaw, Brent D. Body/Power/Identity. The Passion of the Martyrs. *Journal of Early Christian Studies* 4:3 (1996): 269–312.

Sigismund-Nielsen, Hanne. Vivia Perpetua. An Indecent Woman. In *Perpetua's Passions. Multidisciplinary Approaches to the Passio Perpetua et Felicitias,* Jan N. Bremmer and Marco Formisano (eds.), 103–117. Oxford: Oxford University Press, 2012.

Soja, Edward W. *Thirdspace. Journeys to Los Angeles and Other Real-and-Imagined Places.* Oxford: Blackwell, 1996.

Solevåg, Rebecca. *Birthing Salvation. Gender and Class in Early Christian Childbearing Discourse.* Leiden: Brill, 2013.

Stewart, Eric C. New Testament Space/ Spatiality. *Biblical Theology Bulletin* 42 (2012): 139–150.

Tuck, Steven L. Representations of Spectacle and Sport in Roman Art. In *A Companion to Sport and Spectacle in Greek and Roman Antiquity*, Paul Christesen and Donald G. Kyle (eds.), 422–477. Oxford: Wiley Blackwell, 2013.

Veyne, Paul. *Bread and Circuses. Historical Sociology and Political Pluralism*, Brian Pearce (trans.). London: Penguin, 1980.

Veyne, Paul. The Roman Empire. In *A History of Private Life*, Philippe Ariès and George Duby (eds.), vol. 1, *From Pagan Rome to Byzantium*, Paul Veyne (ed.), Arthur Goldhammer (trans.), 5–234. Cambridge, MA: Belnap, 1992.

Webb, Ruth. *Ekphrasis Imagination, and Persuasion in Ancient Rhetorical Theory and Practice.* Surrey: Ashgate Publishing, 2009.

Werner, Michael M. and Bénédicte Zimmermann. Beyond Comparison. *Histoire Croisée* and the Challenge of Reflexivity. *History and Theory* 45 (2006): 30–50.
Werner, Michael M. and Bénédicte Zimmermann. Vergleich, Transfer, Verflechtung. Der Ansatz der *Histoire Croisée* und die Herausforderung des Transnationalen. *Geschichte und Gesellschaft* 28 (2002): 607–636.
Woolf, Greg. *Becoming Roman. The Origins of Provincial Civilization in Gaul*. Cambridge: Cambridge University Press, 2000.

Alfred Schmid

Die Macht des Schicksals?

Zur antiken Vorgeschichte des Reichsapfels

English abstract: The Power of Fate? The Ancient History of the Imperial Orb. Power, destiny, and the globe are a somewhat unusual association. They were nevertheless parts of a conceptual unit that had very real political connotations in antiquity. The possibility of combining power with the globe (in forging the concept of a "global" power) was provided by Greek metaphysics, in the context of which the world was newly conceived as a providential unity. It was a cosmos dominated by divine spirit or soul where the accidental was overwhelmed by fatal substance, the material by the spiritual, the ephemeral by the durable: The world was one. As such it was a "conspiration" of divine agencies residing in the outer spheres of the world's divine shell.

The "world view" of the shared concepts described can be illustrated by ancient astrology, as well as the dignity it achieved in the eyes of the imperial elites of Hellenistic, and especially Roman, society. The belief of the first emperor (and most of his successors) in astrology is revealing. The fatality of the world was incarnate in the perfect geometrical form of its divine sphericity. This fatality was astrologically transmittable to the human realm, i. e. to the forces chosen by providence to dominate the globe (and this providential stature of imperial power was much esteemed by the "provincialized" elites: it made the bare fact of subjugation deeply meaningful). It may be remembered that the iconography of global enterprise (as by the actual UN or environmental agencies) is still dependent on the imagery of the Roman empire. Chosen by fate or a horoscope to save the world from insignificance, the emperor's image could hold the globe in his own hand. Even Christ, depicted as "cosmocrator", was his successor in this respect. So it seems fair to conclude that all later conceptions of "global" power and organization are influenced by, and dependent on, the imago of cosmic order that shaped the form of the world in antiquity serving as paradigm of order and justice for all mankind.

DOI 10.1515/9783110418750-015

„er hat eine Vision von Europa, ein echtes Zivilisationsprojekt. Sein größtes Vorbild ist im Grunde genommen Kaiser Augustus“
(Michel Houellebecq, Unterwerfung, 2015)

„fuck autonomy“
(ebd.)

Macht, Schicksal und der Globus sind drei recht heterogen erscheinende Konzepte, wobei immerhin die „Macht des Schicksals“ schon als Operntitel[1] Verwendung fand. Ungewöhnlich mag eher der Globus in dieser Kombination erscheinen, obschon wir uns mittlerweile an „globale“ Perspektiven von Politik und Geschichte gewöhnt haben.

Im Folgenden ist darzulegen, wie die Vorstellung von der kugelförmigen Einheit des Weltganzen mit einem neuen „teleologischen“ Verstehen der Welt zusammenhing. Sie hatte mit spezifischen Sinn-Annahmen über die Welt zu tun – ja sie sollte diese, in der griechischen Metaphysik weitreichend theoretisch ausgeführten, Sinn-Annahmen gerade verkörpern[2] und sozusagen zur „Realexistenz“ werden lassen.

Es soll hier also die Entstehung der Vorstellung von der Fatalität der Welt beleuchtet werden, die im hellenistischen Einzugsbereich bis zur Konzeption und weitgehenden Akzeptanz von Astrologie in der römischen Kaiserzeit[3] führte – dieser Vorstellungsbereich entwickelte sich also synchron zur Bildung von wegweisend imperialen Formationen.[4] Dann ist die Frage zu stellen, worin deren Affinität zu Macht bestehen und wie sie sich realisieren konnte. Zum Schluss soll auf das „Globale“ als Signet für den unbegrenzten Zugriff politischer Ordnungs- (und Überlegenheits-) Konzepte etwas näher eingegangen werden.

Die Fatalität als weiträumig angewandte Verständnishilfe spielt in hellenistisch-römischer Zeit eine große Rolle. So habe, laut Joseph Manning, etwa der Historiker Polybios (200–120 v. Chr.) auch den politischen Erfolg der Ptolemaier der Tyche (oder der *fortuna*, was ziemlich auf dasselbe hinauskam) zugeschrieben.[5] Und

1 „La forza del destino“ (von G. Verdi).

2 Alfred Schmid. Das ptolemäische Weltbild. In *Mensch und Raum von der Antike bis zur Gegenwart* (= Colloquia Raurica 9), Antonio Loprieno (Hrsg.), 127–149. München und Leipzig: K. G. Saur, 2006.

3 Alfred Schmid. *Augustus und die Macht der Sterne. Antike Astrologie und die Etablierung der Monarchie in Rom.* Köln und Weimar: Böhlau Verlag, 2005; Frederick Cramer. *Astrology in Roman Law and Politics*. Philadelphia, Pennsylvania: The American Philosophical Society, 1954.

4 Augustus war nicht bloß der erste römische Kaiser, sondern zugleich der erste astrologiegläubige Kaiser.

5 Joseph Manning. *The last Pharaohs. Egypt Under the Ptolemies, 305–30BC*. Princeton: Princeton University Press, 2010, 31 A 5.

derselbe Polybios hat bekanntlich Roms imperiale Hegemonie mit einem Konzept von Weltgeschichte verbinden wollen: durch diese Hegemonie habe nun die ganze Geschichte ein Ziel und einen Körper erhalten,[6] und es liegt nahe, in diesem Körper die Welt (oder die *oikumene*) zu sehen. Begriffe wie *pronoia* (Vorsehung), *fatum, tyche, fortuna, heimarmene* und Verwandtes[7] hatten damals Konjunktur, und diese „Konjunktur" hat neben dem Historischen[8] auch das Politische erreicht.[9]

Man könnte zwar, etwa im Hinblick auf die Rolle der schicksalspinnenden Moiren bei Homer oder die vielfachen Klagen über die Kontingenz der Welt in der archaischen Lyrik der Griechen, im Hinweisen auf Fatalität überhaupt einen Grundzug antiker Bildung (und Götterglaubens) erkennen wollen. Aber das hieße, den entscheidenden Schritt zu einer Vorstellbarkeit der Welt verkennen, die seit dem 4. Jhdt. v. Chr. eine fundamental teleologische Verankerung von größter theoretischer Finesse erhielt, welche fast zweitausend Jahre lang die Wahrnehmungen zu lenken vermochte.[10]

Zentrales Dokument oder Manifest des Paradigmenwechsels zu einem Weltbild fataler Destiniertheit ist der platonische *Timaios*, der um die Mitte des 4. Jhdts. v. Chr. geschrieben worden ist. An diesbezüglich konstruktiver Bedeutung ist ihm wohl nur das 12. Buch der aristotelischen *Metaphysik* zur Seite zu stellen. Dabei steht ausser Frage, dass sich Aristoteles in seinen kosmosfrommen Erwägungen in der Nachfolge Platons gesehen hat.[11] – Im *Timaios* (einem Dialog, dessen Rahmenhandlung an die Nacht im Piräus anschließt, in welcher das berühmte

6 Polybios 1, 3, 4–1, 4, 2. Polybios schreibt, dass seit Rom Karthago besiegt hatte und bald darauf zur Vormacht der hellenistischen Welt wurde, die früher isolierten historischen Ereignisse einen körperartigen *(somatoeides)* Zusammenhang gewonnen hätten; die Geschehnisse Italiens und Afrikas (Lybiens) würden in diejenigen von Griechenland und Asien verwickelt und alles erhalte seitdem die Tendenz auf ein Ziel (*telos*) hin. – Roms Aufstieg zur Hegemonie eröffnete für Polybios somit die oikumenische (oder ‚globale') Einheit der Geschichte.

7 Etwa zum *kairos* in hellenistischer Zeit siehe Chaniotis, Angelos. The Ityphallic Hymn for Demetrios Poliorketes and Hellenistic Religious Mentality. In *More than Men, less than Gods. Studies on Royal Cult and Imperial Worship*, P. Iossif Panagiotis et al. (Hrsg.), 157–196. Leuven: Peeters Publishers, 2011.

8 Neben Polybios ist an den Historiker, Philosophen und Forscher Poseidonios zu denken (vorab an dessen, in ihrer Authentizität umstrittene, Bezeichnung des Historikers als „Diener der Vorsehung" nach dem Proöm Diodors).

9 Vom Glück des Sulla oder Caesars zur *fortuna* des Augustus, die dieser auch durch ein Element seines Horoskops illustrierte: Schmid, *Augustus*.

10 Dazu Schmid, *Augustus*, 119–202; Schmid, *Weltbild*.

11 Aristoteles soll eine Epitome des Timaios zum Eigengebrauch verfertigt haben (ebd. Schmid, *Weltbild*, 129); zum Einfluss des Timaios noch auf Zenon, den Gründer der Stoa: David Sedley. The School, from Zeno to Arius Didymus. In *The Cambridge Companion to the Stoics*, Brad Inwood (Hrsg.), 12. Cambridge: Cambridge University Press, 2003.

Gespräch der *Politeia* stattfand) soll die Möglichkeit der idealen Ordnung nun konkret veranschaulicht werden; es ist dabei der *clou* der Argumentation, dass die Antwort auf die ursprünglich politische Frage nach der Realexistenz idealer Ordnung in einem langen Monolog über die Entstehung des Kosmos besteht. Die Welt als der göttliche, kugelförmig die Welt ein- und abschließende Himmel[12] wird zum Nachweis dafür, dass das Ideale Erscheinung werden kann, womit ein Paradigma für die politische Utopie existiert, das ihre Möglichkeit sichern soll.

Die Welt ist Inkarnation oder Erscheinung einer demiurgischen Anstrengung, die von der Unordnung zur Ordnung führte. Sie ist als unerreichtes Vorbild idealistischer Theoretisierung die realisierte „Phänomenologie des Geistes" – als göttlich gerundeter *kosmos* das inkarnierte Ideal. Die Welt ist ein beseeltes, intelligentes Lebewesen (weil das Beseelte vollkommener ist als das Unbeseelte) und sie ist durch göttliche *pronoia* (Voraussicht) entstanden.[13] Dabei ist die Kugelform – also die „Globalität" – dieser Welt ein Ausdruck ihrer Göttlichkeit,[14] deren Replik quasi die Erdkugel ist. Sie bezeichnet oder verkörpert auch die Einmaligkeit und Einzigkeit der Welt[15] und sie ist theoretisches Postulat gewesen noch bevor man versucht hat, sie empirisch zu belegen.[16]

Für Aristoteles wird die Kugelwelt zum physikalischen Unterbau seiner „Wirklichkeits"-Theologie: die ewig bewegten Sphären garantieren die Priorität, ja den Sieg des Wirklichen über das Mögliche (denn ihre primäre „Wirklichkeit ist am meisten Bewegung")[17] – in ihnen erscheint sozusagen die erste Wirklichkeit selbst. – Es ist klar, dass in dieser himmlischen ‚Primär-Welt' nichts „aufs Geratewohl" oder rein zufällig bewegt wird (*Metaphysik* 1071b 35f.). Man möchte sagen, dass diese göttliche Schale der Welt[18] alles Sinnlose und Zufällige mit einem Mantel teleologischer Sinnhaftigkeit, reinster Intelligibilität, substantiel-

---

12 Zur Synonymität von von *kosmos* und *ouranos* („Himmel" und „Welt") in Platons Timaios schon André Festugière. *La revelation d'Hermes Trismegiste II: Le Dieu cosmique*. Paris: Librairie Lecoffre, 1949, 128, Anm. 2; 244, Anm. 4.

13 Platon, *Timaios* 30b–c.

14 Die Welt als sichtbarer Gott der auch der „eingeborene" heißt (*monogenes*: ebd. 31b; 92c).

15 Platon, *Timaios* 31a

16 Michel-Pierre Lerner. *Le monde des sphères, vol, 1: Genèse et triomphe d'une représentation cosmique*. Paris: Les Belles Lettres, 1996, 18: *„le produit d'une construction mentale"*.

17 Aristoteles, *Metaphysik* 1047a 30–32. – Das deutsche Wort „Wirklichkeit" ist überhaupt ein übersetztes „Kunstwort" des Aristoteles (*energeia*), und zwar laut Georg Picht. Der Begriff der Energeia bei Aristoteles. In, ders. *Hier und jetzt: Philosophieren nach Auschwitz und Hiroshima*, Bd. 1, 289. Stuttgart: Klett-Cotta, 1980 als Verdeutschung der Latinisierung von *energeia* als *actualitas* durch den mittelalterlichen Mystiker Meister Eckhardt.

18 Die Sterne sind Götter (dazu Filip Karfik. *Die Beseelung des Kosmos. Untersuchungen zur Kosmologie, Seelenlehre und Theologie in Platons Phaidon und Timaios*. München und Leipzig: Saur,

ler Lebendigkeit und Idealität umgibt. – Und diese Basis-Anschauungen griechischer Metaphysik wurden im Hellenismus Teil allgemeiner Bildung, zum einen durch die Erfolge griechischer Astronomie und Astrologie (die mit diesen theologischen Axiomen stets kompatibel blieben) und zum anderen auch dadurch, dass einflussreiche Denkschulen (wie vorab die Stoa) von dieser Weltvorstellung providentieller Zusammenhänge stark geprägt worden sind.

Diese Teleologie war einigermaßen spektakulär, sie war das revolutionäre und zugleich sichtbare Modell erfolgreicher Geometrisierung als Rationalisierung der Welt. In ihr war der Kosmos, also vorab Himmel und Gestirne, zum Organ oder zur Erscheinungsform des neuen philosophischen Vorsehungs-Glaubens geworden. Der Sphärenkosmos wurde zur sichtbaren und physikalisierten Apparatur der neuen Teleologie; er war die inkarnierte „Rettung der Phänomene".[19] Dabei sollte man den polemischen Horizont des theoretischen Postulats nicht außer Acht lassen: ein teleologisches Verständnis von Verursachungen ist sinnvoll, wo es andere, als sinnlos, zufällig oder willkürlich wahrgenommene Kausalmodelle gibt. Francis Cornford hob in seinem Timaios-Kommentar den Gegensatz von *„reason and necessity"* hervor:[20] offenbar gab es Weltmodelle, die durch eine sinn- (oder ziel-)lose Nezessität reguliert wurden, etwa durch „Natur und Zufall".[21] Die Teleologie der griechischen Metaphysik hatte selber konträre Auffassungen zur Voraussetzung: Weltvorstellungen evolutionistischer und jedenfalls nicht teleologischer Art, wie den Atomismus eines Demokrit.[22]

Gerade das „Zwingende" ist bereits vor Platon ein wichtiges Konzept vorsokratischer Vorstellungen von ‚natürlicher' Anordnung gewesen,[23] wogegen aber nunmehr als weise die Entdeckung von intelligent-beseelter, selbstbewegter Kausalität gilt – während die Ursächlichkeit in Dingen, die über andere durch Zwang bewegt werden, von minderer Art sei.[24] – Die neue Teleologie, welche die

---

2004, 117ff.; 139ff. und sonst) und die Himmelssphären erste Substanzen, die nach inspirierter Meinung die wahren Götter sind (Aristoteles, *Metaphysik* 1074b 10ff.)

19 Zu dieser angeblich platonischen Parole (zit. nach Simplicius, *In Aristotelis de Caelo* 488, 21–24 Heiberg) Lerner, *Monde*, 21ff.; Schmid, *Augustus*, 162ff.

20 Francis Cornford. *Plato's Cosmology. The Timaeus of Plato*. Cambridge: Routledge, 1937, 163ff.

21 Cornford, *Plato's Cosmology*, 167.

22 Zu nicht-teleologischen griechischen Auffassungen über die Welt und die Herkunft der Kultur Thomas Cole. *Democritus and the Sources of Greek Anthropology*. Philadelphia: American Philological Association, 1967 und Christian Utzinger. *Periphrades Aner. Untersuchungen zum ersten Stasimon der Sophokleischen „Antigone" und zu den antiken Kulturentstehungstheorien*. Göttingen: Vandenhoeck & Ruprecht 2003.

23 Zum Zwingenden („anankisch") z. B. Demokrit DK 68 A 39; Leukipp DK 67 B 2; Parmenides DK 28 B 8 26–31, B 10; Heraklit DK 22 B 11, B 94.

24 Cornford, *Plato*, 172, zu Platon, *Timaios* 46e

Welt an den ‚kosmischen Rändern' mit neuen weltkonstruktiven Intelligenzen bevölkerte, ist nur als (anti-moderne) Polemik gegen die ‚moderne' Verfügbarkeit einer phänomenalisierten Welt[25] verständlich, die auch einem Milieu politischer Modernität[26] (mit Bürger- und Redefreiheit sowie geregelter Partizipation) entsprungen ist. Welt war (in dieser etwa durch die Sophistik artikulierten ‚Moderne'), was das bürgerliche Subjekt autonom ermächtigter Kollektive dafür hielt oder wahrnahm. Daher ist es sicher nicht zufällig, dass erstens diese neuen Paradigmen von Ordnung und Intelligibilität in den Rahmen einer kritischen Debatte um die angemessene Verfassung der *polis* (und bei Platon greifbar: einer erbitterten Kritik an Demokratie) gehören, und dass sie zweitens in einer Zeit entstanden sind, in welcher Monarchen erfolgreich als neue Garanten von Ordnung, seit Alexander[27] mit wegweisend imperialer Tendenz, auftreten konnten.

So gesehen musste die neue Fatalität der Welt (die sich in der idealen Kugelform der Sphären und damit in ihrer „Globalität" realisierte) einen gewissen Appeal auf monarchische Formen politischer Ermächtigung ausüben: sehen sich doch Monarchen in aller Regel als Vertreter himmlischer Mächte oder Mandate und damit auch als die erwählten Organe der Vorsehung.

Im südfranzösischen Nîmes (und an vielen anderen Orten) kann man heute noch (an der sog. *„porte d'Auguste"*) eine Inschrift lesen, die mit dem Titel beginnt:[28] Imp. Caesar Divi F. Augustus.[29] Das ist ein zu dieser Zeit vierstelliger Name, wobei das zum ‚Kaiser-Vornamen' gewordene „Imperator" (ursprünglich ein übertragbarer Ehrentitel, den man im Triumphjahr führen durfte) eine trium-

---

25 Gemeint sind Regeln und Gesetzmäßigkeiten von ‚physikalischen' Verläufen, die durch keine vernünftige oder beseelte Intention gelenkt wurden: das Intendieren war durch das politisch-autonome Bürger-Subjekt *de facto* monopolisiert worden. – Das Problem des Zwanges stellte sich dann zwar aufs Neue etwa bei der Astrologie, doch dies war die Folge des Übergreifens von ‚naturaler' Kausalität auf die Sphäre menschlicher Freiheit selbst (die sich modern etwa bei der Gentechnologie wieder stellt).

26 Zu dieser antiken ‚Modernität' (oder Proto-Moderne), die sich seit dem 4. Jhdt. v. Chr. *ex negatione* in einer ‚anti-modernen' intellektuellen Entwertung von Demokratie und generell in der Suche nach autoritativen Transzendentalien der Ordnung zeigte, die auch das Natur-Konzept kosmologisch-providentiell ‚umgeschrieben' hat, verweise ich auf das 7. Kapitel in Alfred Schmid. *Die Geburt des Historischen aus dem Geiste der Politik. Zur Bedeutung frühgriechischer Geschichtsschreibung, mit einem Seitenblick auf Altchina*. Basel: Schwabe, 2016.

27 Zu dessen Erziehern bekanntlich Aristoteles gehörte.

28 Sie stammt aus dem Jahr 16/15 v. Chr., was sich durch Angabe des Konsulatsjahrs und der „tribunizischen Gewalt" eindeutig verifizieren lässt; siehe Dominique Darde und Victor Lassalle. *Nîmes antique*. Paris: Imprimerie Nationale, 1993, 54–55.

29 Folgt Angabe des 11. Konsulats und der achten *tribunicia potestas* mit dem Hinweis, dass der Erwähnte der „Kolonie" (Nemausus) Mauern und Stadttor gab.

phale Potenz als bleibende persönliche Eigenschaft insinuierte. Der Name Caesar weist auf den toten Adoptivvater hin, der Gallien erobert hatte und vor seiner Ermordung die vielleicht effizienteste Militärmaschine der Vormoderne unter seine Kontrolle gebracht hatte. Er war zum Staatsgott (*Divus*) erhoben worden und daher ist Augustus *divi filius* („Sohn des Vergöttlichten"). Und vor allem ist er „Augustus", ein Name, den ihm der Senat 27. v. Chr. als Ehrung verlieh, und der ebenfalls „erhabene" und gewinnende, ihn mit dem Göttlichen verbindende Eigenschaften assoziiert: Das *augurium augustum* war das glückverheissende Vogelzeichen gewesen, das dem Romulus als Gründer der siegriechen Potenz Roms zuteil geworden sein soll.

Die Titulatur kreist um gewinnende, glückverheißende Eigenschaften, die über das Auspizien-Ritual auch mit der Amtsmacht der höchsten römischen Beamten zusammenhing, die als *imperium* bezeichnet wurde. Die Auspizien, unter denen römische Heere kämpften, hingen an dem *imperium* ihrer Feldherrn oder Ober-Feldherrn; mit ihm war sozusagen rituell eine *felicitas* verbunden:[30] eine Art Garantie, zu siegen, die mit der göttlichen Fürsorge der Götter für Rom zusammenhing.

Es ist für das gewählte Beispiel wichtig, dass Augustus als Monarchie-Gründer in einer Gesellschaft mit anti-monarchischer Tradition für den Umfang seiner Kommando-Potenz keine Basis in dieser Tradition fand (in deren Formen er und seine Mit-Aristokraten gedrillt waren). Er musste sich das Recht, Heere außerhalb Roms zu kommandieren, noch regelmäßig als ein „prokonsularisches Imperium" vom Senat verleihen lassen (wie etwa 17 v. Chr. auf fünf Jahre).[31] Um dieses Kommando zu rechtfertigen, musste er entweder Kriege führen, provinzialisierte Gebiete „befrieden" (lat. *pacare*) oder auch neue, reichserweiternde Operationen vorbereiten.[32] Nebst allerlei zivilisierenden Maßnahmen in ehemals „barbarischen" Gebieten,[33] zu denen u. a. der Bau von Theatern und Amphitheatern, Foren und Tempeln oder eben auch Stadtmauern gehörte. Aber dies auf der Basis militärisch-organisatorischer Potenz provinzialisierend ausgreifende Gebaren war älter als die Monarchie, die sich nun unter Augustus endgültig anbahnte, obschon es, bei Caesar wie Augustus, auch ein fundierendes Element bei der

---

30 Erik Wistrand. *Felicitas imperatoria*. Göteborg: Acta Universitatis Gothoburgensis, 1987.

31 Dietmar Kienast. *Augustus. Prinzeps und Monarch*. Darmstadt: Wissenschaftliche Buchgesellschaft, 1999, 119–120.

32 Damals in Südfrankreich vermutlich eine Groß-Operation gegen Germanien (Kienast ebd.); zum Bedeutungswandel von *imperium* (das nun einen Raum als Einflussbereich bedeuten konnte) in jener Zeit J. S. Richardson. Imperium Romanum: Empire and the language of Power. *The Journal of Roman Studies* 81 (1991): 1–9.

33 Greg Woolf. *Becoming Roman: The Origins of Provincial Civilization in Gaul*. Cambridge: Cambridge University Press, 1998.

Bildung monarchischer Macht gewesen ist. Es ist schon früh realistisch wahrgenommen worden, etwa vom bereits eingangs erwähnten Griechen Polybios: „Viele der römischen Entscheidungen sind nämlich jetzt von dieser Art:[34] dass sie durch Ausnutzen des Unverstands anderer die eigene Vormacht vergrößern und herstellen, geschickt operierend und zugleich den Anschein erweckend, den Verlierern eine Wohltat zu erweisen." (Pol. 31,10,7).[35]

Polybios bemühte sich, dem Aufstieg Roms zur Supermacht (den er selbst miterlebte) einen Sinn abzugewinnen. Der besteht etwa darin, dass es nun nur noch *eine* (umgreifende) „Geschichte" geben kann,[36] also eine universale oder eben eine Weltgeschichte. Die Welt verkörpert und illustriert (bildlich als Globus) dann gerade die Einheit von Geschichte[37], denn die Welt wird gleichsam ihr (providentieller)[38] Körper. Wenn somit die Geschichte (durch Rom) von nun an „ein Ziel" haben solle, dann ist das auch eine fatale Dimension: Rom richtet das historische Geschehen (damit auch Vergangenheit und Zukunft) *in toto* auf sich aus.[39] Und es war gerade für die Verlierer verlockend, die notorisch siegreiche Potenz in einen providentiellen Rahmen zu stellen. Diesen Rahmen konnte nur Rom hergeben, und nur als ‚globales Format'.[40]

---

34 Bezieht sich auf die Regelung eines Zwists um die Thronfolge in Ägypten zwischen zwei Ptolemaiern, wobei die Römer gegen das Recht den Jüngeren bevorzugen: Rom entscheidet hier als Vormacht, obwohl Ägypten durchaus noch keine römische Provinz ist.

35 Augustinus, selbst römischer Bürger, sprach später von Roms *libido dominandi* (*De Civitate Dei* 1,30) und sah das als Laster (*vitium*). Livius ist in seiner *praefatio* stolz darauf, die *memoriae* des führenden Volkes auf Erden (*principis terrarum populi*) zu schreiben; nach Velleius Paterculus (1,12, 5) hatte Rom schon Karthago zerstört aus *invidia imperii*. – Sallust, Parteigänger Caesars und zeitweilig Provinzstatthalter in Nordafrika, sprach ganz allgemein von der *natura mortalium avida imperi*, einer natürlichen menschlichen Herrschgier (*Bellum Iugurthinum* 6,3).

36 S. o. Anm. 6.

37 Der hier vielleicht naheliegende Einwand, dass nach einer heute fast kanonischen These Reinhart Kosellecks der Begriff „Geschichte" (als „Kollektivsingular") erst modern sei, wird, unter anderem, schon durch die oben erwähnte Polybiosstelle widerlegt. Ausführlicher zur Kritik dieser These Schmid, *Die Geburt des Historischen*, Kap. 8.

38 Die Angleichung von Geschichte und kosmischer Vorsehung wird dann offenbar bei dem einflussreichen Stoiker Poseidonios (einem Zeitgenossen Ciceros) vollzogen.

39 Vergleichbar ist hier etwa die Ansicht von Jerry Bentley. World History and Grand Narrative. In *Writing World History 1800–2000*, Benedikt Stuchtey, Eckhardt Fuchs (Hrsg.), 51. Oxford: Oxford University Press, 2003, wo die Globalgeschichte als logische (d. h. auch: fatale) Folge von Evolution seit der Steinzeit angesehen wird; dabei seien drei Faktoren ursächlich: „rising human population, expanding technologicl capacity, and increasing interaction between peoples of different societies."

40 Rom sah sich in diesem Format: schon seit Sulla sei „in Rom der Gedanke der Weltherrschaft und die Gleichsetzung des *imperium Romanum* mit dem *orbis terrarum* eine Selbstverständlichkeit geworden" (Kienast, *Augustus*, 335).

Abb. 1: Denarius, P. Cornelius Lentulus Spinter 74 B.C., Crawford 1974 Nr. 397.

Jedenfalls taucht schon im 1. Jhdt. v. Chr. der Globus als politisches Konzept auf: aufschlussreich ist eine Münze, auf der eine menschliche Figur (als „Genius des römischen Volks" identifizierbar) den Fuss auf eine kleine, etwa tennisballgroße Kugel setzt (Abb. 1)[41] die offenbar die Weltkugel bedeuten muss. Nach der Schlacht bei Actium, als Octavian (der noch nicht Augustus hieß) seinen letzten großen Rivalen Marcus Antonius besiegt hatte und dadurch laut Manilius zum *rector Olympi* aufstieg,[42] ließ er eine Münze prägen auf der eine heroisch nackte Gestalt (offensichtlich ihn darstellend) ihren Fuss auf eine etwa wasserballgroße Kugel setzt (Abb. 2)[43] in der man vielleicht sogar den Himmelsglobus erkennen sollte.[44] Bekannt sind auch Münzen, auf denen der Capricorn (als das Sternzeichen des sog. „Glückspunkts" im Horoskop des Augustus)[45] eine Weltkugel zwischen den Klauen hält (Abb. 3).[46]

Bei Vergil wird schon deutlich, wie Zeitgenossen des Augustus die römische Hegemonie als etwas Vorbestimmtes wahrnehmen konnten, wobei die Person des *princeps* dafür den ‚Aufhänger' oder das Organ hergab: schon in mythischer Vorzeit wurde Anchises in der Unterwelt die Vision des späteren Augustus, des Gründers eines Goldenen Zeitalters, zuteil (*hic vir, hic*).[47] Dieser wartete an einem zeiterhabenen Ort als prädestinierter auf seine Realexistenz (die daher nur das Ein-

41 Michael H. Crawford. *Roman Republican Coinage*. Cambridge: Cambridge University Press, 1974, Nr. 397.

42 Manilius I 916.

43 Denar (Harold Mattingly, Edward E. Sydenham. *Roman Imperial Coinage*, vol. 1. London: Spinks, 1923, Nr. 256).

44 Zu dieser Debatte Schmid, *Augustus*, 248–249.

45 Schmid, *Augustus*, 37–54. – Hier ist zu beachten, dass der Globus auch zur Ikonographie der *fortuna* (die ‚die Welt regiert') gehörte.

46 *Roman Imperial Coinage*, vol 1, 2nd ed., Nr. 125.

47 Vergil, *Aeneis* VI 791–3: *Hic vir, hic est, tibi quem promitti saepius audis/ Augustus Caesar, divi genus, aurea condet/ saecula*: „Dies aber, dies ist der Mann, der so oft dir verheissen, Augustus Caesar aus der Götter Geschlecht, der das Goldene Zeitalter gründet".

Abb. 2: Denarius, Augustus 32–29 B.C., RIC I (2nd ed.) Nr. 256.

Abb. 3: Aureus, Augustus 17 – 16 B.C., Colonia Patricia, RIC I (2nd ed.) Nr. 125.

lösen einer fatalen *matrix* war). Und das Statuenprogramm des neuen Augustusforums mit dem Tempel des Mars Ultor kann als Hinweis darauf gelesen werden, dass auch der *princeps* selber die Geschichte als einen ‚Prozess' ansehen wollte, der in ihm und seiner epochalen Dimension kulminierte.[48] Schließlich wurde im Jahr 17 v. Chr. die Epochalität der neuen Monarchie, die zur verbindlichen politischen Form der ersten dezidiert globalen Macht geworden war, öffentlich im Rahmen eines alten religiösen Rituals (der *ludi saeculares*)[49] inszeniert. Eine neue Epoche hatte begonnen, und sie konnte sogar als ein „goldenes Zeitalter", d. h. als eine geschichtsbesiegende, Geschichte ausrichtende oder gar beendende wahrgenommen werden. Zudem wies vorab das gewollte Operieren des Princeps mit astrologischer Symbolik (als ein Hinweisen auf „natale" Erwähltheit)[50] darauf hin, dass der „erste Kaiser" sich selber als bestimmt durch die übergeordneten ‚Supermächte' der alles verfügenden Himmelssphären interpretieren wollte. Er sah sich somit als ‚Schnittstelle' zwischen dem Fatum, das im kosmosfrommen astrologischen Format ein weltumgreifend ‚globales' war, und der Politik (als dem Bereich der dezidiert menschlichen Selbstorganisation).

---

48 Schmid, *Augustus*, 385ff. (mit weiterer Literatur).

49 Alfred Schmid. Epoche als Ritual. Anmerkungen zu den augusteischen Säkularspielen. In *Die Wahrnehmung des Neuen in Antike und Renaissance*, (= Colloquium Rauricum 8), Achatz v. Müller, Jürgen v. Ungern-Sternberg (Hrsg.), 90–103. München und Leipzig: Saur, 2004.

50 Schmid, *Augustus*, *passim*.

Im Zeichen der erwählten Person (die dabei selbst zur ‚Verkörperung' globaler Supermacht wurde) war die ‚prinzipale Gesellschaft' das Organ einer himmlischen Superstruktur, die in der primärbeseelten[51] Weltschale zur theophanen[52] Erscheinung wurde. Damit war sie ‚geschichtsmächtig': als (mundane) Fatalität ist ja Politik sozusagen erhaben über Geschichte, sie bändigt sie und ihre – selber fatalen – Kontingentien. Das Fatum ist mächtiger als Geschichte, wie der „ptolemäische" Sphärenkosmos, der das Geschichtliche als das exklusiv Menschliche zum Reich sublunarer Akzidenzien relegierte.

In Anbetracht der frommen Assoziationen, die der Globus – als primärer Himmelsglobus – hier erweckte, stellt sich die Frage nach seinem politischen Gehalt noch einmal. – War er wirklich das Signet himmelsfrommer Bescheidung des Menschlichen, die sich etwa in der Meinung des Aristoteles ausdrücken konnte, der Mensch sei „nicht das Beste, das es in der Welt gibt"?[53] – Ist der Globus das Zeichen von finaler Begrenzung (und nur in dieser Begrenzung sinnhafter) menschlicher Potenz, oder im Gegenteil von deren endgültiger Ermächtigung?

Das Problem besteht hier nicht zuletzt darin, dass der Globus eine theoretisch-astronomische, in der Antike auch eine theologische Grösse war – ein Bezug zu politisch-konkreten, historisch fassbarem Tun war nicht gegeben (d. h. ein ‚globales Volk' war immer eine nachträgliche Konstruktion).[54] Der Globus eignete sich allerdings gerade (als ‚Verkörperung' des unkonkreten Zusammenhängens von allem) zum Signet eines nicht mehr begrenzbaren Zugriffs, einer politischen Potenz des „Handelns", das nach Hannah Arendt grenzenlos sein soll.[55] Ihm entspräche eine Ambition, die nur durch die Grenzen der Welt zum Stehen gebracht oder stabilisiert werden kann. Der Globus wäre dann das Symbol des *imperium sine fine*, (der „Herrschaft ohne Grenze") das nach Vergil[56] den Römern (von den Göttern als den fatalen Mächten) geschenkt oder zugeteilt worden sei. Und als Ikone für grenzenlose Ambition fungiert der Globus (als Erdglobus) bis heute – sehr bekannt ist die Filmszene, in der Charlie Chaplin als Adolf Hitler mit dem Erdball spielt (Abb. 4); im Gangsterfilm „Scarface" kann man in der Empfangshalle des exilkubanischen Drogenbosses einen Globus mit dem Motto: *„the world*

---

51 Alfred Schmid. Der Himmel als Präsenz des Ursprungs. Intentionen antiker Physiko-Theologie. In *Anfang und Ursprung*, Emil Angehrn (Hrsg.), 67ff. Berlin und New York: de Gruyter, 2007.

52 Schmid, *Ursprung*, 64.

53 Aristoteles, *Ethica Nicomachea* 1141a 21ff. (die beseelten göttlichen Himmelskörper sind besser).

54 Besonders beliebt war der Versuch, dem ‚imperialen Volk' den Sitz in der Weltmitte zuzuschreiben (einige Beispiele bei Schmid, *Augustus*, 318ff.)

55 Arendt, *Vita Activa*, 193ff; 239ff. und sonst.

56 *Aeneis* I 279.

Abb. 4: „The Great Dictator“, Charles Chaplin, 1940.

Abb. 5: „Scarface“, Brian De Palma, 1983. **Abb. 6:** Logo des Umweltgipfels von 1992.

is yours“ als Skulptur erkennen. (Abb. 5) Aber auch der als „Erdgipfel“ bezeichnete Umweltgipfel von Rio hatte einen Globus zum „Logo“, den eine menschliche Hand umfasste (Abb. 6), und den Auftraggebern des Logos dürfte kaum bekannt gewesen sein, dass sie hier eine imperiale Ikonographie benutzten: die Hand, die den Globus hält, ist zuerst die Hand eines Kaisers gewesen[57] – und erst später konnte sie die Hand eines Gottes, dann Christi (als *kosmokrator* mit dem Reichsapfel) und endlich die einer abstrahierten, die Welt umgreifenden Menschheit sein.

Das ideale Modell globalen Zusammenhalts der Welt, das durch die gött-iche Schale ihrer Wirklichkeit begrenzt wurde, die zugleich unendlich geistvoll, ja

57 Der erste Mensch, der mit dem Globus in der Hand abgebildet wurde, war aber wohl der römische General Pompeius: Claude Nicolet. *Space, Geography, and Politics in the Early Roman Empire*. Ann Arbor: University of Michigan Press, 1991, 38.

der „Geist“ selber [58] war, hatte aber einen Haken, der es allerdings zur imperialen Verwendung umso geeigneter erscheinen ließ. Der Globus war ein Modell, in dem die Welt sozusagen eingeschlossen war in die Kugel ihrer Vorstellbarkeit. Als diese Kugel war die Welt gefangen in ihrer Oberfläche (astronomisch genauer: in einem System konzentrischer Oberflächen).[59] Sie ist damit als ‚endgültiges‘ Phänomen gleichsam unter das ‚Auge Gottes‘ gebracht – d. h. eigentlich in das Wahrnehmungsfeld einer absolut autonomen Subjektivität.[60] Damit wäre der Globus das Symbol des imperialen Anspruchs eben dieses Subjekts: nämlich seiner Autonomie gegenüber der Welt, seiner Weltlosigkeit, die sich in der Betrachtung von Himmelsgloben gerade als welterhaben inszenierte.[61] Der Anspruch zeigt sich als Zugriff einer Rationalisierung, die aus dem ‚Extramundium‘ jener kollektiven Autonomie heraus, die sich politisch ihre eigene Welt geschaffen hatte, die Welt zu ‚ihrer‘ Erscheinung sozusagen ‚herunterformatierte‘: zu dem, was dem politisch autonomen Subjekt erschien, was unter seine Wahrnehmbarkeit fiel.[62] Das war die Objektivierung der Welt zu einer verfügbaren „Natur“, von der sich der menschlich-kulturelle und politische Bereich so sehr trennte, dass ihm sein Partizipieren an Welt zweitrangig wurde. Und dies Partizipieren wurde zugleich erheblich behindert dadurch, dass dem „archimedischen Punkt“[63] seiner Autonomie keine konkrete Ortschaft der Welt mehr als verbindliche Entsprechung genügen konnte.[64]

---

58 Aristoteles, *Metaphysik* XII 7.

59 Schmid, *Weltbild*.

60 Diese Konsequenz muss man ziehen in Anbetracht antiker Himmelsgloben, die in der Tat ‚von außen‘ angeordnet sind, d. h. dass die Sternbilder alle seitenverkehrt dargestellt worden sind, nicht wie sie ein irdischer Betrachter als das ‚Obere‘ wahrnehmen würde. Wir betrachten sie *„wie von ausserhalb des Himmelsgewölbes, auf die imaginäre Aussenhaut des Himmelsgewölbes blickend.“* (Ernst Künzl. *Himmelsgloben und Sternkarten*. Darmstadt: Konrad Theiss Verlag, 2005, 65).

61 Für Platon und Aristoteles war die Konstruktion des ‚globalen‘ Kosmos noch der (anti-moderne) Versuch der theoretischen Begrenzung einer demokratischen, athenisch-maßlosen, *libido dominandi* gottlos menschlicher Autonomie gewesen. Doch in der geometrisierbaren Phänomenalität der neuen Himmelsvorstellung war dann die *libido dominandi* dieser menschlich-elitären Autonomie als Voraussetzung enthalten.

62 Siehe Platon, *Theaitetos* 167c: *„Denn was jedem Staate* (πόλις) *schön und gerecht erscheint* (δόκῇ), *das ist es ihm ja auch, solange er es dafür erklärt“* (ταῦτα καὶ εἶναι αὐτῇ ἕως ἂν αὐτὰ νομίζῃ); vgl. 172a: *„was in diesen Dingen der Staat für eine Meinung fasst und daran festhält als gesetzmässig, das ist es nun auch für jeden in Wahrheit, und in diesen Dingen ist um nichts weiser ein einzelner als der andere noch ein Staat als der andere“* (Übers. Schleiermacher; wird als Ansicht des Protagoras referiert).

63 Descartes, Meditationes II 2.

64 Michael Hardt, Antonio Negri. *Empire*, Cambridge Mass.: Harvard University Press, 2000, 91 heben zwar mit Recht die Problematik des Menschen *„that stands separate from and above nature“* hervor. Sie halten das für eine Folge der Metaphysik, aber das scheint mir falsch zu sein: die Ab-

Was diese Welterhabenheit[65] hervorgebracht hatte, das war die politische Kultur der griechischen Poleis gewesen, doch fehlte diesen seit dem 4. Jhdt. v. Chr. ganz offensichtlich das organisatorisch-militärische Potential, um dieser Erhabenheit durch imperiale Anstrengung ersichtlichen Nachdruck oder „Realexistenz" zu verschaffen. Der Anspruch auf Weltüberlegenheit[66] dieser Autonomie (in der Moderne auch durch umfassend institutionalisierte „Wissenschaft" zum Ausdruck kommend), die schon die Welt theoretisch ‚unter ihr Auge' gebracht hatte, bedurfte offenbar der monarchischen Entrepreneurs (wie zuerst Philipps und Alexanders).[67] – Das wäre die pragmatische Seite der neuen Attraktivität von Monarchen seit dem 4. Jhdt. v. Chr. in Griechenland, die dann auch auf Rom übergegriffen hat.

Der Geist grenzenloser Expansion[68] (als Entsprechung grenzenloser Rationalisierung, als Objektivierung, Verfügbarmachung oder Verdinglichung der Welt) hat sich dann als globaler Anspruch in der Ikonographie des Christus als Kosmokrators über das Mittelalter hinweg gerettet (in frommer und heilsgeschichtlicher Einkleidung), aber er trat seit der Aufklärung als unbegrenzt kolonialisierendes,

trennung des menschlich kulturellen Bereichs von einem anderen, rein phänomenal aufgefassten Bereich der „Natur" ist schon so alt wie die griechische Antithese von *nomos* und *physis* (Belege bei Felix Heinimann. *Nomos und Physis*. Basel: Reinhardt, 1945) und sie ist von der griechischen Metaphysik schon übernommen worden. Platon wollte sie durch sein kosmosfrommes Uminterpretieren der *physis* im *Timaios* entschärfen (Schmid, *Augustus*, 125 mit älterer Lit.), was ihm aber letztlich nicht gelingen konnte: der Natur-Begriff setzt die (politische) Autonomie einer menschlichen Sphäre (und damit deren Weltlosigkeit) immer voraus. Die Metaphysik hat diese Autonomie bloss ‚verinnerlicht', essentialisiert und wohl auch moralisiert. – Zum griechischen Natur-Begriff und seiner substantiellen Phänomenalität ausführlich Schmid, *Die Geburt des Historischen*, Kap. 3.

65 Sie ist auch von Hannah Arendt (dort öfter auch als „Weltlosigkeit" bezeichnet) hervorgehoben worden: Hannah Arendt, *Vita activa*. München: Piper, 1981, 246: „Lange bevor wir gelernt hatten, die Erde zu umkreisen [...] hatten wir den Globus in unsere Wohnung gebracht, um gleichsam symbolisch die Erde wie einen Ball in die Hände zu schliessen oder vor den Augen kreisen zu lassen."

66 Er hat sich nicht nur bei Isokrates auch als Überlegenheit gegenüber einer „barbarischen" Umwelt wahrgenommen, die ihrerseits einer kolonialisierenden Zivilität seitens der ‚selbstbewussten Kulturmacht' bedürftig schien.

67 Zur organisatorisch-logistischen Überlegenheit Philipps von Makedonien meinte Demosthenes: „Vor allem war er der unumschränkte Gebieter über seine Gefolgschaft, und das ist im Krieg das Allerwichtigste. Zudem hatten sie ihre Waffen stets in Händen, war er mit Geldmitteln wohl versehen, konnte tun, was immer ihm richtig schien ohne es in Beschlüssen vorzulegen oder öffentlich beratschlagen zu lassen, ohne Rücksicht auf Denunzianten oder Scheu vor Klagen wegen Ungesetzlichkeit, niemandem rechenschaftspflichtig war er unabdingbar der Chef, Kommandeur und Herr über alles und alle." (*Über den Kranz* 18, 235–236).

68 Man kann hier an Jacob Burckhardts berühmte Definition von (an sich böser) „Macht" erinnern: „Sie ist kein Beharren, sondern eine Gier und eo ipso unerfüllbar, daher in sich unglücklich und muss daher andere unglücklich machen." (Jacob Burckhardt. *Ästhetik der bildenden Kunst. Über das Studium der Geschichte* (=JBW 10). München und Basel: C.H. Beck Verlag, 2000, 205).

und nun auch reflektiert merkantiles Welt-Projekt wieder ans Licht. Dazu bleibt eine Menge zu sagen,[69] doch soll hier nur noch kurz auf die antike Rolle der Monarchen zurückzukommen sein.

Die römische Monarchie eignete sich offensichtlich, um einer sehr bewusst organisierten[70] Tatsache der ‚Provinzialisierung' der Welt einen Sinn zu verschaffen. Dieser sollte ein fataler sein – Polybios wie Poseidonios können das Bedürfnis von ‚Unterlegenen' belegen, in der organisatorisch-militärischen Übermacht auf einen fatalen Faktor gestoßen und von einem globalen Geschick erfasst worden zu sein. Nicht nur lag darin der Trost einer Solidarität mit allem, das durch eine „kosmische Sympathie"[71] (die zugleich ein Instrument der Vorsehung war) verbunden war. Man war hier Bürger einer fatalen Weltlichkeit, mit der der Monarch (oft etwa noch durch solare Attribute erhellt) besonders verwandt war als die Verbindungsstelle zwischen Welt und Gesellschaft, die der König in monarchischen oder „kompakten"[72] Gesellschaften darstellen oder verkörpern muss.

Die im Prinzip grenzenlos expansive Gewalt der damals grössten Militärmaschine der Welt, deren hegemonialen Auftrag abzulehnen einem Einspruch gegen eine Vorsehung gleichkam, die den himmlischen Sphärenglobus zum ‚Organ' oder als ihren Götter-Apparat hatte, war weltmächtig, weil sie sich mit dem ordnenden Potential kosmischer Hierarchien assoziieren ließ. Widerstand dagegen war zwecklos, oder war, frei nach Kafka, schon eine Krankheit, die nur im Bett geheilt werden konnte. – In allen Provinzen konnte der Fatalität der Macht auch in Tempeln für die Herrscher (oft zusammen mit der Göttin Roma) gehuldigt werden. Und dieser Kult war ein konkretes soziales Ereignis, das auch die Spitzen der Gesellschaft involvierte und ihnen Sichtbarkeit verschaffte. Das war die konkretere Seite des Umgangs mit der neuen provinzialisierenden Weltmächtigkeit. Abgehobener konnte die Potenz der ‚hegemonialen Sozietät' etwa durch den

---

69 Der modernen Globalisierung von Macht gilt Hardt, Negri, *Empire*.

70 So wurde zur Zeit des Augustus offenbar in Autun (Gallien) eine Schule für die Söhne der vornehmen Gallier gegründet, um diese dem Einfluss der Druiden zu entziehen (Kienast, *Augustus*, 495–496). Das Ausmaß an ‚zivilisierenden' Bauten und Urbanisierung lässt sich etwa im Rhonetal von Lyon bis Arles noch heute an monumentalen Überresten wahrnehmen.

71 Dieser stoische Begriff der *sympatheia* meinte ein ‚holistisches' Zusammenhängen aller Dinge. Es wurde auch mit natürlichen Phänomenen wie dem Einfluss des Mondes auf die Gezeiten und nicht zuletzt mit Astrologie (und prinzipiell der Möglichkeit von mantischen Voraussagen) illustriert. Man kann darin eine stoische Weiterentwicklung der platonisch-aristotelischen Physiko-Theologie sehen. Der Klassiker dazu, stark auf Poseidonios orientiert, ist Karl Reinhardt. *Kosmos und Sympathie. Neue Untersuchungen über Poseidonios* München: Beck, 1926.

72 Zu diesem Begriff Eric Voegelin, *Ordnung und Geschichte*, Bd. 1. München: Fink, 2002, 19ff. (einleitend von Jan Assmann), 41ff.

Dichter Manilius artikuliert werden, der als Zeitgenosse den Augustus als Kosmokrator beschrieb, welcher *socio tonante*[73] (d. h. als Gefährte des Jupiter, hier als Himmelsherrscher gedacht) den Himmel regierte. – Das war keine Undenkbarkeit mehr angesichts eines Himmels, dessen Gestirne sich für Manilius *velut imperio*[74] bewegten: mit der Ungehindertheit römischer Amtsmacht.

Literaturverzeichnis

Arendt, Hannah. *Vita Activa*. München: Piper, 1981.

Bentley, Jerry. World History and Grand Narrative. In *Writing World History 1800–2000*, Benedikt Stuchtey, Eckhardt Fuchs (Hrsg.), 173–196. Oxford: Oxford University Press, 2003.

Burckhardt, Jacob. *Ästhetik der bildenden Kunst. Über das Studium der Geschichte* (=JBW 10). München und Basel: C.H. Beck Verlag, 2000.

Chaniotis, Angelos. The Ityphallic Hymn for Demetri's Poliorketes and Hellenistic Religious Mentality. In *More than Men, less than Gods. Studies on Royal Cult and Imperial Worship*, P. Iossif Panagiotis et al. (Hrsg.), 157–196. Leuven: Peeters Publishers, 2011.

Cole, Thomas. *Democritus and the Sources of Greek Anthropology*. Philadelphia: American Philological Association, 1967

Cornford, Francis. *Plato's Cosmology. The Timaeus of Plato*. Cambridge : Routledge, 1937.

Cramer, Frederick. *Astrology in Roman Law and Politics*. Philadelphia: The American Philosophical Society, 1954.

Crawford, Michael H. *Roman Republican Coinage*. Cambridge: Cambridge University Press, 1974.

Darde, Dominique und Lassalle, Victor. *Nîmes antique*. Paris : Imprimerie Nationale, 1993.

Festugière, André. *La revelation d'Hermes Trismegiste II: Le Dieu cosmique*. Paris: Librairie Lecoffre, 1949.

Hardt, Michael und Antonio Negri. *Empire*, Cambridge Mass.: Harvard University Press, 2000.

Heinimann, Felix. *Nomos und Physis*. Basel: Reinhardt, 1945.

Karfik, Filip. *Die Beseelung des Kosmos. Untersuchungen zur Kosmologie, Seelenlehre und Theologie in Platons Phaidon und Timaios*. München und Leipzig: Saur, 2004.

Kienast, Dietmar. *Augustus. Prinzeps und Monarch*. Darmstadt: Wissenschaftliche Buchgesellschaft, 1999.

Künzl, Ernst. *Himmelsgloben und Sternkarten*. Darmstadt: Konrad Theiss Verlag, 2005.

Lerner, Michel-Pierre. *Le monde des sphères, vol, 1: Genèse et triomphe d'une représentation cosmique*. Paris: Les Belles Lettres, 1996.

Manning, Joseph. *The last Pharaohs. Egypt Under the Ptolemies, 305–30BC*. Princeton: Princeton University Press, 2010.

73 Manilius I 800.

74 Manilius I 493; Eric Voegelin hat (in Bezug auf die pseudo-aristotelische Schrift *De mundo*) von einer „imperialen Deformation" der Welt gesprochen, ja der Kosmos werde hier zum *„imperial giant"* (Eric Voeglin. *Order and History* IV. Baton Rouge: Louisiana State University Press, 1974, 35; 200).

Mattingly, Harold und Edward Sydenham. *Roman Imperial Coinage* , vol. 1. London: Spinks, 1923.
Nicolet, Claude. *Space, Geography, and Politics in the Early Roman Empire*. Ann Arbor: University of Michigan Press, 1991.
Picht, Georg. Der Begriff der Energeia bei Aristoteles. In ders., *Hier und jetzt: philosophieren nach Auschwitz und Hiroshima*, Bd. 1, 289–307. Stuttgart: Klett-Cotta, 1980.
Reinhardt, Karl. *Kosmos und Sympathie*. München: Beck, 1926.
Richardson, J. S. Imperium Romanum: Empire and the language of Power. *The Journal of Roman Studies* 81 (1991): 1–9.
Schmid, Alfred. *Augustus und die Macht der Sterne. Antike Astrologie und die Etablierung der Monarchie in Rom*. Köln und Weimar: Böhlau Verlag, 2005.
Schmid, Alfred. Das ptolemäische Weltbild. In *Mensch und Raum von der Antike bis zur Gegenwart* (= Colloquium Rauricu 9), Antonio Loprieno (Hrsg.), 127–149. München und Leipzig: K. G. Saur, 2006.
Schmid, Alfred. Der Himmel als Präsenz des Ursprungs. Intentionen antiker Physiko-Theologie. In *Anfang und Ursprung*, Emil Angehrn (Hrsg.), 61–83. Berlin und New York: de Gruyter, 2007.
Schmid, Alfred. Epoche als Ritual. Anmerkungen zu den augusteischen Säkularspielen. In *Die Wahrnehmung des Neuen in Antike und Renaissance*, (= Colloquium Rauricum 8), Achatz v. Müller und Jürgen v. Ungern-Sternberg (Hrsg.), 90–103. München und Leipzig: K. G. Saur, 2004.
Schmid, Alfred. *Die Geburt des Historischen aus dem Geiste der Politik. Zur Bedeutung frühgriechischer Geschichtsschreibung, mit einem Seitenblick auf Altchina*. Basel: Schwabe, 2016.
Sedley, David. The School, from Zeno to Arius Didymus. In *The Cambridge Companion to the Stoics*, Brad Inwood (Hrsg.), 7–32. Cambridge: Cambridge University Press, 2003.
Utzinger, Christian. *Periphrades Aner. Untersuchungen zum ersten Stasimon der Sophokleischen „Antigone" und zu den antiken Kulturentstehungstheorien*. Göttingen: Vandenhoeck & Ruprecht, 2003.
Voegelin, Eric. *Order and History* IV. Baton Rouge: Louisiana State University Press, 1974.
Voeglein, Eric. *Ordnung und Geschichte*, Bd. 1. München: Fink, 2002.
Wistrand, Erik. *Felicitas imperatoria*. Göteborg: Acta Universitatis Gothoburgensis, 1987.
Woolf, Greg. *Becoming Roman: The Origins of Provincial Civilization in Gaul*. Cambridge: Cambridge University Press, 1998.

Abbildungsnachweise

Abb. 1: acsearch.info: https://www.acsearch.info/search.html?id=1757337 (15.08.2016)
Abb. 2: Mantis. American Numismatic Society: http://numismatics.org/collection/1944.100.39145 (15.08.2016)
Abb. 3: acsearch.info: https://www.acsearch.info/search.html?id=949730 (15.08.2016)
Abb. 4: https://gabelingeber.wordpress.com/2014/09/21/der-grosse-diktator/ (15.08.2016)
Abb. 5: https://hiphoptaughtme.files.wordpress.com/2015/08/world-is-yours-statue.jpg (15.08.2016)
Abb. 6: http://www.fairfaxcounty.gov/nvswcd/images/earthsummit92.jpg (13.07.2016)

Jens Kugele

Ästhetische Formationen der RaumZeit

Religion, Nation und imperiale Imagination

English abstract: Aesthetic Formations of SpaceTime. Religion, Nation, and Imperial Imagination. With the successful Apollo 11 lunar landing on July 20, 1969, the so-called Space Race between the United States of America and the Soviet Empire peaked. In the following years, the Apollo 11 mission found a most prominent place in US memory culture. It also became an important part of the material and ritual design of one of the most prominent sacred memory spaces in the nation's capital: the Washington National Cathedral. This paper takes as its starting point a memorial service held at the National Cathedral in honor of Neil Armstrong and sheds light on the role of religion in the context of national and imperial order. A critical reading of Hardt and Negri's theoretical engagement with religion discusses their interest in the anti-imperial potential associated with their notion of religion as well as their interest in processes of collective imagination. In a second step, this paper draws on theoretical concepts of "sacred space" to introduce the Washington National Cathedral as a space interweaving the religious field and the national collective imagination. In a third part, the paper focuses on the material and ritual design of the National Cathedral employing the concept of "aesthetic formation" (Birgit Meyer) to highlight its potential for a national cultural imagination.

1 Raumfahrt, imperialer Wettstreit und sakraler Raum

Als US-Präsident John F. Kennedy am 12. September 1962 eine Rede an der Rice University in Texas hält, wendet er sich nicht nur an die versammelten Gäste, sondern zugleich an die gesamte US-amerikanische Nation und die internationale Weltöffentlichkeit. Unter dem Beifall seiner Zuhörer erläutert er seine Vision eines ambitionierten Raumfahrt-Projekts, für das er im Jahr zuvor mit berühmt gewordenen Worten vor dem US-Kongress geworben hatte:

DOI 10.1515/9783110418750-016

> I believe that this nation should commit itself to achieving the goal, before this decade is out, of landing a man on the moon and returning him safely to Earth.[1]

Weniger als sieben Jahre später, am 20. Juli 1969, gelingt diese hoch ambitionierte nationale Entdeckungsreise und mit ihr der ersehnte Erfolg im symbolisch aufgeladenen raumzeitlichen Wettlauf gegen das Sowjetimperium. Mit der erfolgreichen Apollo-11-Mission gelingen nicht nur die Landung eines bemannten Raum-Shuttles und Neil Armstrongs erste Schritte auf der zuvor lediglich imaginierten und wissenschaftlich kartographierten Mondoberfläche. Sondern mit dem berühmt gewordenen Bild der wehenden US-Flagge auf dem Mond, direkt neben dem Landeplatz der Apollo 11, wurde darüber hinaus und für alle Welt sichtbar eine territoriale Markierung im imperialen Wettstreit des westlichen Einflussbereichs gegenüber der sowjetischen Union gesetzt.

Die Geschichte des nationalen Projekts der Exploration des unbekannten Raums im Weltall, des unbekannten Raums auf dem Mond sowie der erfolgreichen, sicheren Rückkehr zur Erde fand ihren äußerst prominenten Ort in den Narrativen der US-amerikanischen Erinnerungskultur. Unter anderem wurde die Mission auch wichtiger Bestandteil der materiellen und rituellen Gestaltung eines besonders prominenten sakralen Erinnerungsorts in der Hauptstadt der USA: der Washington National Cathedral. (Abb. 1) Ein Auszug aus Kennedys Rede wurde in dieser National-Kathedrale beispielsweise im Rahmen eines Gedenkgottesdienstes eingespielt, der anlässlich des Todes von Neil Armstrong 2012 gehalten wurde. Dieser Gottesdienst sowie der sakrale Raum der Kathedrale sollen im Folgenden als Fallbeispiel dienen, um den Blick auf die Rolle von Religion im Kontext nationaler und imperialer Ordnung zu richten. In einem ersten Schritt wird der Beitrag zentrale Aspekte der Religion in Hardt und Negris Theorie herausstellen, die den theoretischen Bezugspunkt des vorliegenden Bandes bildet. Wie zu zeigen sein wird, verbindet sich Hardt und Negris Interesse an Religion zum einen mit ihrem Interesse an anti-imperialen Potenzialen, allerdings in einem weiteren und langfristigeren Sinne auch mit ihrem Interesse an Prozessen der Gemeinschaftsbildung im Kontext der Kollektivimagination. In einem zweiten Schritt soll die Washington National Cathedral mithilfe des Konzepts des *sakralen Raums* als ein Raum vorgestellt werden, der das religiöse Feld mit der nationalen Kollektivimagination verschränkt. Dabei wird in einem dritten Schritt die ästhetische Gestaltung fokussiert und mit dem Konzept der *ästhetischen Formation* das Potenzial für die nationale Gemeinschaftsbildung aufgezeigt.

1 Andrew Chaikin. *A Man on the Moon*. Alexandria, Va: Time-Life Books, 1994, 15.

Abb. 1: Frontseite der National-Kathedrale (Foto: Jens Kugele).

2 Religion in Hardts und Negris Empire

Michael Hardt und Antonio Negri richten in ihrem neo-marxistischen Werk *Empire*[2] den Blick besonders dort auf das Themenfeld Religion, wo sie ihre Vorstellung von Moderne und Nationalstaatlichkeit in den Wurzeln der europäischen Geschichte erläutern und Prozessen der Gemeinschaftsbildung nachgehen. Im Folgenden sollen fünf zentrale Bausteine die Grundlage für unsere Auseinandersetzung mit ihrer Sicht von Religion und Imperium bilden.

---

2 Michael Hardt, Antonio Negri. *Empire*. Cambridge und London: Harvard University Press, 2000.

2.1 Religion und Moderne

Hardt und Negri verwenden den Begriff der Religion zum ersten Mal explizit im Kapitel *Two Europes, Two Modernities.*[3] Die Passage ist eingebettet in einer Charakterisierung von „Moderne“ als „Krise“, die zwei „Modi der Moderne“ vorstellt: den Modus eines radikalen revolutionären Prozesses auf der einen Seite und den Gegen-Modus, der den neuen Kräften den Krieg erklärt, um gleichzeitig eine übergeordnete Macht zu etablieren, welche diese neuen Kräfte im Zaum halten soll.[4] Hardt und Negri, die diese wettstreitenden Prozesse der Revolution und Gegenrevolution in der Epoche der Renaissance verorten, sehen beide Modi der Moderne mit „Religion“ verbunden. Im ersten Fall, da die von ihnen beschriebenen neuen Kräfte und Dynamiken alle gesellschaftlichen Bereiche beeinflussen und somit auch das religiöse Feld:

> This modernity destroys its relations with the past and declares the immanence of the new paradigm of the world and life. It develops knowledge and action as scientific experimentation and defines a tendency toward a democratic politics, posing humanity and desire at the center of history. From the artisan to the astronomer, from the merchant to the politician, in art as in religion, the material of existence is reformed by a new life.[5]

Während die Autoren in diesem ersten Modus der Moderne auch das religiöse Feld durch eine geistige Revolution beeinflusst sehen, kommt ihrer Ansicht nach der Religion gleichzeitig auch im zweiten Modus der Moderne eine zentrale Rolle zu. In diesem gegenrevolutionären Modus werde das „neue Bild der Menschheit“, wie es die revolutionären Kräfte vertreten, von der Gegenrevolution auf eine „transzendente Ebene verpflanzt“.[6] Dies geschehe im Bemühen, die Stoßkraft der revolutionären Dynamik abzulenken und das weltverändernde Potenzial der Wissenschaft zu relativieren:

> This is the second mode of modernity, constructed to wage war against the new forces and establish an overarching power to dominate them. It arose within the Renaissance revolution to divert its direction, transplant the new image of humanity to a transcendent plane, relativize the capacities of science to transform the world, and above all oppose the reappropriation of power on the part of the multitude. The second mode of modernity poses

3 Hardt, Negri, *Empire*, 69–93.
4 Hardt, Negri, *Empire*, 74.
5 Hardt, Negri, *Empire*, 74.
6 Hardt, Negri, *Empire*, 74.

> a transcendent constituted power against an immanent constituent power, order against desire. The Renaissance thus ended in war – religious, social, and civil war.[7]

Religion ist in dieser Sicht also sowohl beeinflusst von den neuen Kräften der Moderne als auch instrumentalisiert vonseiten der Gegenrevolution, um letztlich zu einer der Hauptarenen zentraler Konflikte der Moderne zu werden.

2.2 Religion und Nationalstaatlichkeit

Im direkten Anschluss an die oben genannte Passage zur Moderne als Krise findet sich die zweite ausführlichere Auseinandersetzung Hardts und Negris mit dem Begriff der Religion. Im Kapitel *Sovereignty of the Nation-State*[8] wird hier die Ausformung des Nationalstaats im Kontext einer Macht-Maschinerie betrachtet, die in Reaktion auf die Krise entstanden sei: „As European modernity progressively took shape, machines of power were constructed to respond to its crisis, searching continually for a surplus that would resolve or at least contain the crisis."[9] Der europäische Nationalstaat wird von den Autoren hier zwar nicht als direkte Konsequenz des zweiten Modus der Moderne vorgestellt, jedoch als eine parallel verlaufende Entwicklung, die schließlich auf jenen gegenrevolutionären Kräften aufbaut, um einen perfekten Mechanismus zur Wiederherstellung von Ordnung und Kontrolle zu konstruieren: „The second approach centers on the concept of nation, a development that presupposes the first path and builds on it to construct a more perfect mechanism to reestablish order and command."[10]

Als Garant für Stabilität, Frieden und Kontrolle sozialer Entwicklungen sei der Einfluss der patrimonialen Monarchie so weit gegangen, dass letztlich sogar die Religion der territorialen Gebietsherrschaft des Souveräns untergeordnet werden musste: „‚Cujus regio, ejus religio' – or really, religion had to be subordinated to the territorial control of the sovereign [...]. Even religion was the sovereign's property."[11] Entscheidend für die Folgezeit ist in Hardt und Negris Augen eine Transformation des absolutistischen und patrimonialen Modells, das über die bürgerlichen Revolutionen in England, Amerika und Frankreich hinaus fundamentale Charakteristika in die folgenden Jahrhunderte übersetzte.

7 Hardt, Negri, *Empire*, 74.
8 Hardt, Negri, *Empire*, 93–114.
9 Hardt, Negri, *Empire*, 93.
10 Hardt, Negri, *Empire*, 93.
11 Hardt, Negri, *Empire*, 94.

Die Transformation bestehe in einem graduellen Prozess, der die theologische Grundlage des territorialen Patrimonialrechts durch eine neue, ebenfalls transzendente Grundlage ersetzte:

> The transformation of the absolutist and patrimonial model consisted in a gradual process that replaced the theological foundation of territorial patrimony with a new foundation that was equally transcendent. The spiritual identity of the nation rather than the divine body of the king now posed the territory and population were conceived as the extension of the transcendent essence of the nation.[12]

Hardt und Negri sehen folglich im Konzept der modernen Nation den Erben des patrimonialen Körpers des vormaligen monarchischen Staates. Das Konzept der Nation habe diesen patrimonialen Körper in neuer Form wiedererfunden, aufbauend auf einer stabilisierenden nationalen Identität auf der Basis kultureller, integrierender Identität und biologischer Kontinuität von Blutsverwandtschaft, territorialer Kontinuität und linguistischer Gemeinsamkeit. Im Konzept der Nation sehen Hardt und Negri gleichsam eine ideologische Abkürzung (*ideological shortcut*), der die Konzepte der Souveränität und der Moderne vom Widerstreit der Krise zu befreien suchte. Im Rekurs auf Benedict Anderson heben die Autoren in diesem Kontext die Rolle der kollektiven Imagination hervor, als einer aktiven Herstellung einer Gemeinschaft von Bürgern im nationalen Gewand. Besonders kritisch am Nationenkonzept sehen die Autoren sein Potenzial zum ideologischen Missbrauch und verweisen auf die Entwicklungen des neunzehnten und zwanzigsten Jahrhunderts, als die Nation als alleiniges Modell galt, Moderne und Entwicklung zu vereinbaren. Hardt und Negri basieren ihre Kritik auf die Gedanken Rosa Luxemburgs, deren wichtigsten Beitrag sie in ihrem Argument sehen, Nation sei nicht kompatibel mit demokratischer Organisation. Denn nationale Souveränität in Verbindung mit nationaler Mythologie besetze letztendlich das Terrain der demokratischen Organisation, indem mittels der Mobilisierung einer aktiven Gemeinschaft die alte Macht territorialer Souveränität in modernisierter Gestalt erneuert werde.[13] Religion wird diesem Prozess vollkommen unterordnet, ist aber dennoch nicht Teil des „ideological shortcut“. Deshalb wird es möglich, dass sie entweder von Nationalstaaten zur Legitimation genutzt oder aber gegen ihre imperialen Ansprüche ins Feld geführt werden kann.

12 Hardt, Negri, *Empire*, 95.
13 Hardt, Negri, *Empire*, 96.

2.3 Religion und Imperialismus

Eine dritte Passage gibt Aufschluss über die Rolle von Religion in Hardts und Negris Analyse des europäischen Imperialismus. Auch diese Passage befindet sich wie die beiden oben erläuterten in dem Abschnitt ihres Werkes, der sich Fragen der Souveränität widmet. Im Kapitel *The Dialectics of Colonial Sovereignty*[14] stellen die Autoren im Anschluss an ihre Gedanken zum Nationalstaat die koloniale Souveränität als weiteren unzureichenden Versuch vor, die Krise der Moderne zu bewältigen. Während die Kolonialzeit mit einem der europäischen Moderne intrinsischen Rassenkonflikt einhergehe, müsse das Zeitalter europäischer Entdeckungen auch in seinem utopischen Moment gewürdigt werden. Nicht nur die zunehmende Kommunikation zwischen Räumen und Menschen in der Folgezeit, sondern auch ein Glaube an die universelle Freiheit und Gleichheit der Menschheit, wie die Autoren ihn im revolutionären Gedankengut der Renaissance ausmachen, trete hier auf die globale Bühne.[15] Hieran schließt sich auch die folgende explizite Hoffnung Hardt und Negris an:

> This utopian element of globalization is what prevents us from simply falling back into particularism and isolationism in reaction to the totalizing forces of imperialism and racist domination, pushing us instead to forge a project of counterglobalization, counter-Empire. This utopian moment, however, has never been unambiguous. It is a tendency that constantly conflicts with sovereign order and domination. We see three exemplary expressions of this utopianism, in all its ambiguity, in the thought of Bartolomé de Las Casas, Toussaint L'Ouverture, and Karl Marx.[16]

Am Beispiel des Missionars Las Casas suchen die Autoren die Ambivalenz des Gleichheitskonzepts und der Rolle von Religion im kolonialen Kontext zu veranschaulichen. Postulierte Las Casas zwar einerseits die Gleichheit der Menschen auch für die indigene Bevölkerung der spanischen Kolonie, so fußte sein Gleichheitskonzept doch auf einem problematischen Eurozentrismus. Dieser ging lediglich insofern von einer Gleichstellung aus, als er in der indigenen Bevölkerung potenzielle Europäer bzw. potenzielle Christen sah:

> Las Casas cannot see beyond the Eurocentric view of the Americas, in which the highest generosity and charity would be bringing the Amerindians under the control and tutelage of the true religion and its culture. The natives are undeveloped potential Europeans.[17]

---

14 Hardt, Negri, *Empire*, 114–137.
15 Hardt, Negri, *Empire*, 115.
16 Hardt, Negri, *Empire*, 115–116.
17 Hardt, Negri, *Empire*, 116.

Die Ambivalenz, die Hardt und Negri im Kern der kolonialen Projekte ausmachen, ist somit eng mit dem religiösen Feld verbunden: Das utopische Element der Globalisierungsidee, welches das revolutionäre Gedankengut der Renaissance sogar in der kolonialen Idee aufscheinen lässt, sei untrennbar mit der Kehrseite eines folgenschweren, christozentrischen Universalanspruchs des kolonialen Europas verbunden.

2.4 Religion, Fundamentalismus und Postmoderne

Eine vierte Passage zur Religion thematisiert den Fundamentalismus als Symptom der Veränderungen im ausgehenden zwanzigsten Jahrhundert. In diesem Zusammenhang argumentieren Hardt und Negri gegen eine Interpretation des Fundamentalismus, die ihn auf den Versuch einer rückwärtsgewandten Wiederherstellung einer *vormodernen* Welt reduziert. Vielmehr treten sie für eine Lesart ein, die den Fundamentalismus als Phänomen der *Postmoderne* sieht und somit als machtvolle Ablehnung gegenwärtiger historischer Prozesse.[18] In diesem Sinne sehen Hardt und Negri im Fundamentalismus ein Symptom des historischen Übergangs zum „Empire“ – ähnlich den ihm verwandten Phänomenen wie etwa der postmodernen und postkolonialen Theorie.[19] Als solche postmoderne Bewegung sei der Fundamentalismus immer auch anti-westlich motiviert:

> Certainly, powerful segments of Islam have been in some case ‘anti-Western’ since the religion’s inception. What is novel in the contemporary resurgence of fundamentalism is really the refusal of the powers that are emerging in the new imperial order. [...] we might think of it as the first postmodernist revolution.[20]

Hier lässt sich eine Ambiguität erkennen, mit der Hardt und Negri die Rolle von Religion im Kontext (inter-) nationaler Konstellationen betrachten. Zwar wählen sie besonders in ihren Passagen zum Fundamentalismus sehr zurückhaltende, zumeist deskriptive Formulierungen, doch klingt vor dem Hintergrund ihrer klaren normativen Befürwortung des „Counter-Empire“ das revolutionäre Potenzial an, das sie der (islamischen) fundamentalistischen Kraft zuschreiben.

18 Hardt, Negri, *Empire*, 146.
19 Hardt, Negri, *Empire*, 147.
20 Hardt, Negri, *Empire*, 149.

2.5 Die Figur des Armen und ihre Apotheose

In einem kursiv gesetzten Epilog zum Abschluss des Kapitels *Symptoms of Passage* stellen die Autoren die Figur des Armen vor und feiern sie als exemplarisch für das historische Subjekt, als Figur des unterdrückten, ausgeschlossenen und ausgenutzten, nicht-lokalisierbaren Fundament der „Multitude" und letztlich als Fundament der Möglichkeiten der Menschheit im allgemeinen.[21] Die prophetische Kraft, die der Figur des Armen auch in unterschiedlichen religiösen Traditionen zugeschrieben wird, betonen Hardt und Negri dabei in Rekurs auf Niccolò Macchiavelli, und sehen in ihr gar die Grundlage für eine „Göttlichkeit der Menge der Armen". Doch diese Göttlichkeit ist in den Augen der Autoren nicht auf eine transzendente Ebene gerichtet, sondern zeichnet sich durch eine Innerweltlichkeit aus:

> The divinity of the multitude of the poor does not point to any transcendence. On the contrary, here and only here in this world, in the existence of the poor, is the field of immanence presented, confirmed, consolidated, and opened. The poor is god on earth.[22]

Diese Apotheose der Figur des Armen sehen Hardt und Negri als gegenwärtige Nachfolge einer längst vergangenen Illusion eines transzendenten Gottes. Macchiavelli steht Hardt und Negri auch Pate in ihren Gedanken zum Untergang des Römischen Imperiums, der sich mit der aufkommenden christlichen Religion verbinde:

> Christian religion is what destroyed the Roman Empire by destroying the civic passion that pagan society had sustained, the conflictual but loyal participation of the citizens in the continuous perfecting of the constitution and the process of freedom.[23]

Eine Analyse der Kräfte, die einst eine fundamentale Krise des imperialistischen Europas auslösten, sehen die Autoren insofern als zentral an für ihr Projekt, als sie gleichzeitig Ansatzpunkte für eine neuzugestaltende gesellschaftliche Zukunft des *Counter-Empire* liefern. Der Fall des untergehenden Römischen Reichs ist in ihren Augen dabei ein lehrreiches Beispiel, denn die treibenden Kräfte der Krise werden hier zum Fundament einer neuen Welt. Die zuvor undifferenzierte Masse werde im Zuge eines Vergemeinschaftungsprozesses zu einer produktiven Kraft:

---

21 Hardt, Negri, *Empire*, 156.
22 Hardt, Negri, *Empire*, 157.
23 Hardt, Negri, *Empire*, 373.

> The agents of the crisis of the old imperial world became foundations of the new. The undifferentiated mass that by its simple presence was able to destroy the modern tradition and its transcendent power appears now as a powerful productive force and an uncontainable source of valorization. A new vitality, almost like the barbaric forces that buried Rome, reanimates the field of immanence that the death of the European God left us as our horizon.[24]

2.6 Religion und Counter-Empire

Hardt und Negri widmen ihr letztes Kapitel der *Multitude* und ihrem politischen Macht-Potenzial (*The Multitude against Empire*). Die beiden Epitaphe, die dieses letzte Kapitel einleiten, stellen einen direkten Zusammenhang her zwischen dem Thema Religion im Zitat aus dem „ältesten Systemprogramm des deutschen Idealismus"[25] und dem Thema Widerstand im Zitat von Deleuze und Guattari. Hardt und Negri fokussieren somit am Ende ihrer Schrift die Hoffnung auf Widerstand in Verknüpfung mit dem religiösen Feld.

Die Figur des Armen, die Hardt und Negri wie oben besprochen als Ideal des *Counter-Empire* eingeführt haben, nehmen die beiden Autoren im Schlussabschnitt des gesamten Buches noch einmal auf und verbinden sie erneut mit dem Themenfeld der Religion. In beispielhafter Weise habe in ihren Augen Franziskus von Assisi die Verbindung der Figur des Armen mit der Hoffnung auf eine neue Gesellschaft verkörpert. In seinem bewussten Leben in Armut sehen sie den Legendenstoff, der das künftige Leben kommunistischer Militanz verdeutlichen könne.[26] Indem sie diese Heiligenfigur als Vorreiter ihrer Idee eines neuen Kommunismus herausstellen, setzen sie nicht nur ihre Vorstellung einer „materialen, innerweltlichen Religion" ins Bild, sondern nutzen erneut die Bilder- und Figurenwelt der religionsgeschichtlichen Tradition, um ihrer Vision eine Legitimations- und Motivationskraft zu verleihen.

Mit Blick auf die Rolle der Religion in Hardt und Negris historischen Streifzügen, in ihrer Theorie des Empire sowie in ihrer Vision eines revolutionären Potenzials sind in unserem Kontext zwei Dinge festzuhalten. Hardt und Negris Blick auf die europäische Religionsgeschichte als Christentumsgeschichte ist

---

24 Hardt, Negri, *Empire*, 376.

25 Christoph Jamme, Heinrich Schneider. *Mythologie der Vernunft – Hegels „Ältestes Systemprogramm des deutschen Idealismus"*. Frankfurt am Main: Suhrkamp, 1984. Für eine Diskussion der Autorschaftsfragen siehe auch: Terry Pinkard. *Hegel. A Biography*. New York: Cambridge University Press, 2000, 682–683. Pinkard schreibt die Ideen Hölderlin zu und räumt Hegel einen möglichen Beitrag ein.

26 Hardt, Negri, *Empire*, 413.

selbstverständlich in einer Weise verengt, die weder historische Differenzierung noch seriöse Betrachtung anstrebt.[27] Es sind vielmehr die Zusammenhänge von Macht, politischer Souveränität, Individuen, normativen Ordnungen und kollektiver Imagination, die Hardt und Negri in ihrem Narrativ der europäischen Religions- und Nationsgeschichte in zugespitzter Weise herausstellen. Ihre Pointierungen werfen Fragen auf nach der Rolle und dem Ort von Religion im Kontext der imperialen, raumzeitlichen Ordnungen. Hier möchte mein Beitrag ansetzen und das Argument in die Diskussion des vorliegenden Bandes einbringen, dass sich in diesem Zusammenhang sakrale Räume als hilfreiche Untersuchungsobjekte erweisen. In Bezug auf die Arbeiten der Religionsethnologin Birgit Meyer werde ich sakrale Räume im Allgemeinen und die Washington National Cathedral als Ort der *Aesthetic Formation* vorstellen, welche raumzeitliche Ordnung, Materialität und soziale Praxis vereinen und somit Nuklei der Kollektivimagination darstellen.

3 Sakraler Raum der nationalen Erinnerung

Die Washington National Cathedral stellt einen monumentalen und äußerst prominenten Sakralraum dar, der im Laufe des zwanzigsten Jahrhunderts im Zentrum der raumzeitlichen Nationalimagination der USA, der Hauptstadt Washington D.C., entstand. Errichtet an einer der höchsten Stellen der Stadttopographie überragt die National Cathedral als sakraler Erinnerungsraum die nationalen Denkmäler und Erinnerungsorte Washingtons. (Abb. 2) Bereits der Blick auf die Planungs- und Entstehungsphase der Kathedrale führt uns die Vermessung und Kartierung des Raums als Akt der Imagination in Verbindung mit der hiermit untrennbar verschränkten materialen Gestaltung vor Augen.

Der französischstämmige Stadtarchitekt Pierre L'Enfant entwarf Washington D.C. vor etwas mehr als 220 Jahren im Jahr 1791.[28] Hier an neutraler Stelle zwischen Nord- und Südstaaten sollte auf bisherigem Sumpfgebiet ein neues Zentrum nationaler Einheit entstehen. Bereits L'Enfants frühe Entwürfe sehen

---

27 Für eine nuancierte Perspektive auf die pluralistische Europäische Religionsgeschichte siehe: Hans G. Kippenberg et al. *Europäische Religionsgeschichte. Ein mehrfacher Pluralismus*. Göttingen: Vandenhoeck & Ruprecht, 2009. Für eine Perspektive auf die vielfältigen Christentümer siehe: Christoph Auffarth et al. *Religion auf dem Lande. Entstehung und Veränderung von Sakrallandschaften unter römischer Herrschaft*. Stuttgart: Franz Steiner, 2009.

28 Nicholas R. Mann, *The Sacred Geometry of Washington, DC: The Integrity and Power of the Original Design*. Somerset: Green Magic, 2006.

Abb. 2: Blick von der National Cathedral über Washington DC (Foto: Jens Kugele).

mit dem Präsidentensitz oder auch dem Kapitol städtebauliche Kernelemente des imaginierten nationalen Zentrums vor. Grundbestandteil war allerdings auch ein Gebäude, das als spirituelles Zentrum der US-amerikanischen Nation an prominentester Stelle anvisiert war. Im Laufe des zwanzigsten Jahrhunderts sollte die Idee einer rituellen Stätte für nationale Anlässe wie öffentliche Gedenkgottesdienste oder Begräbnisfeiern tatsächlich zur Errichtung einer Kirche als „nationales Gebetshaus“ führen.[29]

In den 1890er Jahren, einhundert Jahre nach L'Enfants Entwurf, verband sich dessen frühere Idee eines National-Schreins mit den Bemühungen der episkopalen Kirche, ihre Stellung in den USA durch ein monumentales Bauwerk weit über die Generationengrenzen hinweg zu dokumentieren. Der neu gegründeten *Protestant Episcopal Cathedral Foundation* gelang es, mit dem Hügel im Nordwesten Washington D.C.s, dem Mount St. Alban, einen der höchsten geographischen Punkte der entstehenden Hauptstadt zu erwerben.[30] Wenige Jahre später begann

29 Richard Greening Hewlett. *The Foundation Stone. Henry Yates Satterlee and the Creation of Washington National Cathedral.* Rockville: Montrose Press, 2007.

30 Hewlett, *The Foundation Stone*, 65, 96; Anne-Cathrine Fallen. *Washington National Cathedral. The Washington National Cathedral Guidebooks.* Washington, D.C.: Washington National Cathe-

1907 die physische Konstruktion der *Cathedral Church of St. Peter and St. Paul*, die ausgewählte Ereignisse der Nationalgeschichte in Form von Statuen, Fensterbildern, Reliefs und rituellen Objekten erinnert und von Beginn an den Beinamen *Washington National Cathedral* trägt.[31]

Zwischen der Grundsteinlegung am 29. September 1907 und dem Abschluss am 30. September 1990 wurde mit der National Cathedral ein enorm ambitioniertes Konstruktions-Projekt im doppelten Sinne durchgeführt. Die physische Konstruktion des Bauwerks geht damit Hand in Hand mit der diskursiven Konstruktion einer US-amerikanischen Nationalgeschichte, deren Fortschreibung sich über die unterschiedlichen Bauphasen hinweg nachzeichnen lässt. Über 60 Millionen US-Dollar an Spenden wurden im Laufe des zwanzigsten Jahrhunderts für das Projekt gesammelt, und nach wenig mehr als 80 Jahren kann sich die National Cathedral stolz zu den weltweit sechs größten Kathedralen zählen.[32] Geleitet von der episkopalen Gemeinde, doch erklärtermaßen „offen für alle Glaubensrichtungen" ist somit ein monumentales „Haus des Gebets" entstanden, das vom höchsten geographischen Punkt im Nordwesten der Stadt aus die Hauptstadt der Vereinigten Staaten von Amerika mitsamt ihren Regierungsgebäuden und ihrer nationalen Erinnerungslandschaft überschaut.

Im monumentalen Innern der imposanten neogotischen Kathedralen-Konstruktion begegnet dem Besucher in Form medialer Repräsentation historischer Ereignisse eine zweite Form der Konstruktion, nämlich die einer US-amerikanischen Nationalgeschichte. Motive der kunstvoll gestalteten Glasfenster setzen ausgewählte Themen der US-Historie ins Bild wie die Unabhängigkeitserklärung, den amerikanischen Arbeiter oder auch die US-Luftwaffe; lebensgroße Statuen der ehemaligen US-Präsidenten George Washington und Thomas Jefferson zieren die Seitenschiffe; Büsten, Wandteppiche, Stuhlkissen und ausgestaltete Nischen ehren herausragende Figuren der nationalen Historie.

---

dral., 2001, 24; Margaret Bergen Davis et al. *Living Stones. Washington National Cathedral at 100*. Washington, D.C.: Washington National Cathedral, 2007, 6, 48.

31 Vgl. für eine kritische Betrachtung des Kathedralen-Projekts in Washington Hans-Georg Lippert, Alte Ordnung für die Neue Welt – Betrachtungen zur Kathedrale von Washington. In Joachim Klose, Klaus Morawetz (Hrsg.), Aspekte der Zeit – Zeit-Geschichte, Raum-Zeit, Zeit-Dauer und Kultur-Zeit. Münster: Lit, 2004, 198–220.

32 Fallen, *Washington National Cathedral*, 94.

3.1 Kultort, Kultplatz, sakraler Raum

Im allgemeinsten religionswissenschaftlichen Sinne lässt sich die National Cathedral und ihre Umgebung des Mount St. Alban zunächst als *Kultort* beschreiben. Christoph Elsas' Begriffsbestimmung fasst Kultort dem Heiligtum gleich als „natürliche oder künstlich geformte Stätte dauerhafter religiöser Verehrung".[33] Die Grundaspekte des Sakralen, des Ortes und der auf Dauer hin feststellbaren sozialen Praxis ergänzt Elsas im Weiteren durch die Aspekte des Jenseits-Verweises, der kulturellen Narrative sowie der sozialen Kohärenzstiftung. Als spezielle Form des Kultorts lässt sich Elsas zufolge der Kultbau begreifen, dessen „zeichenhafte und inszenatorische Ausgestaltung" durch architektonische Mittel wie „Steinsetzung, Hütte, Gebäude" vorgenommen wird.[34] Unser Blick richtet sich im Zusammenhang mit Kultorten dementsprechend nicht allein auf *gegebene* Orte, sondern darüber hinaus auf in verschiedener Weise *gestaltete* Orte, Gebäude und Räume.

Als den Kultorten eng verwandter Begriff werden Kultplätze in der Forschungsliteratur zumeist aufgefasst als (1) besondere oder abgesonderte Orte bzw. Plätze, an (2) bedeutungsvoller Stelle, welche entweder durch ihre natürliche, geographische bzw. strategische Lage oder durch ihre Funktion als Träger kollektiven Gedächtnisses Bedeutung erhält, und (3) schließlich als Orte religiöser ritueller Handlungen.

Mit Blick auf das Areal, auf dem die National Cathedral errichtet wurde, lässt sich bereits vor Baubeginn von einem Kultplatz in mehrfacher Hinsicht sprechen. Als Mount St. Alban war das Gelände im Nordwesten Washingtons bereits Teil einer Kirchenanlage; mit der feierlichen Einweihung des Friedenskreuzes durch Bischof Satterlee in Anwesenheit des US-Präsidenten Theodore Roosevelt und zahlreicher einflussreicher Gäste wurde das Areal und mit ihm das neu anvisierte Kathedralen-Projekt vom Kultplatz lokaler Prägung zum Kultplatz nationaler Bedeutung. Ab diesem Zeitpunkt sind mit dem Mount St. Alban folglich zentrale Charakteristika verbunden, die in Jürgen Wolfs Worten Kultplätzen im Allgemeinen eigen sind: „Kultplätze sind heilige Orte, die sich durch ihre besondere Lage oder eine mit einem bestimmten Ort verbundene Erinnerung auszeichnen und an denen wiederholt religiöse Riten vollzogen wurden."[35] In Verschränkung des

---

33 Cristoph Elsas. Kultort. In *Handbuch religionswissenschaftlicher Grundbegriffe*, Band 4, Hubert Cancik, Burkhard Gladigow, Karl Heinz Kohl (Hrsg.), 32. Stuttgart: Kohlhammer, 1998.

34 Elsas, Kultort, 32.

35 Jürgen Wolf. Kultplatz. In *Metzler Lexikon Religion*. Band 2, Christoph Auffarth, Jutta Bernard, Hubert Mohr (Hrsg.), 287. Stuttgart: J. B. Metzler, 1996.

gegebenen Ortes und des gestalteten Kultortes kommt der Trias aus Absonderung, Bedeutungszuschreibung und sozialer Praxis auch im Fall des Mount St. Alban zentrale Bedeutung zu, die sich mit kulturellen Narrativen und sozialer Kohärenzstiftung verbindet.[36]

Um die komplexe raumzeitliche Konstellation der National Cathedral und ihre ästhetische Gestaltung zu fassen, erweisen sich theoretische Perspektiven als fruchtbar, die in den vergangenen Jahren im Kontext des *Spatial Turn* nicht nur eine verstärkte Aufmerksamkeit gegenüber Räumen und Orten nach sich zogen, sondern im Gegensatz zu früheren Konzeptionen von Raum und Ort vielmehr einen kultur- und religionswissenschaftlichen auf *sakrale* Räume Blick geschärft haben. In den Worten der Kulturwissenschaftlerin Doris Bachmann-Medick kommen hierdurch sowohl soziale Produktion von Raum als auch neue Konfigurationen in den Fokus:

> [E]mphasis is placed on the social production of space as a complex and often contradictory social process, a specific localization of cultural practices, a dynamic of social relations that points to the mutability of space. This view of the configurability of space through capital, labor, economic restructuring, social relations and social conflicts has been further reinforced by the transformation of cities and landscapes as the result of unequal global developments rooted in the spatial division of labor.[37]

Dementsprechend wird auch ein kritisches Verständnis räumlicher Zusammenhänge im Allgemeinen geschärft und in Bachmann-Medicks Augen eine Grundlage geschaffen für einen spezifischen „Denk-Modus", den in kultur- und sozialwissenschaftlichen Studien selbst ausgemacht werden kann und Einblicke in ihre epistemologischen wie auch politischen Dimensionen schenkt.[38]

Überdies sieht Bachmann-Medick im *Spatial Turn* das Potenzial für einen selbstreflektierten Umgang der Kulturwissenschaft mit dem eigenen sprachlichen Instrumentarium – insbesondere mit dem oft metaphorisch verwendeten

---

36 Zu den Aspekten Kult, Ritus und Ritual in diesem Zusammenhang siehe: Bernhard Lang. Kult. In *Handbuch religionswissenschaftlicher Grundbegriffe*, Hubert Cancik, Burkhard Gladigow, Karl-Heinz Kohl (Hrsg.), 475–488. Stuttgart: Kohlhammer, 1998; Bernhard Lang. Kult/Ritual. In *Metzler Lexikon Religion*, Band 2, Christoph Auffarth, Jutta Bernard, Hubert Mohr (Hrsg.), 267–274. Stuttgart: Metzler, 1999. Archäologische Perspektive bietet E. Heinrich. Kulthaus. In *Reallexikon der Assyrologie und der Vorderasiatischen Archäologie*, Band 6, Erich Ebeling, Dietz O. Edzard, Bruno Meissner, Wolfram v. Soden, Ernst Weidner (Hrsg.), 319–323. Berlin und New York: de Gruyter, 1981.

37 Doris Bachmann-Medick. *Cultural Turns. New Orientations in the Study of Culture*. Berlin: de Gruyter, 2016, 214.

38 Bachmann-Medick, *Cultural Turns*, 214.

Raum-Begriff in kulturwissenschaftlichen Studien.[39] Hier bietet sich eine Chance, auch im Dialog mit der Geschichtswissenschaft Fragen nachzugehen, die etwa Susanne Rau in ihrer Einführung in die historische Raumforschung formuliert;[40] nämlich die Bestimmung von Raumtypen und Konfigurationen, die Analyse von Raumdynamiken und von subjektiven Konstruktionen von Räumen in Wahrnehmung, Erinnerung und Repräsentation sowie eine Analyse von Raumpraktiken in Verbindung mit der Körperforschung, der Geschichte der Globalisierung und der *Material Culture*.

Auf die soziale Konstruktion von Räumen und sakralen Räumen im Besonderen wurde in der religionswissenschaftlichen Forschung verschiedentlich hingewiesen. Jonathan Z. Smith beleuchtet diese soziale Konstruiertheit sakraler Räume im Zuge seiner Ritualtheorie. Rituale in diesen Räumen sind Smith zufolge nicht als Ausdruck oder als Antwort auf ein gegebenes *Sakrales* zu sehen; vielmehr würden Dinge bzw. Menschen durch Rituale erst *sakral gemacht*: „Ritual is not an expression of or a response to ‚the Sacred'; rather, something or someone is made sacred by ritual (the primary sense of *sacrificium*)."[41]

Kategorien wie die des Göttlichen und Menschlichen, des Sakralen und Profanen sind, wie Smith uns zeigt, transitive Kategorien in dem Sinne, dass sie der Kartierung (*maps*) und Kennzeichnung (*label*) dienen, jedoch nicht substanziell zu verstehen seien. Objekten selbst wohnt demnach keine Sakralität inne, diese wird ihnen immer erst durch den sakralen Ort verliehen[42]: „The *sacra* are sacred solely because they are used in a sacred place; there is no inherent difference between a sacred vessel and an ordinary one."[43]

Die Verbindung zwischen abgesondertem Raum, sozialer Handlung und narrativer Institutionalisierung ist auch in den Arbeiten der US-amerikanischen Forscher David Chidester und Edward T. Linenthal von zentraler Bedeutung.[44] Im

39 Bachmann-Medick, *Cultural Turns*, 236.

40 Susanne Rau. *Räume – Konzepte, Wahrnehmungen, Nutzungen*. Frankfurt a. M.: Campus, 2013.

41 Jonathan Z. Smith. *To Take Place. Toward Theory in Ritual*. Chicago und London: University of Chicago Press, 1987, 105.

42 Während Jonathan Z. Smith die zentrale Rolle des Ortes für das Studium von Ritual und Religion im weiteren Sinne fokussiert, ist Grimes im Gegenzug um eine Differenzierung bemüht, indem er einen handlungstheoretischen Kontrapunkt setzt und auf die soziale Produktion des Ortes verweist, siehe: Ronald L. Grimes. Jonathan Z. Smith's Theory of Ritual Space. *Religion*, 29:3 (1999): 261–273.

43 Smith, *To Take Place*, 106.

44 David Chidester, Edward T. Linenthal. *Introduction. American Sacred Space*. Bloomington und Indianapolis: Indiana University Press, 1999, 1–42.

Unterschied zu zahlreichen Arbeiten zur US-amerikanischen Religion richtet sich Chidesters und Linenthals Perspektive weit darüber hinausgehend auf Räume, die auf den ersten Blick eventuell nicht als klassische sakrale Räume betrachtet würden. Sakraler Raum erscheint in ihrem Ansatz als Arena von Zeichen und Symbolen, dessen kontinuierliche Produktion eine soziale Arbeit erforderlich macht und der somit zugleich innerhalb eines ökonomischen und politischen Geflechts auch Dominanzbestrebungen, Umdeutungsversuchen, Aktionen der Desakralisierung, Entweihung, Entortung und Umcodierung ausgesetzt ist – ja, letztlich erst durch diese soziale Realität wird. Insofern sind ein Kultort sowie ein sakraler Raum immer auch ein Erinnerungsort, der in seiner politischen Aushandlung vergangene Zeit mit gegenwärtiger Zeit verschränkt.[45]

Auf die Verschränkung von physisch gestaltetem und kulturell imaginiertem Raum weist mit besonderem Nachdruck der Sozialgeograph Edward Soja hin. Als dezidiertes Ziel seiner theoretischen Überlegungen in seinem Werk *Thirdspace* nennt Edward Soja eine Veränderung unserer Denkweise bezüglich Raum und Räumlichkeit.[46] Dabei fordert Soja nicht eine komplette Abkehr vom bisherigen Denken, sondern eine Öffnung und Erweiterung mit Blick auf räumliche und geographische Imagination. Soja argumentiert dabei für eine kreative Rekombination und Erweiterung (*recombination and extension*), die zweierlei Raumperspektiven aufs Engste verbindet: zum einen die Raumperspektive auf die *reale, materielle* Welt und zum zweiten die Raumperspektive auf *imaginierte* Repräsentationen dieser Räumlichkeit:

> Thirdspace too can be described as a creative recombination and extension, one that builds on a Firstspace perspective that is focused on the 'real' material world and a Secondspace perspective that interprets this reality through 'imagined' representations of spatiality.[47]

Die postulierte konzeptionelle Verschränkung in der Betrachtung sowohl imaginierter als auch materieller Dimensionen von Kultur bildet auch den Kern der religionsethnologischen Arbeiten Birgit Meyers. Ihr Konzept der *Aesthetic Formation* erweist sich als besonders ertragreich für unsere Betrachtungen zur Rolle von sakralen Räumen im Kontext der Kollektivimagination und soll im Folgenden skizziert werden.

45 David Lowenthal. Past Time, Present Place. Landscape and Memory. *The Geographical Review* 65:1(1975): 1–36. 1975.

46 Edward William Soja: *Thirdspace. Journeys to Los Angeles and Other Real-and-Imagined Places.* Oxford: Blackwell, 1996.

47 Soja, *Thirdspace*, 6.

4 *Imagined Communities* und Mediale Vermittlung

Ebenso wie Hardt und Negri ist auch Birgit Meyer an Prozessen der Gemeinschaftsbildung interessiert. Dieses Interesse entwickeln sie jeweils mit explizitem Bezug auf die Arbeit Benedict Andersons, der besonders für Meyers Ansatz der ästhetischen Formationen eine wichtige Grundlage darstellt. Die im wegweisenden Band *Aesthetic Formations*[48] zusammengetragenen Forschungsergebnisse basieren auf ethnographischen Studien in Afrika, Asien, Lateinamerika und den karibischen Inseln, die den Fokus auf den Zusammenhang von Medien (-technologie) und Religion legten. Der Medienbegriff geht dabei jedoch dezidiert über den konventionellen Bereich hinaus und schließt nicht nur Film, Fernsehen, Video, Radio, Audiokassetten oder das Internet mit ein, sondern lenkt den Blick beispielsweise auch auf den Körper als medialen Raum. Insgesamt wird in der Zugangsweise Religion nicht als Objekt der Repräsentation gefasst, sondern selbst als Ort und Form der *Mediation.*

In ihrer programmatischen Einleitung setzt sich Birgit Meyer kritisch mit dem Konzept der *imagined community* auseinander, den Benedict Anderson Anfang der 1980er Jahre in die Diskussion um politische Gemeinschaften im Kontext der Nationalstaatsidee einbrachte. Aufbauend auf Andersons Ansatz führt Meyer den Begriff der ästhetischen Formation ein, der für ein Verständnis von Religion als Medien- und Mediatisierungs-Praxis (*practice of mediation*) argumentiert. Diese Medien-Praxis sieht Meyer um einzelne sinnlich wahrnehmbare und gestaltete Formationen (*sensational forms*) organisiert. Durch ihren Zugang sucht Meyer das Verhältnis von Medien und Religion neu zu verorten und nicht mehr länger als Bereich einer abstrakt verstandenen Imagination und Virtualität zu fassen. Vielmehr sollen die materiellen Dimensionen im Zusammenhang von Medium und Religion ernst genommen werden und die sinnlich wahrnehmbaren Aspekte kollektiver Imagination mit einbezogen werden.[49]

Meyers Auseinandersetzung mit Benedict Andersons imaginierten Gemeinschaften würdigt einige der zentralen Leistungen seines Konzepts, die auch für unseren Kontext von Bedeutung sind. Andersons Interesse gilt der Gemeinschaftsbildung jenseits der direkten interpersonalen Kommunikation und betont die medial vermittelten Imaginationen, die räumliche Entfernungen zwischen Mitgliedern einer Gemeinschaft durch ein Gefühl des Zusammenseins überwindet. Diese Gemeinschaftsbildung in der Verbindung von Individuen, kollektiven

---

48 Birgit Meyer. *Aesthetic Formations: Media, Religion, and the Senses*. New York: Palgrave Macmillan, 2009.

49 Meyer, *Aesthetic Formations*, 4.

Vorstellungen und medialer Vermittlung ist nicht nur Thema Hardt und Negris, sondern auch Meyers Hauptinteresse an Andersons Arbeit. Mit Verweis auf den Zirkulationsbegriff Latours reformuliert Meyer den medial vermittelten Aspekt der Andersonschen Gemeinschaftsbildung als stetigen Prozess: „In this understanding, a community is not a preexisting entity that expresses itself via a fixed set of symbols, but a formation that comes into being through the circulation and use of shared cultural forms and that is never complete […].“[50] In dieser Abkehr von einer Konzeptualisierung von Gemeinschaft als statische, gegebene Einheit, die sich auf symbolischer Ebene repräsentieren ließe, liegt der fundamentale Beitrag in Andersons Arbeit für Meyers Ansatz. Medialer Vermittlung kommt somit nicht nur eine funktionale, stabilisierende Rolle zu, sondern sie spielt eine konstitutive Rolle in jeder Gemeinschaftsformierung.

In einer Parallele zu Hardt und Negri stellt auch für Birgit Meyer Andersons Arbeit zur imaginierten Gemeinschaft den Ausgangspunkt der eigenen Überlegungen zum Verhältnis von Religion, Medien, Materialität und Gemeinschaft dar. Andersons Interpretation der frühen Nationalidee in ihrem engen Verhältnis zur Religion fokussiert einen Zusammenhang, der sowohl für Hardt und Negri als auch für Meyer von zentraler Bedeutung ist, wenngleich Anderson den Aufstieg der Nationalstaaten im historischen Kontext des Niedergangs großer religiöser Gemeinschaften untersuchte und das Hauptaugenmerk sowohl in *Empire* als auch in *Aesthetic Formations* auf dem (post-) modernen Kontext liegt.

> While Anderson was interested in the implications of the shift from premodern large-scale religious imagined communities to the rise of nationalism and the nation-state as the prime module of the modern era, this volume explores the rise and appeal of religious, mass mediated communities at a time in which the nation-state is found to be somewhat in disarray (partly because it faces the emergence of religious communities that challenge the capacity of the secular state to maintain the primacy of the nation over religious identities).[51]

Alle vier AutorInnen teilen dabei einen Grundgedanken: Die Basis der machtvollen Idee einer nationalen Gemeinschaft besteht aus starken Gefühlen der Zugehörigkeit, Bindung und Hingabe und stabilisiert sich im Vertrauen der Bürger in die Wahrhaftigkeit ihrer Fiktionalität („citizens' cofindence in the truthfulness of its fictionality“).[52]

Während Meyer Andersons Grundinteresse an der Formierung von Gemeinschaft teilt, grenzt sich ihr Ansatz deutlich von Andersons Sicht auf Sprache als

---

50 Meyer, *Aesthetic Formations*, 4.
51 Meyer, *Aesthetic Formations*, 4.
52 Meyer, *Aesthetic Formations*, 5.

Medium und Zeichensystem ab. Im Unterschied zu Andersons konstruktivistischem Verständnis ist Meyers Zugang an der Frage interessiert, auf welche Weise die abstrakten Imaginationen für die Menschen wiederum real werden können:

> While the usefulness of constructivist approaches is beyond doubt, certainly when it concerns the unmasking of power claims that posit essentialized truths, in order to understand how imaginations become real for people we must move a step further [...].[53]

Die materielle Kultur rückt insofern ins Zentrum der Betrachtung, als sie den Ort und die Form darstellt, in welcher individuelle und kollektive Vorstellungen erst im Kontext gelebter Umgebung sinnlich erfahrbar werden:

> [I]maginations are required to become tangible outside the realm of the mind, by creating a social environment that materializes through the structuring of space, architecture, ritual performance, and by inducing bodily sensations [...]. In brief, in order to become experienced as real, imagined communities need to materialize in the concrete lived environment and be felt in the bones.[54]

4.1 *Aesthetic Formations*

Meyers Plädoyer gilt daher einer verstärkten Aufmerksamkeit auf mediale Aspekte der materiellen Kultur im Kontext der Imaginations-Forschung: „More attention needs to be paid to the role played by things, media, and the body in actual processes of community making."[55]

Ausgehend von Artistoteles' Begriff der *aisthesis* argumentiert Meyer dabei für einen erweiterten Ästhetikbegriff, der nicht auf die Lehre vom Schönen im Bereich der Künste beschränkt bleibt, sondern die körperliche Fähigkeit bezeichnet, deren Grundlage das psychische Vermögen darstellt, Objekte in der Welt mittels der menschlichen fünf Sinnesmodi wahrzunehmen und gleichzeitig spezifische sinnliche Konstellationen als Ganzes:

> Imaginations, though articulated and formed through media and thus 'produced,' appear as situated beyond mediation exactly because they can be – literally – incorporated and embodied, thus invoking and perpetuating shared experiences, emotions, and affects that

53 Meyer, *Aesthetic Formations*, 5.
54 Meyer, *Aesthetic Formations*, 5.
55 Meyer, *Aesthetic Formations*, 6.

> are anchored in, as well as triggered by, a taken-for-granted lifeworld, a world of, indeed, *common sense*.[56]

Meyers Ansatz reiht sich somit ein in die Forschungsbeiträge der vergangenen Jahre, die einen neuen Akzent in der Religionswissenschaft setzen möchten, indem sie einer Tendenz zur „Dematerialisierung" des Religionskonzepts entgegenwirken.[57] Der Fokus wird im Zusammenhang von Religion, Materialität und Medialität nun, wie Jeremy Stolow es formulierte, ein Verständnis von Religion *und* Medien („religion and media") hin zu einem Verständnis von Religion *als* Medium („religion as media") wandeln.[58]

Um Gemeinschaftsbildung in ihren Prozessen analytisch zu fassen, schlägt Meyer den Begriff der Formationen (*formations*) anstelle der Gemeinschaften (*communities*) vor. Einerseits grenzt sich Meyer hiermit auch von Zygmunt Baumans Begriff der *aesthetic community* ab. Andererseits soll durch den umfassenderen und dynamischeren Charakter des Begriffs der Formationen noch deutlicher Abstand genommen werden von einem Verständnis der Gemeinschaft als einer statischen, geschlossenen sozialen Gruppe. Formationen sollen in Meyers Ansatz sowohl soziale Einheiten bezeichnen als auch gleichzeitig Prozesse des Formens und Formierens:

> The term aesthetic formation, then, highlights the convergence of processes of forming subjects and the making of communities – as social formations. In this sense, 'aesthetic formation' captures very well the formative impact of a shared aesthetics though [*sic*] which subjects are shaped by tuning their senses, inducing experiences, molding their bodies, and making sense, and which materializes in things [...].[59]

Zentren der *Aesthetic Formations* bilden in Meyers Ansatz *Sensational Forms*, ein Konzept, das auf die Art und Weise abzielt, in der religiöse mediale Vermittlung Menschen adressiert, mobilisiert und ästhetisch form(ier)t. *Sensational Forms* versteht Meyer als etablierte Modi des Evozierens und Organisierens unterschied-

---

56 Meyer, *Aesthetic Formations*, 7. Hervorhebung im Original.

57 Siehe hierzu auch Birgit Meyer, Jojada Verrips. Aesthetics. In *Key Words in Religion, Media and Culture*, David Morgan, 20–30. London und New York: Routledge, 2008; Birgit Meyer et al. The Origin and Mission of Material Religion. *Religion* 40:3 (2010): 207–211. 2010; Lucia Traut, Annette Wilke. *Religion – Imagination – Ästhetik. Vorstellungs- und Sinneswelten in Religion und Kultur*. Göttingen: Vandenhoeck & Ruprecht, 2015; Jens Kugele, Katharina Wilkens. Relocating Religion(s) – Museality as a Critical Term for the Aesthetics of Religion. Special Issue. *Journal of Religion in Europe* (2011). Leiden und Boston: Brill.

58 Jeremy Stolow. Religion and/as Media. *Theory, Culture and Society* 22:4 (2005), 119–145.

59 Meyer, *Aesthetic Formations*, 7.

licher Zugänge zu einer transzendenten Dimension. Das Entscheidende mit Blick auf den Aspekt der Gemeinschaftsbildung ist dabei die Verbindung zwischen den einzelnen Individuen im Kontext religiöser Machtstrukturen, Symbole und normativer Ordnungen:

> These are relatively fixed, authorized modes of invoking and organizing access to the transcendental, thereby creating and sustaining links between believers in the context of particular religious power structures. Sensational forms shape both religious content (beliefs, doctrines, sets of symbols) and norms.[60]

5 *Aesthetic Formation* und die National Cathedral

Aus diesem Blickwinkel lassen sich einige Besonderheiten in dem oben erwähnten Gedenkgottesdienst beleuchten, der am 13. September 2012 zu Ehren Neil Armstrongs in der National Cathedral gehalten wurde. Zu Beginn der etwas mehr als einstündigen Feier tragen Elitesoldaten der US-Navy (Neil Armstrong selbst war einst Mitglied der Navy) die amerikanische Flagge in die Stille des Kathedralen-Raums und schreiten entlang der versammelten Besucher, die ihre Ehrfurcht durch körperliche Gesten bekunden: Alle Anwesenden erheben sich von ihren Sitzen und richten den Blick auf die Flagge, während zahlreiche unter ihnen symbolisch die rechte Hand auf ihr Herz legen und durch die Geste patriotische Verbundenheit zeigen.[61] Die Gastgeberschaft der episkopalen National Cathedral wird durch die folgende Prozession der Mitglieder des Cathedral-Chors und des Navy-Chors verdeutlicht, die in weithin sichtbarer Höhe das trinitäre Arrangement aus einem goldenen von zwei Kerzen begleiteten Kreuz tragen.[62] Während die US-Flagge unter Schweigen getragen wurde, umrahmt hier nun als Eingangslied ein „Halleluja" den Einzug, in das die Gottesdienst-Besucher lautstark mit einstimmen.

Der damalige Interims-Dekan der National Cathedral, Francis H. Wade, eröffnet die Feier mit Worten, die gleich zu Beginn nationale Bedeutung und religiöse Vorstellungswelt miteinander verschränken:

---

60 Meyer, *Aesthetic Formations*, 13. An anderer Stelle fasst Meyer die hier erwähnten „Inhalte" wie etwa Überzeugungen, Regeln, Symbole sowie Normen mit dem Begriff des „Regimes" zusammen: Meyer, Verrips, *Aesthetics*, 20–30.

61 Washington National Cathedral: In Celebration of the Life of Neil Armstrong. DVD. 2012. Min 5:30 f.

62 Washington National Cathedral: In Celebration of the Life of Neil Armstrong. DVD. 2012. Min 5:50 f.

> Good morning; grace and peace to you from God our father. My name is Frank Wade, I am the interim dean of this Cathedral and it's my honor to thank you for allowing Washington National Cathedral to fulfill part of its mission as a spiritual home for the nation. It is important in times like this to have places like this where we can in fact hold before God our grieve, our joy, our thanksgiving and our hope. For all of us as a nation have participated no matter how vicariously in the great explorations of Neil Armstrong and his companions and so it is important for us as a nation, as a community, as a people to gather here in this place to consider the mysteries of creation, of life, of death, also to give thanks for a life well lived and for service boldly rendered. That's what we will be doing in this time, and I thank you for sharing in it. May I call your attention to the order of service in front of you.[63]

Was folgt, ist eine kurze liturgische Passage, die im fließenden Übergang zu den oben zitierten Eröffnungsworten des Dekans nun die Versammelten als rituelle Gemeinschaft in die performative Gestaltung der religiösen Dimension des Anlasses mit einbezieht:

> [*Dean Wade*] Glorify God, all the works of God, [*Gemeinde*] and give praise and give honor forever [*Dean Wade*] In the high vault of heaven glorify God [*Gemeinde*] In praise and in honor forever [*Dean Wade*] God of grace and glory, you create and sustain the universe in majesty and beauty. We thank you for all in whom you have planted the desire to know your creation and to explore your work and your wisdom. Lead us, like them, to understand better the wonder and mystery of creation, through Christ, your eternal word through whom all things are made.[64]

Sowohl auf sprachlicher wie auch auf ritueller Ebene wird in Dean Wades Eröffnung eine grundlegende Verbindung kreiert zwischen dem Raumfahrtprogramm, Neil Armstrongs Leben im Kontext einer nationalen Bedeutung sowie der religiösen Vorstellungswelt der episkopalen Gastgeberinstitution, in welche hier die Idee einer Exploration des Weltraums als Exploration der göttlichen Wunder eingebettet wird.

Ein Blick in das Filmmaterial, das die gesamte Nation an diesem Gedenkgottesdienst teilnehmen ließ und als DVD im Gift Shop der Kathedrale erhältlich ist, verdeutlicht weitere Dimensionen der Zeremonie im Kontext der ästhetischen Formation. Im Wechsel der Kamera-Einstellungen zwischen der totalen Perspektive von oben in den Kirchenraum, den Naheinstellungen mit Fokus auf Dean Wade sowie auf Gesichter der Ehrengäste in der ersten Reihe manifestiert sich eine zusätzliche mediale Ebene in der ästhetischen Inszenierung. Manche Schnitte und Perspektivwechsel sind dabei an den Sprecherwechseln orientiert

63 Washington National Cathedral: In Celebration of the Life of Neil Armstrong. DVD. 2012. Min 8:15 f.

64 Washington National Cathedral: In Celebration of the Life of Neil Armstrong. DVD. 2012. Min 9:40 f.

und erleichtern lediglich ein Nachvollziehen des Ablaufs, indem sie jeweils diejenigen Akteure des Rituals ins Bild setzen, die eine aktive Rolle einnehmen. So wechselt beispielsweise die Kameraperspektive an der Stelle vom Fokus auf Dean Wade auf die erste Reihe (einschließlich Carol Armstrong sowie der Apollo-11-Crew-Mitglieder Buzz Aldwin und Michael Collins), an welcher die Gemeinde in der liturgischen Passage im Chor Antwort gibt. An einer prominenten Stelle jedoch leistet die Kameraperspektive bereits in dieser kurzen Eröffnungsphase des Gottesdienstes eine zusätzliche Funktion, da sie die ästhetische Inszenierung des nationalen Raumfahrtprogramms im Rahmen des Gedenkgottesdienstes mit der ästhetischen Inszenierung des nationalen Raumfahrtprogramms in einem der prominentesten Kirchenfenster der National Cathedral verschränkt: Während Dean Wades Worten: „Lead us, like them, to understand better the wonder and mystery of creation, through Christ, your eternal word through whom all things are made", wechselt die Kameraperspektive auf das sogenannte „Space Window" im südwestlichen Seitenschiff der Kathedrale. Das Kirchenfenster, das die erfolgreiche Apollo-11-Mission ins Bild setzt, zählt zu einer der herausragenden Besucher-Attraktionen der Kathedrale. Die Kamera schwenkt just in dem Augenblick vom „Space Window" über die Südseite des Kirchenschiffs hinweg, als Dean Wade im Chor mit den Versammelten mit einem „Amen" den rituellen Abschluss der Eröffnung bekräftigt. Der nun folgende Perspektivwechsel zurück auf die Totale mit Blick auf den Altarbereich zeigt Dean Wade beim Verlassen des Pults. Dean Wade gibt hiermit im wörtlichen Sinne Raum für eine weitere Stimme, die nun das Zentrum einnimmt – allerdings ausschließlich im akustischen Sinne: Über die Lautsprecheranlage der Kathedrale ist die Stimme des ehemaligen US-Präsidenten John F. Kennedy zu vernehmen, die jetzt den Kirchenraum erfüllt. Zu hören sind Ausschnitte aus Kennedys historischer Rede an der Rice University, wo er exakt fünfzig Jahre zuvor den Studierenden vor Ort wie auch gleichzeitig der gesamten Nation sein Raumfahrtprogramm als nationales Projekt im (nuklear) strategischen Wettstreit mit dem Sowjet-Imperium erläutert:

> Those who came before us made certain that this country rode the first waves of the industrial revolution, the first waves of modern invention, and the first wave of nuclear power, and this generation does not intend to founder in the backwash of the coming age of space. We mean to be a part of it – we mean to lead it. [*begeisterter Applaus der damaligen Zuhörer zu hören*] For the eyes of the world now look into space, to the moon and to the planets beyond, and we have vowed that we shall not see it governed by a hostile flag of conquest, but by a banner of freedom and peace.[65]

65 Washington National Cathedral: In Celebration of the Life of Neil Armstrong. DVD. 2012. Min 10:15 f.

Zwar werden in den eingespielten Tonaufnahmen einige der Passagen in Kennedys Rede ausgelassen, welche das Raumfahrtprojekt im Kontext eines präemptiven Vorgehens skizzieren und die Notwendigkeit betonen, als erste Nation den Weltraum zu besetzen, um die Gefahr einer nuklearen Bedrohung vom Weltall aus durch die Sowjetische Union zu bannen. Doch stellen die Formulierungen im obigen Auszug das Apollo-Projekt eindeutig in den Zusammenhang des imperialen Wettstreits, der Eroberung sowie einer internationalen Friedens- und Freiheits-Mission der USA. Der eingespielte Ausschnitt endet mit der Schlusspassage der Rede, in welcher Kennedy dem nationalen Projekt nicht nur eine menschheitsgeschichtliche sondern auch eine heilsgeschichtliche Dimension zuschreibt: „And, therefore, as we set sail we ask God's blessing on the most hazardous and dangerous and greatest adventure on which man has ever embarked. Thank you. [*stürmischer Applaus der damaligen Zuhörer in Houston*]"[66]

Schlussbetrachtungen

Entstehungsgeschichte, sakrale Monumentalität und Nutzung der National Cathedral zeigen den prominenten Platz dieses sakralen Raumes in der US-amerikanischen Hauptstadt und ihre engagierte Rolle innerhalb der US-amerikanischen Kollektivimagination. Unter Verweis auf Pierre L'Enfant, den visionären Stadtplaner Washington D.C.s, präsentiert sich die National Cathedral explizit als legitime Realisierung seiner oben erwähnten Idee, an prominenter Stelle in der US-Hauptstadt ein nationales Gebetshaus zu errichten, das allen Bürgern, über die Glaubensgrenzen hinweg, offen steht und nationalen Zwecken dienen soll. Besonders die aktiv ausgestaltete Verschränkung des politischen Feldes mit dem religiösen Feld, welche in der National Cathedral nach dem Vorbild der Westminster Abbey in Großbritannien Tradition hat, wirft die Frage nach der Trennung von Staat und Kirche auf, die in der US-amerikanischen Verfassung eingeschrieben ist. Tatsächlich lässt sich mit Blick auf die kontinuierliche Selbstfinanzierung der National Cathedral von einer Trennung zwischen *Staat* und *Kirche* sprechen. Für unseren Kontext der Kollektivimagination und Fragen der (imperialen) Souveränität ist jedoch von größerer Bedeutung, dass hier gleichzeitig eine zentrale Verbindung von *Nation* und *Religion* festzustellen ist. Für ihre semiotische Ver-

66 Washington National Cathedral: In Celebration of the Life of Neil Armstrong. DVD. 2012. Min 10:40 f.

schränkung bietet die National Cathedral eine öffentliche Bühne der *Aesthetic Formation.*

Über die Verschränkung nationaler und religiöser Dimensionen hinaus, die bereits im Namen der National Cathedral angelegt sind, schafft der sakrale Raum einen Ort für die rituelle Gestaltung nationaler Erinnerungskultur. Als außergewöhnlicher Ort nationaler Erinnerung und Repräsentation US-amerikanischer Geschichte konstruiert, versammelt, erzählt und kanonisiert die National Cathedral Aspekte nationaler Kollektivimagination. Welche Rolle die Kathedrale dabei in der nationalen Gemeinschaftsbildung der USA, aber auch in der internationalen Wahrnehmung der USA spielt, ist nicht zuletzt mit der Art und Weise verbunden, in der zeremonielle Ereignisse wie Gedenkgottesdienste, Inaugurationen oder nationale Gebete für die US-Truppen im Krieg gestaltet werden. Die komplexe Verbindung von nationaler Souveränität und nationaler Mythologie stellt Erinnerungskultur und kollektive Legitimation von gültigen Narrativen vor Herausforderungen. Hardts und Negris Gedanken zur Rolle von Religion im Zusammenhang mit der aufkommenden Nationalstaatsidee sensibilisiert wie oben dargestellt[67] für den Ort des demokratischen Diskurses in diesem Zusammenhang. Diese Fragen übertragen sich zudem auf die internationale Ebene. Wenn die Zeremonie zu Ehren Neil Armstrongs beispielsweise das nationale Raumfahrtprojekt in eine kosmologisch-heilsgeschichtliche Sinngebung kleidet, findet hierdurch eine sakrale Aufladung einer nationalen Mission statt. Der Politikwissenschaftler Herfried Münkler sieht hierin gar ein Signum imperialer Souveränität. Staaten und Hegemonialmächte, so Münkler, agierten charakteristischerweise ihrem eigenen Selbstverständnis nach *in* der Zeit, „während Imperien für sich in Anspruch nehmen, dass sie über den Zeitlauf entscheiden. Stärkster Ausdruck dessen ist die sakrale Aufladung der imperialen Mission."[68]

Wie Münkler mit Blick auf die Geschichte der Imperien feststellt, provozierte die ausgeprägte Selbstsakralisierung eines Imperiums schon immer starke antiimperiale Reaktionen. Den Kampf des Islamismus gegen das *American Empire*[69] sieht Münkler als größte Herausforderung, die dezidiert gegen die Mission der USA gerichtet ist:

---

67 Siehe den obigen Abschnitt 2.2 Religion und Nationalstaatlichkeit.

68 Herfried Münkler. *Imperien. Die Logik der Weltherrschaft – vom Alten Rom bis zu den Vereinigten Staaten*. Berlin: Rowohlt, 2008, 138–139.

69 Für eine Diskussion der neuartigen Aspekte im amerikanischen Imperium des zwanzigsten und einundzwanzigsten Jahrhunderts siehe Münkler, *Imperien*, 228 ff. Zu einer differenzierten Diskussion des Imperiumsbegriffs im Kontext der USA siehe Michael Hochgeschwender. Die USA – ein Imperium im Widerspruch, *Zeithistorische Forschungen* 3:1 (2006), 55–76.

> Das dabei zu beobachtende Grundmuster, nach dem eine auf quasi-religiöse Gewissheiten zurückgreifende Imperialrhetorik antiimperiale Gegenrhetoriken provoziert, die sich ihrerseits auf religiöse Gewissheiten stützen, kennzeichnet auch die gegenwärtige Debatte über den Status und die Macht des *American Empire*: Je stärker die US-Politik in quasi-religiösen Gewissheiten kommuniziert wird, etwa wenn sie ihre Gegner als satanisch oder dämonisch bezeichnet, desto stärker übernehmen jene Kräfte den Part des wichtigsten antiimperialen Gegenspielers, die ebenfalls mit religiösen Gewissheiten aufzuwarten vermögen. Es kommt nicht von ungefähr, dass diese Rolle seit geraumer Zeit dem Islamismus zugefallen ist. Daran wird sich trotz Chinas wirtschaftlichem Erstarken vorerst nichts ändern. Der Islamismus ist die wirkmächtigste Herausforderung des amerikanischen Empire, weil er dessen Mission bestreitet und die USA seinerseits als den ‚großen Satan' bezeichnet.[70]

In diesem Sinne ist die mediale Inszenierung nationaler Vergangenheit sowie die konkrete Ausgestaltung der *Sensational Forms* wie etwa im sakralen Raum der National Cathedral nicht nur von Bedeutung für die US-amerikanische Kollektivimagination und Gemeinschaftsbildung, sondern ebenso für die symbolisch aufgeladene Arena der internationalen Politik in einer globalisierten Welt.

Literaturverzeichnis

Auffarth, Christoph. *Religion auf dem Lande: Entstehung und Veränderung von Sakrallandschaften unter römischer Herrschaft*. Stuttgart: Franz Steiner, 2009.

Bachmann-Medick, Doris. *Cultural Turns: New Orientations in the Study of Culture*. Berlin: de Gruyter, 2016.

Chaikin, Andrew: *A Man on the Moon*. Alexandria, Va: Time-Life Books, 1994.

Chidester, David und Edward T. Linenthal. *American Sacred Space*. Bloomington und Indianapolis: Indiana University Press, 1995.

Davis, Margaret Bergen: *Living Stones. Washington National Cathedral at 100*. Washington, D.C.: Washington National Cathedral, 2007.

Elsas, Cristoph. Kultort. In *Handbuch religionswissenschaftlicher Grundbegriffe*, Band 4, Hubert Cancik, Burkhard Gladigow, Karl Heinz Kohl (Hrsg.), 32–43. Stuttgart: Kohlhammer, 1998.

Fallen, Anne-Cathrine. *Washington National Cathedral. The Washington National Cathedral Guidebooks. Washington*, D.C.: Washington National Cathedral, 2001.

Grimes, Ronald L. Jonathan Z. Smith's Theory of Ritual Space. *Religion* 29:3 (1999): 261–273.

Hardt, Michael und Antonio Negri. *Empire*. Cambridge und London: Harvard University Press. 2000.

70 Münkler, *Imperien*, 149. Hervorhebung im Original. Münklers Blick auf den islamischen Fundamentalismus weist dabei zwar Parallelen zur Analyse Hardts und Negris auf, ist allerdings klar in seinem historisch-deskriptiven Blick verortet ohne Hardt und Negris zweifelhafte Andeutungen einer ambivalenten Faszination gegenüber der revolutionären Kraft des Islamismus.

Heinrich, E. Kulthaus. In *Reallexikon der Assyrologie und der Vorderasiatischen Archäologie*, Band 6, Erich Ebeling, Dietz O. Edzard, Bruno Meissner, Wolfram v. Soden, Ernst Weidner (Hrsg.), 319–323. Berlin und New York: de Gruyter, 1981.

Hewlett, Richard Greening: *The Foundation Stone. Henry Yates Satterlee and the Creation of Washington National Cathedral*. Rockville: Montrose Press, 2007.

Hochgeschwender, Michael. Die USA – ein Imperium im Widerspruch. *Zeithistorische Forschungen* 3:1 (2006), 55–76.

Jamme, Christoph und Heinrich Schneider. *Mythologie der Vernunft – Hegels „Ältestes Systemprogramm des deutschen Idealismus"*. Frankfurt am Main: Suhrkamp, 1984.

Kippenberg, Hans G., Jörg Rüpke und Kocku von Stuckrad (Hrsg.). *Europäische Religionsgeschichte: Ein mehrfacher Pluralismus*. Göttingen: Vandenhoeck & Ruprecht, 2009.

Kugele, Jens und Katharina Wilkens (Hrsg.). Relocating Religion(s) – Museality as a Critical Term for the Aesthetics of Religion. Special Issue. *Journal of Religion in Europe* (2011). Leiden, Boston: Brill.

Lang, Bernhard. Kult. In *Handbuch religionswissenschaftlicher Grundbegriffe*, Hubert Cancik, Burkhard Gladigow, Karl-Heinz Kohl (Hrsg.), 475–488. Stuttgart: Kohlhammer, 1998.

Lang, Bernhard. Kult/Ritual. In *Metzler Lexikon Religion*, Band 2, Christoph Auffarth, Jutta Bernard, Hubert Mohr (Hrsg.), 267–274. Stuttgart: Metzler, 1999.

Lippert, Hans-Georg. Alte Ordnung für die Neue Welt – Betrachtungen zur Kathedrale von Washington. In *Aspekte der Zeit – Zeit-Geschichte, Raum-Zeit, Zeit-Dauer und Kultur-Zeit*, Joachim Klose, Klaus Morawetz (Hrsg.), 198–220. Münster: Lit, 2004.

Lowenthal, David. Past Time, Present Place. Landscape and Memory. *The Geographical Review* 65:1 (1975): 1–36.

Mann, Nicholas R. *The Sacred Geometry of Washington, DC: The Integrity and Power of the Original Design*. Somerset: Green Magic, 2006.

Meyer, Birgit. *Aesthetic Formations: Media, Religion, and the Senses*. New York: Palgrave Macmillan, 2009.

Meyer, Birgit, David Morgan, Crispin Paine und S. Brent Plate. The Origin and Mission of Material Religion. *Religion* 40:3 (2010): 207–211.

Meyer, Birgit und Jojada Verrips. Aesthetics. In *Key Words in Religion, Media and Culture*, David Morgan (Hrsg.), 20–30. London und New York: Routledge, 2008.

Münkler, Herfried. *Imperien. Die Logik der Weltherrschaft – vom Alten Rom bis zu den Vereinigten Staaten*. Berlin: Rowohlt, 2005.

Pinkard, Terry. *Hegel. A Biography*. New York: Cambridge University Press, 2000.

Rau, Susanne. *Räume – Konzepte, Wahrnehmungen, Nutzungen*. Frankfurt a. M.: Campus, 2013.

Smith, Jonathan Z. *To Take Place. Toward Theory in Ritual*. Chicago und London: University of Chicago Press, 1987.

Soja, Edward William: *Thirdspace. Journeys to Los Angeles and Other Real-and-Imagined Places*. Oxford: Blackwell, 1996.

Stolow, Jeremy. Religion and/as Media. *Theory, Culture and Society* 22:4 (2005), 119–145.

Traut, Lucia und Annette Wilke. *Religion – Imagination – Ästhetik. Vorstellungs- und Sinneswelten in Religion und Kultur*. Göttingen: Vandenhoeck & Ruprecht, 2015.

Wolf, Jürgen. Kultplatz. In *Metzler Lexikon Religion*. Band 2, Christoph Auffarth, Jutta Bernard, Hubert Mohr (Hrsg.), 287–288. Stuttgart: J. B. Metzler, 1996.

Washington National Cathedral: *In Celebration of the Life of Neil Armstrong*. DVD. 2012.

Abbildungsnachweise

Abb. 1: Frontseite der National-Kathedrale (Foto: Jens Kugele).
Abb. 2: Blick von der National Cathedral über Washington DC (Foto: Jens Kugele).

Cartographies of the Imperial Age

Sebastian Dorsch and Iris Schröder

Introduction
Spatiotemporalities of Cartographic Empire-Building

The history of the close relationship between cartography and Empire has often been told. In fact, mapping and empire building seem to be intrinsically linked with one another. In this vein, historians of cartography have pointed to "Empire" as "a cartographic construction", underscoring that "modern cartography is the construction of modern imperialism."[1] However, this said, the many ways in which cartography and empires actually related to one another still calls for further exploration. While some have argued that cartography was mainly a European affair used by imperial agents as some sort of useful governmental tool, others have stressed the importance of looking more closely into the many colonial situations on the ground where we might potentially discover the collective character of cartographic knowledge production.[2] While there is an ongoing controversy about the need to introduce non-European actors into the history of cartography, all in all, scholars tend to agree on the importance of cartographic images that considerably contributed to many empires' imaginary, as well as concurring that maps can be seen as an important form of a visual, foremost imaginary appropriation of the world.[3]

In addition to this, in recent years, Empire as a political stance has come to the forefront. In this context, the book *Empire* (2000), by Michael Hardt and Antonio Negri, has generated widespread interdisciplinary discussions.[4] Studying the post-modern (i.e. post-Cold War) world order Hardt and Negri maintained that

1 Matthew H. Edney. The irony of imperial mapping. In *The imperial map: Cartography and the mastery of empire*, James R. Akerman (ed.), 11–45. Chicago: University of Chicago Press, 2008, 45.
2 See for the first mentioned interpretation: Matthew H. Edney. *Mapping an Empire. The geographical construction of British India, 1765–1843*. Chicago: University of Chicago Press, 1997; Graham Burnett. *Masters of All They Surveyed: Exploration, Geography, and a British Eldorado*. Chicago: Chicago University Press, 2000; for the second one Kapil Raj. *Relocating Modern Science: Circulation and the Construction of Knowledge in South Asia and Europe* 1650–1900. Basingstoke: Palgrave Macmillan, 2007 as well as for an early call to reconsider the manifold interactions between "metropole and the colony": Frederick Cooper, Ann Laura Stoler (ed.). *Tensions of Empire: Colonial Cultures in a Bourgeois World*. Berkeley: University of California Press, 1997.
3 See for example: Ricardo Padrón. *The spacious word. Cartography, literature, and empire in early modern Spain*. Chicago: University of Chicago Press, 2004.
4 Michael Hardt, Antonio Negri. *Empire*. Cambridge: Harvard University Press, 2000.

DOI 10.1515/9783110418750-017

Empire is now turning into a new global form of sovereignty. By now, they claim, there is a decentered, deterritorializing and progressively expanding Empire that stands in sharp contrast to the 'old' European imperial powers, which were based on centered, territorialized regimes.[5] While Hardt and Negri insisted on the new character of contemporary Empire, historians have challenged this juxtaposition, especially the suggested interpretation of the so-called older imperial regimes. New scholarship, especially focussing on the British case, has thus pointed to the "Empire project" underlining the need to see the bricolage character of the imperial endeavour on the whole.[6] Moreover, in an even bolder version of world history, the term Empire itself seems to be fading away, leaving only "imperial formations" behind which, however, are most often sustained by mapping endeavours.[7]

It goes without saying that the many stakes alluded to here should generate further research. Therefore, the following chapters on cartography and empire will not just be influenced by Hardt/Negri, but rather the other way around, they claim that new historical Empire studies can advance their theory in many ways. In this particular section, we focus on the spatiotemporality of Empires in the making, drawing our attention especially to nineteenth century imperial settings. In order to advance a more nuanced understanding of the questions at stake, we collected three studies from different historical contexts: Bernardo Michael analyses demarcation practices before and during the Anglo-Gorkha War (1814–1816) between the East India Company and the regime of Gorkha (today Nepal). Then, Stephen Walsh's chapter points to spatiotemporal groundings of the production of mass data driven inductive science in mapping the High Arctic island of Jan Mayen by Austro-Hungarians in the 1880s. Finally, yet importantly, Alrun Schmidtke examines how armchair geographer Bruno Hassenstein as a non-governmental actor in Gotha, Germany, tried (and failed) to map a distant empire, Japan, in the last decades of the nineteenth century.

Bernardo Michael's chapter focuses on the tensions between everyday governance practices and linear mapping strategies in the frontier area between regions claimed by the East India Company, as well as by Indian and the Gorkhali kingdoms. Confronted with long-standing disputes between local Gorkhali and

5 In this concept, Empire has no spatial or temporal boundaries – as a biopower, Empire is dedicated to "perpetual and universal peace outside of history". Only the creative power of what they term the Multitude is or will be able to contradict the holistic spatiotemporal Empire, see Hardt and Negri, *Empire*, xv.

6 John Darwin. *The Empire Project: The Rise and Fall of the British World-System, 1830–1970*. Cambridge: Cambridge University Press, 2009.

7 Burbank and Cooper, *Empires*.

Indian actors in the beginning of the nineteenth century, Company officials tried to re-organise the region's "territorial illegibility" and its fiscal system. By analysing everyday directives, records etc., Michael demonstrates that the local officials established a fixed principle, that of limitation between the actors on the ground, which fixed territorial claims in space and time. By implementing this principle, officials aimed to produce a-historical, linear bounded "absolute spaces", "containers" as Michael points out. Contrary to many studies, he emphasizes that it was not the European centre that introduced linear order in the "periphery", but local agents confronted with local disputes. Imperial mapmakers in the later decades of the nineteenth century could build on these daily practices of making territories legible – and they could make historical practices vanish in their maps through strategies of spatiotemporal fixation.

Building off Paul Carter, Stephen Walsh conceptualizes maps/mapping as "spatial interviews" between invaders/consumers and nature. He proposes Western Science as an "empire". In the latter half of the nineteenth century, new practices of mass data inductive science such as geomagnetism attempted to transform the exploration of wild environments into a laboratory science, i.e. turn the adventures of strong, but subjective, men into a "real modern" science which proliferated reliable knowledge – or as Stephen Walsh calls it, allegedly "objective geographic truth". With the removal of local and historical contexts as characteristic of Western imperial cartography, and the inscription of "objective" knowledge into maps, cartographic actors could produce putatively "objective", unchangeable images of the world in a widely apprehensible manner and, consequently, spatiotemporal certainty within the interested public sphere. At the same time, Walsh demonstrates how the arctic island Jan Mayen was transformed from a void *terra nullius* into something 'meaningful': into a scientifically dominated "historical space". Walsh's chapter clearly spells out that by taking the localized and historicized actors of mapping into account, by interviewing their environments, we can broaden the Hardt/Negri conception of empire and cartography.

Alrun Schmidtke, in her contribution, deals with the production of the Atlas of Japan (1885/87) by Bruno Hassenstein and thus points to specific spatial and temporal characteristics of private, non-governmental imperial practices. Hassenstein's employer – the publishing house Perthes in Gotha, Germany, one of Europe's cartographic "hot spots" of that time – was confronted with the speed of the market, with changing demands and new products by competitors. Additionally, the scientific claims of Hassenstein (with his doctoral degree) and those of Perthes (as a publishing house renowned for its high quality products) were interfering. As Schmidtke shows, these temporal obstacles have to be analysed in direct connection with the spatial ones: Hassenstein's concept of his atlas, as a product that could easily be exported to the mapped country, failed. It foundered due to distance. Hassenstein as armchair geographer invested (too) much time in establishing a "transimperial network"[8] for acquiring and translating his distant "sources". As a form of cultural othering of the "extra-European," he thought that he could apply the same imperial cartographic techniques to Japan that he had used before with colonial Africa. Hence, he neglected established systems of transcription from Sino-Japanese to Latin characters and produced a confusing, and scientifically excessive, map that could not meet the daily practical demands of its expected consumers, the *Oyatoi gaikokujin* ("contract foreigners") in Japan. "Imperial repertoires"[9] had to suit to the spatiotemporal environments.

The often supposed global character of science, as well as its alleged objectivity which is often taken as a simple given, especially in cartography, connects seemingly well to the Hardt/Negri thesis of a decentered and unbounded Empire "out of history." However, the chapters of the following section openly challenge the Hardt and Negri's concepts in several ways. Drawing on "critical cartography" the authors posit multiple centres with multiple actors, stressing the manifold media of empire-building. They claim that actors in specific historical *loci of enunciation* (Walter Mignolo),[10] confronted with specific spatiotemporal contexts, had specific interests in producing cartographic world-images as well as power relations. In this vein, the following section suggests to consider these actors as translocal producers of empires.

References

Burbank, Jane and Frederick Cooper. *Empires in world history. Power and the politics of difference*. Princeton: Princeton University Press, 2010.

Burnett, Graham. *Masters of All They Surveyed: Exploration, Geography, and a British Eldorado*. Chicago: Chicago University Press, 2000.

Cooper, Frederick and Ann Laura Stoler (ed.). *Tensions of Empire: Colonial Cultures in a Bourgeois World*. Berkeley: University of California Press, 1997.

Darwin, John. *The Empire Project: The Rise and Fall of the British World-System, 1830–1970*. Cambridge: Cambridge University Press, 2009.

Edney, Matthew H. *Mapping an empire. The Geographical construction of British India, 1765–1843*. Chicago: University of Chicago Press, 1997.

8 We take this notion from Bernhard C. Schär. *Tropenliebe. Schweizer Naturforscher und niederländischer Imperialismus in Südostasien um 1900*. Frankfurt: Campus Verlag, 2015.

9 Burbank and Cooper, *Empires*, 3.

10 See for example in: Walter Mignolo. *Local histories/global designs. Coloniality, subaltern knowledges, and border thinking*. Princeton: Princeton University Press, 2000.

Edney, Matthew H. The irony of imperial mapping. In *The imperial map: Cartography and the mastery of empire*, James R. Akerman (ed.), 11–45. Chicago: University of Chicago Press, 2008.

Hardt, Michael and Antonio Negri. *Empire*. Cambridge: Harvard University Press, 2000.

Mignolo, Walter. *Local histories/global designs. Coloniality, subaltern knowledges, and border thinking*. Princeton: Princeton University Press, 2000.

Padrón, Ricardo. *The spacious word. Cartography, literature, and empire in early modern Spain*. Chicago: University of Chicago Press, 2004.

Raj, Kapil. *Relocating Modern Science: Circulation and the Construction of Knowledge in South Asia and Europe*, 1650–1900. Basingstoke: Palgrave Macmillan, 2007.

Schär, Bernhard C. *Tropenliebe. Schweizer Naturforscher und niederländischer Imperialismus in Südostasien um 1900*. Frankfurt: Campus Verlag, 2015.

Bernardo A. Michael

The Spatial Anxieties of Everyday Colonial Rule and the History of Cartography: Connecting the Dots[1]

> "... the British possessions, ..., are classed by governments and revenue or judicial divisions, must be considered only *approximations to truth* on a subject, in its nature of great difficulty, and which has hitherto not received from the Indian authorities the degree of attention which its *practical value* and importance is entitled to" (emphasis added).
> *(John Crawfurd[2])*

> "... this rediscovery and rewriting of the imperial past and postcolonial present should be done in explicit connection with a sense of the lived geographies of empire..."
> *(Neil Smith[3])*

Introduction

Most of the research for this paper was conducted in connection with an examination of the territorial disputes surrounding the Anglo-Gorkha war (1814–1816) and formed part of a larger study on statemaking and space.[4] There, I endeavored to reveal how the extraction of tribute and taxes along with the arrangement of landed tenures produced the discontinuous and dispersed territories that made up the shared agrarian frontier of the English East India Company and the Himalayan kingdom of Gorkha (present-day Nepal). Such landed relationships produced shifting patterns of land-use that ultimately impacted the layout and organization of territory. Consequently, lands frequently shifted back and

---

1 I am grateful to Ishita Bannerjee-Dube, Raymond Craib, Saurabh Dube, Matthew H. Edney, Sebastian Dorsch, Iris Schröder who along with the organizers of the conference on Imperial Space-Time at the University of Erfurt, in Germany in October 2014 have sustained a long drawn conversation on space, cartography and statemaking. Thanks are also due to Ambika Arvindd for her editorial assistance. The usual caveats apply.

2 John Crawfurd *Description of India, 1822–23*. IOR/Mss Eur D457a and 457b, *IOR, APAC, BL*. The quote is from D 457a, 177–178 (emphasis mine).

3 Neil Smith. Geography, Empire, and Social Theory. *Progress in Human Geography* 18 (1994): 491–500, 491.

4 See Bernardo A. Michael. *Statemaking and Territory: Lessons from the Anglo-Gorkha War (1814–1816)*. London: Anthem Press, 2012.

DOI 10.1515/9783110418750-018

forth between agents attached to different states with no consideration of linear boundaries or continuity of territory. Patches of land belonging to both states along the Anglo-Gorkha frontier could come under the extractive levies of some local magnate or petty chief in one year, remain waste in another, and revert to cultivation in the third, only this time in the hands of some new source of political authority. As a result, the layout and organization of administrative divisions such as a *pargana* or *tappa* along the Anglo-Gorkha borderlands began to break up leaving them with intermixed bodies possessing fuzzy and shifting boundaries. Such territorial illegibility was unacceptable for the British and in the ensuing war the Gorkhalis were defeated in 1816. It is interesting to note that immediately after the war, the British began demarcating a linear boundary between the two states, one that was now capable of representation on modern maps.

The implications of such spatial dynamics for the history of cartography deserve additional comment. Recent histories of empires have examined the role of cartography in representing and (re)ordering the territories of states. Maps typically seek to represent, bound, mark, and organize space or territory. While such studies have made impressive examinations of the role of maps, surveyors, and surveying institutions, they have also left us with the impression that states were just "put" on maps through metropolitan initiative.[5] Sometimes little is known of the initiative taken by local agents, official and otherwise, as they jostled to organize and represent the territories they inhabited or administered. The initiative of such agents presented a terrain of activity that affirmed or confounded metropolitan visions of territory. At times, and despite the absence of reliable maps and facilities for undertaking formal surveys, colonial officials would attempt to organize their territories under their jurisdiction in ways that would reproduce the neat lines of a modern map. Their spatial agency needs to be given greater recognition in the history of cartography.

Admitting such a plurality of voices in the history of cartography calls for discerning "cartographic" agency within new contexts that might be distinct from the customary activities of surveyors, their collaborators, and surveying departments or the ideological and symbolic functions of maps. One way of doing this would be to trace the connections between cartography and considerations of spatiality emanating from other dispersed arenas of governance. That is, carto-

5 See for instance the chapter "Census, Map, and Museum" in: Benedict Anderson. *Imagined Communities*. London: Verso, 1991; Michael Biggs. Putting the State on the Map: Cartography, Territory and European State Formation. *Comparative Studies in Society and History* 41 (2): 374–405; and Joseph W. Konvitz. *Cartography in France, 1660–1848: Science, Engineering, and Statecraft*. Chicago: The University of Chicago Press, 1987.

graphic agendas can be potentially informed by spatial impulses that originate in arenas of social life far removed from the institutional framework and agendas of states and their surveying departments. Such a view examines questions of space to establish connections between histories of mapmaking and the activities of human agents inhabiting diverse social landscapes. For instance, cadastral mapping may be viewed as the cartographic component of the larger social arena within which various forms of agrarian entitlements are defined, organized, and regulated.[6] Like Ariadne's thread of yore that connected the labyrinth to the outside world, we need to join the dots between maps, and the wider socio-spatial conditions of their production, in previously untreated ways.

Perceiving the social tissue that links the mapmaker's room to the deeper spatial impulses and anxieties lying elsewhere allows us to write thicker accounts of cartographic history that assign greater agency to (especially) non-elite subjects whose activities shaped the organization of territories and their subsequent representation on maps. "Worlding" maps, in this fashion, builds on the pioneering work of scholars like Arthur Robinson, David Woodward, and John Brian Harley who argued that maps are not the mere products of technical skill, but are also awash with the forces of culture, power, and history.[7] Cartographic history's task as a theoretically sophisticated and ethically informed practice was to excavate the hidden agendas, silences, and relations of power that informed the production of maps. In this connection, it is then appropriate to recall that the geography of the state was not simply represented on maps; rather its territorial details were worked out in a piecemeal and uncoordinated fashion at the level of everyday governance. In South Asia, the early colonial state's geography was realized through all kinds of patchy and shifting initiatives on the ground, often without the help of accurate maps. In other words, colonial mapmaking was not an overdrawn homogenous metropolitan project in-the-making that was then thrust on a passive landscape. The work of surveying departments was complemented, confounded, and sometimes even preceded by the work of numerous (sometimes unwitting) agents on the ground. In focusing on these themes, this paper seeks to add to the growing literature on the history of cartography in South Asia that have detailed how modern cartography as a new form of global power

---

6 I am grateful to Matthew H. Edney for his comments and suggestions that have helped me clarify this point.

7 While the literature on the subject is too vast to cite here, see the following: a collection of J. B. Harley's work can be found in: Paul Laxton. *The New Nature of Maps: Essays in the History of Cartography*. Baltimore: The John Hopkins University Press, 2002. See also: Matthew Edney. Putting 'Cartography' into the History of Cartography: Arthur H. Robinson, David Woodward, and the Creation of a Discipline. *Cartographic Perspectives* 51 (2005), 14–29.

resulted in the reorganization of state spaces.[8] Official correspondence from the early colonial period reveals the presence of deep seated spatial anxieties about the structure, layout, boundaries, and internal divisions of the Company's recently acquired territories. Lacking the resources and information to produce maps or conduct expensive surveys, colonial officials devised various responses that had a map-like quality to them because they attempted to install a particular vision of territorial organization within the routines of every rule without taking recourse to any formal surveying and mapmaking. This turned out to be a long-drawn, complicated, and uneven process. The following sections will examine evidence from the routines of everyday governance, far removed from the surveyor's offices, where colonial officials reveal a territorial design that was based on a distinct vision of the geography of the state. It also alerts us to the possibility of sidestepping some of the homogenizing and dichotomizing tendencies within the literature on the history of cartography in order to reveal new and valuable counterpoints. Such a move has methodological implications as well, calling for an examination that is inclusive of a wider range of local and regional archival materials, and not confined to the mere study of maps and survey records lying in national repositories.

1 A "Principled" Production of Space

Ever since the English East India Company acquired territory in India in the eighteenth century, colonial governance produced deep seated spatial or territorial anxieties. During the early decades of colonial rule, the Company state possessed little or no information of the actual layout, internal divisions, and boundaries of its territories. For instance, indigenous administrative divisions such as the *pargana* tended to possess shifting and discontinuous bodies that were hard to discern without survey and mapping. Company officials struggled to make sense of the intermixture, overlaps, and fluid boundaries that characterized the organization and layout of these *pargana* divisions. Such an experience of terri-

8 See for instance: Ian Barrow. *Making History, Drawing Territory: British Mapping in India, c. 1756–1905*. New Delhi: Oxford University Press, 2003; Lucy Chester. *Borders and conflict in South Asia: The Radcliffe boundary commission and the partition of Punjab*. Manchester University Press, 2009; and: Matthew H. Edney. *Mapping an Empire: The Geographical Construction of British India, 1765–1843*. Chicago: University of Chicago Press, 1997. For a powerful and widely debated account of the reorganization of social life by new forms of global power see Michael Hardt, Antonio Negri. *Empire*. Cambridge, MA.: Harvard University Press, 2000.

torial anxiety was widespread. In 1818, George Dowdeswell, the vice president of the Company's governing council at Calcutta echoed the sentiments of many company officials when he observed: "We have almost *daily experience* of the very defective state of our geographical information in many parts of our own territories and of the serious inconveniences and embarrassments arising from this cause in the conduct of *ordinary* affairs of the administration."[9] This problem was accentuated during the early decades of colonial rule, when lack of personnel and finances prevented the East India Company from surveying its territories and producing detailed local maps. However, this paucity of information and maps did not prevent colonial administrators from taking various measures aimed at keeping their territorial jurisdictions coherent, continuous, and clearly demarcated. Lacking the resources and information to produce maps or conduct expensive surveys, colonial officials devised responses that had a map-like quality to them because they attempted to install a particular territorial vision within the routines of everyday rule without taking resorting to any formal surveying and mapmaking initiatives.

Such a vision began to crystallize ever since the Company acquired territories in 1765. Company officials increasingly held a view of the state as a territorially coherent and continuous entity. Such a vision found expression in official discourse. Take for instance the representations of the Himalayan state of Gorkha (present-day Nepal) in colonial documents. Captain Knox, the Company's Resident at Kathmandu (1802–1803) described the Gorkhali state as being "weak and *unconnected*," while William Kirkpatrick in his late-eighteenth century account of the Government of Nepal found it hard to "determine the actual force of the *machine* of government."[10] Such a desire to see a unified and connected picture of the state, and its geography is evident in the instructions given by the Secretary John Adam to Thomas Rutherford, one of the Company's agents on the Western sections of the Anglo-Gorkha frontier. Rutherford was instructed to organize such information under certain heads – roads, rivers, positions, defiles etc. – "to aid inquiry and the *reduction* of desultory information into some *connected form.*"[11] This proclivity to represent the state as a territorially unified entity was not the monopoly of European states and their agents. Any state, colonial or indigenous,

---

9 Cited in: Edney, *Mapping*, 119. The emphasis is mine.

10 Captain Knox to Secretary to Government, *FS* Consl. 30 June 1802, no. 43, *NAI*. William Kirkpatrick. *An Account of the Kingdom of Nepaul*. Reprint, Delhi: AES Publications, 1969 [1811], 196–197. The emphasis is mine.

11 John Adam to T. Rutherford, *FS* Consl. 16 August 1814, no. 11, *NAI*. Emphasis mine. Rutherford was the Company's civil surgeon in the town of Moradabad. He also held private trading interests in the hills.

with centralizing tendencies could have potentially developed such a coherent vision of its territories.[12] In the case of England, the idea of a territorially unified state perhaps found early expression in the sixteenth century County Atlases of Christopher Saxton. Saxton's atlases not only contained individual county maps, but also a general map of England showing all its counties colorfully delineated. Such cartographic images of England were widely disseminated among the country's elite and common folk through paintings, tapestry work, and even playing cards.[13] But, like in India, it was only in the nineteenth century that the English state was able to take concrete measures to reorganize its poorly demarcated and ill-defined internal administrative divisions. The process by which this happened and its connections to territorial rearrangements being worked out in the colonies merits further exploration, but remains outside the scope of this paper.[14]

In the early nineteenth century, the English East India Company undertook silent spatial rearrangements along its frontier with the Himalayan kingdom of Gorkha (present-day Nepal), without attempting formal surveying and mapping operations. The territories of the Company state, Gorkha, and numerous petty principalities lay mutually entangled along this frontier because of an older regime of political, taxation, and tenurial relationships. Over time, a host of disputatious territorial claims had emerged between these parties. The Company, for its part, was very desirous of preserving the coherence and continuity of its territories, and suspicious of any territorial claims that confounded this vision. To achieve this without the assistance of modern maps, company officials formulated certain guidelines or principles that were designed to preserve the geographical unity of their territories. While such principles did not arise out of formal mapping exercises, they were nevertheless instrumental in preserving their vision of territory, a vision that would only become visible in maps created many decades later.

Such moves are evident all along the Anglo-Gorkha frontier prior to the outbreak of war in 1814. In the Western Himalayas for instance, when the Company

---

12 For the case of China and Siam see: Laura Hostetler. *Qing Colonial Enterprise: Ethnography and Cartography in Early Modern China*. Chicago: University of Chicago Press, 2001; and: Thongchai Winichakul. *Siam Mapped: The History of the Geo-Body of a Nation*. Honolulu: University of Hawaii Press, 1994.

13 See: Victor Morgan. The Cartographic Image of 'The Country' in Early Modern England. *Transactions of the Royal Historical Society (Sixth Series)* 29 (1979): 129–154.

14 For details see: David Fletcher. The Ordnance Survey's Nineteenth Century Boundary Survey: Context, Characteristics and Impact. *Imago Mundi* 51 (1999): 131–146; Bernardo A. Michael. Making Territory Visible: The Revenue Surveys of Colonial South Asia. *Imago Mundi* 59:1 (2007): 78–95.

took control of Gorkhali territories, it was faced with large number of competing territorial claims put forward by the kingdoms of the region. The situation was complicated by the Gorkhali invasion of the region (from the 1790s) which had produced substantial rearrangements in the structure of political relations, along with the payments of tribute and taxes. Realizing that some of these disputes were of a long-standing nature that could never be easily resolved, company officials attempted to formulate some fixed "principle" against which all such competing territorial claims could be "adjusted." These concerns surface in the instructions given to Col. David Ochterlony, the English commander-in-charge of the Western Himalayas:

> Without some *fixed principle* for the decision of such questions, it would be extremely embarrassing and difficult if not practicable to undertake the adjustment of the multiplied and conflicting claims of territory arising out of the disturbed state of the country for a long series of years, and the *frequent transition* of particular districts from the authority of one principality to that of another. The period of the Gorkhali invasion is in all respects the most convenient from being of so modern a date as to render evidence early attainable while with reference to our public declaration and the avowed objects of our interference in the affairs of the Hill Chiefs, namely the restoration of things to the state in which they were at that time, we might be perhaps justly considered to be precluded from taking cognisance of claims originating in the transactions of an antecedent period.[15]

By taking the Gorkhali invasion of the region as the cut-off point for determining the political status of a kingdom, and ignoring all territorial claims prior to the Gorkhali invasion, the Company was able to fix the fluctuating territorial claims of these kingdoms not only in time, but also in space or territory. The formulation of such "principles" was not an entirely neutral and innocent affair. Underlying such "principles" was a particular vision of territory that sought to cut through the complicated claims and relationships that left most territories mutually entangled. It is for this reason that Company officials sought to preserve the *status quo* of older political, taxation, and tenurial relationships as they existed just prior to the Gorkhali invasion. Such a simplification allowed the Company to reduce the number of competing claims to territory whose history could never be fully investigated. Such claims invariably generated highly convoluted geographies producing territories that were discontinuous, overlapping, and constantly shifting. Thus, the "principle" of possession at the time of the Gorkhali invasion, had

15 Letter from John Adam, Secretary to Government to Col. David Ochterlony, *FS* Procs. 2 December 1815, no. 4, *NAI*. The emphasis is mine.

a map-like quality to it, as it tried to articulate boundaries and reify the territories of the kingdoms of the Western Himalayas.

The results, however, were not always favorable to the British. For instance, again during the Anglo-Gorkha war, the British instigated the hill state of Kulu to take over some hill posts from the Gorkhalis. However, after the war, the Kulu raja began to claim these posts by virtue of conquest and ancient possession. Much to their shock, their rights were not recognized by the Company, as it went against the principles established by government to determine contested claims to territories. The principle (discussed earlier) was the restoration of the status quo to the pre-war state of affairs, that is, as existing in 1814. Prior to the Anglo-Gorkha war, these Gorkhali posts had been in the hands of the states of Basahar and Kumharsein. Thus, the posts were now given to the pleasantly surprised rulers of these states. What adds more irony to all this was the fact that prior to the Gorkhali invasion of this area in the first decade of the eighteenth century, Basahar and Kumharsein had forcibly wrested these very posts from the Kulu raja! Needless to say, the logic of the British "principles" must have baffled the Kulu raja who saw himself as the rightful owner of these posts.[16]

Another reordering principle with distinct spatial effects was the Company's steadfast refusal to occupy and govern lands in the hills. This "principle of limitation," was clearly motivated by spatial concerns.[17] First enunciated in 1813, it declared to Gorkhali authorities that, "no interference should take place in the proceedings of Nipaul in the hills and on the other hand the Nipaulese authority should on no account be extended below the hills."[18] The aim of this "principle" was to put a halt to the interminable territorial disputes that frequently arose between the Gorkhali commanders in the hills and their counterparts in the plains of North India. That is, hill and plain chieftains were asked to renounce all claims in each other's territories. This was tantamount to a naturalization of the territories of hill and plain kingdoms, and ignoring the age-old claims they had exercised in each other's territories. Pragmatic considerations possibly underwrote principles such as the "principle of limitation," the logistical problems involved in governing these hill territories, and the desire to retain them as a convenient buffer between the Company and the Chinese to the north. As far as Gorkha goes, it could also be argued that by 1814, the Company had come to view the Himala-

16 For details, see letter of Thomas Metcalf, Resident at Delhi to J. Adams, Secretary to Government, *FS* Consl. 16 December 1815, no. 3, *NAI*.

17 The term "principle of limitation" was coined by the Nepali historian Ludwig F. Stiller. See: Ludwig F. Stiller, The Principle of Limitation. *The Nepalese Perspective* 9 (1974): 4–21.

18 Cited in Stiller, The Principle, 6.

yas as the natural location of the Gorkhali state, with the foothills of the Himalayas forming the Company's natural frontiers. Such arbitrarily determined "fixed principles," too had a map-like quality to them. They allowed Company officials to organize and make legible in the form of a fixed structure, the territories under their control without resorting to formal surveying and mapmaking. The pragmatic needs of everyday governance demanded such responses which it was hoped would simplify the complicated territorial disputes and transfers that peppered the geography of its frontier. During the first decades of colonial rule, given the shortage of surveying and mapping resources, the British formulated such "principles" of legibility and simplification in response to questions of everyday governance. These "principles" tried to uphold a vision of territory as coherent, contiguous, and well-defined, a vision that would find ultimate expression in the colonial maps produced in succeeding decades.

2 Revenue Records and the Ordering of Territory

In precolonial South Asia, states possessed well-established traditions of record keeping. Detailed records were maintained pertaining to almost every aspect of general administration, revenue administration, trade, intelligence gathering, and social communication.[19] Rajasthani and Marathi documents for instance, provided detailed information on the revenue accounts of individual villages, soil type, area, proprietary rights, crops, and so on.[20] Fiscal-Administrative divisions such as *parganas* (sub-districts), and *mauzas* (villages) remained the basic territorial units in these records.[21]

However, divisions such as *parganas* were not objectified as territorially coherent units. Rather, they emerged in the form of detailed village lists showing land-related information, such as measured area, status (taxed or exempt), soil type, crop output, rate of taxation, lands added to the *mauza* or *pargana*, and

19 For historical treatment of information gathering and social communication in south Asia see: Christopher Bayly. *Empire and Information: Intelligence Gathering and Social Communication in India, 1780–1870*. Cambridge: Cambridge University Press, 1996.

20 For details see: Swaraj Prakash Gupta. *Agrarian System of Eastern Rajasthan*. Delhi: Manohar, 1986, 317–334; Swaraj Prakash Gupta. Khasra Documents in Rajasthan. In *Medieval India: A Miscellany*, vol. 4. Bombay: Asia Publishing House, 1977, 168–176.

21 See for instance, Satish Chandra and Swaraj Prakash Gupta. The Jaipur Paragana Records. *Indian Economic and Social History Review* 3 (1966): 303–315. Chandra and Gupta note that such *pargana* records from Rajasthan, despite their detailed nature possess gaps which "do not permit the continuous study of any pargana" (309).

so on. However, this was done without any reference to preserving the layout and territorial continuity of the concerned *pargana*. Representing the *pargana* as a *whole*, or getting a fix on it in terms of a continuous block of territory do not appear to have figured in the calculations of the record keepers. This is all too evident even in a statistically rich document as the *Ain-i-Akbari* (1595).[22]

It might be recalled that the *Ain* provides detailed statistics on the *sarkars* and *parganas* of the Mughal Empire – their social composition, revenue yields, crops grown, and even their measured area figures (*arazi*). It is the reliability of the *arazi* figures that concerns us. Even today, scholars are divided in their opinions about the exact nature of the *arazi* figure, whether they represented the entire area (including cultivated lands, cultivable wastes and uncultivable wastes), or only the net cultivable area (including cultivable waste), or only seasonal data pertaining to the winter (*kharif*) or summer harvests (*rabi*). At the same time, vast tracts of the Mughal Empire remained unsurveyed. Irfan Habib has estimated that fifty per cent of the Mughal Empire remained unsurveyed even at the time of Aurangzeb's reign. Of this, forty three per cent of unsurveyed lands lay in Bihar, and about thirty six per cent in Awadh.[23] The Ain offers no *arazi* statistics at all for the entire *suba* of Bengal and parts of the *suba* of Bihar.[24] While the technicalities of the debate are beyond the scope of this work, it is unlikely that these figures refer to the total land area of the empire's *mahals*. The cost and labor of surveying along with resistance from local elites would have dissuaded Mughal officials from undertaking detailed surveys of the empire's lands, especially uncultivable waste lands (including forests, hills, and other wastelands). Their prime concern may have been to estimate the land that was under cultivation or had the potential to be cultivated, and ensure the taxation of both. The *Ain* like other contemporary records pertaining to agricultural production and revenue assessment, was not so concerned with drawing up a complete topographical profile of a division such as the *pargana*, as it was with assessing its revenue-yielding potential. The revenue-yielding potential of a fiscal division in turn was never constant – given the constant back and forth slippage between cultivation and waste land. Francis Buchanan-Hamilton, during his Mysore survey of 1800, commented on the nature of such precolonial taxation surveys:

---

22 See: Abul Fazl. *Ain-i-Akbari*, H. S. Jarrett (trans.), vol. 2. Calcutta: The Asiatic Society, 1949, 186/167.

23 Irfan Habib. *The Agrarian System*. Bombay: Asia Publishing House, 1963, 4.

24 Shireen Moosvi. *The Economy of the Mughal Empire, c. 1595*. Delhi: Oxford University Press, 1987, 64, footnote 34.

> The principal native officer here says that people are now employed in measuring the lands which belong to all the villages in this lately acquired division of Major Mcleod's district. The measurement, however, will by no means be complete; as large hills and wastes are not included within the boundaries of any village and will not be comprehended in the accompts. Even within the village boundaries it is only the lands that are considered arable, or as capable of being made so, that are actually measured, steep and rocky places are taken by conjecture.[25]

The underlying premise of Buchanan's statement is that for a survey to be *complete* it must take into account *all* the topographical features, both within and beyond the boundaries of a village – something indigenous officials were not terribly concerned about. Also underlying this statement was a distinct vision of territory – that would be protracted by combining distinctly demarcated divisions. Predecessor regimes such as the Mughals were never really capable of, or even interested in, surveying the *entire* body of their territories. To do so would have called for a vision of the state possessing a naturalized geography that needed to be surveyed as a *whole* and not merely limited to the survey of politically important or revenue yielding lands.

Even if the Mughal elite possessed such a spatial vision, they would have been unable to concretize this vision given the decentralized sources of power and varying levels of authority and power that mediated the Mughal state's control over land. Thus, precolonial revenue records did not insist on the maintenance of spatial order in the way a modern map would have. Precolonial administrative and revenue records were more in the nature of lists of villages showing the larger fiscal divisions and sub-divisions they were attached to. Frequently, villages or parts thereof were attached or detached from their parent *parganas* keeping in line with changes in revenue collection arrangements and shifts in political allegiances. The need to preserve the territorial continuity of a fiscal division or sub-division rarely concerned the preparers of these documents. In fact, the resources of a village or its part could be attached to *parganas* that lay separated from them by the territories of intervening *parganas*. There is some evidence from the colonial record to support this.

In 1793, J. Lumsden, the Collector of Saran district discovered that the *taluqa* of Sangrampur belonged to the *tappa* of Mando (*pargana* Majhowa, *sarkar* Champaran). What attracted the Collector's attention was that Sangrampur was physically separated from its parent *tappa* of Mando, by the intervening *tappas* of Dowlatta and Belwa (see Map 1). On further inquiry, it was found that in 1782, this *taluqa* fell in *tappa* Mando when it was granted by Jugal Kishor Singh, the raja of

25 Cited in Moosvi. *Economy*, 44, footnote 9.

Bettiah to a Bengali speculator, Jagmohan Mukherji and his dependents. Then in 1788 and 1789, there was no mention of this village in the assets of *tappa* Mando. The local *quanungo's* clarified that this was so because the *taluqa* of Sangrampur had during these years become attached to the *tappa* of Soogaon which itself lay separated from it (*taluqa* Sangrampur) by the intervening *tappas* of Hurni Tand and Owlaha.[26] Lumsden discovered a similar dynamic taking place in the case of a number of other villages.[27] In another instance from the district of Gorakhpur, the raja of Gorakhpur possessed a *taluqa* consisting of three whole *tappas* (in *pargana* Haveli Gorakhpur) and a few detached villages spread over the *parganas* of Bhowapar, Sylhet, and Dhuriapar. Between 1685 and 1728, the rajas *taluqa* villages were included in the general accounts of the *pargana* of Haveli Gorakhpur. But in 1729, one Atma Ram who was the *diwan* or head of the revenue department, sequestered these villages from the accounts of the several *parganas* and constituted them into a separate *mahal* (revenue yielding unit). In this manner, the *taluqa* of the raja was detached from the fiscal resources of four *parganas* to constitute a separate revenue-paying unit.[28] This move may have been initiated by the raja of Gorakhpur in order to free him from the influence of local officials and landed magnates competing for control over those *parganas*. Thus, what we are observing is an established practice of sequestering certain lands from the fiscal accounts of administrative units. The reasons for doing this were often connected with the formation of new revenue engagements, and conferment of new land rights, or the result of routine contests between landed magnates. When making such rearrangements, no attention seems to have been paid to preserving the territorial compactness of the concerned division.

Company officials were quick to notice this "confusion" and "disorganization" in the revenue records. They initiated certain steps to "remedy" the situation. One of the first steps they took was to systematize information on various issues pertaining to land rights, revenue administration, and record-keeping. For instance, in 1773, J. I. Kieghly, the Collector of Tirhut District (along the Anglo-Gorkha frontier) began to reorganize the accounts of each *pargana* under his

---

26 See letter of J. Lumsden, Collector of Saran District to BOR, May 1793, in *Letters Sent to BOR Register*, 13 February 1793–5 July 1793, *District Record Room*, Chapra, North Bihar. See also Charles James Stevenson-Moore. *Final Report on the Survey and Settlement Operations in the Champaran District, 1892–1899*. Calcutta: Bengal Secretariat Press, 1900, 30.

27 See Letter of J. Lumsden. Collector of Saran District to BOR, May 1793. In *Letters Sent to BOR Register*, 13 February 1793–5 July 1793, *District Record Room*, Chapra, North Bihar.

28 See: Syed Najmul Raja Rizvi. Srinet Zamindars of Sarkar Gorakhpur Under the Mughals. *Uttar Pradesh Historical Review* 7&8 (1993–1995): 35–41.

Map 1: *Tappa* divisions of *pargana* Majhowa, in *sarkar* Champaran showing detached portions of *tappas* Belooa, Jaffarabad, Mando, Mudhwal. The *taluqa* of Sangrampur lies between the *tappas* of Sonwal and Dowlatta. Mapwork by Bernardo Michael. This map is not drawn to scale.

charge, in order to make the "Purgunnnah as complete as possible."[29] Between 1776 and 1778, the Company set up a special office, known as the Amini Commission, to conduct one of the most detailed investigations into the revenue records of its provinces. The aim was to make sense of the mass of unwieldy data on revenue accounts by systematizing them. Its sponsor, Governor-General Warren Hastings, explained that the Commission's mandate was to "collect these different accounts and methodize them for our guidance in forming a new settlement is one of the principal objects of the temporary office which I have proposed."[30] The commission in its report noted that the records of each district "underwent considerable and annual alterations, having in many instances been augmented and

29 *Proceedings of the Controlling Committee of Revenue at Patna*, 1773, vol. 7, 193, *BSA*.

30 The complete report of the Amini Commission can be found in Richard Bury Ramsbotham. *Studies in the Land Revenue History of Bengal, 1769–87*. London: Oxford University Press, 1926. See especially page 80.

in others diminished..."[31] The Amini Commission was headed by English officers who deputed *amins* (surveyors) to collect these records. However, after two years of collecting such records (17761778), the Commission was scrapped after failing to make any geographical sense of these records and resolve the nagging question of the actual location, boundaries, layout, and organization of the Mughal *pargana* divisions.[32]

There is more evidence concerning the Company's increasing awareness that revenue records needed to be reorganized and systematized on new lines. In 1789, the Board of Revenue instructed Archibald Montgomerie, the Collector of Saran district that when making revenue settlements, care was to be taken to show clearly in the revenue accounts the addition or separation of territory into fiscal units (such as *tappas* or *parganas*).[33] There was also recognition, at least within official circles, that the internal divisions of the Company's territories needed to be preserved from frequent changes in their size and composition. In 1791, the Amended Code of Regulations for the Decennial Settlement for Bengal and Bihar instructed the Collector of every district in Bengal and Bihar to establish a fixed arrangement of divisions such as *parganas*, *tarafs*, *taluks*, and *kismuts* within each *zamindari*. The *zamindars* and *taluqdars* were not to make any alterations to these divisions, either in their extent, designation, or disposition. These regulations also stipulated that the revenue accounts were to be systematized according to *parganas*, with the accounts of the internal divisions of a *pargana* being inserted into the account register of *that pargana*, and not any other.[34] Attempts were also made to delineate clear-cut territorial jurisdictions for officials in the North West provinces. In 1815, when the Company decided to allow the reappointment of *patwaris* (village accountants) in the North West provinces, it added the caveat that these *patwaris* should either be appointed to separate villages or any number of *contiguous* villages.[35] Later, in 1820, it was decided that every *pargana* would have three registers for recording information

---

31 Ramsbotham, *Studies*, 123.

32 For an insightful account that situates the Amini Commission's activities within a wider inter-regional and international context of politics and finance see: Robert Travers. The Real Value of the Lands: The Nawabs, the British and the Land Tax in the Eighteenth-Century Bengal. *Modern Asian Studies* 38:3 (2004): 517–558.

33 Instructions to Archibald Montgomerie, Collector of Saran from BOR, January 1789, Letters Received Register, 1788–1789, *Saran Collectorate Records, BSA*.

34 Amended Code of Regulations for the Decennial Settlement for Bengal and Bihar, September 1791, in *Procs. of the BOR*, 23 September 1791, no. 17, *WBSA*. The emphasis is mine.

35 Government to Board of Commissioners, 12 August 1815, Regulations of 1815, section 4, *Pre-Mutiny Records*, Letters Received from the Government to BOR, 1810–1816, *UPSA*.

pertaining to estates (*mahals*), rent-free lands (*lakhiraj*), and villages (*mauzas*).[36] By the 1840s, the arrangement of records witnessed further systematization. The *Circular Orders by the Sudder Board of Revenue, North West Provinces Addressed to Commissioners of Revenue on Records and Registration* (1840) had this to say about the organization of records.

> In preparation of Reports intended for the Boards Office, or for transmission to Government from your divisions, you are requested to adopt the arrangement of collectorates which is given in the Appendix. This geographical arrangement has been enjoined by orders of government to ensure facility of reference.[37]

> The records are to be kept parganawar and mauzawar...A separate shelf or space is to be set apart for each pergunnah, and the name of the pergunnah is to be clearly and durably written both on the outside of the door or the press and on front of the shelf or shelves on which the records of that pergunnah may be arranged.[38]

Such measures highlight the Company's determination to carefully reiterate the geography of its territories in the organization of its records. Therefore, even prior to the mapping exercises of nineteenth century, company officials put a halt – *in the revenue record* – to the back and forth movement of villages from one *tappa* or *pargana* to another.

To conclude, the Company since its very inception as a territorial power in north India, in 1765, had begun to grapple with the realities of revenue collection and its representation in the extant revenue records. Company officials soon came to realize that these records were appallingly deficient in enabling them to grasp the contours, layout, and internal divisions – in a word – the *complete* architecture of the territories they had acquired. It took time for this realization to dawn on them, and our discussion in no way exhausts, the contests, dissonances, contradictions, and ambiguities that may have marked the slow progress of the Company's projects in this area. In fact, historicizing the production of these arrangements within the Company's early revenue records constitutes an entire study in its own right, which falls beyond the scope of this paper. Indeed, this exploration has only served to indicate that there were rearrangements being made within the internal

36 Extracts from the Proceedings of the Committee of Records, 6 August 1820, *Pre-Mutiny Records*, Letters Received from the Government to BOR, January–August 1820, *UPSA*.

37 *Circular Orders by the Sudder Board of Revenue, North West Provinces Addressed to Commissioners of Revenue on Records and Registration* (Calcutta: Baptist Mission Press, 1840), Section 4, Commissioners Records, 263.

38 *Circular Orders*, Section 5, Arrangement of Collectors Records and Distribution of Business, 267.

organization and physical arrangement of revenue records, and these rearrangements must not be treated as unproblematic or inconsequential, or mere exercises in administrative "reform." Rather, they announced the insertion of a distinct territorial order into the maintenance of colonial state's revenue records.

3 Reordering Territory: Thanas, Sub-divisions, and Districts

Ever since the acquisition of the provinces of Bengal, Bihar, and Orissa in 1765, the East India Company took steps to establish an administrative apparatus over its conquered territories.[39] But this was easier said than done, as the Company's districts were constituted out of the older Mughal *pargana* divisions with their discontinuous bodies and boundaries. Consequently, these spatial features crept into the emerging geographical structure of the Company's own districts giving rise to much confusion and dispute. Company officials were aware of these spatial issues and tried to remedy the situation through a process of territorial reorganization. Perhaps, one of the earliest observations on the spatial characteristics of the Company's districts came from Charles Bentley the Collector of Chittagong. In 1772, Bentley was asked to give his opinion on the Government's new policy of establishing farming contracts to collect revenue. Bentley in his comments, and without the aid of modern maps, noted that in the district under his charge not a single *chakla* or *pargana* "stood entire," making it difficult to determine the total resources of the district. Bentley tried to resolve this (unsuccessfully) by dividing Chittagong into nine new *compact* revenue zones, each called a *chakla*.[40] Colo-

39 Descriptive accounts of the Company's administrative system include the following: Animesh Chakrabarti. The Origin of Subdivisions in Bengal. *Bengal Past and Present* 47 (1978), 36–64; Niranjan Dhar. *The Administrative System of the East India Company in Bengal, 1714–1786*, 2 vols. Calcutta: Eureka Publishers, 1964; Bankey Bihari Misra. *The Central Administration of the East India Company, 1773–1834*. Manchester: Manchester University Press, 1959; Bankey Bihari Misra. *The Judicial Administration of the East India Company in Bengal, 1765–1782*. Delhi: Motilal Banarasidass, 1961. These works approach the question of administrative geography from the perspective of government policy and administrative rationalizing and not in terms of a *longue duree* view of spatial transformation. The constitutive role of *pargana* divisions in the formation of modern administrative divisions remains undiscussed.

40 Cited in: Alamgir Muhammad Serajuddin. *The Revenue Administration of the East India Company in Chittagong, 1761–1785*. Chittagong: University of Chittagong, 1971, 60–61. See also pages 58–76 for details.

nial officials stationed on the ground shared Bentley's spatial anxieties and tried to institute territorial changes that, often without the help of maps, would render their districts compact and well defined.

Another notable reorganization of districts took place in 1786 under the Governor-Generalship of John Macpherson (1785–1786). During this time, the Company's territories continued to rest on a received territorial structure consisting of *parganas*, *zamindaris*, and *taluqs* that remained largely undefined, scattered and intermixed. Moreover, neither were detailed surveys undertaken nor were accurate maps available for the conduct of daily administration. Thus, on the advice of the Court of Directors (in London), 35 new districts or Collectorships were formed with each having a minimum rent roll of five lakh rupees. These were later reduced to 23 on the recommendations of John Shore, the then President of the Board of Revenue.[41] Shore's Minute of 13 March 1787 called for the reorganization of territorial divisions, making them more compact and with clearer boundaries. He also called for the "re-annexation of *Talooks* or smaller portions of land to the principal divisions from which they have been separated, or to which they are contiguous."[42] Again, in 1789, he would observe that a good idea of the revenues of the province of Bengal could be gained

> … by assuming as a *ground-work* the present distribution of the country into collectorships, after such correction as it may admit, and by obtaining through the different collectors, an account of the distribution of the revenue upon pergunnahs and villages throughout their respective jurisdictions.
>
> To perform this, we must first determine the number of pergunnahs in each zamindarry, and prescribe a form of arrangement for them, which shall be established by law, imposing penalties for every unauthorized alteration. I do not by this mean to prevent a zemindar changing the subdivisions of a pergunnah, but to require only that the constituent subdivisions shall *always* stand under the same pergunnah, where they are originally placed, and that they shall *not* be transferred to any other. Thus, supposing a pergunnah to be portioned out into five or more terfs or kismuts, comprehending the *whole*, it shall stand in the accounts as *one* article only, comprizing so many divisions.[43]

---

41 Misra, *The Central Administration*, 156–157; Radha Kumud Mookerji. *Indian Land System: Ancient, Medieval and Modern*. Alipore: Bengal Government Press, 1940, 74–75.

42 See Sir John Shore's Minute of 13 March 1787, cited in W. K. Firminger (ed). *The Fifth Report From the Select Committee on the House of Commons on the Affairs of the East India Company*, 3 vols., Calcutta: R. Cambray & Company 3 (1917): 734–736. See especially page 735.

43 Minute of Sir John Shore, dated 18 June 1789, respecting the Permanent Settlement of the Land in the Bengal Provinces, cited in Walter K. Firminger. *The Fifth Report*, 1–145. The quotation is from page 100 (emphasis mine).

Meanwhile, new territorial arrangements were instituted by Governor-General Cornwallis in his rules of 7 December 1792. According to those rules, the various districts were divided into *thanas* or police jurisdictions of about twenty to thirty miles square in extent. Considerations of economy, revenue collection, internal security, administrative convenience, and spatial compactness determined the location and layout of these new *thana* divisions. In dividing his district into *thanas*, each Magistrate was to see that the principal towns and markets were generally situated in the center and duly served by the new *thanas*.[44] The *sarkar* of Gorakhpur which was initially divided into revenue divisions called *tahsils*, which were later renamed *thanas* or police divisions. Thus, the district of Gorakhpur was divided into one *sadr* (central/main) police *thana* and nine compact subordinate *muffasil thanas* with a *daroga* appointed in charge of each, and ten *chaukis* or inferior stations, each under a *jamadar*.[45]

However, these *thana* divisions, though set up with the intention of being compact, were in reality patchy, irregularly shaped, and intermixed in many places. This was because the entire framework of the *thana* jurisdictions was constructed out of the pre-existing framework of the *pargana* divisions. Commenting on this the British surveyor Francis Buchanan Hamilton who had undertaken a topographical survey of the region in 1809, noted that a number of *thanas* had detached segments, separated from the parent body, by intervening stretches of territory of neighboring *thanas*. In Buchanan Hamilton's discussion of the layout each *thana* of Gorakhpur we find a profusion of highly spatialized terms and phrases like – scattered, detached, long narrow jurisdiction, long and narrow and winds around, compact division, petty division, centrical, ill arranged, and so on (see Map 2). His description of the *thana* of Maghar (Gorakhpur district) exemplifies this. He observed that the *thana* of Maghar had an "irregular straggling jurisdiction, exceedingly ill arranged having nearly in its middle, or intermixed with it, the division of Bakhira, and containing two detached portions of Bangsi." To resolve this spatial irregularity, Hamilton suggested that, "If these portions were joined to the two divisions and Bakhira made the residence of the officers of police, the jurisdictions would be tolerably compact, of a reasonable size, and its capital tolerably near the center."[46] On the internal divisions of Gorakhpur he would observe that "most unexampled carelessness and confusion has arisen," where the different shapes and sizes of divisions were "subdivided by intervening portions of those, which are adjacent and many have the residence of their

44 Misra, *The Central Administration*, 332.

45 Misra, *The Central Administration*, 335, 352–353.

46 Francis Buchanan-Hamilton, *The Gorakhpur Report*, 1: 297–298, Eur Mss. D 91, *IOR, APAC, BL*.

officers in situations that are exceedingly inconvenient." Within Gorakhpur, the main police divisions (*thanas*) and their subdivisions (*chaukis*) too varied in size. Once again, for Hamilton, this was an anomaly that needed rectifying. Hamilton, did not comment on the complicity of the older *pargana* divisions in creating the discontinuous and patchy bodies of the *thanas*.

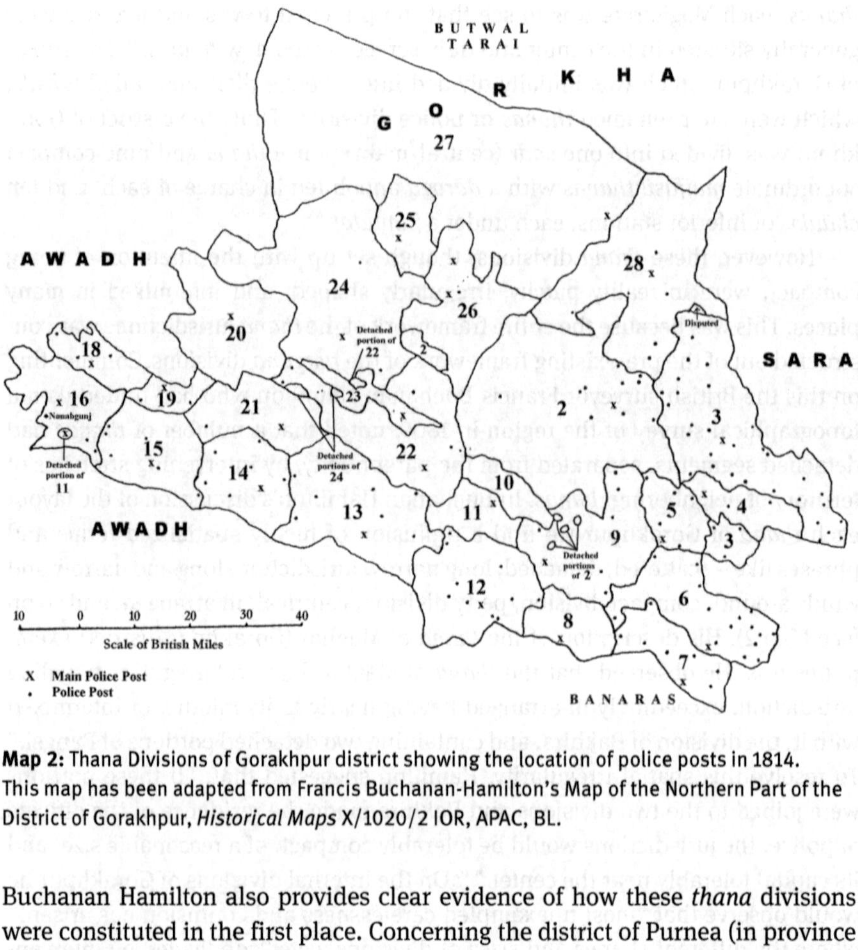

Map 2: Thana Divisions of Gorakhpur district showing the location of police posts in 1814. This map has been adapted from Francis Buchanan-Hamilton's Map of the Northern Part of the District of Gorakhpur, *Historical Maps* X/1020/2 IOR, APAC, BL.

Buchanan Hamilton also provides clear evidence of how these *thana* divisions were constituted in the first place. Concerning the district of Purnea (in province of Bihar), he had this to say:

> The sub division into Thanahs [*thanas*] has been made with as little care as in Rongopur [Rangpur district, north Bengal]. Their jurisdictions are miserably intermixed, of very unequal sizes, and population and the residence of the officers has often without any apparent reason, been placed very far from a central position. In forming these jurisdictions the maps never seem to have been consulted and the guide seems to have been a

> report of some native (munshi [scribe] or amin [surveyor]) who knew nothing of the country. I saw a curious specimen of the ideas of such people in a plan of Thanah Dimiya, which was furnished from the public office and represented Dimiya as placed in the centre of a square district, with villages extending all around to nearly equal distances. Whereas the division is in reality somewhat semi-circular, with at projection at each angle, and Dimiya, until carried by the Kosi [river], stood exactly on its bank, which formed the boundary, and it was much nearer to the south angle than to the northern extremity.[47]

In Buchanan Hamilton's account of the indigenous "plan" of the *thana* of Dimiya, we get a rare insight of a local perspective on these new divisions, one which totally missed the physical topography of the *thana*. This was "curious" to him as he was more concerned with the actual layout of the *thana* than with some imaginary conception of it. It might be suggested that Buchanan Hamilton in examining the layout of *thanas* was actually trying to comprehend them as distinct parts of the broader geographical frame of the state, a perspective local officials did not share. Colonial officials hoped to reorder older Mughal *pargana* divisions in order to provide the foundational material on which the Company would erect the framework of its new administrative divisions for policing (*thanas*) and revenue collection (*tahsils*). Every effort was made to render these new *thana* divisions compact with well-defined bodies, which taken together would comprehend a *whole* district. The new administrative divisions would be "absolute" spaces, produced out of the sanction of the state, and devoid of impress of local interventions. These spaces were absolute–an empty field, a container, a co-ordinate system of discrete locations on the earth that could be comprehended as a whole.[48]

The establishment of *thanas* gave way to other problems. For instance, the jurisdictions of these *thanas*, did not always coincide with the jurisdictions set up for law courts or for the collection of revenue. Once again, Francis Buchanan Hamilton's observations on the Gorakhpur district are revealing. He notes:

> The extent of the jurisdictions for trying petty debts, and for the collection of revenue are totally different from those of the police establishment, and from each other, which occasions a good deal of trouble to the subject and some of the former are of such enormous extent as to afford very little relief either to suitors or witnesses, so that the only complaints

---

47 Francis Buchanan-Hamilton. *An Account of the District of Purnea in 1809–10*. Patna: Bihar Orissa Research Society, 1928, 4.

48 I use the term "absolute" space as suggested by Neil Smith and Cindi Katz as "our naively assumed sense of space as emptiness." See: Neil Smith and Cindi Katz. Grounding Metaphor: Towards A Spatialized Politics. In *Place and the Politics of Identity* Michael Kieth and Steve Pile (eds.). New York: Routledge, 1993, 67–83, esp. 75.

> likely to be made to such courts are those of litigious wealthy men, who wish to harass the poor into compliance, with whatever they choose to demand.[49]

In the years following Hamilton's survey, steps were taken to remedy the situation. According to Regulation 11 of 1831, the revenue and police functions of the *tahsildar* (collector of revenue) were made co-extensive, and the jurisdictions of the several *thanas* were made to correspond. However, the problem of intermixture and detached segments (outlined in previous paragraphs) continued to persist. This as we know was largely due to the "irregular" nature of *pargana* divisions on which these *thana* divisions were constructed. The *pargana* boundaries needed to be "adjusted" if the *thana* divisions were to become compact.

Consequently, the boundaries of *parganas* were steadily adjusted to facilitate this convergence and render the districts more compact. In 1793, a large number of territorial transfers took place between various Collectorships. Villages and *mahals*, either whole or in part, that lay detached from their parent *parganas*, were transferred between various districts, so as to render the concerned districts more compact.[50] Such territorial transfers would continue to take place, for instance, in the provinces of Bihar and Bengal, in a piecemeal fashion, over the next century. In the first decade of the nineteenth century, similar maneuvers were initiated in the Company's newly acquired territories of the Ceded and Conquered Provinces in North India. Concerning these attempts to adjust the police jurisdictions of the *thanas* of Azamgarh, Jaunpur, Ghazipur (Banaras Division), the Magistrate of Azamgarh had this to report to the Commissioner of Ghazipur – "The great irregularity of the Pergunnah boundaries made this [the adjusting of *thana* jurisdictions] very perjudicial, but as the Pergunnah boundaries are now *adjusted* and *rounded* off, the objection ceases. The Thanah boundaries of course, correspond with the *new* Pergunnah boundaries" (emphasis added).[51] In the meantime, in 1845, another administrative unit was created between the district and the *thana* called the Subdivision. Thus, the District of Champaran (Bihar) was divided into two sub-divisions, Motihari and Bettiah which were further sub-divided into *thana* divisions. By the 1870's the civil, criminal, and revenue jurisdic-

---

49 Francis Buchanan Hamilton, *The Gorakhpur Report*, 3 vols., 1: 10–11, Eur Mss D 91, *IOR, APAC, BL.*

50 For details see Procs. of BOR, 15 May 1793, *BSA*; Procs. of BOR 16 August 1793, *BSA*; Printed Index to the Procs. of the BOR, 1793, *WBSA*.

51 Letter of Magistrate of Azamgarh to the Commissioner, Ghazipur, 8 November 1835, *GDR*, basta 27, Vol. 57, *RSA*.

tions were made coterminous. These *thana* divisions became the basic template on which other divisions such as *tahsils* were pegged.

However, it was not only the *thana* and *tahsil* divisions that were introduced and periodically reorganized. The same happened with the jurisdictions of the law courts, called *munsifi* divisions. They too were adjusted, rounded off and then made coterminous with the *thana* divisions of the districts.[52] Villages, which belonged to the criminal jurisdiction of one district and the fiscal jurisdiction of another, were exchanged. On the converse, *thanas* located in the fiscal and criminal jurisdiction of one district but in the civil jurisdiction of another were reshuffled. Take the case of *thana* Raniganj. This *thana* straddled the civil jurisdiction of Birbhum district (in Bengal), the criminal jurisdiction of Bankura district (also in Bengal), and the revenue jurisdiction partly of Burdwan district and partly in Manbhum district. The conflicts in jurisdiction resulted in people fighting cases in as many as three districts simultaneously. The *thana* of Raniganj was subsequently rearranged.[53] In this manner *thana, munsifi,* and fiscal jurisdictions were rearranged both within and across administrative divisions and sub-divisions to neatly overlap with the boundaries of districts. By 1875, the colonial state was making serious efforts to render permanent the boundaries of these emerging districts. For example, government clearly enunciated the principle that the boundary of a district was to be a permanent one and distinct from the boundary of an estate paying revenue. Every district contained lands belonging to estates, which belonged to the revenue rolls of some other district. This circumstance was to be treated as separate and distinct, from the permanent boundary of the district. In this manner, by the late nineteenth century, the policy of objectifying the colonial state's internal divisions through territorial exchanges between *thanas* and their continuous compaction would go on from the 1790 to well into the twentieth century (see Map 3).[54] The process by which the structure of colonial territories was fashioned out of the older *pargana* divisions constitutes one of the great silent spatial settlements conducted by the British in South Asia.

---

52 See P/242, *Bengal Judicial Consultations*, Procs. Of the Revenue Branch, Jurisdictions and Boundaries, 1–131, *IOR, APAC, BL*. See especially 12–14.

53 See Procs. of the Lt. Governor of Bengal (Revenue Department), March 1875, 155–160, P/ 242 *Bengal Judicial Consultations, IOR, APAC, BL.*

54 The Bengal Judicial Consultations are full of such instances, which in some cases run into hundreds of pages. See the following series in the *Bengal Judicial Consultations*, Judicial, 1887–1905, P/5408–5410 (for 1898); Revenue (1874–1886 and 1906–1936), P/242, P/236, P/910, P/1319, P/1641, P/1836, P/2024, P/2238, P/2487 (all for 1874–1886), *IOR, APAC, BL.*

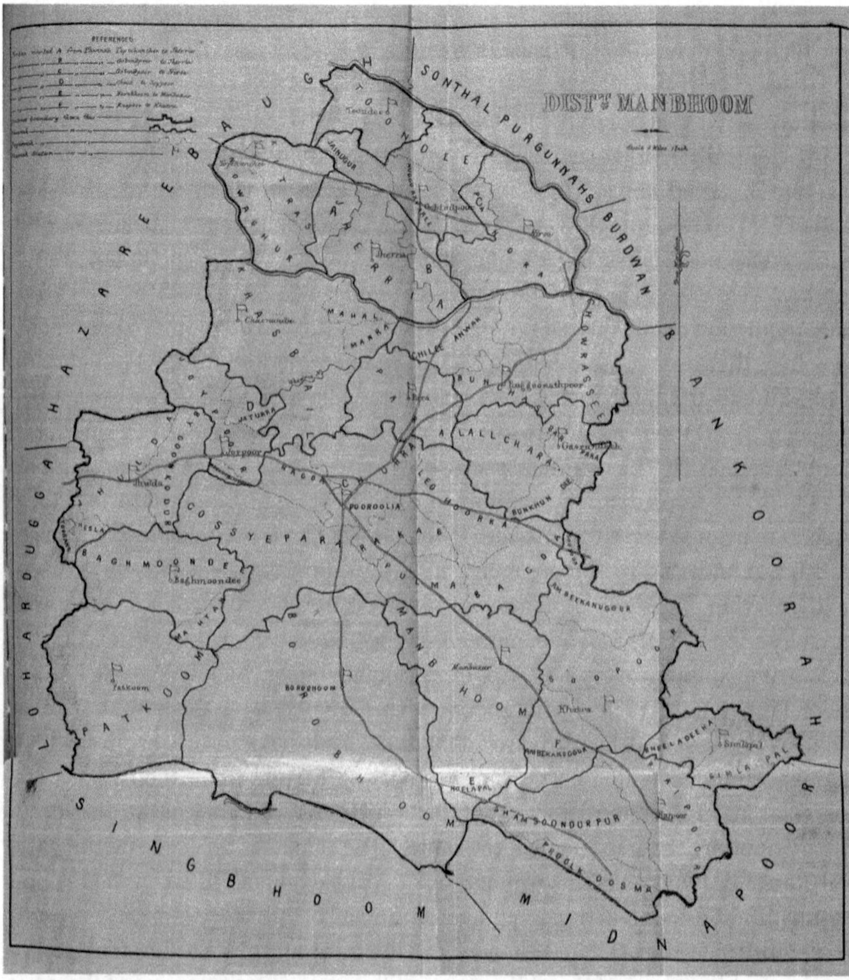

Map 3: Rearrangement of Thana boundaries in District Manbhoom. Map of District Manbhoom [Manbhum], 39 × 34 cm, scale = eight miles to one inch, north at top. Reproduced with permission from the British Library c British Library Board. Source: Procs of the Lt. Governor of Bengal: Revenue Branch (Jurisdictions and Boundaries), Bengal Judicial Consultations P/242, March 1875, pp. 92–3, India Office Records, Asia, Pacific & Africa Collections, British Library.

Conclusion

In South Asia, from the very inception of colonial rule, the English East India Company took steps to reorder its territories, to fit an emergent spatial vision of a clearly bounded state possessing continuous and clearly bounded territory. This vision found its ultimate expression in modern surveys and maps. However, the impulse to represent an ordered and well-bounded state first surfaced as the British sifted through the realities of everyday governance, long before formal surveys and mapmaking were undertaken in the mid nineteenth century. In this regard, the creation of imperial space was not undertaken merely through the medium of colonial surveys and mapmaking; rather it was silently encoded in the lived realities and rituals of daily rule. Through directives issued to its officials, the reorganization of records, and the reordering of administrative jurisdictions, colonial officials tried to fashion a spatial template for the state – that was continuous, clearly bounded, and perfectly consistent in all its parts. This was a long-drawn, complicated, and often ill-coordinated process. Such research also brings to the fore a range of hitherto unacknowledged human agents, both elite and subaltern, who while operating in contexts that were far removed from the mapmakers office became important intermediaries in the construction and circulation of maps, visions of territory, and geographical information.[55]

In any case, the preexisting territorial divisions such as the *pargana* were transformed in this process. These *pargana* divisions and their constituent *tappas* witnessed all sorts of pruning, grafting and reshuffling exercises that changed their contours forever, and ultimately prepared the way for their almost unnoticed demise. The actual spatial layout of the *parganas* would be made visible for the first time through the revenue surveys of the 1840s when the first *pargana* maps were compiled by the British.[56] For a brief moment, the British were able to obtain a snapshot of the territories they had inherited from indigenous states, before they slipped into historical obscurity. In 1838, Martin Montgomery published the journal of the British traveler and surveyor Francis Buchanan Hamilton pertaining to his travels through the district of Gorakhpur. Martin chose to leave out Buchanan-Hamilton's detailed observations on the confusing and highly disproportionate features of Gorakhpur's *thana* or police divisions. While no reason was ascribed for this omission, Martin may have seen no practical use in providing details that would only have confused his readers. After all, his stated goal was "to present a clearer view of the actual *frame-work* and *anatomy* of society in

55 For more on the subject of circulation in colonial networks see: Kapil Raj. *Relocating Modern Science: Circulation and the Construction of Knowledge in South Asia and Europe, 1650–1900*. London: Palgrave, 2007; and: Neil Safier. *Measuring the New World: Enlightenment Science and South America*. Chicago: The University of Chicago Press, 2008.
56 For details see: Michael, Making Territory Visible.

the East, than anything they saw or heard during their sojourn in Hindostan."[57] Martin, like many before and after him, would narrate this story of colonial territorial transformation as an enlightened one that distilled order from the chaos and confusion of the East. It also entailed a moment of forgetting where the historical forces at work in the constitution of indigenous territorial divisions such as the *pargana* would be hid from view, initially in the documentary record, and ultimately in modern maps. However, it is useful to remember that the British Empire in South Asia at this time, and indeed until its very demise, despite the obsession to create clearly defined imperial spaces – remained a territorial entity whose contours were shaped not just by the drawing of maps, but also by the drawing up of law codes, bureaucratic routines, redirection of tax flows and the redefinitions of tenurial rights. At the same time the process remained an incomplete one. The continued presence of enclaves, dispersed territories, straggling corridors of imperial control and shared sovereignty were a reminder of the flip side of the neat lines, order, and rationality that have traditionally been viewed as the hallmarks of Empire.[58]

This article suggests that at the heart of empire lay a tremendous concern for things spatial – in rearranging territory, relationships, objects, and bodies. The spatial anxiety around territory made itself visible not only in the surveying and mapping projects of the state, but also in the routines and policies of everyday governance. At times, these routines and policies predated and overlapped the introduction of the detailed colonial surveying and mapmaking initiatives of the nineteenth century. Together, they retailed an ordered, simplified and legible image of the territories of the colonial state. Such an effect was not an actual structure, but the powerful "metaphysical effect of practices [of spatial organization, temporal arrangement, functional specification, supervision and surveillance] that make

57 See: Martin Montgomery. *The History, Antiquities, Topography and Statistics of Eastern India*, 3 vols., London: William H. Allen & Company, 1838. See especially vol. 2, i. Emphasis mine. However, these details have not been deleted from the printed versions of Hamilton's reports on Patna-Gaya, Shahbad and Purnea that came out ninety years later, in 1928.

58 For detailed explorations of this theme see: Lauren Benton. *A Search for Sovereignty: Law and Geography in European Empire, 1400–1900*. Cambridge: Cambridge University Press, 2010; Sudipta Sen. *Distant Sovereignty: National Imperialism and the Origins of British India*. New York: Routledge, 2002; and: Laura Ann Stoler. On Degrees of Imperial Sovereignty. *Public Culture* 18:1 (2006): 125–46. For a timely study of the English East India Company set against the evolving backdrop of the transition from early-modern to modern forms of state, sovereignty and political power see: Philip Stern. *The Company-State: Corporate Sovereignty and the Early Modern Foundations of the British Empire in India*. New York: Oxford University Press, 2011.

such structures appear to exist [independent of society]."[59] This has some implications for our studies on the history of cartography. For one, it would appear that the vision that a state's territories were continuous, well bounded, and internally structured predated the arrival of detailed maps of the subcontinent. This vision of territory was a panoptic one, viewed from above, as if from nowhere, creating a framework of territorial spaces that was absolute – a taken-for-granted container. Such an Icarian vision might have developed not just within England, but may have been simultaneously forged within the context of the emerging colonial state in South Asia.[60] In the end, the story of the spatial redrawing of the colonial state needs to be understood against a wider background of the development of territorial sovereignty in Europe and the discontinuous and shared character of imperial rule at home and within the colony. This points to a more complicated and uneven story with possible counterflows that would decenter neat teleologies of an easy and innocent transfer of notions of territorial sovereignty from the metropole to the periphery. More research needs to be undertaken if we are to understand the distinct genealogies of these projects and the existence of any mutually informing connections, along with the mediation of local and regional forces that ultimately resulted in the creation of the dispersed "geo-bodies" of the British state and its colony in India.

Glossary

APAC	Asia Pacific Africa Collection
BL	British Library, London
BOR	Board of Revenue
BSA	Bihar State Archives, Patna
Col.	Colonel
Consl.	Consultation
Eur. Mss.	European Manuscripts
FP	Foreign Political
FS	Foreign Secret

---

59 See: Timothy Mitchell. The Limits of the State: Beyond Statist Approaches and their Critics. *American Political Science Review* 85 (1991): 77–96, 94.

60 This Icarian ideal suggests that imperial power installs an "off the ground" perspective using media such as cartography while denying or ignoring local experiences and practices. This view is expounded in: Paul Carter. *The Road to Botany Bay: An Essay in Spatial History*. London: Faber & Faber, 1987.

GCR	Gorakhpur Collectorate Records
GDR	Gorakhpur Divisional Records
IESHR	Indian Economic and Social History Review
IOR	India Office Records
Lt.	Lieutenant
NAI	National Archives of India, Delhi
Procs.	Proceedings
PIHRC	Proceedings of the Indian Historical Records Commission
RSA	Regional State Archives, Allahabad
UPSA	Uttar Pradesh State Archives, Lucknow
WBSA	West Bengal State Archives, Kolkata
Vol.	Volume

References

Anderson, Benedict. *Imagined Communities*. London: Verso, 1991.

Barrow, Ian. *Making History, Drawing Territory: British Mapping in India, c. 1756–1905*. New Delhi: Oxford University Press, 2003.

Bayly, Christopher. *Empire and Information: Intelligence Gathering and Social Communication in India, 1780–1870*. Cambridge: Cambridge University Press, 1996.

Benton, Lauren. *A Search for Sovereignty: Law and Geography in European Empire, 1400–1900*. Cambridge: Cambridge University Press, 2010.

Biggs, Michael. Putting the State on the Map: Cartography, Territory and European State Formation. *Comparative Studies in Society and History* 41 (2): 374–405.

Buchanan-Hamilton, Francis. *An Account of the Districts of Bihar and Patna in 1811 & 1812*. Patna: Bihar Orissa Research Society, 1928.

Buchanan-Hamilton, Francis. *An Account of the District of Purnea in 1809–10*. Patna: Bihar Orissa Research Society, 1928.

Carter, Paul. *The Road to Botany Bay: An Essay in Spatial History*. London: Faber & Faber, 1987.

Chakrabarti, Animesh. The Origin of Subdivisions in Bengal. *Bengal Past and Present* 47 (1978): 36–64.

Chandra, Satish and Satya Prakash Gupta. The Jaipur Pargana Records. *Indian Economic and Social History Review* 3 (1966): 303–315.

Chester, Lucy. *Borders and conflict in South Asia: The Radcliffe boundary commission and the partition of Punjab*. Manchester: Manchester University Press, 2009.

Circular Orders by the Sudder Board of Revenue, North West Provinces Addressed to Commissioners of Revenue on Records and Registration. Calcutta: Baptist Mission Press, 1840.

Dhar, Niranjan. *The Administrative System of the East India Company in Bengal, 1714–1786*, 2 vols. Calcutta: Eureka Publishers, 1964.

Edney, Mathew H. *Mapping an Empire: The Geographical Construction of British India, 1765–1843*. Chicago: University of Chicago Press, 1997.

Edney, Mathew H. Putting 'Cartography' into the History of Cartography: Arthur H. Robinson, David Woodward, and the Creation of a Discipline. *Cartographic Perspectives* 51 (2005): 14–29.
Fazl Abul. *Ain-i-Akbari*, H. S. Jarrett (trans.), vol. 2. Calcutta: The Asiatic Society, 1949.
Fletcher, David. The Ordnance Survey's Nineteenth Century Boundary Survey: Context, Characteristics and Impact. *Imago Mundi* 51 (1999): 131–146.
Firminger, Walter Kelly (ed.). The Fifth Report From the Select Committee on the House of Commons on the Affairs of the East India Company, 3 vols. Calcutta: R. Cambray & Company, 1917.
Gupta, Satya Prakash. *Agrarian System of Eastern Rajasthan*. Delhi: Manohar, 1986.
Gupta, Satya Prakash. Khasra Documents in Rajasthan. In *Medieval India: A Miscellany*, vol. 4, 168–176. Bombay: Asia Publishing House, 1977.
Habib, Irfan. *The Agrarian System of Mughal India*. Bombay: Asia Publishing House, 1963.
Hardt, Michael and Antonio Negri. *Empire*. Cambridge, MA: Harvard University Press, 2000.
Hostetler, Laura. *Qing Colonial Enterprise: Ethnography and Cartography in Early Modern China*. Chicago: University of Chicago Press, 2001.
Kirkpatrick, William. *An Account of the Kingdom of Nepaul*. Reprint, Delhi: AES Publications, 1969 [1811].
Konvitz, Joseph W. *Science, Engineering, and Statecraft*. Chicago: The University of Chicago Press, 1987.
Laxton, Paul. *The New Nature of Maps: Essays in the History of Cartography*. Baltimore: The John Hopkins University Press, 2002.
Michael, Bernardo A. Making Territory Visible: The Revenue Surveys of Colonial South Asia. *Imago Mundi* 59:1 (2007): 78–95.
Michael, Bernardo A. *Statemaking and Territory: Lessons from the Anglo-Gorkha War (1814–1816)*. London: Anthem Press, 2012.
Misra, Bankey Bihari. *The Central Administration of the East India Company, 1773–1834*. Manchester: Manchester University Press, 1959.
Misra, Bankey Bihari. *The Judicial Administration of the East India Company in Bengal, 1765–1782*. Delhi: Motilal Banarasidass, 1961.
Mitchell, Timothy. The Limits of the State: Beyond Statist Approaches and their Critics. *American Political Science Review* 85 (1991): 77–96.
Mookerji, Radha Kumud. *Indian Land System: Ancient, Medieval and Modern*. Alipore: Bengal Government Press, 1940.
Montgomery, Martin. *The History, Antiquities, Topography and Statistics of Eastern India*, 3 vols. London: William H. Allen & Company, 1838.
Moosvi, Shireen. *The Economy of the Mughal Empire, c. 1595*. Delhi: Oxford University Press, 1987.
Morgan, Victor. The Cartographic Image of 'The Country' in Early Modern England. *Transactions of the Royal Historical Society (Sixth Series)* 29 (1979): 129–154.
Raj, Kapil. *Relocating Modern Science: Circulation and the Construction of Knowledge in South Asia and Europe, 1650–1900*. London: Palgrave, 2007.
Ramsbotham Richard Bury. *Studies in the Land Revenue History of Bengal, 1769–87*. London: Oxford University Press, 1926.
Rizvi, Syed Najmul Raja. Srinet Zamindars of Sarkar Gorakhpur Under the Mughals. *Uttar Pradesh Historical Review* 7 & 8 (1993–1995): 35–41.

Safier, Neil. *Measuring the New World: Enlightenment Science and South America*. Chicago: The University of Chicago Press, 2008.
Sen, Sudipta. *Distant Sovereignty: National Imperialism and the Origins of British India*. New York: Routledge, 2002.
Serajuddin, Alamgir Muhammad. *The Revenue Administration of the East India Company in Chittagong, 1761–1785*. Chittagong: University of Chittagong, 1971.
Smith, Neil and Cindi Katz. Grounding Metaphor: Towards A Spatialized Politics. In *Place and the Politics of Identity*, Michael Kieth and Steve Pile (eds.), 67–83. New York: Routledge, 1993.
Smith, Neil. Geography, Empire, and Social Theory. *Progress in Human Geography* 18 (1994): 491–500.
Stern, Philip. *The Company-State: Corporate Sovereignty and the Early Modern Foundations of the British Empire in India*. New York: Oxford University Press, 2011.
Stevenson-Moore, Charles James. *Final Report on the Survey and Settlement Operations in the Champaran District, 1892–1899*. Calcutta: Bengal Secretariat Press, 1900.
Stiller, Ludwig F. The Principle of Limitation. *The Nepalese Perspective* 9 (1974): 4–21.
Stoler, Laura Ann. On Degrees of Imperial Sovereignty. *Public Culture* 18 (2006): 125–146.
Travers, Robert. 'The Real Value of the Lands': The Nawabs, the British and the Land Tax in the Eighteenth-Century Bengal. *Modern Asian Studies*, 38:3 (2004), 517–558.
Winichakul, Thongchai. *Siam Mapped: The History of the Geo-Body of a Nation*. Honolulu: University of Hawaii Press, 1994.

Maps

Map 1: *Tappa* divisions of *pargana* Majhowa, in *sarkar* Champaran showing detached portions of *tappas* Belooa, Jaffarabad, Mando, Mudhwal. The *taluqa* of Sangrampur lies between the *tappas* of Sonwal and Dowlatta. Mapwork by Bernardo Michael. This map is not drawn to scale.

Map 2: Thana Divisions of Gorakhpur district showing the location of police posts in 1814. This maps has been adapted from Francis Buchanan-Hamilton's Map of the Northern Part of the District of Gorakhpur, *Historical Maps* X/1020/2 IOR, APAC, BL

Map 3: Rearrangement of Thana boundaries in District Manbhoom. Map of District Manbhoom [Manbhum], 39 × 34 cm, scale = eight miles to one inch, north at top. Reproduced with permission from the British Library © British Library Board. *Source*: Procs of the Lt. Governor of Bengal: Revenue Branch (Jurisdictions and Boundaries), Bengal Judicial Consultations P/242, March 1875, pp. 92–3, India Office Records, Asia, Pacific & Africa Collections, British Library.

Alrun Schmidtke

Mapping a Distant Empire: Bruno Hassenstein's *Atlas of Japan* (1885/87)[1]

Introduction

In 1887, an extravagant cartographic project reached its final stages. In the vicinity of the Thuringian Forest in the duchy Saxe-Coburg and Gotha a German-language atlas on a single foreign country was being published without that country's participation and amidst diverging agendas on issues from linguistic and scientific ambitions to monetary and political aspirations. The atlas depicted "the land of the rising sun" in the far East of the continent of Asia: Japan. The cartographic monograph *Atlas von Japan* [*Atlas of Japan*] showed Japan's topography with many large and smaller islands, its long coastal line, and obscure denominations of towns, rivers, mountain ranges, peaks, and lakes romanised into Latin letters.[2]

In the interpretation of the atlas' maker and his supporters, Japan had only just opened its harbours to foreign people and for international trade, starting infrastructural programmes and educational reforms as well as organising structured, well-equipped land surveying projects based on western scientific standards.[3] Many of these projects relied on technological and knowledge trans-

---

1 Research for this article was made possible by Fritz Thyssen Foundation (Herzog Ernst Scholarship programme) and conducted at the *Forschungsbibliothek* and *Forschungszentrum Gotha* (Erfurt University). This article has profited from discussions with and input from organisers and participants of the conference SpaceTime of the Imperial. I am most thankful to Alice Goff and Alexander Blum for linguistic revisions and Sven Ballenthin, Nils Güttler, and Petra Weigel for their invaluably helpful practical, material, and conceptual insights into the Perthes Collection.

2 Bruno Hassenstein. *Atlas von Japan. Sieben Blätter im Massstabe von 1:1000000 und eine Übersichtskarte im Massstabe von 1:7500000* [Atlas of Japan. Seven sheets in on a scale of 1:1,000,000 and one general chart on the scale of 1:7,500,000]. Gotha: Justus Perthes, 1885/1887.

3 The political shift that ended Japan's isolation politics occurred towards the end of the Edo era. Modernisation was a key political ambition in the Empire of Japan, starting with the Meiji era in 1868. See: Andrew Gordon. *A modern history of Japan. From Tokugawa times to the present.* New York and Oxford: Oxford University Press, 2003. For surveying projects in imperial or nation-state contexts, see: D. Graham Burnett. *Masters of all they surveyed. Exploration, geography, and a British El Dorado.* Chicago, IL and London: University of Chicago Press, 2000. Kapil Raj. *Relocating modern science. Circulation and the construction of knowledge in South Asia and Europe, 1650–1900.* Houndmills: Palgrave Macmillan, 2007. David Gugerli and Daniel Speich. *Topografien der Nation. Politik, kartografische Ordnung und Landschaft im 19. Jahrhundert* [topographies of the nation. 19th-century politics, cartographic regime, and landscape]. Zurich: Chronos, 2002. Mat-

DOI 10.1515/9783110418750-019

fers, with the Japanese government employing thousands of foreigners into the process of modernisation.[4] When these foreigners, the so-called *Oyatoi*, experienced the inaccessibility of local topographical information – difficulty in laying their hands on maps in general and the inability to read complex Sino-Japanese toponyms –, some of them sought support from a by then well-known and well-established publishing house in Europe: Justus Perthes' Geographische Anstalt, situated right at the foot of the castle hill in Gotha, specialising in geographical and cartographical publications.[5]

Although demands for romanised maps of the Japanese islands date back to the 1850s in Perthes' archives, the project of a comprehensive map of Japan only took off towards the end of the 1870s when the employment of large numbers of highly paid European, American and Asian foreigners in Japan was well established and requests for romanised maps peaked. Judging from this demand and the general employment rates of German scientific and engineering staff hired by the Japanese administration, a cartographer in Gotha ventured out to create a scientific product for export, the large-sized *Atlas von Japan*.

The publishing process of the atlas as a specific object within the history of cartography in the Age of Empire[6] is central to my study. The aim is to interpret a rich set of previously unknown sources from a publisher's archive as a result of a "transimperial network"[7] which—in its spatial extent—was neither local nor global: The atlas' maker was something of an armchair cartographer from Gotha, busily demanding contributions from German and British travellers to Japan, seeking occasional help from Japanese students staying in Germany. To the cartographer, mapping the distant Empire of Japan meant to collate an assortment of sources and thereby promoting himself as a capable member of the European cartographic community able to handle a language as difficult as Japanese and to

---

thew H. Edney. *Mapping an empire. The geographical construction of British India, 1765–1843.* Chicago, IL and London: University of Chicago Press, 1997.

4 Hazel Jones. *Live machines: Hired foreigners and Meiji Japan.* Tenterden, Kent: Paul Norbury, 1980.

5 Its name affix "Geographische Anstalt" roughly translates into "geographical institute." In the latter half of the 19th century it was considered a distinguished research facility, sometimes even praised as a university-like institution. See John George Bartholomew. The philosophy of map-making and the evolution of a great German atlas. In *Scottish Geographical Magazine* 18:1 (1902): 34–39.

6 Eric Hobsbawm. *The age of empire: 1875–1914.* London: Weidenfeld & Nicolson, 1987.

7 Bernhard C. Schär. *Tropenliebe: Schweizer Naturforscher und niederländischer Imperialismus in Südostasien um 1900* [love of the tropics: Swiss natural scientists and Dutch imperialism in South East Asia around 1900]. Frankfurt on the Main: Campus Verlag 2015 [Globalgeschichte, vol. 20], 20.

afford the expensive collection of data for such an undertaking. On a conceptual scale, his publishing endeavour was part of a "culture of colonialism," helping to promote a hierarchical perception of the world,[8] actively contributing to a larger project of cultural othering, including scholarly fields like cartography and geography. This becomes strikingly clear when comparing the visual characteristics of Hassenstein's maps with his Japanese source materials: Brown-coloured mountain sides and green lowlands dominate the atlas' sheets, whereas his main source map, a large-sized Japanese wall map, emphasises roads connecting inhabited areas.[9] At the same time of copying the Japanese source map, the cartographer diminishes the original in his *Mémoire* – the map commentary – by denying its quality.

Furthermore, the making of the atlas depended on "networks nurtured by imperial contacts,"[10] albeit not an imperial project per se: Japan had become the Empire of Japan in 1868, being itself a colonial power.[11] Hassenstein's transimperial network consisted of people from different empires with varied political, economic and scientific agendas, as will be explicated in this article. In studying sources from publisher's archives, active participants and prospective secondary profiteers of empire come to life, providing empirical confirmation of imperial practices. By drawing on these sources, this paper seeks to illuminate in what respect cartographic endeavours might have been inspired by imperial practices, themselves not being carried out by government agencies. As will be shown, the publishing house was mainly oriented towards a readership perceived as a sales market. By serving a print culture in which reading habits favoured dichotomous presentations of culture, the published *Atlas of Japan* was both product and producer of a transimperial network.

8 Schär, *Tropenliebe*, 15.

9 冨士見十三州輿地全圖 (*Fujimi jusanshu yochi no zenzu,* literally "map of those thirteen provinces from which Mount Fuji can be viewed") with Hassenstein's handwriting and cartographic triangulations is kept at Perthes Collection, FB Gotha, SPK 547$112045243.

10 Jane Burbank and Frederick Cooper. *Empires in world history. Power and the politics of difference.* Princeton, N.J.: Princeton University Press, 2010, 2.

11 Nadin Hee. *Imperiales Wissen und koloniale Gewalt: Japans Herrschaft in Taiwan 1895–1945* [imperial knowledge and colonial violence: Japan's rule over Taiwan 1895–1945]. Frankfurt am Main: Campus Verlag 2015 [Globalgeschichte, vol. 11].

1 The setting

The book format of an atlas depicting one particular country came to be a frequent goal of nation states, which published this specific format for educational purposes, sometimes as a by-product of detailed land surveying with the construction of maps acting as a means for governing smaller and larger pieces of land.[12] The *Atlas of Japan* published in 1885 and 1887 was much different in terms of aim and scope. No government was involved in the process of its making, nor was it authored or published by anyone familiar with the mapped country's geography. The atlas' author Hassenstein went abroad only twice in his life, travelling to Berlin and to Venice, respectively.[13] The publishing firm Perthes had not initiated the making of the atlas and was hardly equipped to ensure all stages of its production as becomes obvious in the varied files pertaining to the process of the atlas' publication.[14] Records include sketches and drafts for the maps, extensive data collection, original sources that were collated in the making of the atlas' map sheets, hundreds of letters to and from Hassenstein, and the journals he kept in relation to his cartographic work.[15]

Perthes came to be Europe's hot spot for cartographic publishing in the second half of the 19th century.[16] Within professional and popular publishing,

12 Burnett, *Masters*. Raj, *Relocating*. Gugerli and Speich, *Topografien*. Edney, *Mapping*.

13 Incidentally, he went to Venice to attract supporters for his atlas project. See: *Personalakte Hassenstein, Bruno* [personal file Hassenstein, Bruno]. Forschungsbibliothek Gotha (FB Gotha), Sammlung Perthes Archiv (SPA) Familien- und Firmenarchiv [family and company archive] (FFA) P 206.

14 In the following, I am going to present recently uncovered sources from the publisher's archives of Perthes which were made accessible to the public with grants from Thuringia's Ministry of Education and several German cultural foundations in 2003. The Perthes Collection is held by Erfurt University Library and stored at the Perthes Forum in Gotha. Sources from the Perthes Collection under consideration for this study include the map collection (SPK), the publisher's archive (SPA), the publisher's library (SPB), and the family and company collection (FFA). Most of the archive has been organised according to geographical location. Thus, structural questions towards publishing procedures remain difficult to answer. See: Nils Güttler. *Das Kosmoskop. Karten und ihre Benutzer in der Pflanzengeographie des 19. Jahrhunderts* [the cosmoscope. Maps and their users in 19th-century plant geography]. Göttingen: Wallstein, 2014.

15 The journal kept for the *Atlas of Japan* is: Bruno Hassenstein. *Kartographische Notizen, Heft 27: 1880–1885. Japan* [cartographic notebook no. 27: 1880–1885]. FB Gotha, SPA Redaktionsarchiv Petermanns geographische Mitteilungen [editorial archive of the journal *Petermanns geographische Mitteilungen*] (PGM) 138/9a.

16 Georg Jäger. Kartographischer Verlag [cartographic publishing]. In *Geschichte des deutschen Buchhandels im 19. und 20. Jahrhundert* [history of the German book trade in the 19th and 20th century], Georg Jäger (ed.). Frankfurt on the Main: Buchhändler-Vereinigung, 2001.

the firm had several strong pillars that helped establish it internationally despite competitors and political turmoil.[17] Besides the publishing arm of their *Almanach de Gotha*, a directory of royalty and nobility, their cartographic ventures were the strongest in atlas publishing and in their popular monthly, now known as *Petermanns geographische Mitteilungen*.[18] The idea for the making of a stand-alone publication on the cartographic appearance of Japan goes back to the founder of the successful Perthes journal. Started in 1855, this journal had been designed to come out monthly, always including at least one original map as stated in the preface of the first issue:

> Unsere "Mittheilungen" sollen sich dadurch von ähnlichen Schriften unterscheiden, dass sie auf sorgfältig bearbeiteten und sauber ausgeführten Karten das Endresultat neuer geographischen Forschungen [sic] zusammenfassen und graphisch veranschaulichen. Nie wird deshalb eine Nummer unserer Schrift ausgegeben werden, ohne eine oder mehrere Karten-Beilagen, und diese werden mit besonderer Rücksicht darauf entworfen werden, dass sie allen Besitzern von Stieler's Hand-Atlas, Berghaus' Physikalischem Atlas, und anderen aus der Anstalt hervorgegangenen Kartenwerken ein fortlaufendes leicht zugängliches Supplement in handlicher Form gewähren. [Our Transactions shall be distinct from other similar publications for their carefully executed maps which will sum up and graphically present the results of recent geographical research. Never will there be an issue without the addendum of one or more maps, which will be designed as a progressive, easily accessible, and handy supplement with special reference to the owners of *Stieler's Hand-Atlas*, *Berghaus's Physical Atlas*, and other publications from our institute.][19]

---

17 Jürgen Espenhorst. *Petermann's planet. A guide to German handatlases and their siblings throughout the world, 1800–1950*. Schwerte, Germany and Falls Church, VA: Pangaea, 2003. Heinz Peter Brogiato. Baedeker und Stieler. Die Rolle des Verlagswesens zwischen Popularisierung und Professionalisierung der Geografie im 19. Jahrhundert [Baedeker and Stieler. On the role of publishing between the popularisation and professionalisation of geography in the 19th century]. In *Wissenschaftsverlage zwischen Professionalisierung und Popularisierung* [academic publishers between professionalisation and popularisation], Monika Estermann, Ute Schneider (ed.), 77–114. Wiesbaden: Harrassowitz, 2007 [Wolfenbütteler Schriften zur Geschichte des Buchwesens, vol. 41].

18 Its first issue's complete title read *Mittheilungen aus Justus Perthes' Geographischer Anstalt über wichtige neue Erforschungen auf dem Gesamtgebiete der Geographie* [transactions of Justus Perthes' Geographical Institute on important new research in the whole field of geography] which was changed to *Dr. A. Petermann's Mitteilungen aus Justus Perthes' Geographischer Anstalt* in 1879 and to *Petermanns geographische Mitteilungen* in 1938, commonly shortened to PGM. Its last issue was published in 2004. Franz Köhler. *Gothaer Wege in Geographie und Kartographie* [Gotha's ways in geography and cartography]. Gotha: Haack, 1987.

19 August Petermann. Vorwort [preface]. In *Mittheilungen aus Justus Perthes' Geographischer-Anstalt uber wichtige neue Erforschungen auf dem Gesamtgebiete der Geographie* 1:1 (1855): 1–2, 2.

Its founder August Petermann had used his international network to promote the Perthes journal and was very careful to aim at both a professional and at a popular readership.[20] Whenever maps became too large, whenever research or public interest in a certain area increased, Petermann published supplements, which gave ample space for larger accounts of journeys, research fields, or surveys. When he died in 1878, he left the editorial team with a rough sketch of what special issues might be prepared for publication next.

A worn sheet of paper gives evidence that Petermann had conceptualised a comprehensive map of Japan [fig. 1].[21] It is kept within the correspondence about the making of the atlas and bears the handwriting of Bruno Hassenstein, a Perthes cartographer who was trained by Petermann and who succeeded him in overseeing the journal's cartographic department.[22] This plan shows a map of Japan that has been drawn over in several layers. The key explicates the design: Petermann's projected map sheets would have mapped Japan in seven segments in a landscape mode. His apprentice's first design consisted of 18 north-oriented maps, which Hassenstein discarded in favour of his second design that was spread out on only eight map sheets. These sheets were to be published as supplement to the Perthes journal, *Dr. A. Petermann's Mitteilungen aus Justus Perthes' Geographischer Anstalt* (*Petermann's Mitteilungen)*.[23] This issue was never published. Instead, the atlas came to be a stand-alone publication in two volumes, published in 1885 and 1887 with a low print-run of 300 copies, and only after the cartographer had acquired funds for covering the costs of printing.[24]

---

20 Imre Josef Demhardt. *Der Erde ein Gesicht geben: Petermanns Geographische Mitteilungen und die Anfänge der modernen Geographie in Deutschland; Katalog zur Ausstellung der Universitäts- und Forschungsbibliothek Erfurt/Gotha im Spiegelsaal auf Schloß Friedenstein in Gotha, 23. Juni bis 9. Oktober 2005* [putting a new complexion on earth. Petermanns Geographische Mitteilungen and the beginnings of modern geography in Germany. Catalogue of an exhibition held by Erfurt University Library and Research Library Gotha from June 23 to October 9, 2005 in the "Spiegelsaal" at Friedenstein Castle]. Gotha: Univ. Erfurt 2006 [Veröffentlichungen der Forschungsbibliothek Gotha, Vol. 42]. Philipp Felsch. *Wie August Petermann den Nordpol erfand* [how August Petermann invented the North Pole]. Munich: Luchterhand, 2010.

21 Bruno Hassenstein. *Drei Entwürfe für eine Specialkarte von Japan* [three concepts of a special map of Japan], *1878/1879*. FB Gotha, SPA ARCH PGM 150/1, 18–18a.

22 For Hassenstein's biography see: Friedrich Ratzel. Br. Hassenstein †. In *Dr. A. Petermann's Mitteilungen aus Justus Perthes' Geographischer Anstalt* 48 (1902): 1–5. Demhardt, *Erde*, 71–72.

23 This plan was pursued at least until early 1882, i.e. three years after the first drafts. Bruno Hassenstein. *Letter to Johannes J. Rein, November 10, 1881*. FB Gotha, SPA ARCH PGM 150/3, 515a–515d.

24 Besides the correspondence, the atlas' publication time may be reconstructed from Perthes' advertisements: *Empfehlenswerte Weihnachtsgeschenke aus dem Gebiete der Kartographie und*

Fig. 1: Hassenstein's former supervisor and instructor August Petermann had made plans for a comprehensive map of Japan to be published as a supplement to the journal *Mittheilungen aus Justus Perthes' Geographischer Anstalt über wichtige neue Erforschungen auf dem Gesamtgebiete der Geographie* [transactions of Justus Perthes' Geographical Institute on important new research in the whole field of geography]. — Bruno Hassenstein. *Drei Entwürfe für eine Spezialkarte von Japan* [three concepts of a special map of Japan], 1878–1879 (Forschungsbibliothek Gotha, Perthes Collection, SPA ARCH PGM 150/1, 18–18a).

When Hassenstein became chief technical officer responsible for all production levels of the journal's maps, he had to monitor the manufacturing stages for at least 24 maps annually, supplements not included.[25] Hassenstein's job as cartographer at Perthes was to make sense of travellers' notes, their maps and data collected during their expensive and time-consuming voyages of exploration.[26] While drawing maps, keeping track of distances, terrain, and specific landscapes during the course of a journey was a well-established duty of the traveller on the road, the cartographers still attempted to at least harmonise incoming maps before publication.[27] Producing maps for the Perthes journal meant to visualise data according to the standards of the journal and its readers. Frequently, the areas that were to be mapped were hardly known at all to Perthes' readership. Supplements covered regions such as Australia, Africa, or distant islands that had not been publicly available on maps before.[28] Hassenstein himself had finished his apprenticeship with maps depicting travels of exploration in Africa.[29] The situation was very different for Japan. Although it had only been accessible as a travel destination from the 1850s onwards, large and detailed maps of the islands existed already.[30] However, they were in Japanese, hardly available

---

Geographie [recommendable cartographical and geographical Christmas presents]. FB Gotha, SPA ARCH PGM 138/9c, 425 [*Almanach de Gotha* 1885, insert]. "Eine große neue Karte von Japan" [a large new map of Japan], newspaper clipping. In *Weser Zeitung*, December 13, 1885. FB Gotha, SPA ARCH PGM 138/9c, 423. Scholarly reviews of the atlas were first published in 1886, for example: Franz M. Hilgendorf. Hassenstein, Bruno: Atlas von Japan. Sieben Blätter im Massstabe von 1 : 1 000 000 und eine Uebersichtskarte i. M. von 1 : 7 500 000, I. Abth. Süd- und Central-Japan. Gotha 1885 [Rezension] [review]. In *Verhandlungen der Gesellschaft für Erdkunde zu Berlin* 13:4 (1886). Hassenstein's last notes on the subject are recorded in Bruno Hassenstein. *Zeitleiste: Japan-Atlas. 25.4.1886–26.11.1887* [timeline: Atlas of Japan. April 25, 1886–November 26, 1887]. FB Gotha, SPA ARCH PGM 138/9a, 111.

25 Jan Smits. *Petermann's maps. Carto-bibliography of the maps in Petermann's Geographische Mitteilungen 1855–1945*. 't Goy-Houten: Hes & De Graaf, 2004.

26 A model of the map production for *Petermanns geographische Mitteilungen* has been designed by Bruno Schelhaas and Ute Wardenga: Bruno Schelhaas and Ute Wardenga. 'Inzwischen spricht die Karte für sich selbst.' Transformation von Wissen im Prozess der Kartenproduktion ['By now the map is self-evident'. Knowledge transformation during map production]. In *Die Werkstatt des Kartographen. Materialien und Praktiken visueller Welterzeugung* [the cartographer's shop. Materials and practices of a visual production of the world], Petra Weigel, Steffen Siegel (eds.). Munich: Fink, 2011, 89–107.

27 See Güttler, *Kosmoskop*.

28 The example of Australia is covered in: Demhardt, *Erde*, 61–70.

29 Ratzel, Hassenstein.

30 A contemporary account of more recent maps that were also relevant to Hassenstein, is given by Erwin Knipping. Ueber eine neue Karte von Japan und ihre Quellen [*on a new map of Japan*

to anyone in Japan both for a general lack of distribution and for linguistic difficulty.[31] Petermann's letters from travellers headed for Japan include the repeated entreaty to publish reliable, up-to-date, sufficiently detailed and readable maps of the Japanese islands, i.e. graphically and linguistically normalised designs.[32] Travellers experienced great difficulty in obtaining maps that were of any help in preparing for a journey to Japan or in attempting to familiarise themselves with Japan's topography even when they were able to read a little Japanese.[33]

It was Perthes' excellent connections in geographical circles that made the compilation of a romanised map of Japan feasible. Geographers approached the firm at least as early as 1859 with requests for romanised maps of Japan.[34]

and its sources]. In *Mitteilungen der Deutschen Gesellschaft für Natur- und Völkerkunde Ostasiens* II, 11 (1876), 20–24. A broader collection of geographical and cartographical material published at the time is: Friedrich von Wenckstern. *A bibliography of the Japanese empire. Being a classified list of all books, essays and maps in European languages relating to DAI NIHON (Great Japan) published in Europe, America and in the East from 1859–93 A.D. To which is added a facsimile-reprint of Léon Pagès, Bibliographie Japonaise depuis le XVe siècle jusqu'à 1859*. London: Kegan Paul/ Trench/ Trübner, 1895. A modern account of Japanese cartography is given in: Namba Matsutaro (ed.). *Old maps in Japan*. Osaka: Sogensha, 1973, and: Marcia Yonemoto. *Mapping early modern Japan. Space, place, and culture in the Tokugawa period (1603–1868)*. Berkeley and Los Angeles, CA: University of California Press, 2003. European cartographic publications of the Japanese islands are presented in: Lutz Walter. *Japan mit den Augen des Westens gesehen: gedruckte europäische Landkarten vom frühen 16. bis zum 19. Jahrhundert* [Japan in the eyes of the West: printed European maps, 16th–19th century]. Munich: Prestel, 1993 [exhibition catalogue].

31 Hassenstein made a similar observation when asking Japanese students in Leipzig to transcribe certain Japanese toponyms, which more often than not they were unable to recognise. See, for example: Bunjiro Koto. *Note regarding the translation of a map legend. 1882*. FB Gotha, SPA ARCH PGM 284, 110.

32 One example is: Ferdinand Freiherr von Richthofen. *Letter to August Petermann, April 20, 1873*. FB Gotha, SPA ARCH PGM 150/2, 220.

33 Von Richthofen recalled older maps as being "a peculiar blending of truth and poetry" ("eine eigenthümliche Mischung von Wahrheit und Dichtung"), in Richthofen, *April 20, 1873*, 220. A critique of specific maps was proposed in Hassenstein's *Atlas of Japan*, see: Bruno Hassenstein. Kartographische Bemerkungen über Entwurf und Ausführung der Karten [cartographic remarks on the design and realisation of the maps]. In: Hassenstein, *Atlas*, 2–4, 2–3.

34 *File PGM 150/2, 1859–1900*. FB Gotha, SPA ARCH PGM 150/2, 219–327. The completeness of this correspondence can only be assumed, as there are no records of incoming and outgoing letters as sometimes is the case in publishers' archives.

2 The cartographer's zeal

Hassenstein's plan, dated October 1879, was designed around the time that he published his very first Japanese map in *Petermann's Mitteilungen* [fig. 2].[35] This

Fig. 2: Hassenstein's first map depicting Japan was reordered frequently, leading him to great expectations towards the success of his Atlas of Japan. The map was based on a Japanese source with a similar cartographic section, namely the popular tourist map 富士見十三州輿地全圖 (Fujimi jusanshu yochi no zenzu, 'map of those thirteen provinces from which Mount Fuji can be viewed'), copies of which were owned by Hassenstein and several of his correspondents. – Bruno Hassenstein. "Die Umgegend der Bai von Tokio und des Vulkans Fuji-no-Yama. Mit Unterstützung von Prof. J. Rein und Tadashi Sanda gezeichnet von B. Hassenstein" [the environments of Tokyo Bay and of the volcano Fuji-no-yama. Drawn by B. Hassenstein with the assistance of Prof. J. Rein and Tadashi Sanda]. In Dr. A. Petermann's Mitteilungen aus Justus Perthes' Geographischer Anstalt 25:X (1879): plate 19 (Forschungsbibliothek Gotha, SPK 547-111978386).

35 Bruno Hassenstein. Die Umgegend der Bai von Tokio und des Vulkans Fuji-no-Yama. Mit Unterstützung von Prof. J. Rein und Tadashi Sanda gezeichnet von B. Hassenstein [the environments of Tokyo Bay and of the volcano Fuji-no-yama. Drawn by B. Hassenstein with the assistance of Prof. J. Rein and Tadashi Sanda]. In *Dr. A. Petermann's Mitteilungen aus Justus Perthes' Geographischer Anstalt* 25:X (1879): plate 19. As the journal's map editor, Hassenstein supervised the publication of about 24 maps annually for the regular issues alone.

first map depicted the Greater Tokyo area and Mount Fuji, an all-time favourite religious and secular tourist destination on the large main island of Japan, Honshu.[36] Mount Fuji's popularity becomes clear from the sheer existence and incredibly high numbers of societies in Tokyo alone whose members were devoted to admiring the iconic volcano from afar: In the first half of the 19th century about 800 of these clubs in Tokyo were dedicated to viewing Mount Fuji.[37] Moreover, the centre of Japan's main island Honshu was exactly the area that was of interest to many foreigners living in Japan.[38]

Hassenstein's first map was published as an addendum to an essay by German professor of geography Johannes Justus Rein.[39] This was their first cartographic collaboration and Hassenstein insisted on drawing a new map because he was dissatisfied with the quality of the map Rein had submitted.[40] Hassenstein's map "Die Umgegend der Bai von Tokio und des Vulkans Fuji-no-Yama" took almost a year to complete from design to print.[41] Records in the Perthes Collection show that there were three main obstacles in the preparation of a new map of an area that was relatively well known to westerners compared to other parts of Japan: First, Perthes' library department held few original sources of Japan's cartography and geography, hence the loan of specific maps consumed the cartographer's time.[42] Secondly, Hassenstein was unable to read Japanese and had to acquaint himself with the syllabic alphabets Hiragana and Katakana

---

36 Hassenstein, Umgegend. H. Byron Earhart. *Mount Fuji. Icon of Japan*. Columbia, SC: University of South Carolina Press, 2011 [Studies in comparative religion].

37 Andrew Bernstein. Whose Fuji? Religion, region, and state in the fight for a national symbol. In *Monumenta Nipponica* 63:1 (2008): 51–99.

38 Kristin Meißner. Responsivity within the context of informal imperialism. Oyatoi in Meiji Japan. In Journal of Modern European History 14:2 (2016): 268–289.

39 Johannes J. Rein. "Der Fuji-no-yama und seine Besteigung" [Mount Fuji and its ascent]. In *Dr. A. Petermann's Mitteilungen aus Justus Perthes' Geographischer Anstalt* 25:X (1879): 365–376. For biographical information on Johannes Rein, see: Sebastian Conrad and Matthias Koch. Einleitung [introduction]. In *Johannes J. Rein. Briefe eines deutschen Geographen aus Japan 1873–1875* [Johannes J. Rein. Letters from a German geographer in Japan, 1873–1875], Sebastian Conrad and Matthias Koch (eds.), 25–93. Munich: Iudicium, 2006 [Monographien aus dem Deutschen Institut für Japanstudien, vol. 40].

40 Johannes J. Rein. *Letter to Bruno Hassenstein*. September, 26 1879. FB Gotha, SPA ARCH PGM 150/1, 90–91.

41 Johannes J. Rein. *Letters to Bruno Hassenstein and Ernst Behm regarding the publication of the essay "Der Fuji-no-yama und seine Besteigung."* November 1878–September 1879. FB Gotha, SPA ARCH PGM 150/1.

42 Hassenstein made use of maps from Rein's private collection and sea charts held by the Prussian Admiralty in Berlin. See: Bruno Hassenstein. Bemerkungen über die hauptsächlich benutzten Quellen [notes on principal sources]. In Hassenstein, Umgegend.

as well as with Sino-Japanese Kanji characters.[43] Thirdly, Hassenstein decided to base his map on a popular map with Sino-Japanese characters that were difficult to transcribe even for Japanese with high linguistic skills.[44]

Despite these hindrances, Hassenstein decided to move forward with Petermann's plan of a comprehensive map of Japan. A strong motivation for Hassenstein was his idea that a cartographic venture the size of an atlas would suffice for a German university to award him a doctoral degree. In July, 1879, he addressed this issue in a letter to Rein:

> P.S.: In einem meiner nächsten Briefe werde ich mir erlauben, Sie mit einer Frage zu behelligen, die mich schon lange beschäftigt, und in Bezug zu meinem Vorwärtskommen in der hiesigen Anstalt steht: nämlich eine Anfrage über die Möglichkeit der Erlangung des Doktordiploms von der Universität Marburg für mich, einen Autodidakten. Könnten Sie mir schon jetzt, damit ich klar sehe, irgendein Formular über die zur Erlangung des Doktor-Diploms nötigen Bedingungen verschaffen? Ich würde Ihnen sehr dankbar sein! [P.S. In one of my next letters I will be so bold as to trouble you with a question that has been bothering me for quite some time now, concerning advancement in this company. This is an enquiry on the awarding of a doctorate degree from the University of Marburg for myself, an autodidact. So that it is clear to me, would you be able to send me some form regarding all necessary preconditions for the awarding of such a diploma? I would be very grateful!][45]

Hassenstein's request of a doctoral degree addressed to a German professor of geography seemed far-fetched even at the time. Johannes Rein replied vaguely, carefully avoiding explicit concessions. Yet, when Georg August University of Göttingen celebrated its 150th anniversary in 1887, another colleague of Rein's

---

43 This is evident in letters as well as in pencilled-in transcriptions on maps in Perthes' map collection (SPK) and in books stored in the library's section on Japan (SPB Japan). See, for example: August Pfizmaier. *Über japanische geographische Namen* [on Japanese toponyms]. Wien: Karl Gerold's Sohn, 1875/1876. FB Gotha, SPB Japan No 28 (SPB 8° 3100.00028). Murray S. Day. *Report of the trigonometrical survey of the island of Hokkaido for 1875*. New York: Francis Hart & Co., 1876. FB Gotha, SPB Japan No 37 (SPB 4° 3100.00370).

44 The colourful map 冨士見十三州輿地全圖 *[fujimi jusanshu yochi no zenzu, map of those thirteen provinces from which Mount Fuji can be viewed]* bears Hassenstein's handwriting and cartographic triangulations. FB Gotha, SPK 547$112045243. As given in the map's title, Tadashi Sanda helped with the transcription. Tadashi Sanda. *Letter to Bruno Hassenstein, August 27, 1879*. FB Gotha, SPA ARCH PGM 150/1, 39–40v. According to Rein and Ernest Mason Satow, British expert on Japan, Sanda was employed at the Japanese ambassador's office in London. Ernest Mason Satow. *Letters to the editors of Petermanns geographische Mitteilungen, 1879–1883*. FB Gotha, SPA ARCH PGM 356.

45 Bruno Hassenstein. *Letter to Johannes J. Rein, July 17, 1879*. FB Gotha, SPA ARCH PGM 150/1, 113–113c, 113c.

and a former co-worker of Hassenstein's at Perthes, Hermann Wagner, supported the presentation of an honorary doctoral degree for the Gotha cartographer.[46] Wagner praised Hassenstein as "proven expert on the geography of Africa" in his letter of recommendation and recognised the cartographer's efforts for a comprehensive map of Japan.[47]

In contrast to this public valuation of his work and his placement at Perthes' Geographische Anstalt, Hassenstein's desire for advancement at Perthes through the publication of an outstanding piece of cartography was bitterly disappointed. Instead of appreciating his tenaciousness in pursuing the complicated mapping of a distant empire, his superiors at the publishing house condemned his ignorance of cost-effectiveness and his misjudgement of the market.[48] Hassenstein was harshly criticised for his unauthorised allocation of his own as well as his assistants' working hours, and his costly ordering of maps, transcriptions and translations without prior permission. When asked to sum up his expenses since the beginning of his mapping project in 1881, he had to give estimates because his records had never before included any cost reports.[49]

However, although the publisher disapproved of the final atlas, the beginning of the project was much more promising for Hassenstein. He had intended to compile his maps for a supplement of the Perthes journal. These were typically published in a print-run of about 2,000 copies, less than a standard issue of the journal, but much more than the final print-run of the *Atlas of Japan*, which only reached a meagre 300 copies.[50] These supplements addressed topics

---

46 Archival material pertaining to the university's mass presentation of honorary academic degrees is held by the university archives. For Hassenstein's case, see: Hermann Wagner, Christian August Volquardsen and Bruno Hassenstein. *Ehrenpromotion Bruno Hassenstein, 1887* [honorary doctorate Bruno Hassenstein, 1887]. Georg August University Göttingen, University Archives, Dekanat der Philosophischen Fakultät [arts faculty, dean's office], vol. 173b, 57a–57d.

47 "Er ist heute der beste Kenner afrikanischer Geographie [...]." Hermann Wagner. *Letter of recommendation for Bruno Hassenstein, [1887]*. Georg August University Göttingen, University Archives, Dekanat der Philosophischen Fakultät [arts faculty, dean's office], vol. 173b, 57a [1–3], 57a [2].

48 His personal file contains some references to this instance: See Perthes, *Personalakte*. Reference to a similar later incident is made in: Bruno Hassenstein. 7.11.1890, Vormittag 10–11 Uhr. Scene in P.s Vorkomptoir [November 7, 1890, morning, 10–11. Scene in Perthes's office]. In *Nachlass Bruno Hassenstein* [Bruno Hassenstein papers]. Staatsbibliothek Berlin Preußischer Kulturbesitz, Handschriftenabteilung, box 4.

49 Bruno Hassenstein. Conto für Japan-Arbeit [expense account for the Atlas of Japan]. [1882]. In Bruno Hassenstein. *Kartographische Notizen, Heft 27: 1880–1885. Japan*. FB Gotha, SPA ARCH PGM 138/9a, 140–144.

50 Justus Perthes' Geographische Anstalt. *Auslieferungsbuch 1869–1894* [account book on deliv-

and areas that were of greater interest to the journal's audience or the general public, but that were too extensive for inclusion in the journal itself. They were not included in the subscription, thus always addressing special-interest readerships with varying results. Japan seemed to attract a satisfying number of readers, as supplement no. 59 showed: Johannes Rein's and Erwin Knipping's accounts of one of the major routes in Japan, the road *Nakasendo*,[51] sold 2,239 copies in its first year, with an additional 89 copies sent out for review purposes.[52] Accordingly, Hassenstein imagined a comprehensive series of maps that was to be widely circulated and publicised via the journal's subscribers as well as through word-of-mouth recommendations by colleagues like Rein and Knipping.

Knipping, a German meteorologist, had been recruited by the Japanese government in 1871 and had been working as *Oyatoi* in Japan when he undertook his own mapping project.[53] The meticulously hand-coloured map consisted of six sheets that were to be collated and mounted after distribution.[54] *Stanford's library map of Japan compiled by Erwin Knipping* was published in 1879,[55] but its London-based publisher had stopped distribution in the same year,[56] presum-

eries, 1869–1894]. FB Gotha, FFA Verlagsauslieferungsbuch (VAB). Bruno Hassenstein. *Note on sales of the first volume.* November 1886. FB Gotha, SPA ARCH PGM 138/9a, 222.

51 At the time, it was one of two standard routes connecting the old capital Kyoto in South West Honshu with the modern capital Tokyo on the East coast of the island. See: Jilly Traganou. *The Tokaido road. Travelling and representation in Edo and Meiji Japan.* London and New York, NY: Routlege Curzon, 2004.

52 Perthes, *Auslieferungsbuch.*

53 Knipping was also under contract for regular summarising work for Perthes in the 1880s. A biographical sketch is included in his memoirs. Erwin Knipping: *In japanischen Diensten. Zwei Jahrzehnte eines preußischen Meteorologen in der ersten Hälfte der Meiji-Zeit (1868–1912)* [In the service of Japan. A Prussian meteorologist's two decades in the first half of the Meiji period (1868–1912)], Löhne: Cass 2014, 87–88.

54 The copies in the Perthes Map Collection (SPK) are still separate. See Erwin Knipping. *Stanford's library map of Japan. Principally compiled from Japanese documents by E. Knipping.* London: Stanfords, 1879 (FB Gotha, SPK call numbers 547$112555659, 547$112555624, 547$112554768, 547$112554792, 547$112555268, and 547$112555292).

55 Knipping, *In japanischen Diensten*, 86–87.

56 "New Maps of Japan" [newspaper clipping]. In *Japan Daily Herald, November 1877* (FB Gotha, SPA ARCH PGM 138/9a, 138). The review dates back to 1877 and was written on the basis of a photograph. Knipping presumably used this early review of his manuscript map to promote its publication by European publishing houses. There are no records pertaining to the publishing of *Stanford's library map of Japan*. The publisher's history has been summarised by Peter Whitfield. *The mapmakers. A history of Stanfords.* London: Compendium, 2003.

ably because of low demand, high costs and the effort required for colouring the sheets.[57]

However, Knipping and Hassenstein shared the same target group, namely the many Europeans, westerners and other non-Japanese citizens who stayed in Japan as so-called *Oyatoi gaikokujin* ("contract foreigners"). Employed by the Japanese government, the *Oyatoi* were recruited to support Japanese modernisation efforts such as educational reforms and infrastructural developments.[58] Drawing on the frequent reorders of the Greater Tokyo map, Hassenstein assumed the *Oyatoi* and other businessmen taking up trade in Japan would want to buy maps of their new temporary home.[59] All of them were expecting to be staying in Japan for only a limited time, spanning from three months to several years. Usually, they remained far from fluent in Japanese, thus preferring maps with romanised characters.[60]

The records of Hassenstein's enterprise give a fine example of the pitfalls in transnational publishing ventures in the imperial age: Not only was there international competition, but the preconditions for each publishing endeavour differed from one another. Some were endorsed by governments, others were private scholarly or corporate economic projects; some were collated from original source materials and with the linguistic help of locals, others relied on vague transcriptions and on the memories of foreign travellers.[61] Hassenstein and Knipping's biggest obstacle during the mapping of the islands was in obtaining and translating sources. Cartographic prints were mainly available in tourist contexts,

57 See: Nils Güttler. Unsichtbare Hände. Die Koloristinnen des Perthes Verlags und die Verwissenschaftlichung der Kartographie im 19. Jahrhundert [invisible hands. The female colourists at Perthes and the 19th-century scientification of cartography]. In *Archiv für Geschichte des Buchwesens* 68 (2013): 133–153.

58 Jones, *Machines*. Meißner, *Responsivity*.

59 Bruno Hassenstein. *Letter to Johannes J. Rein, March 1, 1880*. FB Gotha, SPA ARCH PGM 150/3, 490a–490c. Judging from another letter to Hilgendorf, Hassenstein had also imagined a German imperial engagement in Japan. Bruno Hassenstein. *Letter to Franz M. Hilgendorf, November 18, 1885*. FB Gotha, SPA ARCH PGM 282, 36–37.

60 Foreigners were not the only ones who longed for simplification of the complex Sino-Japanese characters. The Japanese ministry of education employed multiple task forces, one of which was to figure out how Japanese place names were to be transcribed. Transcription in a straightforward syllabic alphabet was to be the first step, not only helping foreigners but also native Japanese to better understand the reading of certain characters in different contexts. For a contemporary account see: Pfizmaier, *Namen*. For a historical analysis see: Nanette Gottlieb. *Kanji politics. Language policy and Japanese script*. London: Kegan Paul International, 1995.

61 Hardly anything can be said about circulation figures, distribution patterns, and readerships. Figures for print runs are available from Perthes' account book of deliveries. Perthes, *Auslieferungsbuch*. For Knipping's publisher Stanford's no such archival source has been recorded. See: Whitfield, *Mapmakers*.

aimed at those who could read Japanese and displaying very unusual visual characteristics for a European eye.[62] Existing maps with Latin characters and of Japanese origin were frequently either regarded as imprecise, criticised for their scale as being far too large for practical purposes, or they were outdated by the 1870s due to administrative reforms and infrastructural developments.[63] According to Perthes' correspondents, a modern scientific survey and mapping project of Japan was not started until the 1880s,[64] and the contracting authority, Japan's ministry of education, published some results, but they were difficult to obtain in Gotha.[65] This added to Hassenstein's conviction that his mapping project would find great numbers of purchasers, as Perthes' publications were already being ordered and sent out across the globe.[66]

3 The disillusionment

With the progression of the project, the maps of the Japanese islands got more and more expensive. In 1882, all Perthes employees were asked to report on existing projects and Hassenstein compiled a list of his expenses.[67] As it seems, it was not until then, about three years after starting the project, that Hassenstein first totalled the overall costs of his effortful atlas. He had already spent what would be

62 Kären Wigen. *A malleable map. Geographies of restoration in central Japan 1600–1912*. Berkeley and Los Angeles, CA: University of California Press, 2010.

63 Traganou, *Tokaido*. Constantine Nomikos Vaporis. *Breaking barriers. Travel and the state in early modern Japan*. Cambridge, MA: Harvard University Press, 1994.

64 A forthcoming collection of essays on the history of cartography in Japan includes one section on "Modern maps for Imperial Japan," covering Japanese cartographic projects. See: Kären Wigen, Sugimoto Fumiko and Cary Karacas (eds.). *Cartographic Japan. A history in maps*. Chicago, IL and London.: University of Chicago Press, 2016 [forthcoming].

65 Knipping had to turn down requests from Perthes employees for photographs of the new Japanese maps. Although he was working for a government agency, he was unable to get hold of the newest material: "Es ist hier in Japan nicht immer möglich, Karten oder Drucksachen zu bekommen, die von der Regierung veröffentlicht werden, wenn man nicht mit den Autoren oder den Beamten der betreffenden Abtheilung persönlich bekannt ist." [Here in Japan one can rarely get one's hand on maps or any prints that are published by the government if one is not personally acquainted with the authors or officials involved.] Erwin Knipping. *Letter to Ernst Behm, June/July 1877*. FB Gotha, SPA ARCH PGM 450, 47–48.

66 See: Jörg Dünne. Portable Media und Weltverkehr. Die Taschenatlanten des Perthes Verlags [portable media and world traffic. Perthes' pocket atlases]. In Weigel and Siegel, *Werkstatt*, 185–203.

67 Hassenstein, *Conto*.

about double as the annual salary of a civil servant. Not wanting to let his efforts go to waste, Hassenstein continued despite being told to focus on his other duties in the company. Long periods of sickness took turns with his growing frustration with the firm's new practice. Hassenstein expressed his regrets that his teacher's legacy was being wasted. He felt that Perthes publications should emphasise critical interpretation and the cartographers' explanatory *Mémoire*, which he regarded as scientifically virtuous.[68] Instead, he felt that economic aspects were superseding scientific standards his employer's reputation was built upon.[69] In keeping with Hassenstein's scientific claims, the *Atlas of Japan* starts with a lengthy description of its sources, design and linguistic decisions.[70] The exact course of events in the downsizing of his projected supplement is not recorded. In 1880, Hassenstein wrote to Rein that he had been asked to adjust his design to Perthes' atlas maps.[71] Two years later, the publication format of the journal was back on the table, this time altered to three supplements for a large map on Japan.[72] Finally, Hassenstein's *Atlas of Japan* got published in two volumes with a very low print-run as he had to finance the printing himself.[73]

The final publication consisted of two volumes, one of which was published in 1885, the other in 1887. Both consist of an introduction, explaining the mapmaker's methods, sources, and resources, and four maps.[74] After having finished his map on central Japan for Rein's essay in *Petermann's Mitteilungen*, Hassenstein started the production of those map sheets that were easier to draw.[75] Especially data for the northern main island Hokkaido was scarce. Thus, the first publication included the map sheets for the chain of islands Ryukyu and other southern islands belonging to Japan. The second map sheet was titled "Nagasaki," the third "Kioto," and the last section of the first volume depicted the central part of

68 Bruno Hassenstein. *Redaktionelle Korrespondenz 1888–1890, Hefte II, III, IV, V, VI, VII* [editorial correspondence 1888–1890, notebooks II, III, IV, V, VI, VII]. FB Gotha, SPA ARCH PGM 138/8, no. VII, 11–14v.

69 Bartholomew, Map-making.

70 Hassenstein, *Kartographische Bemerkungen*.

71 Bruno Hassenstein. *Letter to Johannes J. Rein, March 13, 1880*. FB Gotha, SPA ARCH PGM 150/3, 460–460c.

72 Bruno Hassenstein. *Letter to Johannes J. Rein, June 9, 1882*. FB Gotha, SPA ARCH PGM 150/3, 443–443c.

73 Bruno Hassenstein. *Letter to Johannes J. Rein, August, 8 1885*. FB Gotha, SPA ARCH PGM 150/3, 386–386c.

74 Hassenstein, *Atlas*.

75 The following reconstruction of the publication process is mainly based on letters from Hassenstein that are kept with the Johannes Rein Papers in the Perthes Collection. FB Gotha, SPA ARCH PGM 150.

Honshu, including Mount Fuji and Tokyo. The second volume depicted northern Honshu, western and eastern Hokkaido. The last sheet shows Japan's prefectures and districts with telegraph and railway lines, and steamboat connections.

With the exception of the general chart, Hassenstein chose Japanese toponyms instead of translations for topographical denominations such as "river," "bay" or "mountain." Earlier maps would have given Japanese names and their translations in the same place or they would have randomly alternated styles.[76] Hassenstein wanted his maps of Japan to be widely understandable and adaptable for selling abroad, which showed some understanding of his employer's publishing style since Perthes had been familiar with multi-language publications, for example with their *Stieler* handatlases.[77] The atlas' cartographic design was modern: lowlands were coloured in green, mountains and higher terrain in brown. It included visualised data such as the population of cities and the depths of sea around the coastline. By incorporating Japan's administrative features the mapmaker demonstrated his efforts to publish an up-to-date account, since the subdivisions had only been reformed in 1878.[78]

As a cartographer, Hassenstein lamented Japan's political independence and its status as an Empire. His "cartographic remarks on the design and realisation of the maps" include detailed explanations for his selection of sources, his acquisition of unpublished material from the Prussian Admiralty, the project's supporters and translators, and his choices for the rendering of topographical data. The section on language is especially noteworthy. In it Hassenstein explained his situation at the start of the project:

> Zu den schwierigsten Aufgaben der Kartographie außereuropäischer Länder gehört die konsequente und korrekte Durchführung der geographischen Nomenklatur [...]. Wenn es sich um die kartographischkompilatorische Darstellung eines Landes handelt, welches sich wie etwa Indien, Algier, die Philippinen u. a. im Besitz einer europäischen Macht befindet, deren Pflicht es ist (oder vielmehr sein sollte), für eine einheitliche Namenschreibung in den abhängigen Gebieten zu sorgen, so ist dem Kartenzeichner die Notwendigkeit sprachlichen Studiums erspart; er hat sich einfach an die offizielle Schreibart zu halten, und es

---

76 Walter, *Japan*. This was true for his first map, too, where for example Tokyo Bay is named as "Tokio-Ura oder Yedo-Bai der Europäer." See: Hassenstein, Umgegend.

77 See: Espenhorst, *Petermann's planet*. Dünne, Portable Media. Apart from prospective sales, some sponsors made their support conditional on the internationalisation of the publication. See quotes from Ernest Mason Satow in Johannes J. Rein. *Letter to Bruno Hassenstein, March 23, 1880*. FB Gotha, SPA ARCH PGM 150/3, 461–462.

78 Bruno Hassenstein. *Kartographische Notizen. Heft 25. Japan: 1880.* [Cartographic notebook no. 25. Japan: 1880.] FB Gotha, SPA ARCH PGM 138/10a, 2–41.

genügt, dieselbe unter dem Kartentitel durch ein kleines Wörterverzeichnis und Notizen über die Aussprache kurz zu erläutern.

[The consistent and correct execution of geographical nomenclature is one of the most complicated tasks for the cartographic representation of non-European countries. If compiling a cartographic design of a country that is owned by a European power, such as India, Algier, or the Philippines, they are (or should be) obliged to ensure a consistent designation; in this case the cartographer is free from having to attend to linguistic studies, he just has to adhere to the designation and it suffices to explain this in a short table beneath the map's heading, including some notes on pronunciation.][79]

Due to the fact that no national guideline for the transcription of Japanese characters existed, Hassenstein had to choose between other available systems. After lengthy discussions with Bunjiro Koto and Isao Ijima, overseas students from Japan and Hassenstein's part-time translators of Japanese maps,[80] he combined two competing systems of transcription, resulting in confusion especially for his English-speaking audience.[81] Thus, Hassenstein's choice to romanise Japanese characters was highly questionable with regards to his claim to produce a truly internationally oriented publication.[82] Moreover, the translation of sources and the homogenisation of topographical names was the main reason why this publishing venture took almost ten years before its finalisation, during which time Hassenstein shifted from praising the atlas as his "favourite piece of work"[83] to describing it as his "interminable sea serpent [...] dominating [his] life"[84].

Contemporary reviewers rated Hassenstein's atlas as remarkable. The cartographic skill he had exhibited with this publication was received as outstanding. The Royal Geographical Society's reviewer concluded: "[T]he maps

79 Hassenstein, Kartographische Bemerkungen, 3.

80 Bunjiro Koto was a student of geology in Leipzig and Munich, Isao Ijima studied zoology in Leipzig. Bunjiro Koto. *Letters to Bruno Hassenstein. 1881–1882*. FB Gotha, SPA ARCH PGM 284. Isao Ijima. *Letters to Bruno Hassenstein. 1881–1884*. FB Gotha, SPA ARCH PGM 138/10 II and PGM 284.

81 He combined a German system of transcription based on Karl Richard Lepsius's "standard alphabet" with James Curtis Hepburn's romanisation system, a transcription style still in use today.

82 Johannes J. Rein. *Letter to Bruno Hassenstein, March 23, 1880*. FB Gotha, SPA ARCH PGM 150/3, 461–462.

83 Bruno Hassenstein. *Letter to Erwin Knipping, April 19, 1882*. FB Gotha, SPA ARCH PGM 450, 28.

84 Bruno Hassenstein. *Letter to Franz Hilgendorf, November 18, 1885*. FB Gotha, SPA ARCH PGM 282, 36–37.

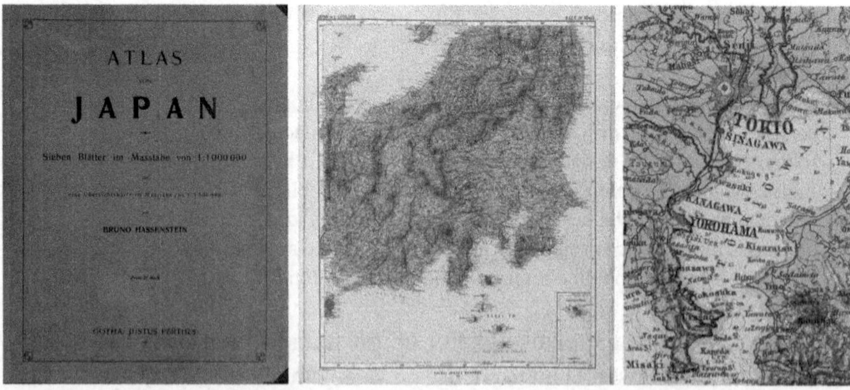

Fig. 3: Cover page, page view and detail from Bruno Hassenstein's Atlas of Japan as preserved in the Perthes Collection (Forschungsbibliothek Gotha, SPB Japan No 60). — Bruno Hassenstein. Atlas von Japan. Sieben Blätter im Massstabe von 1:1000000 und eine Übersichtskarte im Massstabe von 1:7500000 [Atlas of Japan. Seven sheets on a scale of 1:1,000,000 and one general chart on the scale of 1:7,500,000]. Gotha: Justus Perthes, 1885/1887, cover, plate IV, and detail of plate IV.

are beautiful specimens of cartography."[85] As was to be expected from the warnings of his supporters who had actually been to Japan, his representation of toponyms came to be criticised for its arbitrary dealing with the complexity of the Japanese language [fig. 3]. Especially his treatment of similar syllables like じ ("ji"), ち ("chi"), and し ("shi") was viewed to be unhelpful for using the maps in Japan. Basil Hall Chamberlain, author of a widely circulated and frequently reprinted British guidebook on Japan, concluded:

> It is only to be regretted that the author [i.e. Hassenstein, A.S.] should have considered the usual method of spelling Japanese not good enough for him, and should accordingly have evolved a new one from the depths of his inner consciousness. Are not Japanese names hard enough already? What traveller – what resident even – can be expected to recognise, say, the town of Choshi under the disguise of 'Tsosi' or Hachijo under 'Fatsidjio'?[86]

All in all, Chamberlain recommended the use of Greater Tokyo area maps to foreigners travelling through Japan. A map depicting all islands was not regarded as necessary for visitors and this might be part of the answer why Hassenstein's

---

85 John Coles. Japan. – Atlas von –. In *Proceedings of the Royal Geographical Society and monthly record of geography, new monthly series* 9:8 (1887): 526–530, 530.

86 Basil Hall Chamberlain. *All things Japanese. Being notes on various subjects connected with Japan for the use of travellers and others.* 2nd ed. London: Kegan Paul/ Trench/ Trübner 1891, 281.

atlas was publicised so reluctantly by his employer Justus Perthes' Geographische Anstalt.

Perthes faced increasing rivalry from other producers of atlases and from newly established scientific journals that emphasised their more direct links to geography as a discipline at university.[87] Perthes' cartographers had to find new ways of making themselves indispensable, whereas the firm as a whole was seeking to strengthen its thriving branches.[88] This specialisation did not include the promotion of special-interest publications such as map sheets of East Asia. Apparently, Bernhard Perthes was right with his decision. The sales figures for Hassenstein's atlas were very low. Its price of 12 Goldmarks made it an expensive booklover's edition hardly attractive for being taken on a long journey. Furthermore, its large format – four times the size of the handy journal pages – made it unsuitable for its supposed main target group of westerners travelling in Japan. Its scarce dispersion today indicates that it probably never came to be used as a standard reference work for foreigners staying in Japan.[89]

The distance of the armchair cartographer from his mapped target had manifold consequences: Being well acquainted with the relatively loose rendering of African toponyms in European cartography, he gravely underestimated the linguistic challenges of a written language already familiar to his probable critics. Instead, he disregarded his supporters' suggestions for the transcriptions of Japanese geographical names. Adding to the project's unfortunate design was the Japanese Empire's own cartographic agenda, which did not rely on a strictly limited target audience. To the contrary, it served the administration's existing need for exact surveying as a foundation for governance. At the same time, the author of the German atlas on the Japanese islands was experiencing a shift in his employer's publishing strategy. When geographical research became an established discipline at German universities, cartographical endeavours came to be increas-

87 Brogiato, "Baedeker."

88 Justus Perthes' Geographische Anstalt. *Justus Perthes in Gotha, 1785–1885*. Gotha: Justus Perthes, 1889. Justus Perthes' Geographische Anstalt. *Haupt-Katalog* [main catalogue]. Gotha: Justus Perthes, 1915. Justus Perthes' Geographische Anstalt. *Haupt-Katalog* [main catalogue]. Gotha: Justus Perthes, 1935.

89 Besides the Perthes Collection at the Research Library Gotha, copies are kept in Göttingen, Marburg, and Munich. Libraries in Paris and London have several copies and there is one in Oxford. Outside of Europe, only one hard copy at the University of Chicago Library has been catalogued by the database WorldCat. Hassenstein also recorded the receipt of one copy for the library of Yokohama's German club.

ingly regarded as an auxiliary science.[90] The publisher had previously relied on its journal's reputation for his marketing in academic circles as well as with a popular readership. When topics became more specialised with the growing professionalisation of geography, the publisher had to adapt to decreasing subscription numbers. Consequently, a complex series of maps with a lengthy explanatory *Mémoire* was not thought to support the journal's reorientation towards a large readership, especially when the mapped country was out of reach – geographically, linguistically, and politically.

Conclusion

Book culture was a vital venue for the enactment of empire and the inclusion of imperial repertoires into people's practices: Instead of travelling to Japan as an explorer, the cartographer Bruno Hassenstein adopted imperial practices from his office desk in Gotha. Just like foreigners could be employed by the Japanese government for infrastructural and educational purposes, Hassenstein conceptualised his atlas as a product of knowledge that could easily be exported to its mapped country. This article highlighted the making of this unfortunate publishing project, showing how its maker tried to transfer a publishing routine he formerly applied to colonial African spaces to the islands of the Empire of Japan.[91] The initial excitement of Dr. Bruno Hassenstein to be acquainted with Europe's finest experts in Japanese studies could neither turn into a personal nor into monetary profit.

The publisher's main obstacles for the pursuit of publications like Hassenstein's *Atlas of Japan* were temporal and spatial: Publication speed mattered just as much as the choice of spaces to be mapped, as competition and demand varied considerably. Unable to control the outcome of costly commissions for translations, transcriptions, and the acquisition of source material from a distant empire

90 The early 20th-century revival is marked by the publication of: Max Eckert. *Die Kartenwissenschaft: Forschungen und Grundlagen zu einer Kartographie als Wissenschaft* [map science. Research and principles of cartography as science]. Berlin and Leipzig: de Gruyter, 1921/1915 [vols. 1 and 2].

91 While still working for Perthes until his death in 1902, Hassenstein carried out mapping jobs for other publishers. His papers at the Staatsbibliothek Berlin Preußischer Kulturbesitz offer insights into his engagement for Hans Meyer: Hassenstein provided the maps for Meyer's travel accounts to Mount Kilimanjaro. Besides provoking his employers to dismiss him from duty, this publishing job could be interesting in a conceptual way, for it probably shows the same attitude towards a mapped region, but with different material preconditions.

not under colonial rule, the publisher chose to focus on publications with reliable results. The journal issues as well as the multi-language editions of Perthes' *Stieler Hand-Atlas* needed to be published on time and with reasonable detail, accommodating their audience's expectations in order to reach the company's sales target. The private cartographic venture undertaken by Hassenstein differed greatly from government-issued imperial practices of mapping. The Gotha-based cartographer collated various source materials instead of basing his maps on recent government-issued surveys. Unable to read most of his sources, he had to rely on a number of informal supporters, drawing on Imperial Japan's promotion of the studies of its students abroad, which brought Hassenstein's translators to universities in Europe and thus joined his transimperial network.

Hassenstein's plan to publish romanised maps of Japan drew on recent political discussions, leading him to apply his firm's publishing strategies to a new cartographic space. Perthes' position in the imperial publishing landscape becomes strikingly clear in the discussions surrounding the publication of Hassenstein's unsuccessful *Atlas of Japan*. Instead of supporting stand-alone formats as proposed and initiated by Hassenstein, the publishing house further strengthened its publishing palette by focussing on serial publications that were easily adaptable to other languages and national contexts. Perthes tried to avoid high costs of new cartographic works, pooling its resources for securely established publications such as their universal *Stieler Hand-Atlas*. To crown it all, Hassenstein's maps were never even incorporated into the *Stieler*, thus allowing his almost decade-long engagement with Japan completely go to waste.

References

Unpublished sources

Forschungsbibliothek Gotha, Perthes Collection
FFA = Familien- und Firmenarchiv [family and company archives]
SPA = Sammlung Perthes Archiv [archives]
PGM = Schriftleitung Petermann's Geographische Mitteilungen [editorial correspondence for the journal *Petermann's Geographische Mitteilungen*]
SPK = Sammlung Perthes Kartensammlung [map collection]
Georg August University Göttingen, University Archives
Dekanat der Philosophischen Fakultät [arts faculty, dean's office]
Staatsbibliothek Berlin Preußischer Kulturbesitz, Handschriftenabteilung [manuscript division]
Nachlass Bruno Hassenstein [Bruno Hassenstein papers]

Published sources

Bartholomew, John George. The philosophy of map-making and the evolution of a great German atlas. In *Scottish Geographical Magazine* 18:1 (1902): 34–39.

Chamberlain, Basil Hall. *All things Japanese. Being notes on various subjects connected with Japan for the use of travellers and others*. 2nd ed. London: Kegan Paul/ Trench/ Trübner 1891

Coles, John. Japan. – Atlas von –. In *Proceedings of the Royal Geographical Society and monthly record of geography, new monthly series* 9:8 (1887): 526–530.

Day, Murray S. *Report of the trigonometrical survey of the island of Hokkaido for 1875*. New York: Francis Hart & Co., 1876.

Eckert, Max. *Die Kartenwissenschaft: Forschungen und Grundlagen zu einer Kartographie als Wissenschaft*. Berlin and Leipzig: de Gruyter, 1921/1915 [vols. 1 and 2].

Hassenstein, Bruno. *Atlas von Japan. Sieben Blätter im Massstabe von 1:1000000 und eine Übersichtskarte im Massstabe von 1:7500000*. Gotha: Justus Perthes, 1885/1887.

Hassenstein, Bruno. Die Umgegend der Bai von Tokio und des Vulkans Fuji-no-Yama. Mit Unterstützung von Prof. J. Rein und Tadashi Sanda gezeichnet von B. Hassenstein. In *Dr. A. Petermann's Mitteilungen aus Justus Perthes' Geographischer Anstalt* 25:X (1879): plate 19.

Justus Perthes' Geographische Anstalt. *Haupt-Katalog*. Gotha: Justus Perthes, 1915.

Justus Perthes' Geographische Anstalt. *Haupt-Katalog*. Gotha: Justus Perthes, 1935.

Justus Perthes' Geographische Anstalt. *Justus Perthes in Gotha, 1785–1885*. Gotha: Justus Perthes, 1889.

Knipping, Erwin. *In japanischen Diensten. Zwei Jahrzehnte eines preußischen Meteorologen in der ersten Hälfte der Meiji-Zeit (1868–1912)* [In the service of Japan. A Prussian meteorologist's two decades in the first half of the Meiji period (1868–1912)], Lohne: Cass 2014.

Knipping, Erwin. *Stanford's library map of Japan. Principally compiled from Japanese documents by E. Knipping*. London: Stanfords, 1879.

Knipping, Erwin. Ueber eine neue Karte von Japan und ihre Quellen. In Mitteilungen der Deutschen Gesellschaft für Natur- und Völkerkunde Ostasiens II, 11 (1876), 20–24.

Petermann, August. Vorwort. In *Mittheilungen aus Justus Perthes' Geographischer Anstalt über wichtige neue Erforschungen auf dem Gesamtgebiete der Geographie* 1:1 (1855): 1–2.

Pfizmaier, August. *Über japanische geographische Namen*. Wien: Karl Gerold's Sohn, 1875/1876.

Ratzel, Friedrich. Br. Hassenstein †. In *Dr. A. Petermann's Mitteilungen aus Justus Perthes' Geographischer Anstalt* 48 (1902): 1–5.

Rein, Johannes J. Der Fuji-no-yama und seine Besteigung. In *Dr. A. Petermann's Mitteilungen aus Justus Perthes' Geographischer Anstalt* 25:X (1879): 365–376.

Wenckstern, Friedrich von. *A bibliography of the Japanese empire. Being a classified list of all books, essays and maps in European languages relating to DAI NIHON (Great Japan) published in Europe, America and in the East from 1859–93 A.D. To which is added a facsimile-reprint of Léon Pagès, Bibliographie Japonaise depuis le XVe siècle jusqu'à 1859*. London: Kegan Paul/ Trench/ Trübner, 1895.

Literature

Bernstein, Andrew. Whose Fuji? Religion, region, and state in the fight for a national symbol. In *Monumenta Nipponica* 63:1 (2008): 51–99.

Brogiato, Heinz Peter. Baedeker und Stieler. Die Rolle des Verlagswesens zwischen Popularisierung und Professionalisierung der Geografie im 19. Jahrhundert. In *Wissenschaftsverlage zwischen Professionalisierung und Popularisierung*, Monika Estermann, Ute Schneider (ed.), 77–114. Wiesbaden: Harrassowitz, 2007 [Wolfenbütteler Schriften zur Geschichte des Buchwesens, vol. 41].

Burbank, Jane and Frederick Cooper. *Empires in world history. Power and the politics of difference*. Princeton, NJ: Princeton University Press, 2010.

Burnett, D. Graham. *Masters of all they surveyed. Exploration, geography, and a British El Dorado*. Chicago, IL and London: University of Chicago Press, 2000.

Conrad, Sebastian and Matthias Koch. Einleitung [introduction]. In *Johannes J. Rein. Briefe eines deutschen Geographen aus Japan 1873–1875*, Sebastian Conrad and Matthias Koch (eds.), 25–93. Munich: Iudicium, 2006 [Monographien aus dem Deutschen Institut für Japanstudien, vol. 40].

Demhardt, Imre Josef. *Der Erde ein Gesicht geben: Petermanns Geographische Mitteilungen und die Anfänge der modernen Geographie in Deutschland; Katalog zur Ausstellung der Universitäts- und Forschungsbibliothek Erfurt/Gotha im Spiegelsaal auf Schloß Friedenstein in Gotha, 23. Juni bis 9. Oktober 2005*. Gotha: Erfurt University 2006 [Veröffentlichungen der Forschungsbibliothek Gotha, vol. 42].

Dünne, Jörg. Portable Media und Weltverkehr. Die Taschenatlanten des Perthes Verlags. In *Die Werkstatt des Kartographen. Materialien und Praktiken visueller Welterzeugung*. Petra Weigel and Steffen Siegel (eds.), 185–203. Munich: Fink, 2011.

Earhart, H. Byron. *Mount Fuji. Icon of Japan*. Columbia, SC: University of South Carolina Press, 2011 [Studies in comparative religion].

Edney, Matthew H. *Mapping an empire. The geographical construction of British India, 1765–1843*. Chicago, IL and London: University of Chicago Press, 1997.

Espenhorst, Jürgen. *Petermann's planet. A guide to German handatlases and their siblings throughout the world, 1800–1950*. Schwerte, Germany/ Falls Church, VA: Pangaea, 2003.

Felsch, Philipp. *Wie August Petermann den Nordpol erfand*. Munich: Luchterhand, 2010.

Gordon, Andrew. *A modern history of Japan. From Tokugawa times to the present*. New York and Oxford: Oxford University Press, 2003.

Gottlieb, Nanette. *Kanji politics. Language policy and Japanese script*. London: Kegan Paul International, 1995.

Gugerli, David and Daniel Speich. *Topografien der Nation. Politik, kartografische Ordnung und Landschaft im 19. Jahrhundert*. Zurich: Chronos, 2002.

Güttler, Nils. *Das Kosmoskop. Karten und ihre Benutzer in der Pflanzengeographie des 19. Jahrhunderts*. Göttingen: Wallstein, 2014.

Güttler, Nils. Unsichtbare Hände. Die Koloristinnen des Perthes Verlags und die Verwissenschaftlichung der Kartographie im 19. Jahrhundert. In *Archiv für Geschichte des Buchwesens* 68 (2013): 133–153.

Hee, Nadin. *Imperiales Wissen und koloniale Gewalt: Japans Herrschaft in Taiwan 1895–1945*. Frankfurt on the Main: Campus Verlag 2015 [Globalgeschichte, vol. 11].

Hobsbawm, Eric. *The age of empire: 1875–1914*. London: Weidenfeld & Nicolson, 1987.

Jäger, Georg. Kartographischer Verlag. In *Geschichte des deutschen Buchhandels im 19. und 20. Jahrhundert* Georg Jäger (ed.). Frankfurt on the Main: Buchhändler-Vereinigung, 2001.
Jones, Hazel. *Live Machines: Hired Foreigners and Meiji Japan*. Tenterden, Kent: Paul Norbury, 1980.
Köhler, Franz. *Gothaer Wege in Geographie und Kartographie*. Gotha: Haack, 1987.
Matsutaro, Namba (ed.). *Old maps in Japan*. Osaka: Sogensha, 1973.
Meißner, Kristin. Responsivity within the context of informal imperialism. Oyatoi in Meiji Japan. In *Journal of Modern European History* 14:2 (2016): 268–289.
Raj, Kapil. *Relocating modern science. Circulation and the construction of knowledge in South Asia and Europe, 1650–1900*. Houndmills: Palgrave Macmillan, 2007.
Schär, Bernhard C. *Tropenliebe: Schweizer Naturforscher und niederländischer Imperialismus in Südostasien um 1900*. Frankfurt on the Main: Campus Verlag 2015 [Globalgeschichte, vol. 20].
Schelhaas, Bruno and Ute Wardenga. 'Inzwischen spricht die Karte für sich selbst.' Transformation von Wissen im Prozess der Kartenproduktion. In *Die Werkstatt des Kartographen. Materialien und Praktiken visueller Welterzeugung*, Petra Weigel, Steffen Siegel (eds.). Munich: Fink, 2011, 89–107.
Smits, Jan. *Petermann's maps. Carto-bibliography of the maps in Petermann's Geographische Mitteilungen 1855–1945*. 't Goy-Houten: Hes & De Graaf, 2004.
Traganou, Jilly. *The Tokaido road. Travelling and representation in Edo and Meiji Japan*. London and New York: Routlege Curzon, 2004.
Vaporis, Constantine Nomikos. *Breaking barriers. Travel and the state in early modern Japan*. Cambridge, MA: Harvard University Press, 1994.
Walter, Lutz (ed.). *Japan mit den Augen des Westens gesehen: gedruckte europäische Landkarten vom frühen 16. bis zum 19. Jahrhundert*. Munich: Prestel, 1993 [exhibition catalogue].
Whitfield, Peter. *The mapmakers. A history of Stanfords*. London: Compendium, 2003.
Wigen, Kären. *A malleable map. Geographies of restoration in central Japan 1600–1912*. Berkeley/Los Angeles, CA: University of California Press, 2010.
Wigen, Kären, Sugimoto Fumiko and Cary Karacas (eds.). *Cartographic Japan. A history in maps*. Chicago, IL and London: University of Chicago Press, 2016 [forthcoming].
Yonemoto, Marcia. *Mapping early modern Japan. Space, place, and culture in the Tokugawa period (1603–1868)*. Berkeley/Los Angeles, CA: University of California Press, 2003.

Figures

Fig. 1: Hassenstein's former supervisor and instructor August Petermann had made plans for a comprehensive map of Japan to be published as a supplement to the journal *Mittheilungen aus Justus Perthes' Geographischer Anstalt* über *wichtige neue Erforschungen auf dem Gesamtgebiete der Geographie* [transactions of Justus Perthes' Geographical Institute on important new research in the whole field of geography]. – Bruno Hassenstein. *Drei Entwürfe für eine Spezialkarte von Japan* [three concepts of a special map of Japan], 18781879 (Forschungsbibliothek Gotha, Perthes Collection, SPA ARCH PGM 150/1, 18–18a).

Fig. 2: Hassenstein's first map depicting Japan was reordered frequently, leading him to great expectations towards the success of his *Atlas of Japan*. The map was based on a Japanese source with a similar cartographic section, namely the popular tourist map 冨士見十三州輿地全圖 (*Fujimi jusanshu yochi no zenzu*, 'map of those thirteen provinces from which Mount Fuji can be viewed'), copies of which were owned by Hassenstein and several of his correspondents. – Bruno Hassenstein. "Die Umgegend der Bai von Tokio und des Vulkans Fuji-no-Yama. Mit Unterstützung von Prof. J. Rein und Tadashi Sanda gezeichnet von B. Hassenstein" [the environments of Tokyo Bay and of the volcano Fuji-no-yama. Drawn by B. Hassenstein with the assistance of Prof. J. Rein and Tadashi Sanda]. In *Dr. A. Petermann's Mitteilungen aus Justus Perthes' Geographischer Anstalt* 25:X (1879): plate 19 (Forschungsbibliothek Gotha, SPK 547-111978386).

Fig. 3: Cover page, page view and detail from Bruno Hassenstein's *Atlas of Japan* as preserved in the Perthes Collection (Forschungsbibliothek Gotha, SPB Japan No 60). – Bruno Hassenstein. *Atlas von Japan. Sieben Blätter im Massstabe von 1:1000000 und eine Übersichtskarte im Massstabe von 1:7500000* [Atlas of Japan. Seven sheets on a scale of 1:1,000,000 and one general chart on the scale of 1:7,500,000]. Gotha: Justus Perthes, 1885/1887, cover, plate IV, and detail of plate IV.

Stephen A. Walsh

Void into Meaning: Geophysics and Imperial Cartography in the High Arctic

Introduction

This chapter deals with a geographic space that is both very specific and diffuse: the High Arctic, that is, the uninhabited (and uninhabitable) extreme northern reaches of the planet. How did these polar regions become the spaces of competing imperialist projects and lands that could be meaningfully possessed? Up through the Enlightenment, polar exploration was a distinctly utilitarian enterprise and explorers typically saw little use in these frozen latitudes of the world, bereft of peoples, markets, and (for the most part) natural resources that were appreciable at the time.[1] None of this had changed by the First World War, yet polar exploration had become an avenue of global contention as men raced to plant imperial flags in the Earth's most far-flung and inhospitable northern regions. The High Arctic had become a meaningful theater of imperial acquisition and contention. Over this time period, then, what had changed in perceptions of this geographic space to allow such a radical transformation?

Since at least Francis Bacon (1561–1626), exploration and the growth of scientific knowledge about the planet had functioned as important agents of political legitimation. Exploration was, as Dane Kennedy argues, a way to demonstrate and/or claim membership in the exclusive club of first-rank imperial powers. Even when direct territorial aggrandizement was not at stake, exploration was still an assertion of a status at the cutting edge of knowledge production.[2] This meant that science played a key role in interpretations of arctic space.

The rise of mass data driven inductive sciences was a particularly important phenomenon in this regard. Starting in the second half of the nineteenth century, the collection of relatively isolated specimens came to be gradually eclipsed in polar sciences by the aggregation of large amounts of data obtained through increasingly complex and esoteric instrumental observations. This had tremendous repercussions for how humanity considered the utility of the High Arctic. It gave rise to new methods, scientific cultures, types of knowledge and objects

---

1 Glyn Williams. *Voyages of Delusion: The Northwest Passage in the Age of Reason*. London: HarperCollins, 2002, 141–142.

2 Dane Kennedy. Introduction: Reinterpreting Exploration. In *Reinterpreting Exploration: The West and the World* Dane Kennedy (ed.), 2d ed. Oxford: Oxford University Press, 2014.

DOI 10.1515/9783110418750-020

of study. It is my contention that *before* the polar regions could become lands that could be meaningfully possessed in the early twentieth century, they had to become areas of legitimate scientific interest in the nineteenth: the High Arctic had to become scientifically interesting (and therefore apprehensible) before it was socio-politically appropriable. Icebergs, for instance, became objects of glaciological research instead of just obstacles or hazards to trade. Seemingly barren expanses of rock and packed snow became zones teeming with geomagnetic and meteorological data. This transformation tremendously enlarged the amount of "useful" information that scientists could gather in the polar regions.

This chapter focuses on a case study in this development, an expedition conducted by officers and sailors of the Austro-Hungarian Navy to Jan Mayen, a small, barren and uninhabited Arctic island. Here they constructed and ran a year-long scientific research station from 1882 to 1883. This venture was part of a broader international scientific effort, the First International Polar Year, in which eleven countries established fourteen research stations in the high latitudes of the world to collaborate on simultaneous experiments and observations using the same standards and similar instruments for a twelve-month period. This expedition represents part of a turning point for the scientific and geographical appropriation of the Arctic.

Why were these men of the Austro-Hungarian *Kriegsmarine* there in the first place? Polities with immediate Arctic territorial claims (such as Great Britain and Russia), or commercial fleets with significant operations in northern waters (for instance, the United States or the German North Sea ports) had more readily apparent interests in the increase of Arctic knowledge. This was not the case in the Habsburg Monarchy. The black-and-gold flag of the Habsburgs flew at no Arctic trading post, nor was there an Austrian whaling fleet trawling the polar seas. Yet it is precisely Austria-Hungary's lack of direct economic or political interests in the Arctic that makes the link between exploration, science and imperialism especially palpable.

1 Jan Mayen

Jan Mayen is a small isolated island in the northern Atlantic, about 550 kilometers northeast of Iceland. Its landscape is typical of volcanic islands, featuring numerous craters and congealed lava flows. That volcano is the Beerenberg, a peak in the north that dominates the topography of Jan Mayen, reaching 2,277 meters

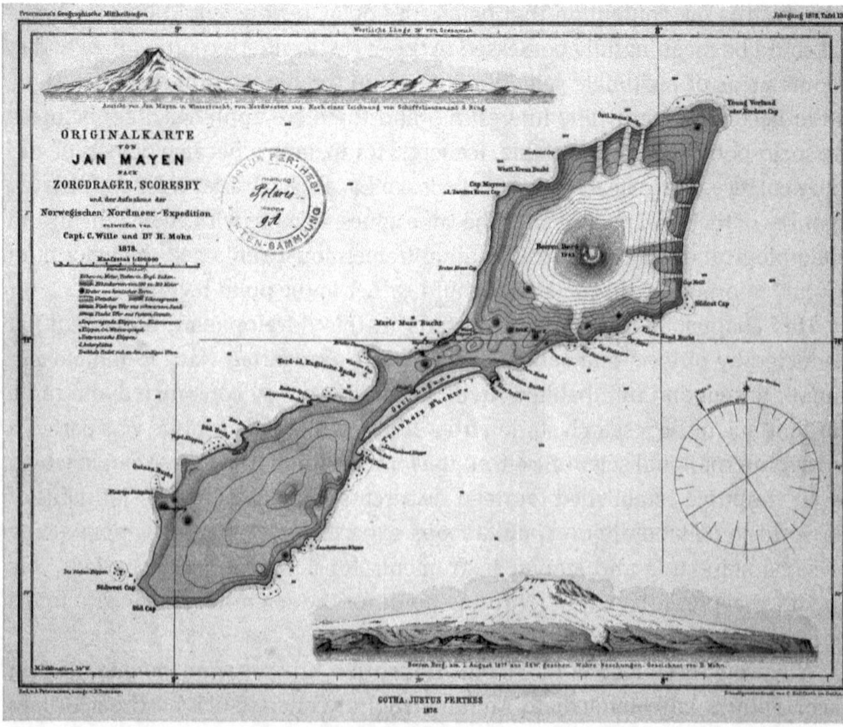

Map 1: C. Wille, H. Mohn. Original Map of Jan Mayen according to Zorgdrager, Scoresby and the Survey of the Norwegian North Sea Expedition.

above sea level.[3] Its rocky terrain offers little nourishment or shelter from the winds of the North Atlantic and its biosphere is consequently highly restricted.

In the early decades of the seventeenth century, Jan Mayen housed summer bases for Dutch whaling cartels, but the whalers quickly exhausted their prey in these waters, and Jan Mayen became largely abandoned by the 1640s. This development mirrored general western interest in Europe's highest waters. As whaling stocks in the European Arctic declined in the eighteenth century, so too did attention paid to the furthest latitudes of Europe. While there was always some limited sealing and fur trapping in areas such as Spitsbergen, the High Arctic offered little in terms of passages or meaningfully appropriable space once whales were no longer plentiful. From the perspective of empire in terms of spatiality and tem-

3 Krzysztof Birkenmajer. Geology of Jan Mayen and Surroundings: An Overview. In *Jan Mayen in Scientific Focus*, Stig Skreslet (ed.), 13–26. Dordrecht: Kluwer Academic Publishers, 2004.

porality, then, we see that in the Early Modern High Arctic, the nexus between the imperial and the availability of natural resources was critical.

Maps of Jan Mayen from this period reflect those resource-based concerns. For example, around 1670, the Dutch cartographer Johann Blaeu published a map of the island that was composed from a nautical perspective, emphasizing the coasts, whaling stations, and surrounding waters, giving little attention to the actual land of the island except for the Beerenberg. Aside from that mountain, early modern depictions of Jan Mayen's interior are largely speculative and abstract, not seriously attempting to represent real places but instead provide a mimetic shorthand for barren, infertile space, a disconnect indicative of a lack of concern.

It was not until the nineteenth century that Jan Mayen once again became an object of documented European interest. This shift began with the English whaler and polar scientist William Scoresby (1789–1857) who briefly visited Jan Mayen in 1817. He included a description of his time there in his 1820 work, *An Account of the Arctic Regions*, a book that has been called "one of the founding classics of polar science."[4]

Scoresby's version of polar science focused on an idea of "improvement" akin to agricultural reformers, with resource extraction (above all, whales) at the forefront.[5] The perspective of Scoresby's science was thus from the deck of a ship. On Jan Mayen, Scoresby and his crew conducted glaciological observations, and calculated the dimensions of various physical phenomena. But they primarily collected geological samples, took compass readings and made a few latitude and longitude calculations. In total, Scoresby spent about one day on the island, after which it seems there was no more worthwhile data to collect. This is not to imply that Scoresby was somehow "missing" important data that more professionalized scientists would have known how to take heed of, but that ideas of what constituted meaningful scientific data were gradually changing.

In 1856, Frederick Temple Blackwood, the Marquess of Dufferin, embarked upon an adventurous yacht cruise to the north, on an itinerary that included Jan Mayen. Blackwood's account of Jan Mayen is marked by romantic imagery of a land where polar ice "glimmered round the vessel like a circle of luminous phan-

4 Anthony Brandt. *The Man Who Ate His Boots: The Tragic History of the Search for the Northwest Passage*. New York: Alfred A. Knopf, 2010, 34. The best academic discussion of Scoresby is Michael Bravo. Geographies of Exploration and Improvement: William Scoresby and Arctic Whaling, 1782–1822. *Journal of Historical Geography* 32 (2006): 512–538. For Scoresby's journey to Jan Mayen, see: William Scoresby. *An Account of the Arctic Regions with a History and Description of the Northern Whale Fishery*, Vol. 1. Edinburgh: Archibald Constable and Co., 1820, 154–169.

5 Bravo. Geographies of Exploration, 518–519.

toms," and his journey indicates how technological improvements were beginning to open the Arctic to privileged, aristocratic adventure.[6] But in terms of scientific endeavor, Blackwood was content with stating, "It was now time to be off. As soon as we had collected some geological specimens, and duly christened the little cove... we walked back to the gig."[7] Such a remark concisely points to both the gradual "scientization" of polar travel and how perfunctory such science frequently was.[8] For all the sublime beauty of Jan Mayen, there was little to attract scientific interest past initial surface geological examinations.

In 1878, a new kind of mission visited Jan Mayen with the Norwegian North Sea Expedition, a venture to study the hydrology of the North Atlantic basin.[9] The expedition's leader was Henrik Mohn, Director of the Norwegian Meteorological Institute. In the summer of 1878, Mohn and his compatriots spent almost a week on and around Jan Mayen, where they conducted cartographic surveys and hydrographical measurements, in addition to collecting zoological, botanical and geological specimens. They also produced by far the most data-rich map of Jan Mayen hitherto rendered.[10] But its scope was still limited. The expedition spent a total of six days in the waters near Jan Mayen but was only able to land on two of them.[11] This is reflected in the map. Unlike most early modern cartographies of the island, this map does not feature mimetic, symbolic/imaginary landscape. Rather, it is an attempt to "scientifically" depict the island, that is, the extent of ice and snow cover, and its topography through the use of contour lines

6 Frederick Temple Blackwood. *Letters from High Latitudes: Being Some Account of a Voyage in 1856 in the Schooner Yacht "Foam" to Iceland, Jan Mayen and Spitzbergen*. London: John Murray, 1857, 209–226. For the development of a "Northern" variety of the Grand Tour in the Romantic era, see: Angela Byrne. *Geographies of the Romantic North: Science, Antiquarianism and Travel, 1790–1830*. New York: Palgrave Macmillan, 2013, 23–27.

7 Blackwood, *Letters from High Latitudes*, 211; 222.

8 I take the term "scientization," for lack of a better term, from: Sverker Sörlin. National and International Aspects of Cross-Boundary Science: Scientific Travel in the Eighteenth Century. In *Denationalizing Science: The Contexts of International Scientific Practice*, Elisabeth Crawford, Terry Shinn and Sverker Sörlin (eds.), 43–72. Dordrecht: Kluwer Academic Publishers, 1993.

9 See: The Norwegian-Atlantic Expedition. *Nature* 14:350 (July 13, 1876): 232.

10 Henrik Mohn. Die Reise der Norwegischen Nordmeer-Expedition nach Jan Mayen. *Petermann's Geographische Mittheilungen* 24 (1878): 228–235; Henrik Mohn and C. Wille. *Originalkarte von Jan Mayen nach Zorgdrager, Scoresby und der Aufnahme der Norwegischen Nordmeer-Expedition* [map]. In *Petermann's Geographische Mittheilungen* 24 (1878), Table 13.

11 Mohn reported first glimpsing Jan Mayen on July 28, 1878 and landing without difficulty. Afterwards, however, the expedition could only reach land on July 30, but circled the island conducting observations and looking for an opportunity to reach shore until August 3rd. Mohn. Die Reise der Norwegischen Nordmeer-Expedition, 228–231.

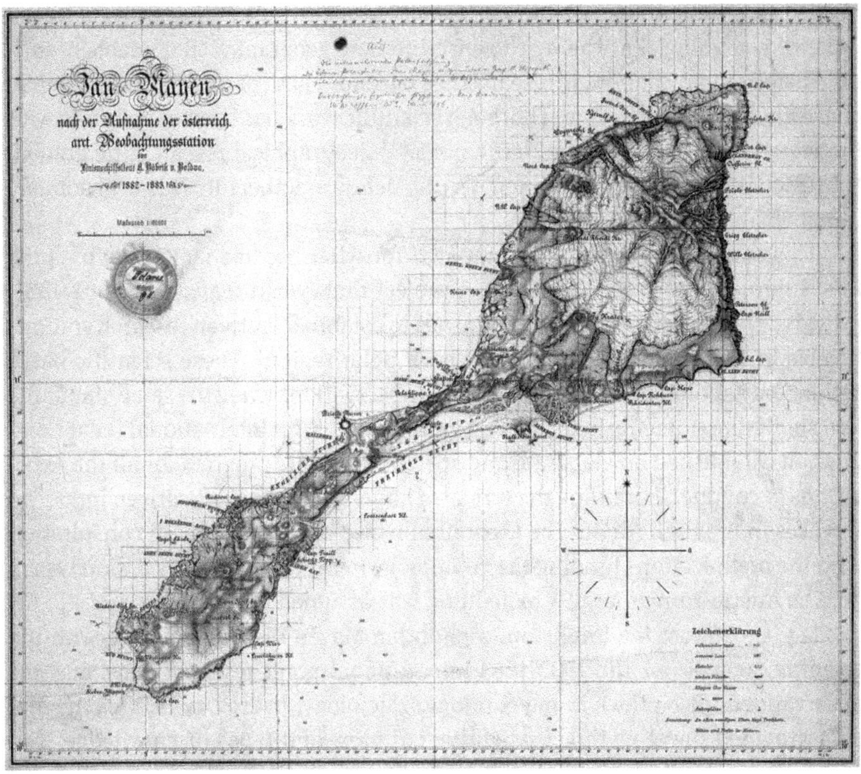

Map 2: Adolf Bóbrik von Boldva. Jan Mayen according to the Survey of the Austrian Arctic Observation Station.

and (to indicate steep slopes) hachure marks. Still, the expedition that produced this map was a mobile and largely maritime affair, and the map features at least as much hydrographical data (such as shoals off the coast) as it does data about the island itself.

2 The First International Polar Year (IPY)

In the meantime, a movement was surfacing that took aim at the geography-centric standard methodology of polar exploration, and argued that the practice needed to be placed on an allegedly firmer model of laboratory-based science. Carl Weyprecht (1838–1881), lieutenant in the Austro-Hungarian Navy and commander of the First Austro-Hungarian North Polar Expedition (1872–74), spearheaded this campaign.[12] Weyprecht dismissed polar exploration as it had been practiced, arguing that the lust to inscribe new geography "has reached such dimensions today that Arctic research has become a kind of international steeplechase toward the North Pole. Thorough scientific work has been supplanted with the mere victory over material difficulties."[13] Geographical research, he argued, "obstructs science and, as so often occurs, nearly smothers it. This is absolutely reprehensible."[14]

This conclusion led him to campaign for what became later known as the First International Polar Year (IPY), a project that, when realized in 1882–1883, provided the platform for eleven countries to establish fourteen research stations (twelve in the Arctic and two in the south polar region). These scientific bases conducted simultaneous experiments and observations using the same standards and similar instruments for a twelve-month period.[15] An International Polar Commission established an elaborate and specific scientific program for all the expeditions to conduct, heralding the arrival of large-scale mass-data driven inductive sciences in the High Arctic. The Commission also decided (albeit in consultation with the organizations funding the ventures) where each national mission was to go. The Austro-Hungarian IPY expedition was assigned Jan Mayen.

Let us contrast the brief stopovers on Jan Mayen discussed earlier with the scientific agenda for the 1882–1883 international polar research mission. New observational disciplines in meteorology, glaciology, hydrology and geomagnetism provided novel methods, scientific cultures, and types of knowledge. The

12 The most extensive modern narrative of the expedition (and the best synthesis) is: Enrico Mazzoli. *Dall'Adriatico ai Ghiacci. Ufficiali dell'Austria-Ungheria con i loro marinai istriani, fiumani e dalmati alla conquista dell'Artico*. Gorizia: Edizioni della Laguna, Museo dell'Antartide, 2003. On the reception of the Austro-Hungarian North Polar Expedition, see: Johan Schimanski and Ulrike Spring. *Passagiere des Eises. Polarhelden und arktische Diskurse 1874*. Cologne, Weimar, Vienna: Böhlau, 2015.

13 Carl Weyprecht. Über die österreichische Nordpol-Expedition. In *Tageblatt der 48. Versammlung deutscher Naturforscher und Ärzte in Graz vom 18. bis 24. September 1875*. J. Frischauf, V. Graber and R. Klemensiewicz (eds.), 39. Graz: Leuschner and Lubensky, k.k. Univ. Buchhandlung, 1875.

14 Weyprecht. Über die österreichische Nordpol-Expedition, 40.

15 Several edited volumes have recently helped ameliorate the scholarly neglect of the International Polar Year. See: Susan Barr and Cornelia Lüdecke (eds.). *The History of the International Polar Years (IPYs)*. Heidelberg: Springer-Verlag, 2010; Roger D. Launius, James Rodger Fleming and David H. DeVorkin (eds.). *Globalizing Polar Science: Reconsidering the International Polar and Geophysical Years*. New York: Palgrave Macmillan, 2010; Jessica M. Shadian and Monica Tennberg (eds.). *Legacies and Change in Polar Sciences: Historical, Legal and Political Reflections on the International Polar Year*. Burlington, VT: Ashgate Publishing Company, 2009.

International Polar Commission instructed that, for meteorology alone, the following observations should be carried out *every hour* for a year: readings from a thermometer fixed at least 1.5 meters above ground, barometric pressure, absolute and relative humidity, wind speed and direction, precipitation, and the formation of clouds in addition to their mass and direction of movement.[16] The geomagnetic program was even more ambitious. In addition to hourly synchronized readings of absolute declination, inclination and magnetic variation, the first and fifteenth days of each month were designated as "term days." This meant that for a twenty-three hour period every expedition would conduct these observations every five minutes simultaneously (according to Göttingen time) and every twenty seconds for one hour.[17] The International Polar Commission also assigned expeditions hourly observations of the *aurora borealis*, optional observations of galvanic currents, hydrographic measurements, and atmospheric electricity, in addition to the collection of zoological, botanical and geological specimens.[18] This expansion of categories of data and types of knowledge marked a fundamental change in how the High Arctic could be conceptualized, no longer just an obstruction or barren expanse, but an object of legitimate scientific enquiry and appropriation. The accumulation of data, in other words, became an end in itself, as opposed to means of discovering navigable passages and natural resources, and/or promoting commerce.

While this shift occurred across many disciplines, for the sake of clarity and space I would like to take geomagnetism as paradigmatic. The study of geomagnetism was the mass data observational inductive discipline *par excellence*

---

16 Protocolle der Verhandlungen der 3. Internationalen Polar-Conferenz in St. Petersburg, 1. Sitzung, August 1, 1881. Nachlaß Emil Wohlgemuth Nr. 4. Kriegsarchiv B/169. Nr. 4. Österreichisches Staatsarchiv. The collected hourly readings can be found in: Adolf Sobieczky. Meteorologie. In *Die internationale Polarforschung. Die österreichische Polarstation Jan Mayen, ausgerüstet durch Seine Excellenz Graf Hanns Wilczek, geleitet vom k.k. Corvetten-Capitän Emil Edlen von Wohlgemuth. Beobachtungs-Ergebnisse*, I. Band, IV Theil, 84–165. Vienna: Kaiserlich-Königliche Hof- und Staatsdruckerei, 1886. A host of other meteorological observations were also conducted, such as earth and sea temperature, but these readings were collected less frequently.

17 Protocolle der Verhandlungen der 3. Internationalen Polar-Conferenz in St. Petersburg, 2. Sitzung, August 3, 1881. Nachlaß Emil Wohlgemuth Nr. 4. Kriegsarchiv B/169. Nr. 4. Österreichisches Staatsarchiv. Magnetic declination is the angle between compass north and true north (as the Magnetic Pole differs from the North Pole). Inclination is the vertical angle made by magnetic instruments (as the Earth's geomagnetic field does not run parallel with the planet's surface).

18 Protocolle der Verhandlungen der 3. Internationalen Polar-Conferenz in St. Petersburg, 3. Sitzung, August 4, 1881. Nachlaß Emil Wohlgemuth Nr. 4. Kriegsarchiv B/169. Nr. 4. Österreichisches Staatsarchiv.

in the nineteenth century. The Earth's magnetic field is subject to almost every imaginable variable, including diurnal variation (change during a twenty-four hour period), secular variation (change during a period of a year or more), and a mutable pattern of geographic variation that is not dependent on the magnetic poles (which also migrate). In the nineteenth century, the only way to take all of these variations into account was through observing and recording as many of them as possible. Consequently, geomagnetism's fundamental research methodology was quantity.

A result of this was that the geographical space that made up Jan Mayen became described and categorized according to data that the discipline of geography did not typically take into account. The enormous number of geomagnetic observations, making up over a hundred pages of tables in the IPY expedition's final report, for example, serve as another kind of thick localized description and taxonomy of a place. But instead of the more familiar geographical vocabulary of mountains, rivers and provinces, it is expressed through the numerical vocabulary of geomagnetic inclination, declination and intensity. Data was not collected from sketches or sextant readings, but from the observation of devices such as a bifilar magnetometer, or declinatorium.

Consequently, the charts of geomagnetic data produced through the use of such instruments are a kind of spatial representation, as their aggregation was an attempt to comprehensively describe Jan Mayen's geomagnetic field, a type of geophysical data significantly defined by space/location (in addition to time), to map a type of geophysical feature of the island that only magnetic devices could discern. In this sense, going by Christian Jacob's broad definition of maps that emphasizes their function rather than formal characteristics, we might understand these collections of geomagnetic observations as a kind of cartography.[19] At the same time, such charts and collections of data can be considered a kind of descriptive geography, a genre that stretches back at least to Strabo.[20]

Despite Weyprecht's condemnations of detailed geographic research, the 1882–1883 Austrian expedition spent a significant portion of its resources and time on Jan Mayen to generate a detailed topographic survey of the island. The

---

19 Christian Jacob. *The Sovereign Map: Theoretical Approaches to Cartography through History*. Tom Conley, trans. Chicago: The University of Chicago Press, 2006, 17–21. For the difficulties of defining a "map," see: Jeremy W. Crampton. *Mapping: A Critical Introduction to Cartography and GIS*. Malden, MA: Wiley-Blackwell, 2010, 42–44.

20 For Strabo and descriptive geography, see: Denis Cosgrove. *Geography and Vision: Seeing, Imagining and Representing the World*. London and New York: I.B. Tauris and Co., Ltd., 2008, 5–8.

rise of mass data inductive science, as exemplified by geomagnetism, had a counterpart in mapmaking more traditionally conceived.

From 1882 to 1883, Adolf Bóbrik von Boldva, the officer in charge of cartography, conducted about half a dozen expeditions to survey land to the north and south of the research station, each lasting approximately one week. Progress was made on foot, or occasionally with a small boat. The men slept in a tent with a rubber floor, preferably placed upon a bed of moss. In the evening, hot pea soup would be served, followed by corned beef and perhaps green peas in addition to cheese and bread baked at the research station. Time for smoking tobacco came at the end of dinner. As the officers planned the next day's surveying, the sailors would attend to chores.[21] In such almost cozy conditions, Boldva and his assistants worked to construct a detailed, survey of Jan Mayen. This description points to an awareness of exploration as science: prescriptive measures, enforced order, and regularized procedures all sought to transform exploration, if not entirely into laboratory science, still into a practice in which reliable knowledge was produced and environmental conditions (that could adversely affect that goal) would be tamed. This meticulous surveying was part of a broader program for how the Austro-Hungarian expedition would engage with the temporality of the future: as a model, a new paradigm for how polar research would operate in times to come, a no longer mobile but localized practice, spatially fixed in laboratory/research stations.[22]

One of Boldva's most significant strategies for claiming legitimacy for his portrayal of Jan Mayen was in the publication of a tremendous amount of "behind the scenes" information concerning the data and calculations he used to produce his map. Equations and formulas are all methods of communication that are intertwined with forms of community and social identity.[23] They can express information, but also assert hierarchies of comprehension. The extent and complexity of the information Boldva provided with his map could equally shroud his cartography in the rhetoric of specialized, elite knowledge as well as elucidate his methodology.

Boldva did not reproduce all of his triangulations. But he did take his reader through the details of constructing one polygon, so that, in his words, "hope-

21 Emil Edlen von Wohlgemuth. Vorbericht. In *Die internationale Polarforschung*, 83.

22 Susan Barr, Louwrens Hacquebord, Erki Tammiksaar and Natal'ya Georgievna Sukhova. The Expeditions of the First International Polar Year, In *The History of the International Polar Years*, Susan Barr and Cornelia Lüdecke (eds.), 35–107. Heidelberg: Springer-Verlag, 2010, 39.

23 Theodore Porter. *Trust in Numbers: The Pursuit of Objectivity in Science and Public Life*. Princeton, NJ: Princeton University Press, 1995, viii.

fully everyone will be able to draw a conclusion about the weightiness of the individual points."[24] Boldva then went into a welter of charts and extremely complex trigonometric calculations, which might initially seem to give lie to his putative audience of "everyone." But he did not direct these equations to such a general audience without reason. For the trigonometric adept, such complex calculations would exemplify a host of salutary virtues, including methodological rigor and precision. For one without such advanced mathematical fluency the "actual" meaning of the calculations would be both unknowable and immaterial. For this reader, Boldva demonstrated his proficiency with advanced, and elite, epistemological tools, and granted his work the authority of arcane, specialized knowledge. The rhetoric of calculation is at least as important as the equations themselves. Legitimacy so claimed was by no means unique to cartography or to the First International Polar Year, but it assumes a particular prominence here because it was used to buttress the claim to "truth" of something readily apprehensible to the public at large, a map.[25]

Boldva's map of Jan Mayen is more a product of a relatively stationary and categorizing perspective than a record of exploration. I borrow this distinction from the influential exploration scholar Paul Carter, who distinguished between the ways an explorer and a surveyor experience geographic space. Unlike the explorer, it is not the goal of the traveling surveyor to find out what lays beyond the horizon. Instead, it was to impose characterizations and taxonomies upon the landscape.[26]

Differing from previous maps of Jan Mayen, the IPY map is not a broad outline of the island's major features. Rather, it is an ostensibly complete visual expression of Jan Mayen's topography. Hardly a millimeter of space exists on the map that is not inscribed with some form of geographic data and placement into the taxonomy of rock, glacier, volcanic sand and solidified lava. Carter asserts that it is the perspective of surveying that leads to appropriation, for the survey is the detailed map's staging ground. "The map was an instrument of interrogation, a form of spatial interview which made nature answer the invader's need for information."[27] Over time, we can see this "spatial interview" grow steadily more intensive, as the representational, mimetic depictions of place that are more

24 Adolf Bóbrik von Boldva, Aufnahme und Beschreibung der Insel Jan Mayen und Beobachtungen über Gletscherbewegung. In *Die internationale Polarforschung*, 11.

25 Boldva, Jan Mayen nach der Aufnahme der österreich. arct. Beobachtungsstation [map]. In *Internationale Polarforschung*, unnumbered table.

26 Paul Carter. *The Road to Botany Bay: An Essay in Spatial History*. New York: Alfred A. Knopf, 1988, 112.

27 Carter. *Road to Botany Bay*, 113.

symbolic than anything else were replaced by intensive surveying informed by mass data. Every centimeter of its geographical space was becoming an object of legitimate scientific interest.

This strategy had a corollary in the general overview of the expedition, written by its leader Emil Edlen von Wohlgemuth. Besides offering the expedition as a model for the future practice of polar exploration, Wohlgemuth also reached into the past, endeavoring to cast Jan Mayen as a space for historical events, both in the environmental and human sense. He took pains to establish the history of the island's discovery, noting that although 1611 was typically given as the year of its discovery, he personally found, at a museum in Bergen, a map that featured Jan Mayen from 1610, and that cartographical depictions of the island grew more accurate over the seventeenth and eighteenth centuries. Human temporality intersected with geological and glaciological temporalities as Wohlgemuth pointed out how it was possible to use those early modern mappings of Jan Mayen (which, he admitted, focused on coastal conditions) to trace the geographical changes, wrought by geology and volcanic eruptions, that had occurred on the island over the centuries. Space and time were thus in a dialogue with each other. Glaciological change had occurred as well, with the early modern Dutch cartography depicting three glaciers, while now five were apparent.[28]

Perhaps the most striking use of history, however, was the inclusion in the IPY results of a diary kept by seven Dutch sailors who had died of scurvy while attempting to winter on Jan Mayen between 1633 and 1634. Like the IPY expedition, this was no accidental overwintering, but a deliberate attempt to study arctic environmental conditions, and its inclusion came with the obvious but unspoken agenda of highlighting the dangers of what the IPY expedition, the next documented attempt to overwinter on Jan Mayen, faced.[29] It also served to situate the IPY expedition in the chronology of polar research, a legacy both stretching into the past and into the future. As Wohlgemuth wrote in the diary of his own expedition as it concluded:

> A year of exhilarating work has faded away! Now the storm, which covers all in this sound of lava (*Lavasund*), will also lay upon this place of honest activities, just as it buried the huts

28 Wohlgemuth, Vorbericht, 64–65.

29 Ein Tagebuch geführt von sieben Seeleuten, welche auf der Insel St. Maurice (Jan Mayen) bei Grönland in den Jahren 1633 bis 1634 überwintern und sämmtlich auf der Insel starben. In *Internationale Polarforschung*, 107–118. That including the diary was part of a deliberate agenda is indicated by its incorporation not just into the expedition's official results, but also in a shorter discussion of the Austrian IPY expedition published before its departure. See *Die österreichische Beobachtungs-Station auf Jan Mayen, 1882–1883*. Vienna: Verlag von Gerold and Co., 1882, 61–79.

> of the Dutchmen, just as it has been accustomed to do for centuries! Overhead the gloomy fog draws in – slowly – gravely – eternally? Gone! Gone! – Not forever, because in the near future the stations will return to life, worthy successors with better instruments will come, just as we took the place of those seven Dutchmen who succumbed to the trials of winter 250 years ago.[30]

As Jan Mayen became a space of legitimate scientific interest, it also became a space for a historical drama that had been ongoing and would continue: the positivist increase of knowledge. What had once been void, or at least dreary, uninteresting space that merited only representational/mimetic depiction, became meaning.

3 Empire in the High Arctic?

Does the idea of imperial appropriation help us understand what was happening on Jan Mayen between 1882 and 1883? When Boldva and his assistants camped out in their rubber-lined tent and measured the angles between mountaintops, were they acting as agents for some kind of Habsburg imperial vision, or an informal appendage of the Austro-Hungarian Empire?[31] At first, the inscription of place names in the map of Jan Mayen seems to suggest something along these lines. With few exceptions, the IPY expedition members had nothing to do with the place names they supposedly discovered. Instead, Count Hans Wilczek (who financed the expedition), back in Vienna, had the last word.[32] Scholars have cited the removal of local context as characteristic of western imperialism, and here we have a distant aristocrat with no experience of the place inscribing its place names.[33] Perhaps the most swaggering manifestation of the imperial in this cartography are the five "peaks" (*Spitze*) which were each christened after a member of the Habsburg dynasty.[34] This geographic category, with clear connotations of superiority, was wholly reserved for imperial blood.[35]

---

30 Wohlgemuth. Vorbericht, 52.

31 On the intersection between cartography and empire, see: Matthew Edney. *Mapping an Empire: The Geographical Construction of British India, 1765–1843*. Chicago: The University of Chicago Press, 1997.

32 Boldva, Aufnahme und Beschreibung, 23.

33 Carter, *Road to Botany Bay*, xiii–xxv.

34 The Habsburgs so honored are Emperor Franz Josef, Empress Elisabeth, Crown Prince Rudolf, his wife, Stéphanie, and Archduke Carl Stephan.

35 Others who received namesakes in mountainous terrain had to be content with the less prestigious, nautically-derived "Topp," such as Wilczek's friend and fellow aristocratic polar explor-

Yet, while not discounting the importance of toponyms, it must be said that an imperial appendage consisting primarily of place names without other forms of a lingering relationship between metropole and supposed periphery, this is imperialism of a highly informal nature. Michael Hardt and Antonio Negri remind us that we need not think of "empire" as a discrete political entity, and indeed, they go so far as to divorce the concept of imperial expansion from the workings of state organs.[36] While the politics of such assertions, in terms of describing actual historical agents and events, might be questionable, their diffuse idea of empire points toward some useful concepts for our purposes. Perhaps the "empire" in question here should be more broadly conceived, as international western science.

But to what extent did Jan Mayen really become cartographically known and fixed into imperial taxonomies as a result of the Austrian expedition? As detailed and precise as Boldva's map might purport to be, it, like all maps, depicts a land that never actually existed. Firstly, Boldva chose to use Mercator's projection, a system that employs parallel lines of longitude. On a globe, of course, lines of longitude are not parallel; they converge at the Poles. This projection therefore increasingly distorts land in the extreme latitudes of the world. Boldva acknowledged that this could result in distortions, especially in coastal formations, and applied corrections according to "numerous detailed measurements" made "only in accordance with [a given phenomenon's] own perception."[37] In other words, he sought to ensure that projection would not cause the land to look different than how he perceived it. Apart from that statement, however, here the curtain closes around Boldva's cartographic calculations. He provided no mathematical equations here to cancel out the untrustworthy subjective nature of the observer, no charts of allegedly objective data, merely the observer's own personal perception. Furthermore, Boldva's "detailed measurements" still did not overcome the inherently metaphorical nature of cartography. Even if Boldva had somehow produced what Lorraine Daston and Peter Galison would term a "mechanically objective" observation of a given phenomenon, he would still have needed to translate that observation into the visual language of the map.[38]

---

er Admiral Max von Sterneck, or with an entire mountain, such as the meteorologist Julius von Hann. This latter category might seem more prestigious at first, but it should be remembered that mountains touch the ground, while peaks float above.

36 Michael Hardt and Antonio Negri. *Empire*. Cambridge, MA: Harvard University Press, 2000, 166–167.

37 Boldva, Aufnahme und Beschreibung, 23.

38 Lorraine Daston and Peter Galison. *Objectivity*. New York: Zone Books, 2007, 121.

The subjectivity of Boldva's Jan Mayen map becomes all the more noticeable as he also admitted that he frequently departed from the "actual" appearance of the landscape. The simple fact that a 1:1 map is impossible means the cartographer inevitably faces questions about what to include and what to omit.[39] The careful attention Boldva paid to Jan Mayen's rocky topography meant that he deliberately excluded all vegetation from the map, privileging the geologist and dismissing the botanist. Driftwood, he wrote, was endemic on Jan Mayen's shores, and therefore only included when in "substantial" amounts. Boldva did not quantify this classification. The cartographic vision depicted on this map is also a seasonal one, limiting its depiction of snow cover to midsummer.[40] This is not to imply that the Austrian map of Jan Mayen was in any way egregiously "false," "subjective" or misleading, but to highlight the potential limits of imperial knowledge and appropriation through systematic surveying and cartographic expression.[41] The road from exploration and mapping to empire is not always a straight one.

Conclusion

The very amount of data generated by the IPY expeditions, as well as detailed published reports of this data, and the exhaustive descriptions of the procedures used to acquire this data, were unprecedented in polar research. They indicate a new level of engagement with the Arctic, a growing perception that the Arctic did not constitute an empty, meaningless space, but a realm teeming with data and legitimate objects of study. As this transpired, the Arctic also became an arena of international scientific competition as well as cooperation, paving the way for imperial rivalries. Even though at around the turn of the twentieth century such rivalries were best known in races to the North and South Pole, the model of the

39 On "cartographic silence," see: John Brian Harley. Silences and Secrecy: The Hidden Agenda of Cartography in Early Modern Europe. In John Brian Harley, *The New Nature of Maps: Essays in the History of Cartography*. Paul Laxton, ed., 84–107. Baltimore: The Johns Hopkins Press, 2001.

40 Boldva, Aufnahme und Beschreibung, 23. The seasonal view was also, implicitly, limited to just two mid-summers, whereas snow conditions could of course also be highly variable year-to-year.

41 On recent scholarship that questions the direct causation between exploration and empire, see: Kennedy, Introduction: Reinterpreting Exploration, in Kennedy, *Reinterpreting Exploration*, 1–18.

IPY, heavily based on the inductive sciences, provided a more enduring legacy for how humanity approached the High Arctic and other uninhabited spaces.

The IPY set the precedent for how polar science was to be conducted in the future – away from the romantic model of a solitary (and famous) explorer squaring off against the forces of nature and toward a model of corporate, coordinated scientific effort. As one polar authority put it in 1935, there was “a new style of exploration which involves the intensive cult of physical science at fixed stations as opposed to the old method of casual journeys through unknown country.”[42] And so the First International Polar Year led to the Second (1932–1933) and the Third, also known as the International Geophysical Year (1957–1958), which provided occasion for the launch of Sputnik. All this points to a significant investment, both financially and intellectually, in the study of the planet’s most frozen and inhospitable regions, a shift that we see manifested in the Austro-Hungarian IPY expedition’s map of Jan Mayen.

The twentieth century was also the timeframe in which formal imperialism in the High Arctic began. The government of Canada began attempting to exercise sovereignty in the most northern regions of North America in 1903, leading to a rise in tensions with the United States that would last decades.[43] In the 1920s, controversies erupted between the Soviet Union and the Scandinavian states over the uninhabited archipelagos north of Europe, such as Franz Josef Land, and those without indigenous populations, such as Spitsbergen (modern-day Svalbard).[44] Disputes over sovereignty in the High Arctic remain to this day.[45]

Although the divvying up of the High Arctic is not settled, the international community formalized the dynamic between empire and appropriation in the Arctic in 1982, with the ratification of the United Nations Convention on the Law of the Sea, passed in part to help govern competing claims upon the north polar ocean. Under this treaty, claims to Arctic space (and its subterranean resources) require the submission of vast amounts of scientific data to an international commission. Modern Arctic conquest does not come from an advancing army, or the planting of a flag, but through the mediation of international science.[46] Bruno

---

42 As quoted by: Janet Martin-Nielsen. *Eismitte in the Scientific Imagination: Knowledge and Politics at the Center of Greenland*. New York: Palgrave Macmillan, 2013, 22.

43 See: Nancy Fogelson. *Arctic Exploration and International Relations, 1900–1932*. Fairbanks: University of Alaska Press, 1992; W. Gillies Ross. Canadian Sovereignty in the Arctic: the ‘Neptune’ Expedition of 1903–4. *Arctic* 29:2 (1976): 87–104.

44 Charles Emmerson. *The Future History of the Arctic*. New York: PublicAffairs, 2010, 81–100.

45 See, for example: Denmark Challenges Russia and Canada over North Pole. *BBC News*, http://www.bbc.com/news/world-europe-30481309. Last accessed March 9, 2016.

46 Emmerson, *Future History*, 83–84.

Latour once famously described cartography as a way to "make domination at a distance feasible."[47] This of course has a significant element of truth. But for the space and time of the High Arctic in the modern (or post-modern) era, geophysics is the official language of empire. While the gathering or construction of scientific knowledge has been associated, at least implicitly, with imperial expansion for centuries, in this particular space and time this connection has become markedly explicit. For this state of affairs to come about, it was necessary that the High Arctic first be seen as apprehensible, as a scientifically worthwhile and legitimate object of inquiry. The efforts of the Austrian expedition on Jan Mayen are a case study of how that process began.

References

Barr, Susan, Louwrens Hacquebord, Erki Tammiksaar and Natal'ya Georgievna Sukhova. The Expeditions of the First International Polar Year, In *The History of the International Polar Years*, Susan Barr and Cornelia Lüdecke (eds.), 35–107. Heidelberg: Springer-Verlag, 2010.

Barr, Susan and Cornelia Lüdecke (eds.). *The History of the International Polar Years (IPYs)*. Heidelberg: Springer-Verlag, 2010.

Birkenmajer, Krzysztof. Geology of Jan Mayen and Surroundings: An Overview. In *Jan Mayen in Scientific Focus*, Stig Skreslet (ed.), 13–26. Dordrecht: Kluwer Academic Publishers, 2004.

Blackwood, Frederick Temple. *Letters from High Latitudes: Being Some Account of a Voyage in 1856 in the Schooner Yacht "Foam" to Iceland, Jan Mayen and Spitzbergen*. London: John Murray, 1857.

Brandt, Anthony. *The Man Who Ate His Boots: The Tragic History of the Search for the Northwest Passage*. New York: Alfred A. Knopf, 2010.

Bravo, Michael. "Geographies of Exploration and Improvement: William Scoresby and Arctic Whaling, 1782–1822." *Journal of Historical Geography* 32 (2006): 512–538.

Byrne, Angela. *Geographies of the Romantic North: Science, Antiquarianism and Travel, 1790–1830*. New York: Palgrave Macmillan, 2013.

Carter, Paul. *The Road to Botany Bay: An Essay in Spatial History*. New York: Alfred A. Knopf, 1988.

Cosgrove, Denis. *Geography and Vision: Seeing, Imagining and Representing the World*. London and New York: I.B. Tauris and Co., Ltd.

Crampton, Jeremy W. *Mapping: A Critical Introduction to Cartography and GIS*. Malden, MA: Wiley-Blackwell, 2010.

Crawford, Elisabeth, Terry Shinn and Sverker Sörlin (eds.). *Denationalizing Science: The Contexts of International Scientific Practice*. Dordrecht: Kluwer Academic Publishers, 1993.

Daston, Lorraine and Peter Galison. *Objectivity*. New York: Zone Books, 2007.

47 Bruno Latour. *Science in Action: How to Follow Scientists and Engineers through Society*. Milton Keynes: Open University Press, 1987, 223.

Denmark Challenges Russia and Canada over North Pole. BBC News, http://www.bbc.com/news/world-europe-30481309. Last accessed March 9, 2016.

Edney, Matthew. *Mapping an Empire: The Geographical Construction of British India, 1765–1843*. Chicago: The University of Chicago Press, 1997.

Emmerson, Charles. *The Future History of the Arctic*. New York: PublicAffairs, 2010.

Fogelson, Nancy. *Arctic Exploration and International Relations, 1900–1932*. Fairbanks: University of Alaska Press, 1992.

Frischauf, J., V. Graber and R. Klemensiewicz (eds.). *Tageblatt der 48. Versammlung deutscher Naturforscher und Ärzte in Graz vom 18. bis 24. September 1875*. Graz: Leuschner and Lubensky, k.k. Univ. Buchhandlung, 1875.

Hardt, Michael and Antonio Negri. *Empire*. Cambridge, MA: Harvard University Press, 2000.

Harley, John Brian. *The New Nature of Maps: Essays in the History of Cartography*. Paul Laxton (ed.). Baltimore: The Johns Hopkins Press, 2001.

Die internationale Polarforschung. Die österreichische Polarstation Jan Mayen, ausgerüstet durch Seine Excellenz Graf Hanns Wilczek, geleitet vom k.k. Corvetten-Capitän Emil Edlen von Wohlgemuth. Beobachtungs-Ergebnisse. Vienna: Kaiserlich-Königliche Hof- und Staatsdruckerei, 1886.

Jacob, Christian. *The Sovereign Map: Theoretical Approaches to Cartography through History*. Tom Conley, trans. Chicago: The University of Chicago Press, 2006.

Kennedy, Dane. (ed.). *Reinterpreting Exploration: The West and the World*. 2d ed. Oxford: Oxford University Press, 2014.

Latour, Bruno. *Science in Action: How to Follow Scientists and Engineers through Society*. Milton Keynes: Open University Press, 1987.

Launius, Roger D., James Rodger Fleming and David H. DeVorkin (eds.). *Globalizing Polar Science: Reconsidering the International Polar and Geophysical Years*. New York: Palgrave Macmillan, 2010.

Martin-Nielsen, Janet. *Eismitte in the Scientific Imagination: Knowledge and Politics at the Center of Greenland*. New York: Palgrave Macmillan, 2013.

Mazzoli, Enrico. *Dall'Adriatico ai Ghiacci. Ufficiali dell'Austria-Ungheria con i loro marinai istriani, fiumani e dalmati alla conquista dell'Artico*. Gorizia: Edizioni della Laguna, Museo dell'Antartide, 2003.

Mohn, Henrik. Die Reise der Norwegischen Nordmeer-Expedition nach Jan Mayen. *Petermann's Geographische Mittheilungen* 24 (1878): 228–235.

The Norwegian-Atlantic Expedition. *Nature* 14:350 (July 13, 1876): 232.

Die osterreichische Beobachtungs-Station auf Jan Mayen, 1882–1883. Vienna: Verlag von Gerold and Co., 1882.

Porter, Theodore. *Trust in Numbers: The Pursuit of Objectivity in Science and Public Life*. Princeton, NJ: Princeton University Press, 1995.

Ross, W. Gillies. Canadian Sovereignty in the Arctic: the 'Neptune' Expedition of 1903–4. *Arctic* 29:2 (1976): 87–104.

Schimanski, Johan and Ulrike Spring. *Passagiere des Eises. Polarhelden und arktische Diskurse 1874*. Cologne, Weimar, Vienna: Böhlau, 2015.

Scoresby, William. *An Account of the Arctic Regions with a History and Description of the Northern Whale Fishery*. Edinburgh: Archibald Constable and Company, 1820.

Shadian, Jessica M. and Monica Tennberg (eds.). *Legacies and Change in Polar Sciences: Historical, Legal and Political Reflections on the International Polar Year*. Burlington, VT: Ashgate Publishing Company, 2009.

Skreslet, Stig (ed.). *Jan Mayen in Scientific Focus*. Dordrecht: Kluwer Academic Publishers, 2004.

Sobieczky, Adolf. Meteorologie. In *Die internationale Polarforschung. Die österreichische Polarstation Jan Mayen, ausgerüstet durch Seine Excellenz Graf Hanns Wilczek, geleitet vom k.k. Corvetten-Capitän Emil Edlen von Wohlgemuth. Beobachtungs-Ergebnisse*, I. Band, IV Theil, 84–165. Vienna: Kaiserlich-Königliche Hof- und Staatsdruckerei, 1886.

Sörlin, Sverker. National and International Aspects of Cross-Boundary Science: Scientific Travel in the Eighteenth Century. In *Denationalizing Science: The Contexts of International Scientific Practice,* Elisabeth Crawford, Terry Shinn and Sverker Sörlin (eds.), 43–72.

Weyprecht, Carl. Über die österreichische Nordpol-Expedition. In *Tageblatt der 48. Versammlung deutscher Naturforscher und Ärzte in Graz vom 18. bis 24. September 1875*. J. Frischauf, V. Graber and R. Klemensiewicz (eds.), 39. Graz: Leuschner and Lubensky, k.k. Univ. Buchhandlung, 1875.

Williams, Glyn. *Voyages of Delusion: The Northwest Passage in the Age of Reason*. London: HarperCollins, 2002.

Figures

Map 1: C. Wille, H. Mohn. Original Map of Jan Mayen according to Zorgdrager, Scoresby and the Survey of the Norwegian North Sea Expedition. *Petermanns Geographische Mittheilungen* 24 (1878), Table 13 (Forschungsbibliothek Gotha, SPK 547-111953316).

Map 2: Adolf Bóbrik von Boldva. Jan Mayen according to the Survey of the Austrian Arctic Observation Station. In *Die international Polarforschung 1882–1883. Die österreichische Polarstation Jan Mayen*. Vol. 1. Vienna: Kaiserlich-Königliche Staatsdruckerei, 1886. Unnumbered table (Forschungsbibliothek Gotha, SPK 547-112213014)

Media Narratives on Königsberg/Kaliningrad: Spatiotemporalities of the Displaced

Holt Meyer and Charlton Payne

Introduction
Temporal and Spatial Displacement: German and Russian Persons, Names and Cities in Königsberg/Kaliningrad

The city of Kaliningrad and the surrounding Kaliningrad district, which exist as such (as a territory under Russian administration) only since 1946, are born of and defined by both spatial and temporal displacement. In the simplest sense, the spatial displacement involves – purely geographically speaking – the territory's shifting from being the most eastern part of the German empire to being the northern section of the most western part of the "Russian Republic" within the Soviet empire. This shift causes a multiple temporal displacement, the two main variants being the official Soviet discovery and/or invention of a history which has the 'ancient Slavic land' "returning home" (a game recently repeated by the Russian government in Crimea), and the expelled former inhabitants, mainly ethnic Germans, also displacing the current day by hallucinating the further existence of East Prussia in prose, documents, poetry and other text types. A novel such as Arno Surminski's *Sommer vierundvierzig oder Wie lange fährt man von Deutschland nach Ostpreußen* (*Summer Forty-Four, or How Long Does It Take to Get from Germany to East Prussia*), published in 1997, attests to the literary afterlife of Königsberg in German writing. Each of these variants of displacement is linked to a specific (nominally post-colonial) imperial model.

This section works through three cases of this spatial and temporal displacement and investigates in which sense it might be termed 'imperial'. Two of the cases involve sign production on the part of the post-1945 Soviet administration and both involve the modeling of space by means of literal shaping and naming. The other one comes from 'the other side' (also in the sense of being a voice of the dead) and retraces the administrative editing of overreaching narratives derived from individual accounts of the forcible displacement of the inhabitants of East Prussia who remained at the end of the "Third Reich" and were in their vast majority expelled from there (in the northern part of East Prussia by the very Soviet administration doing all the modeling, shaping and naming).

***

When hearing the concept of "displacement" in connection with recent European history, the first association might be that of "displaced persons" who lived in

DOI 10.1515/9783110418750-021

camps after WWII and stayed there until they could return home, if that home existed anymore, which it often did not. In his widely read book *Bloodlands*, Timothy Snyder also uses the term "displacement" for the massive movements of peoples as a result of the war "between Hitler and Stalin", remarking, for instance on "forced displacements during and after the war".[1] In Snyder's book, one finds formulations such as the following on "displacement" in this sense, this time referring to the policies of Wilhelm Kube, head of the "General Commissariat White Ruthenia" in the context of general Nazi policy according to which the "Slavs were meant to be starved and displaced".[2]

When Charlton Payne refers to "Post-Imperial Narratives of Displacement" in the title of his paper, he is referring to just this. Payne uses the term "displaced persons" frequently in his contribution, and his interests in the "methodology devised to create testimony by displaced persons in order to craft narratives about the transition to a post-imperial order after the war" takes a "particular focus on depictions of Königsberg/Kaliningrad" as a key area of physical human "displacement".

Timothy Snyder uses the term "displacement" in another sense when he speaks of how Stalinist propaganda during and after the war "displaced east European Jews from their historical position as victims of the Germans, and embedded them instead in an account of an imperialist conspiracy against communism".[3] As in the "German" case of making Germans into victims, resulting in downplaying of the actual victims, namely the Jews and other peoples displaced by the Nazi war machine, here too – mutatis mutandi – Stalinism appeals to the victimization narrative by casting "Russians" (or "peaceful Soviet citizens") in their role as direct and explicit objects of Nazi aggression. This goes so far as to practically eliminate the Jews from official Soviet narratives of Auschwitz. At the same time, late Stalinism appears in a strongly "Russified" version: Its re-signification practices try to render the "Russians" as by nature the real communists. One could ask here if this had a self-delegitimizing effect, or, alternatively, if we are seeing a more fundamental struggle (in the sense of hegemony) to define the terms of the victimization narrative after the war. In any case, multiple shifts of people and their descriptions can be reconstructed (and deconstructed) as a conceptual displacement of the dready physically displaced with the goal of fitting narratives of political programs.

---

1 Timothy Snyder. *Bloodlands. Europe between Hitler and Stalin*. New York: Basic Books, 2010, 331

2 Snyder, *Bloodlands*, 249.

3 Snyder, *Bloodlands*, 376.

In the closing remarks of Bert Hoppe's paper on urban planning in Kaliningrad after the Soviet conquest of the region – planning which aimed to create a miniature copy of Moscow but was never put in to place – we read about another kind of displacement. When Hoppe writes about the "radical plans that ultimately aimed at the creation of an urban clone", what he is describing is a displacement of a town's center from itself and its centuries-old cultural identity. He thus concludes that the center of the town on the edge of the Soviet empire was itself peripheral. The play of the central and the peripheral can be viewed in terms of Lotman's semiosphere, where the semiotic action which is most telling for the center takes place on the periphery.[4] The question arises as to what extent that activity in the "translation zone" or "contact zone" (Pratt[5]) of the periphery, so to speak, impacts the more rigid semiotic structures at the center. The displacement of the center of Kaliningrad adds a new twist to this dynamic, adding in the category of the 'center of the periphery as a copy of the center'.

Hoppe's argumentation is closer to the usage of "displacement" which can be found in Hardt and Negri's *Empire*, for it concerns not only physical, but also conceptual displacement.

Hardt and Negri apply the term "displacement" often as an alternative or development of the Marxist concept of "alienation".

They also apply the term to the area of sovereignty itself (arguably the key term of the entire book), particularly as concerns the nation-state:

> The process of constructing the nation, which renewed the concept of sovereignty and gave it a new definition, quickly became in each and every historical context an ideological nightmare. The crisis of modernity, which is the contradictory co-presence of the multitude and a power that wants to reduce it to the rule of one – that is, the co-presence of a new productive set of free subjectivities and a disciplinary power that wants to exploit it – is not finally pacified or resolved by the concept of nation, any more than it was by the concept of sovereignty or state. The nation can only mask the crisis ideologically, displace it, and defer its power.[6]

> When the nation-state does function as an institution of sovereignty, does it finally manage to resolve the crisis of modernity? Does the concept of the people and its biopolitical displacement of sovereignty succeed in shifting the terms and the terrain of the synthesis between constituent power and constituted power, and between the dynamic of productive forces and relations of production, in such a way as to carry us beyond the crisis?[7]

4 Cf. Juri Lotman, On The Semiosphere, trans. Wilma Clark. *Sign Systems Studies* 33:1 (2005): 205–226.

5 Mary Louise Pratt. Arts of the Contact Zone. *Profession* (1991): 33–40.

6 Michael Hardt, Antonio Negri. *Empire*, Cambridge: Harvard University Press, 2000, 97.

7 Hardt, Negri, *Empire*, 109.

The context of this last argument is the differentiation between nationalism as a context for anti-colonial struggle and nationalism of the "sovereign state". Hardt and Negri state categorically: "As soon as the nation begins to form as a sovereign state, its progressive functions all but vanish."[8] Linking this discussion to the historical configuration at hand, one might ask what displaced persons do to the concept of "the people" in this displacement of sovereignty. The displaced persons under Soviet administration on the way to the Kaliningrad region and the displaced inhabitants of the formerly Nazi-administered East Prussia clearing out and moving to the parts of Germany destined to remain under the limited sovereignty of the individual occupation zones which then became the FRG and the GDR are considered to be part of the sovereign nation based on the "people", framed in a republican-federal or "people's republican" sense. They come from a dissolved time and space, which was organized on the basis of a biopolitical imperialism of the "Volk". In this context we can reconsider Hardt and Negri's concept of "non-place" which allows for a further differentiation of the term "displacement":

> Rule is exercised directly over the movements of productive and cooperating subjectivities; institutions are formed and redefined continually according to the rhythm of these movements; and the topography of power no longer has to do primarily with spatial relations but is inscribed, rather, in the temporal displacements of subjectivities. Here we find once again the non-place of power that our analysis of sovereignty revealed earlier. The non-place is the site where the hybrid control functions of Empire are exercised.[9]

The concept of "temporal displacements of subjectivities" in its relation to the "non-place of power" is developed in the context of the "Empire" of the present which is summed up in the elegant phrase: "our postmodern Empire has no Rome".[10] The hybrid form of current Empire's "monarchic body" causes "temporal displacements" insofar as in the "imperial non-place, in the hybrid space that the constitutional process constructs, we still find the continuous and irrepressible presence of subjective movements" in time.[11]

When we are talking about the anti-imperial empire of the 1940s and the persons who are driven to wander (are "displaced" in the usual sense of the word), we are obviously not dealing with this kind of purely temporal displacement. There is a Rome and its name is Moscow (there are studies demonstrating

8 Hardt, Negri, *Empire*, 109.
9 Hardt, Negri, *Empire*, 319.
10 Hardt, Negri, *Empire*, 317.
11 Hardt, Negri, *Empire*, 319.

that "Moscow as the Third Rome" is much more a projection from Cold War times onto the past than it is an actual doctrine of premodern Russia[12]). Holt Meyer even argues that the spatiotemporal model of Moscow-led late Stalinism tends towards the opposite of "pure temporality", dissolving time itself (in the sense of real historical development) into a rigid teleology, which tends towards pure timeless space, since real events are rendered not only impossible, but unthinkable.

At the same time, the element of (theoretical) global internationality of the anti-imperialist imperial project (which supposedly steers the course of the "displaced persons" we are discussing) puts it at least conceptually beyond classically imperial colonialism based on the conventional military control of space. Those expelled ethnic Germans who go to the Federal Republic, i.e. the "West", are also confronting this purely theoretical international community while trying to be absorbed within a traditionally defined space of the modern state. Projects such as documenting, re-naming, re-locating, re-designing and so on try to harness the constituent intensities set into motion by the movements of displaced persons. The prefix "re-" indicates the temporal displacement enacted in these activities. These efforts to capture subjective movements in new forms of representation, and with a variety of methods and media, try to keep up with the dynamics unleashed by displacement but are themselves caught in a representational logic of control that relies on such categories as "the people" and the nation-state even as it tries to articulate an anti-imperial international order.

Charlton Payne addresses "victimization and loss" in such a manner as to evoke a further, though related, conception of displacement. He states as follows: "The narrative framework of victimization and loss that was so pervasive in the immediate postwar years held an enormous legitimating power of address and appeal, displaying a perlocutionary force that we can describe as itself imperial in its scope." Here he is in effect accounting for a "displacement of displacement", i.e. a conceptual renegotiation on the part of 'displaced Germans' which attempted to shift the blame for the beginning of the massive murderous displacements of the war but "could never entirely cover the traces" of those displacements' "violent origins" (thus the last words of Payne's contribution).

Holt Meyer attempts to address the concept of "displacement" itself, including the two definitions already mentioned (displacement of people and of sovereignty) and their variants and adding in a third: the displacement of time

12 Marshall Poe. Moscow, the Third Rome: the Origins and Transformations of a "Pivotal Moment". *Jahrbücher für Geschichte Osteuropas* 49:3 (2001): 412–429 (In Russian: Izobretenie kontseptsii "Moskva – Tretii Rim". *Ab Imperio. Teoriia i istoriia natsional'nostei i natsionalizma v postsovetskom prostranstve* 1: 2 (2000): 61–86).

itself in a totalitarian chronotope, which attempts to erase history itself. It is this aspect which, on the one hand, unites German "national socialist" and post-war Soviet Stalinist "nationalist socialist" (in Hardt and Negri's terms) strategies in the modeling of time and space, and on the other hand offers a territory on which one can work through the specificity of the latter. Hardt and Negri tend to stress the extreme similarity of the two political models and downplay the differences between Hitler's "national socialism" and what David Brandenberger calls "national Bolshevism".[13]

Meyer takes issue with this and puts forth the claim that the general totalitarian official displacement of time, the replacement of actual social and political history with an ideologically charged screen memory, is completely different when connected to (vulgar) Marxist teleology. This, he notes, is the case even if this teleology is so closely linked to extreme Russian nationalism that it seems to be a kind of echo of the fascism(s) which the USSR claimed to have eradicated from the face of the earth. The pseudo-emancipatory rhetoric of the USSR, which since the mid-30s is linked to an extreme emotionalization and "familiarization" of politics[14] and a vicious, mass-murderous one-man dictatorship, has specific qualities based, among other things, on the fact that "socialist realism" does not just embellish, but actually creates Soviet reality. Meyer demonstrates that the total renaming of toponyms in the northern part of East Prussia, which became the Kaliningrad district, is based not only on extreme nationalist Russification, but also on this general principle of "socialist realism". Precisely the temporary renaming of the small town near Insterburg (now Cherniakhovsk) called Puschdorf after Alexandr Pushkin, who in the late 40s retrospectively became practically a Soviet war hero, is a case in which all of these points come together. Hoppe's study, which addresses not as much naming as imaging (with much less chances of tangible results) and reconstructs a complex dynamic of displacement-to-center, echoes these practices – for Pushkin in the postwar USSR becomes (just after Stalin) the trace of the center par excellence.

All three cases demonstrate that "displacement" in the various meanings sketched out here further the investigation of "imperial spatiotemporality" particularly for the case of anti-colonial imperialism, which in turn seems to be a

13 David Brandenberger. *National Bolshevism: Stalinist Mass Culture and the Formation of Modern Russian National Identity, 1931–1956*. Cambridge: Harvard University Press, 2002.

14 Sabine Hänsgen. Die Familiarisierung des Politischen im Kino der Stalinzeit: Zur Filmkomödie als Genre sowjetischer Populärkultur. In *Neue Politikgeschichte. Perspektiven einer historischen Politikforschung* (= Historische Politikforschung 1), Ute Frevert, Heinz-Gerhard Haupt (eds.), 115–139. Frankfurt, New York: Campus, 2005.

variant or immediate precursor of Hardt and Negri's Empire of the late twentieth century.

References

Brandenberger, David. *National Bolshevism: Stalinist Mass Culture and the Formation of Modern Russian National Identity, 1931–1956*. Cambridge: Harvard University Press, 2002.
Hänsgen, Sabine. Die Familiarisierung des Politischen im Kino der Stalinzeit: Zur Filmkomödie als Genre sowjetischer Populärkultur. In *Neue Politikgeschichte. Perspektiven einer historischen Politikforschung* (= Historische Politikforschung 1), Ute Frevert, Heinz-Gerhard Haupt (eds.), 115–139. Frankfurt, New York: Campus, 2005.
Hardt, Michael and Antonio Negri. *Empire*, Cambridge: Harvard University Press, 2000.
Lotman, Juri. On The Semiosphere, trans. Wilma Clark. *Sign Systems Studies* 33:1 (2005): 205–226.
Poe, Marshall. Moscow, the Third Rome: the Origins and Transformations of a "Pivotal Moment". *Jahrbücher für Geschichte Osteuropas* 49:3 (2001): 412–429 (In Russian: Izobretenie kontseptsii "Moskva – Tretii Rim". *Ab Imperio. Teoriia i istoriia natsional'nostei i natsionalizma v postsovetskom prostranstve* 1: 2 (2000): 61–86).
Pratt, Mary Louise. Arts of the Contact Zone. *Profession* (1991): 33–40.
Snyder, Timothy. *Bloodlands. Europe between Hitler and Stalin*. New York: Basic Books, 2010.

Charlton Payne

Post-Imperial Narratives of Displacement in Germany around 1951

The Schieder Commission's Documentation of Displaced East Prussians and Hans Werner Richter's *Sie fielen aus Gottes Hand*

In the second issue, published in 1954, of the newly founded *Vierteljahrshefte für Zeitgeschichte* (VfZ), Martin Broszat explains the goals and methods of a massive documentation of the millions who fled or were forced out of their homes in "Ost-Mitteleuropa" at the end of WWII.[1] From 1951–54, Broszat was a member of the research team for the ambitious *Documents on the Expulsion of the Germans from Eastern-Central Europe* (*Dokumentation der Vertreibung der Deutschen aus Ost-Mitteleuropa*), a large-scale project initiated and funded by the Federal Ministry for Displaced Persons, Refugees and War Victims (*Bundesministerium für Vertriebene, Flüchtlinge und Kriegsgeschädigte*). The talented young doctoral candidate Broszat, who would go on to become one of Germany's preeminent scholars of the history of National Socialism, was instructed to provide a detailed account of the documentation project's methods for the VfZ by Hans Rothfels, one of the principle scholars commissioned to oversee the documentation when it was formed in 1951 and the main editor of the VfZ since the journal's founding in 1953. Broszat was well-suited for the task, since he was responsible for assessing the viability of the sources under the supervision of the project director, Theodor Schieder.[2]

Both Schieder and Rothfels had a personal connection to Königsberg. After completing his Habilitation under Friedrich Meinecke in 1923, Rothfels became Professor at the Albertus-Universität in Königsberg in 1926. He held that position until he was forced to retire in 1934 due to his Jewish descent. In 1939 he fled via England to the USA. In 1934, Theodor Schieder joined Rothfels in Königsberg in 1934, when he was hired as director of the "East Prussian Department of Postwar History" (*Landesstelle Ostpreußen für Nachkriegsgeschichte*) at the Albertus-Uni-

1 Martin Broszat. Massendokumentation als Methode zeitgeschichtlicher Forschung. *Vierteljahrshefte für Zeitgeschichte* 2 (1954): 202–213. Hereafter cited parenthetically. Here, as with all the other German texts, translations are mine.

2 For Broszat's role in the *Documentation*, see Mathias Beer. Martin Broszat und die Erfahrung der Dokumentation der Vertreibung. In *Martin Broszat, der "Staat Hitlers" und die Historisierung des Nationalsozialismus*, Norbert Frei (ed.), 43–59. Göttingen: Wallstein, 2007.

DOI 10.1515/9783110418750-022

versity. Rothfels was thus part of a recent imperial history situated at the frontier of the old East Prussian Empire and with its sights set on re-establishing and expanding its territories under National Socialist rule. In turn, and in contrast to Schieder, who joined the NSDAP in 1937 and went on to become one of the Reich experts assigned to prepare the war in Poland, Rothfels was also a victim of National Socialism's racially-motivated imperial designs. His personal fate thus intersects with the later desire to integrate the Federal Republic of Germany into a post-imperial world order. For the idea of suffering becomes one of the central topoi of the documentation project spearheaded by Schieder, as Broszat makes very clear in his claims for a post-imperial methodology of documenting effects of the imperial, which is the focus of my analysis in this essay.

I contend that in the case of the documentation of flight and expulsion from Königsberg, the representation of a victimized population of expelled Germans could be used to conceal the immediate Nazi past, discredit the Soviet Union's territorial claims, and prepare the Federal Republic's incorporation by NATO in one fell swoop, so that at the transitional moment in which Königsberg is reset as Kaliningrad (officially renamed on July 4, 1946), the city's recent history as an imperial outpost under the Gauleitung of Erich Koch could be transformed into an emblem of loss and Soviet conquest.[3] In this essay, I am primarily concerned with the methodology devised to create testimony by displaced persons in order to craft narratives about the transition to a post-imperial order after the war. With a particular focus on depictions of Königsberg/Kaliningrad, I will start with a close analysis of Broszat's exercise in methodological reflection on the problem of documenting testimony and then compare it with an alternative effort to create testimony of displacement in Germany at that time, the documentary novel *They Fell from God's Hand* (*Sie fielen aus Gottes Hand*) by Hans Werner Richter. My argument is that, despite all their differences, both documentary projects demonstrate how their efforts to gather testimony for the sake of a post-imperial account of the displacements caused by the imperial era still bear the ineradicable traces of what they strive to eradicate through their documentation. Both show that the act of documenting testimony can also entail concealment and the suppression of alternative voices. As these two efforts to document displacement demonstrate all too well, the documentation of certain testimony of displacement ensured that other testimony would not be documented due to the fact that these documentations were shaped by narratives of how postwar society should view the phenom-

3 I thus agree with and build upon the arguments in Robert G. Moeller. *War Stories. The Search for a Usable Past in the Federal Republic of Germany*. Berkeley, Los Angeles, London: University of California Press, 2001, especially 85.

enon of displacement in a society that wished to establish itself as post-imperial but was unsure of how to (re)write the record of its imperial past.

1 Postimperial Documentation of the Imperial

Rothfels prefaces Broszat's methodological report in the VfZ with a brief statement on the constitution, aims, and findings of the first two volumes documenting the expulsion of the German population from the territories east of the Oder-Neiße. Those first two volumes were published in the fall of 1953.[4] Rothfels' statement consists, for its part, of a lengthy quotation from the last paragraph of the editors' preface to the first volumes of the *Documentation*, in which the editors declare their allegiance as primarily to the scholarly "ethos" of historical research and politically to the tenets of the *Charter of German Expellees* (*Charta der deutschen Heimatvertriebenen*) which in 1950 renounced "vengeance and retribution" for the expulsions. The editors acknowledge that Germany largely contributed to the "disasters" of the last two decades, and they express their hope that their work on the documentation will help to prevent future expulsions and establish a "rearrangement of the relationships among peoples in this space".[5] This new order, born out of "new moral strength" gained through a responsible confrontation with the immediate past, the editors contend, is the precondition for the resolution of the tensions among the different peoples not only of "eastern-central Europe" but of Europe as a whole. Explicitly evoking the notion of suffering that serves as a common denominator of recent European imperial history within the continent and as a topos of the documentation project itself, the editors conclude that the birth of a new European order freed of national tensions and no longer plagued by expulsions would prove that "the unspeakable suffering of our generation does not remain meaningless"[6] The editors of the *Documentation* thus attribute meaning to the act of documentation itself by setting it within a nar-

4 Theodor Schieder. *Dokumentation der Vertreibung der Deutschen aus Ost-Mitteleuropa*. (*Die Vertreibung der deutschen Bevölkerung aus den Gebieten östlich der Oder-Neisse*, vol. I/1–2) Bundesministerium für Vertriebene, Flüchtlinge und Kriegsgeschädigte (eds.), in Verbindung mit Werner Conze, Adolf Diestelkamp, Rudolf Laun, Peter Rassow und Hans Rothfels. Bonn: Bundesministerium für Vertriebene, 1953.

5 "eine Neuordnung der Völkerbeziehungen in dem Raume, der zuletzt ein Inferno der Völker geworden war." Rothfels quoted in Martin Broszat, Massendokumentation als Methode zeitgeschichtlicher Forschung. *Vierteljahrshefte für Zeitgeschichte* 2 (1954): 202–213, here: 202.

6 "das unsagbare Leid unserer Generation nicht ganz sinnlos bleibt", Broszat, Massendokumentation, 202.

rative, told in the future anterior tense, about the recognition of past suffering and resolution of geopolitical conflict as the preconditions for the birth of a new political order guaranteeing peace and justice out of the ruins of war-torn Europe.

Their lofty declaration is ambivalent, to say the least. On the one hand, its renunciation of vengeance and retribution, as well its desire to prevent future expulsions, conjures the image of a world order predicated on peace and justice. On the other hand, while the "rearrangement of the relationships among peoples in this space" could be read as the desire to resolve conflict and affirm international consensus, the vagueness of this "rearrangement" suggests that the current arrangement provides insufficient grounds for international consensus in a new world order.[7] To phrase it differently, I interpret the documentation as invoking an international consensus that, if true to its principles, should not accept the validity or viability of the current arrangement of this particular geopolitical space. What is more, that appeal conjures at least two layers of spatiotemporality. On the one hand, the teleological claim about constituting a future of European peace and justice is underwritten, on the other hand, by the de-localized framework of an already constituted global world order, an empire, in Hardt and Negri's terms, which, as greater than Europe alone, is dedicated to establishing and guaranteeing peace throughout the world – and if necessary through NATO interventions.

The project editors thus evoke a temporality oriented teleologically toward a future which will render meaningful the history of suffering that occurred within a particular geographical space. And yet that future meaning will only be secured in the first place if what transpired with the arrival of Soviet troops is carefully documented and publicly perceived in a way that ensures a moral renewal of the inhabitants of that geopolitical space in particular and of Europe in general. In order to enter into a veritably post-imperial order, the recent imperial past (of suffering) has to be properly reconstructed; however, the reconstruction will only have proven valuable once the post-imperial order has been instituted. Moreover, if this reconstruction is to occur by means of a documentation, it requires a paper trail. However, it is Broszat's task to explicate a methodological problem of documentation and thereby legitimate the solution for it proposed by Schieder's commission of researchers. That methodological problem arises from the fact that there are very few reliable extant sources documenting the suffering of local populations at the hands of imperial political forces. While there are indeed official documents from this NS period, Broszat explains, many are either unavailable to

7 International consensus oriented toward the resolution of conflict is a primary feature of the world order described as empire by Hardt and Negri. (Michael Hardt, Antonio Negri. *Empire*. Cambridge (Mass.), London: Harvard University Press, 2000, 10 and 14.)

the historian of contemporary history or are unreliable due to the saturation of government "jargon" by ideology and propaganda (203). Furthermore, in an age in which the "mass of the population", as Broszat phrases it, either actively participates in events as in democratic regimes or is merely prompted to supply plebeian acclamation for totalitarian regimes, the wider population has in any case become both subject and object of political history (203). Modern historiography, he continues, has to shift its attention to the ways that political events inform the collective fate of a given populace (203). Yet in a decisive move, which is based on the fundamental distinction between democratic and totalitarian forms of participation, this also means emphasizing how groups have *suffered* under the reign of a totalitarian regime. Such groups might have become the subject of history, but they have also had to bear the burdens caused by those who have led and misled them: "as led and misled, as those who have passively suffered the course of history".[8] By arguing in the patois of "seduction" (*Verführung*), of populations being misled in the sense of seduced, Schieder's scholars are latching onto the post-imperial myth behind which the Federal Republic could hide a fresh Nazi past, namely the narrative of a people unlovingly seduced into an abhorrent war by Hitler.[9]

But this argumentative logic leads back once more to the methodological problem of a missing paper trail. Since the existing government documents are, for the aforementioned reasons, inadequate historical sources for the reconstruction of the collective fates of recent history, other sources must be discovered and referenced, and yet such alternative sources do not exist. The teleology of this method thus not only anticipates a future outcome that would confer meaning on recent imperial history from the standpoint of a post-imperial order but also requires the construction of the very documents that are to serve as evidence of collective fates:

> the historian's task has thus expanded to the extent that he not only classifies and verifies extant sources but must himself see to it that, for certain events in the recent past, source material comes into being in the first place.[10]

---

8 "als Geführte und Verführte, als Leidende am Gang der Geschichte beteiligt". Broszat, Massendokumentation, 203–204

9 See Klaus Theweleit. *Pocahontas*, vol. 4, *„You give me fever". Arno Schmidt. Seelandschaft mit Pocahaontas. Die Sexualität schreiben nach WWII.* Frankfurt/Main: Stroemfeld, Roter Stern, 1999, 211.

10 "die Aufgabe des Historikers wird sich insofern zu erweitern haben, als er nicht mehr nur als Sichtender und Prüfender schon vorliegenden Quellen gegenübersteht, sondern selbst dazu beitragen muß, daß für bestimmte Vorgänge jüngst vergangener Epochen überhaupt Quellenmaterial zustande kommt." Broszat, Massendokumentation, 204.

A paper trail of what happened will have to be created by the very scholars who are documenting and thereby providing an account of recent events. The methodological problem for this particular practice of contemporary history (*Zeitgeschichte*) derives of course not only from its teleological argumentation, but also from the pragmatic consequences of trying to account for events which transpired less than ten years ago. At the same time, the documentation project reveals just how intertwined it is with contemporary political concerns in an era and geopolitical regime that wishes to define itself as post-imperial; not surprisingly, the documentation was entrenched in contemporary political designs. As a documentation and research project initiated by Adenauer's *Federal Ministry for Displaced Persons, Refugees and War Victims*, its purpose was to supply the hitherto missing official documentation of suffering (German) populations, as retroactively defined, of course, by the newly founded Federal Republic of Germany which sought to identify itself as part of a post-imperial geopolitical order.

The declared intention of Adenauer's Ministry and the participating commission of historians – many of whom, like Rothfels and Schieder, were directly linked to the outpost of German imperialism in Königsberg – was to supply documentary evidence that the expulsions of Germans from the eastern territories occurred in violation of the Potsdam Accord. In the Potsdam Agreement signed on 2 August 1945, where Attlee and Truman supported the annexation of Königsberg by the Soviet Union, the big three decreed that the transfer of German populations from Hungary, Czechoslovakia, and Poland should be conducted in an "orderly and humane manner."[11] There was even an initial hope among different ministers in Adenauer's regime that the documented material could then be submitted as evidence in support of a challenge to the Oder-Neiße border, which had been established by the Potsdam Agreement, should later negotiations of a peace treaty be conducted.[12] The effort to deliver "clear knowledge of the process of expulsion" on the basis of reliable sources, which first have to be constructed as such,[13] is thus to a significant degree made in the service of territorial claims. Thus, the documentation of displacement and territorial loss harbors remnants of imperial ambitions. Those traces of the imperial past haunt the Federal Republic's appeal to a new European order which defines itself as committed to a human rights

---

11 A thorough, recent account of the expulsions is provided by R. M. Douglas. *Orderly and Humane. The Expulsion of the Germans after the Second World War.* New Haven: Yale University Press, 2012.

12 Eva Hahn, Hans Henning Hahn. *Die Vertreibung im deutschen Erinnern. Legenden, Mythos, Geschichte*. Paderborn: Ferdinand Schöningh, 2010, 460.

13 Broszat, Massendokumentation, 204.

regime predicated on the prevention or alleviation of the suffering of individuals and populations as a result of political conflicts over the spaces and identities of nation-states.

This is a documentation that both wants to create a paper trail of suffering with which the Adenauer regime could wield charges of human rights abuse against the Soviets and for that reason needs at the same time to conceal the traces of German imperial territorial conquest. That tension is reflected in the scholars' procedures of documentation, in the form of a balancing act, held in the posture at least of scholarly reliability, between the "color" of personal testimony, as Broszat calls it, and the verifiable reconstruction of events. It is the task of the contemporary historian of expulsion east of the Oder-Neiße line to ensure that in the project's primary documentary genre of the "report of personal experience" or "personal testimony" (*Erlebnisbericht*), in Broszat's words, the "point of view of authentic representation had to be taken into account and all possible misrepresentation had to be excluded".[14] These "personal testimonies" were reconstructed from diaries, letters, reports and questionnaires (*Befragungsprotokolle*). Due presumably to the extreme nature of the circumstances of expulsion, however, there is a paucity of diaries and letters for the historians to consult, which is why the majority of documents consists of personal testimonies reconstructed from the memory of refugees and eyewitnesses at the behest of the researchers. The refugees and expellees were prompted to either write their own "chronicle-like" reports or to answer specific questions that were then recorded by the researchers. Whereas "self-written reports" supplied the illustrative "color" of originality, the "examinations" (*Vernehmungsprotokolle*) enabled the documenting scholars to establish specific historical details which were unknown to most participants and hence usually left out of the self-written reports.[15] Broszat claims that, despite the dubious quality of protocols of memory, owing to "forgetfulness, embellishment, the creation of legend" that can distort the recollection of events that have transpired as much as seven years before,[16] many of the eyewitnesses were able to recall the time and place of incidents during their escape and expulsion with astonishing detail.[17] The editors of the *Documentation* underscore in their preface to the first volume that the "plasticity of the representation" of a

14 Broszat, Massendokumentation, 205.

15 Broszat, Massendokumentation, 207–208.

16 The circulation of legends around and among the expellees was observed a few years later by Alfred Karasek-Langer and analyzed in his essay: Volkskundliche Erkenntnisse aus der Vertreibung und Eingliederung der Ostdeutschen. *Jahrbuch für Volkskunde der Heimatvertriebenen* 1 (1955): 11–65, esp. 18–25.

17 Broszat, Massendokumentation, 208.

personal testimony is best achieved by precise references to time and place of events.[18] Of course, the opportunity to put these experiences into words also encouraged many to embellish their self-written statements with a high degree of “subjective impulses”, which the documenting scholars sought to regulate based on an assessment of the severity of reported experiences or based on the likelihood of their taking flights of fancy as a matter of occupational habit, as in the case of those refugees and expellees who were professional writers or journalists, or those from other professions with a proclivity for writing, such as teachers or pastors.[19] This mechanism of control obviously did not prevent them from later publishing Hans von Lehndorff’s *Report from East and West Prussia 1945–1947* (*Bericht aus Ost-und Westpreußen 1945–1947*) in 1960, in which the surgeon told of his care for the wounded during the battle of Königsberg in 1945 and of his commitment to his Protestant Christian beliefs. Lehndorff was praised as both a gifted writer and conscientious objector to National Socialism.[20] A year after the publication of his account by the Schieder documentation, his text was published separately and as of 2010 has been reprinted thirty-one times.

Broszat’s exposé presents the documentation as seeking a balance between historical accuracy and subjective emotional resonance, between genuine accounts of suffering and accusation or panegyric,[21] between (potentially overeager) willingness to put an account on paper and reticence, the initiative of those directly affected and the needs of the documenting scholars, but also between exceptional individuals and their stories and representative demographic cross-sections. The desire to create a documentation of representative instances of a general phenomenon motivates the entire enterprise – and it is one of its most dubious features. The editors state this objective clearly in their preface to the first volume, when they write that it is their aim to represent the “entire process of expulsion, in all of its phases, local conditions and variety of occurrences and attendant circumstances that emerged in the process.”[22] This objective renders the “Erlebnisberichte” as testimony in two senses: both in the legal sense of providing evidence, in this case for the argument that the expulsions violated the terms of the Potsdam Accord, and in the broader sense of “the word

18 Schieder, *Dokumentation*, Vol. I.1, III–IV

19 Broszat, Massendokumentation, 205–206.

20 See: Kyrie statt Heil. *Der Spiegel* 35 (1962): 28–30.

21 Broszat, Massendokumentation, 206.

22 “Gesamtvorgang der Vertreibung, d.h. in allen seinen zeitlichen Abschnitten, örtlichen Bedingtheiten und der Vielzahl der dabei auftretenden Erscheinungen und Begleitumstände”. Schieder, *Dokumentation*, Vol. I/1, IV.

of another providing a source of knowledge."[23] The testimonies recorded by the documenters provide knowledge about what happened to individual Germans in Königsberg and elsewhere on the other side of the Oder-Neisse; that knowledge about particular instances of suffering was then packaged to be used in a legalistic sense as evidence for an "entire process of expulsion."

However, the boldest leap occurs when the referential status of this testimony is made to adopt a fictional quality. This happens when the testimony is not only taken as evidence for certain events occurring to specific persons at times and places which could all potentially be verified. The documenters inadvertently risk lending this testimony attributes conventionally identified with the works of narrative fiction by insisting on the representative quality of the narrative reports. Individual testimony is made to stand for the experiences of an entire group and the knowledge to be gained from that testimony wavers between verifiable fact and historical significance. According to the logic of this procedure designed to create testimony in its twofold sense, the accumulation of great numbers of testimonies should, through a process of sorting, assessing, and excluding, result in a reduction of complexity for the sake of creating exemplarity. An array of individual reports are then assembled as cases, to next be sorted and arranged into an account of "the historical fate of individual groups",[24] which in turn becomes representative of the fate of the "Germans in the east" as ultimately yet another example of a century of persecution, ethnic cleansing and mass expulsion.[25] The Germans expelled from Königsberg are thus cast as representatives of not only entire German populations formerly residing in "eastern-central Europe" but of the many populations displaced by geopolitical rearrangements that have taken place there (and elsewhere) in recent history. In his narrative introduction to the first two volumes of the *Documentation* focused on the expulsions east of the Oder-Neisse, Broszat underscores that these expulsions are a "final link in a long chain of rearrangements on the political and ethnographic map of east-central Europe" that he interprets as "the ultimate radicalization and the end of the idea of the nation-state".[26] He thus situ-

23 Leona Toker draws on this distinction in her discussion of the status of Gulag memoirs and novels as testimony. Her discussion about the referential and fictional aspects of testimony raises interesting questions when applied to the testimony crafted by the Schieder commission. (Leona Toker. *Documentation. Return from the Archipelago. Narratives of Gulag Survivors*. Bloomington and Indianapolis: Indiana University Press, 2000, 124.)

24 Broszat, Massendokumentation, 203.

25 Declared, for instance, in: Carl Wingenroth. Das Jahrhundert der Flüchtlinge. *Außenpolitik* 10:8 (1959): 491–499.

26 *Dokumentation der Vertreibung*, Vol. 1/1, quoted in Beer, Martin Broszat und die Dokumentation, 50.

ates the expelled Germans at the end of an imperial order based on the system of nation-states and responsible for numerous fatalities, but also presents them as beacons of a new world order with the potential to guarantee peace and justice.

While this method of documentation enacts a procedure of reduction, it is also productive and self-referential. The leveling of temporal and spatial differences described above is productive of the group identities whose tracks it set out to (dis)cover as those of "expellees" (*Heimatvertriebene*). The method of verification is self-referential, too, because the accuracy of statements is checked by comparing them with the other statements being put on record. Descriptions about the time and place of events are compared with each other to establish their consistency and verifiability.[27] This comparative method runs into the problem of hearsay or of a "later creation of legends" that can retrospectively influence the recollection of events. Broszat explains that such factors are supposedly recognizable in reports through the use of stereotypes or the emergence of strikingly similar accounts of events. Furthermore, since not all events are comparable, some statements cannot be compared with others and hence the veracity of certain statements cannot be confirmed with reference to others, in which case it falls upon the documenters to assess the internal plausibility of a statement, a procedure that Broszat likens to the forensic paradigm of a "criminological investigation".[28] Such a report would then be tested for signs of "conscientiousness" on the manifest level of its rhetorical expression. As examples of lacking credibility, Broszat lists the use of stereotypes, hyperbole, or (as in the case of many local officials) an apologetic tone with regard to one's own actions, or other signs of tendentiousness, but also the use of clichés or the expression of too much pathos could disqualify an account. "Awkwardness of expression",[29] on the other hand, might carry a degree of evidentiary weight for the team of scholarly editors, as in the case of document number six from the first volume, a report by an ethnic German (*Volksdeutsche*) woman (Wanda Dijer) from the district of Sierpe in Poland, which the editors reproduce with its linguistic imperfections (which are impossible to retain in translation):

> until the eighteenth of January, I was in my homeland on my own piece of land in Borkenhagen in the district of Sichelberg. On January 21, 1945, the Russians reached us with our trek in Jaschen. The Poles were also there and stole our belongings that we took with us in our

27 Broszat, Massendokumentation, 210.
28 Broszat, Massendokumentation, 210.
29 Broszat, Massendokumentation, 211.

mule-driven carts, so that me and my two kids, Edith, born on July 4, 1936, and Helmuth, born on May 24, 1942, were left with only a shirt and skirt.[30]

The “naturalness [almost in the sense of ethnicity (C.P.)!] and uprightness” (“Urwüchsigkeit und Biederkeit”) of such reports are said to contain a valuable documentary quality of their own.[31] Leaving aside the questionable use of tainted vocabulary like “naturalness” (“Urwüchsigkeit”) in this context, I can only insinuate that it is itself tendentious for Broszat to cite examples of lacking credibility from reports of events in Yugoslavia and the Sudetenland, whereas his examples of concretely illustrative depictions are unquestionably taken from the volume documenting the region east of the Oder-Neiße line. In any case, Broszat neither includes in his methodological exposé examples of the questions that were posed nor of templates used for protocols or self-written reports.

What image of Königsberg does the Schieder documentation convey? Separated from the rest of the German Reich by the Polish corridor established by the Treaty of Versailles, Königsberg was the urban exclave of Eastern Prussia and hence the easternmost city of the German Reich. That image of an exclave is reproduced in the accounts of Königsberg as an isolated outpost surrounded by attacking Russian forces. Air raids had already destroyed much of the city by the end of August 1944. After January of 1945, the city becomes an island on the eastern front populated by evacuees and refugees as the Soviet army encircles the city. The testimony of Eva Kuckuck from Königsberg, for instance, recounts this sense of being closed in upon: “The ring around Königsberg became tighter and tighter, the sound of canon fire increasingly audible from one day to the next”.[32] Prior to the encirclement, several accounts by inhabitants of other East Prussian cities and villages refer to Königsberg as a bastion of the incompetent leadership of Gauleiter Erich Koch, who refused to organize mass transports of the population to safety after the Russian army had already reached East Prussia in Fall of 1944.[33] Afterwards, Königsberg is depicted as an encampment of injured and threatened refugees desperately seeking to escape over Das Frische Haff, until it becomes the

30 “bis zum 18. Januar war ich in meiner Heimat auf meinem eigenem Grundstik in Borkenhagen, Kr. Sichelberg. Am 21. Januar 1945 erreichten uns die Russen mit unserem Treck in Jaschen. Da waren auch Gleich die Poler da und beraubten unser restliches Hab und Gut, das wir noch mit unseren Geschpanen mit uns führten, so daß ich und meine zwei Kinder, Edith, geb. am 4. Juli 1936, und Helmuth, geb. am 24. Mai 1942, Nur im Hemd und einem Kleit Stehen bliben.” Broszat, Massendokumentation, 212.

31 Broszat, Massendokumentation, 212.

32 Schieder, *Dokumentation*, Vol. I.1, document nr. 21, 82–87, here: 84.

33 For instance, document nr. 5 of Vol. I.1 by the former mayor of Insterburg.

site of mistreatment of the, according to the documentation, 100,000 remaining civilians with the arrival of Russian soldiers on 8–9 April 1945.[34] Finally, the mistreatment of the civilians by the Russian soldiers and administrative authorities in Königsberg (assault, theft, rape, starvation) is documented in the personal testimonies collected in the second book of the first volume.

Post-imperial Empire, as Hardt and Negri describe it, "is not born of its own will but rather is *called* into being and constituted on the basis of its capacity to resolve conflicts."[35] They also point out, however, that it presupposes an international consensus which aims to resolve conflicts for the sake of instating peace and justice in the world. I read the personal testimonies as assembled by Schieder's commission to cumulatively articulate an appeal to an ostensibly already constituted consensus that might discredit the Soviet presence in the formerly German "east," which Königsberg/Kaliningrad emblematizes. Their addressee is that presumed consensus of a post-imperial world order which is at once regarded as already existent and is simultaneously called into being by the various appeals to it. Post-imperial empire also encourages such appeals, because they supply the constituting power on which it feeds for its perpetuation and expansion. As I will show by the end of my analysis of Richter's novel in the next section, this dynamic underlies the nearly imperial reach and efficaciousness of narratives of loss and victimization. Schieder's documentation, I argue, by articulating such an appeal, actively participates in the crafting of a narrative of the transition from an imperial to a post-imperial order. It summons all this testimony to selectively document the effects of imperial violence for a case against those current forces which are represented as prohibiting the emergence of a new, peaceful and just, political order. And yet the Schieder commission's narrative is plagued by its own "forgetfulness, embellishment, and creation of legend" regarding the recent imperial past.

Although the scholarly commission and participating researchers declared it their aim to situate the events reconstructed in their documentation as the flight, persecution, and expulsion of the German population from the areas east of the Oder-Neisse within the events of the previous two decades, the first two volumes restrict their time frame from 1944 to 1950.[36] That their reconstruction starts near the end of the war with the invasion of the Soviet army gainsays their stated inten-

34 See: Schieder, *Dokumentation*, Vol. I.1, document nr. 171.

35 Hardt, Negri, *Empire*, 15.

36 In all fairness, Mathias Beer, who has diligently researched the documentation, remarks that Broszat and his colleagues managed over the ensuing years to better contextualize the expulsions in subsequent volumes. Beer, Martin Broszat und die Dokumentation, 51.

tions. It launches rather a narrative which starts with a string of losses: a lost war, devastated lives and homes, persecution, humiliation, loss of homeland. Started in such a way, the narrative narrows the perspective to the Germans east of the Oder-Neisse, who became protagonists when their suffering began at the end of the war and they were forced to leave their homes with the invasion of the Soviet army. We can certainly read the words of those expelled Germans as a source of knowledge, not just about the events themselves but also about how those events were perceived and retained by individuals. Indeed, we can only imagine what unexpected testimony never made its way into the *Documentation's* narrative and still lurks in the federal archives in Bayreuth ...

2 Narrating the Postimperial as a Novel of Displaced Persons

For other narratives about displaced persons from Königsberg were indeed circulating at the time, and not only informally among families and acquaintances,[37] but also as testimony written in the form of the novel. The example I want to consider here, as a contrasting paradigm to Schieder's *Documentation* of displaced persons, is the novel *Sie fielen aus Gottes Hand* by Hans Werner Richter.[38] Published in 1951, the same year that the work on Schieder's documentation began, Richter's novel is based on personal testimony and interviews with displaced persons held in the transit camp for displaced persons in Hersbruck, near Nürnberg. Richter wrote the novel during 1949–50 after visiting Hersbruck with the journalist Claus Hardt in 1949. Over the course of several months, the two met with refugee commissioners as well as displaced persons in the former concentration camp that had been converted into a transit camp for some of the nearly eleven million displaced persons on German soil at the end of the war. Claus Hardt visited the camp in his role as correspondent for *Spiegel* at the time, while Richter, who became famous in the German literary scene as one of the founders of the Gruppe 47, was also initially there for journalistic purposes.[39]

37 Cf. Karasek-Langer. Volkskundliche Erkenntnisse aus der Vertreibung.

38 Hans Werner Richter. *Sie fielen aus Gottes Hand*. Roman. München: Bertelsmann, 1976 [1951]. Hereafter cited parenthetically. Again, all translations are mine.

39 See Katarzyna Śliwińska. Hans Werner Richters *Sie fielen aus Gottes Hand* oder wie 'ein europäischer Wirrwarr' inszeniert wird. In *'Es sind alles Geschichten aus meinem Leben.' Hans Werner Richter als Erzähler und Zeitzeuge, Netwerker und Autor,* Carsten Gansel, Werner Nell (eds.), 69–81. Berlin: Erich Schmidt, 2011, 69.

In contrast to the narrative framework of the first volume of Schieder's documentation, which starts with the incursion of the Soviet Army in the German-occupied settlements east of the Oder-Neisse, Richter's narrative begins in 1939 with the German invasion of Poland and its setting alternates between several battlefields as well as behind various scenes of war throughout the European landscape. The novel's extensive cast of main characters includes a captain of the Estonian army, a Jewish boy from Warsaw, a chemistry student from Krakau, a young SS-Officer from Luxemburg, a Czech woman married to a German worker, a captured captain of the Spanish Republican Army, a communist army pilot from Leningrad, a wandering young black-market trader from Cairo, a Latvian bartender from Riga, a Ukrainian Red Army lieutenant, a Yugoslavian partisan from a German settlement in Yugoslavia, and a construction worker at the Hersbruck camp. From Königsberg, finally, is an opportunistic German woman who consorts with Nazi officers. In the afterword to the edition reprinted in 1976, Richter explains in retrospect that he aimed to create a panorama of the war which incorporated the perspectives of different refugees later encamped in Hersbruck: "My intention was to combine a selection of the destinies I encountered here in a way that resulted in a human panorama of WWII".[40] Until the characters meet in the former concentration camp of Hersbruck after the end of the war, figural perspectives and locations are not affixed to a single definite geographical region. Richter's narrative seeks to weave together events occurring simultaneously into a panoramic tapestry with a focus on the impressions of the characters experiencing the events. The outbreak of war on the fateful night in 1939 which opens the novel is captured in a panoramic miniature which encapsulates the narrative procedure and frames it within the work on the camp that will become by the end the unifying setting of the novel as a whole:

> In the work camp of the Reich, along a big gravel road not far from Nürnberg, an alarm sounded. The siren raised its wailing voice at the same time that captain Alexander Lewoll set down his shot glass on the table on the Estonian border, as Slomon Galperin turned over in his bed in Warsaw, as lieutenant Pjotr Majsiura disappeared in the dark foyer in Bobruisk, and as Anna Gajek reached for her husband's hands. She cried out in the moment that Francesco Gerdalles, the former captain of the Spanish Red Army, heard the sentry's whistle in the concentration camp in St. Cyprus in Southern France, as Sidi Raumier awoke from his dream in the port of Oran, as Andre Michhalowicz passed through the door of the small inn on the Hungarian-Yugoslavian border, as the chemistry student Hanka Seretzki walked down the street in Krakow, as the apprentice cook Henry Sturm stared in exhaustion at the

40 "Meine Absicht war, die Schicksale, die ich hier antraf, in einer Auswahl so zu bündeln, daß sie ein menschliches Panorama des zweiten Weltkrieges ergaben". Richter. *Sie fielen aus Gottes Hand,* Nachwort.

> headlines of the newspaper that lay at his feet in the Hotel Astoria in Cologne, and Kostja Wagow boarded the train in Leningrad. At this moment, the sustained, plaintive tone of the siren wailed across the roofs of the barracks and wrested sleepers from their beds.[41]

Yet this narrative construction of a panorama relies on a view imagined from outside and above the events. The characters are unaware of each other as the war unfolds and exposes them to contingencies beyond their own powers of control or explanation, but the narrating instance recreates their stories with the benefit of an overview unavailable to the protagonists themselves. In the process, that view approaches the divine perspective which is described at the opening as having abandoned the protagonists. The narrative begins with reference to the invisible hand of God and establishes the, if not omniscient then at least god-like, perspective towards the earth rotating day and night, year upon year, as it then zooms in to depict different European scenes of war.

> The hand which held things could not be seen. The earth spun, one point among billions, and orbited the sun. The sun turned and sped through space, followed by the satellites which encircled it. It seemed as if everything was ordered in accordance with law and goal, measure and rule. The earth stood inclined toward the sun and rotated, the sun glowed and shot its protuberances into space. None of these movements indicated a transformation in the course of things. It was day and night on earth at the same time, one day among the millions and billions of days that had come and gone, and a night like all others over thousands of years.[42]

---

41 "In dem Reichsarbeitsdienstlager, das an einer großen Schotterstraße nicht weit von Nürnberg lag, gab es um diese Zeit Alarm. Die Sirene hob ihre anschwellende Stimme zur gleichen Zeit, als Hauptmann Alexander Lewoll an der estnischen Grenze sein Schnapsglas auf den Tisch setzte, als Slomon Galperin sich in seinem Bett in Warschau herumwarf, als Leutnant Pjotr Majsiura in Bobruisk in dem dunklen Flur verschwand und als Anna Gajek nach den Händen ihres Mannes griff. Sie heulte auf in dem Augenblick, als Francesco Gerdalles, der ehemals rotspanische Hauptmann, in dem Konzentrationslager in Saint Cyprien in Südfrankreich die Trillerpfeife des Postens hörte, als Sidi Raumier im Hafen von Oran aus seinem Traum erwachte, als Andre Michhalowicz aus der Tür des kleinen Gasthofs an der ungarisch-jugoslawischen Grenze ging, als die Chemiestudentin Hanka Seretzki in Krakau auf die Straße lief, als der Kochlehrling Henry Sturm im Hotel Astoria in Köln übermüdet auf die Schlagzeilen der Zeitung starrte, die zu seinen Füßen lag und Kostja Wagow in Leningrad den Zug bestieg. In diesem Augenblick lief der langgezogene, klagende Ton der Sirene über die Dächer der Baracken hin und riß die Schläfer aus ihren Betten." Richter. *Sie fielen aus Gottes Hand*, 19.

42 "Die Hand war nicht sichtbar, die die Dinge hielt. Die Erde drehte sich, ein Punkt unter Milliarden Punkten, und kreiste um die Sonne. Die Sonne drehte sich und jagte durch den Raum, gefolgt von den sie umkreisenden Trabanten. Es schien, als sei alles geordnet nach Gesetz und Ziel, nach Maß und Regel. Die Erde stand der Sonne zugeneigt und rotierte, die Sonne glühte und warf ihre Protuberanzen in den Raum. Keine Bewegung zeigte eine Veränderung im Lauf der

The perspective evokes the metaphysical image of an earth ruled by the firm hand, so to speak, of a cosmic order. What at the outset is taken to be a cosmic order by the inhabitants of that world – "it seemed as if everything was arranged according to law and goal, measure and rule" – unravels into a string of contingencies for the persons who will later be displaced by the war that erupts on this night in 1939. Yet the third-person narrator assumes that empty site imagined as formerly held by God, as the instance capable of shifting among their perspectives and locations, following them but also leading the reader by way of the narrative trajectory to their destination in the camp for displaced persons. The characters, on the other hand, adopt a fatalistic view of the "path of those who are stateless and lost between the nations".[43] For instance, the characters cannot explain why their friends and family have died, why they have fought on one side or the other of warring parties, or even why they have become stateless: "Irene didn't know either why they had all lost their country".[44] While the displaced persons interpret occurrences as inexplicable and their own actions within this terrain of destruction as futile, the narrator reconstructs them as unavoidable destinies in a world ravaged by warring nations, murderous collective identifications, and competing political ideologies: through the narrative composition, which thanks to the auctorial narrator leads them to the camp for displaced persons in Hersbruck, they become emblems of their age.[45]

The narrative thus gives rise to a tension between the novel's documentary or testimonial elements and the imposing auctorial narrator that has been duly criticized in the scholarship on Richter's text. Richter has been criticized for emphasizing futility at the expense of responsibility and agency, for attributing the responsibility for violence to the political irresponsibility of the ruling powers – reiterating thereby a version of the seduced-by-Hitler-story, on the one hand,

Dinge an. Es war Tag auf der Erde und es war Nacht zur gleichen Zeit, ein Tag von den Millionen und Milliarden Tagen, die gekommen und gegangen waren, und eine Nacht gleich allen anderen Nächten in Tausenden von Jahren." Richter. *Sie fielen aus Gottes Hand*, 8.

43 "Weg der Heimatlosen und Verlorenen zwischen den Völkern". Richter. *Sie fielen aus Gottes Hand*, 240.

44 Richter. *Sie fielen aus Gottes Hand*, 414.

45 On the frequent use of the predicate "meaningless" as the clue to the meaning of meaninglessness in the novel, see the excellent analysis by Śliwińska, Hans Werner Richters *Sie fielen aus Gottes Hand*, 76; A similar argument about the narrator's reconstruction of an ordered chaos underlying the destinies is made by Jochen Vogt. '*Sie fielen aus Gottes Hand*' zwischen Sozialreportage und Epochenroman. In '*Es sind alles Geschichten aus meinem Leben.' Hans Werner Richter als Erzähler und Zeitzeuge, Netzwerker und Autor,* Carsten Gansel und Werner Nell (eds.), 83–92. Berlin: Erich Schmidt, 2011, 90.

and eliding significant differences between perpetrators and victims, on the other. Particularly vehement is the criticism of Richter's treatment of the young Jewish boy Slomon Galperin from Warsaw.[46] Accidentally shot near the end of the novel by US military police during a revolt by the camp internees, the dying Slomon sees the "three crosses" hanging outside the concentration camp – "the Christian crucifix, the Orthodox crucifix, and the Jewish Star of David" – enter the barracks and the Star of David comes forth to bow down before him.[47] This bizarre image, with its equation of a crucifix with a Star of David, presumably meant as a symbol of reconciliation,[48] ushers the "return" of Slomon to Israel. In any case, it is a disconcerting metaphor of reconciliation between Jews and Christians in a former German concentration camp. What is more, it illustrates Richter's recurrent criticism of the US occupation of Germany by casting the US Military Police as murderers of the innocent Polish Jew – in a disingenuous identification of US army men as perpetrating the death of Polish Jews. Both the absence of any communist iconography (or rather, the predominance of religious tradition over communism signaled by the Orthodox cross as symbol of the "East") in the camp and Salomon's hallucinatory death sequence imply Richter's own vision of postwar Germany as a "bridge" or third-way between the "liberalistic order" of the USA in the West and the "socialist state" of the USSR in the East.[49] Königsberg is also represented at the scene: Though she hides in shame behind the back of the Estonian Displaced Person Alexander Lewoll, Ingeborg Sänger, from Königsberg, is nevertheless present among the witnesses and mourners at Slomon's death bed.

The occupying US authorities repeatedly criticized in Richter's writings transformed the Hersbruck concentration camp into a transit camp for displaced persons. By making it the final destination for them, however, Richter's novel alienates it once more from its institutional function as a transit camp through

46 See on the final scene of the novel Michael Hofmann. Im Zwielicht des Erlebnisses. Neuanfang und Abwehr von Verantwortung im Nachkrieg. Zu Hans Werner Richter. In: *Literarischer Antisemitismus nach Auschwitz*, Klaus-Michael Bogdal, Klaus Holz, Matthias N. Lorenz (eds.), 147–158. Stuttgart, Weimar: Metzler, 2007, esp. 156–58. For the above-mentioned criticisms, see Śliwińska, Hans Werner Richters *Sie fielen aus Gottes*; Vogt, '*Sie fielen aus Gottes Hand*' zwischen Sozialreportage und Epochenroman; and Klaus Briegleb. Literarische Nachverfolgung. Zu Hans Werner Richters *Sie fielen aus Gottes Hand* (1951). In *Wendezeiten, Zeitenwenden: Positionsbestimmungen zur deutschsprachigen Literatur 1945–1955*, Robert Weninger, Brigitte Rossbacher (eds.), 3–35. Tübingen: Stauffenburg, 1997.

47 Richter. *Sie fielen aus Gottes Hand*, 505.

48 Vogt, '*Sie fielen aus Gottes Hand*' zwischen Sozialreportage und Epochenroman, 92.

49 As Richter's journal programmatically declared in: Deutschland – Brücke zwischen Ost und West. *Der Ruf. Unabhängige Blätter der jungen Generation* 1:4 (1946).

which internees temporarily pass on their way to repatriation. In the process, the camp assumes the meaning of an "existential endpoint" for those who end up there as well as for observers.[50] In a manner similar to Broszat's methodological claim to a complete narration of events, by which Broszat legitimated his historical introduction to the first volume of the *Documentation*, Richter's narrative also aspires to depict an overview of the immediate postwar order. Richter's novel goes about it by way of a "connective parallel arrangement" that casts a wide net, and then hauls all of the narrative strands into the camp in Hersbruck.[51] The symbol of the postwar order, according to the narrative logic of Richter's text, is the camp in all its variations: the concentration camp, the prisoner of war camp, the labor camp, and now the refugee camp.[52] Like the postwar order itself, the refugee camp is a transitional place, which bears traces of the past, embodied by the displaced persons interned in it; but it is also supposed to be a transit point on the way to a future order. By transforming it into a final destination for his diverse cast of displaced persons, however, Richter's novel renders it as a paradigmatic institution of the era.

Königsberg also has a share in the novel's narrative stock. The city on the periphery is present throughout the wanderings followed over the course of the novel as the hometown of Ingeborg Sänger. After the start of the war, she had followed a Hungarian major to Budapest, where she bore his child after he went to fight on the Russian front. With the arrival of the Soviet Army in Hungary, she fled with her child to Vienna: "Not only had she fled from the Russians, but also from the blazing hatred of the Hungarians toward the defeated Germans".[53] Unable to return to Königsberg, she has become "stateless" ("heimatlos"), and in Allied-occupied Vienna, she takes up with an American soldier, Joe from Texas, before

---

50 Vogt, *'Sie fielen aus Gottes Hand*' zwischen Sozialreportage und Epochenroman, 90.

51 "Richter takes on the task of bringing together more than a dozen life-stories into an *overall picture* of an historical situation, of a transitional era" ["Richter stellt sich, wie angedeutet, die Aufgabe, mehr als ein Dutzend jener Lebensläufe zum *Gesamtbild* einer historischen Situation, einer Übergangsepoche zusammenzufügen."] This he does by way of the technique of narrative "Parallelschaltung." (Vogt. *'Sie fielen aus Gottes Hand*' zwischen Sozialreportage und Epochenroman, 87.)

52 For a thorough discussion of the construction of the camp as a paradigm of modernity in postwar German letters, with particular attention to the writings of Andersch and Richter, see Norman Ächtler. Das Lager als Paradigma der Moderne. Der Kriegsgefangenendiskurs in der westdeutschen Nachkriegsliteratur (1946–1966). *Deutsche Vierteljahresschrift für Literaturwissenschaft und Geistesgeschichte* 87:2 (2013): 264–294. Following his desertion from the German army in 1944, Alfred Andersch had been held in a prisoner of war camp in the USA and then returned after the war to Germany to found the Gruppe 47 with Richter.

53 Richter. *Sie fielen aus Gottes Hand*, 297.

heading on to Munich in hope of receiving a visa for the USA. To help her cause, she pretends to be Hungarian. Richter uses Ingeborg Sänger as an example, on the one hand, of the widespread rape of German women by Soviet soldiers; on the other hand, her willingness to opportunistically take up with whichever man promises to care for her in a given situation is a highly gendered take on the seduction narrative which imputes a pragmatic promiscuity to those stateless persons who sympathized or consorted with Nazis. In the Luitpoldkaserne in München, where the many refugees await a decision over whether they will receive official Displaced Person status as well as repatriation, she crosses paths with Alexander Lewoll. As unwanted persons in the former East Prussian and Baltic region now occupied by the Soviets, the two have been forced to seek refuge and, denied passage to the USA, are turned over to the German refugee commission that assigns them to the camp in Hersbruck.

In conjunction with the crisscrossing fates of Ingeborg and Alexander, furthermore, Königsberg has a brief cameo as the location of a Nazi prisoner of war camp. The Estonian captain Alexander Lewoll, namely, before he was assigned to the SS "Estonia" division in 1942, had been imprisoned "together with the Russians near Königsberg" (189). Königsberg thus finds a place on the map of European camps charted by Richter's novel and assigned an historical significance. By way of a briefly mentioned detail from Lewoll's story, the fact that Russian and other soldiers suffered captivity by the Nazis is acknowledged in passing and situated within the narrative of a transition from an imperial order of nation-states at war with one another over geopolitical borders, with brutal designs for populating the European landscape, to a society that still bears traces of its disastrous past in the emergence of two imperial blocks and the prevalent institution of the camp.

Königsberg emerges here as in Schieder's *Documentation* as a contested landmark within a world described by Richter's narrator as still stuck in a state of limbo after the second world war and faced with an incipient new ideological one between two emergent imperial powers:

> The world had not become more peaceful after the war, it had only moved closer together, and the earth's space seemed more clear and manageable than ever before. Two large blocks, powerful empires, loomed along with the threatening storm of a new war of worldviews on the horizon. The wounds of the old war had not yet healed. Millions wandered about homeless and stateless, inundated the countries, searched for a foothold, descended into misery and criminality and were despondent in their forsakenness.[54]

54 "Die Welt war nicht friedlicher geworden seit dem Kriege, sie war nur enger zusammengerückt, und der Raum der Erde schien übersichtlicher und klarer geworden zu sein denn je.

The novel underscores that imperial designs did not disappear after the defeat of the Nazi regime, but were reconfigured under the victorious banners of communism and liberalism. At the end of the novel, the Rabbi who comes to the camp to bless the body of Slomon also consecrates the displaced persons as postwar society's "conscience":

> "Perhaps," the Rabbi continued, as he smiled again, "perhaps many of them are here in this former concentration camp because they still had a conscience. There are forces which continue to have an effect, from the past into the present and perhaps into the future."[55]

The Rabbi presents the future as up for grabs between those malign forces of the past which instituted the camp and the positive forces which were the victims of the camp paradigm and embody the conscience of the present society. In Hardt and Negri's terms, the displaced persons are the multitude within an empire for whom the camp is both the monument of a past it cannot jettison and an institutional pillar.

This is why Richter's particular camp paradigm is highly ambivalent. On the one hand, it is the place where national identities are negated and an (albeit dystopic) European identity emerges: where displacements paradoxically bring together the dispersed people of Europe. In this regard, Richter's text questions the insistence on national identity and the territorial claims asserted out of national interests which underlie Schieder's *Documentation of German Expellees*. A charitable reading of Richter's novel would stress that the panorama includes the often overlooked displacements of the populations of Latvians and Estonians. Alexander Lewoll and Irene Zarinns deliver testimony about this part of the story of displacement. Also depicted are impressions of the conflict between Jews and Arabs in Palestine. Furthermore, in light of recent refugee movement from North Africa and the Middle East to Europe's borders, Richter's novel is quite prescient in its reconstruction of the pathways of the character Sidi Raumier from Cairo. Richter's novel thus, on the one hand, fashions a punctually transnational per-

---

Zwei große Blöcke, gewaltige Imperien, zeichneten sich ab, und das Gewitter eines neuen Weltanschauungskrieges begann drohend am Himmel aufzuziehen. Noch aber waren die Wunden des alten Krieges nicht geheilt. Millionen irrten noch obdachlos und heimatlos umher, überschwemmten die Länder, suchten nach einem Halt, sanken immer mehr in Elend und Kriminalität ab und waren hoffnungslos in ihrer Verlorenheit." Richter. *Sie fielen aus Gottes Hand*, 428.

55 "'Vielleicht', fuhr der Rabbiner fort, und er lächelte wieder dabei, ‚vielleicht sind viele von ihnen hier in diesem ehemaligen Konzentrationslager, weil sie noch ein Gewissen hatten. Es gibt Kräfte, die weiter wirken, aus der Vergangenheit in die Gegenwart und vielleicht in die Zukunft.'" Richter. *Sie fielen aus Gottes Hand*, 508.

spective on war and treats the camp as not only a European but indeed a global phenomenon.

On the other hand, however, Richter's camp paradigm possesses dubious features as well. For one, the movements of the displaced persons who represent the conscience of Europe lead toward Europe's "middle," that is to say, toward Germany. Evoking a notion that closely approximates Richter's view of a defeated Germany as a significant future cultural and geopolitical middle-ground between East and West, Richter's knowing narrator confidently avers:

> Entire nations were condemned collectively and declared guilty, and at the same time the drifters or expellees of all nations gathered in the lap of these nations. Many wandered through Europe and gravitated toward its middle, where there was starvation and misery and the war had left behind a jungle of lawlessness.[56]

Europe's "middle" is filled by "that once hated and combatted and now fully chaotic Germany that was now the most free and unfree country on the earth, administered by foreign military bureaucracies and simultaneously the Eldorado of desperados and refuge for the stateless. Those who fled, fled to Germany."[57] Richter's narrator presents Germany as if it were itself a stateless, and hence lawless, place of congregation for persons without a nation. From this topology it is a small step toward an indiscriminate identification of defeated Germans with all the other "drifters and expellees." In fact, the roofer Krause, who in his heart is a good SPD man, regards all the different internees whom he witnesses entering and leaving the camp over the years as one and the same. He doesn't see any difference between the SS officers who are imprisoned there at the end of the war and all the others who came and went before them: "It was a miserable, wrecked stream of human beings that poured out of the vehicles, and it was no different than that flow of prisoners who were unloaded and then loaded again over the years."[58] He heads a local campaign to have the camp torn down, because he

---

56 "Ganze Völker wurden kollektiv verdammt und für schuldig erklärt, und gleichzeitig sammelten sich die Getriebenen und Vertriebenen aller Nationen im Schoß dieser Völker. Viele irrten durch Europa und strebten seiner Mitte zu, dort, wo Hunger und Elend war und der Krieg einen Dschungel der Gesetzlosigkeit zurückgelassen hatte." Richter. *Sie fielen aus Gottes Hand*, 356.

57 "das einmal so verhaßte und bekämpfte und nun völlig chaotische Deutschland, das jetzt das freieste und unfreieste Land der Erde war, verwaltet von fremden Militärbürokratien und gleichzeitig Eldorado der Desperados und Zuflucht der Heimatlosen. Wenn man flüchtete, flüchtete man nach Deutschland". Richter, *Sie fielen aus Gottes Hand*, 428.

58 "Es war ein elender, zerschlagener Menschenstrom, der sich dort drüben aus den Waggons ergoß, und er unterschied sich durch nichts von jenem Häftlingsstrom, der die Jahre hindurch hier ausgeladen und dann später wieder verladen worden war". Richter, *Sie fielen aus Gottes Hand*, 353.

regards it as a dishonorable relict of an imperial age, in a time when society must instead pave the way towards a new order. His argument for removing the camp puts a different spin on the rabbi's remark about the persistence of social forces: for him, the camp is a "symbol" of the Nazi past and thus represents a negative force (in this case, "tyranny" ["Gewaltherrschaft"]) with the potential to further impact the future ("They attract human beings and are a temptation for every new power") and should therefore be removed.[59] Yet the recourse in his plea to biological metaphors frighteningly similar to Nazi rhetoric reveals just how difficult it was to shake the ideological trappings of the recent past: "The camp lies at the margin of our life, it is growing further and further into our life and threatens our entire civilization. It is a cancer."[60]

Conclusion

With this brief excursus on Richter's novel I intended to show the existence of an alternative narrative pattern, in contrast to Schieder's documentation, for interpreting the testimony of displaced persons around 1951 in Germany. However, these two alternatives prove to be closer to one another than a first glance might suggest, for they both ultimately obey the logic of a paradigm of victimization and loss that dominated self-definitions of Germany's place within the immediate postwar European political order. This framework proved to be highly capable of absorbing both a national and a transnational model, regional and global perspectives, as well as those of victims and perpetrators. The narrative framework of victimization and loss that was so pervasive in the immediate postwar years held an enormous legitimating power of address and appeal, displaying a perlocutionary force that we can describe as itself imperial in its scope. Like an empire that fosters elements of its own potential demise, however, this expansive narrative complex could never entirely cover the traces of its violent origins.

---

59 Richter, *Sie fielen aus Gottes Hand*, 425.

60 "Das Lager liegt am Rande unseres Lebens, es wächst immer tiefer in unser Leben herein und bedroht unsere ganze Zivilisation. Es ist ein Krebsschaden" Richter, *Sie fielen aus Gottes Hand*, 424.

References

Ächtler, Norman. Das Lager als Paradigma der Moderne. Der Kriegsgefangenendiskurs in der westdeutschen Nachkriegsliteratur (1946–1966). *Deutsche Vierteljahresschrift für Literaturwissenschaft und Geistesgeschichte* 87:2 (2013): 264–294.

Beer, Mathias. Martin Broszat und die Erfahrung der Dokumentation der Vertreibung. In *Martin Broszat, der "Staat Hitlers" und die Historisierung des Nationalsozialismus*, Norbert Frei (ed.), 43–59. Göttingen: Wallstein, 2007.

Briegleb, Klaus. Literarische Nachverfolgung. Zu Hans Werner Richters *Sie fielen aus Gottes Hand* (1951). In *Wendezeiten, Zeitenwenden: Positionsbestimmungen zur deutschsprachigen Literatur 1945–1955*, Robert Weninger, Brigitte Rossbacher (eds.), 3–35. Tübingen: Stauffenburg, 1997.

Broszat, Martin. Massendokumentation als Methode zeitgeschichtlicher Forschung. *Vierteljahrshefte für Zeitgeschichte* 2 (1954): 202–213.

Douglas, R. M. *Orderly and Humane. The Expulsion of the Germans after the Second World War.* New Haven: Yale University Press, 2012.

Hahn, Eva and Hans Henning Hahn. *Die Vertreibung im deutschen Erinnern. Legenden, Mythos, Geschichte*. Paderborn: Ferdinand Schöningh, 2010.

Hardt, Michael and Antonio Negri. *Empire*. Cambridge (Mass.), London: Harvard University Press, 2000.

Hofmann, Michael. Im Zwielicht des Erlebnisses. Neuanfang und Abwehr von Verantwortung im Nachkrieg. Zu Hans Werner Richter. In: *Literarischer Antisemitismus nach Auschwitz*, Klaus-Michael Bogdal, Klaus Holz und Matthias N. Lorenz (eds.), 147–158. Stuttgart, Weimar: Metzler, 2007.

Karasek-Langer, Alfred. Volkskundliche Erkenntnisse aus der Vertreibung und Eingliederung der Ostdeutschen. *Jahrbuch für Volkskunde der Heimatvertriebenen* 1 (1955): 11–65.

Kyrie statt Heil. *Der Spiegel* 35 (1962): 28–30.

Moeller, Robert G. *War Stories. The Search for a Usable Past in the Federal Republic of Germany*. Berkeley, Los Angeles, London: University of California Press, 2001.

Richter, Hans Werner. Deutschland – Brücke zwischen Ost und West. *Der Ruf. Unabhängige Blätter der jungen Generation* 1:4 (1946).

Richter, Hans Werner. *Sie fielen aus Gottes Hand*. Roman. München: Bertelsmann, 1976 [1951].

Schieder, Theodor. *Dokumentation der Vertreibung der Deutschen aus Ost-Mitteleuropa*. (*Die Vertreibung der deutschen Bevölkerung aus den Gebieten östlich der Oder-Neisse*, vol. I/1–2). Bundesministerium für Vertriebene, Flüchtlinge und Kriegsgeschädigte (eds.), in Verbindung mit Werner Conze, Adolf Diestelkamp, Rudolf Laun, Peter Rassow, Hans Rothfels. Bonn: Bundesministerium für Vertriebene, 1953.

Śliwińska, Katarzyna. Hans Werner Richters *Sie fielen aus Gottes Hand* oder wie 'ein europäischer Wirrwarr' inszeniert wird. In *'Es sind alles Geschichten aus meinem Leben.' Hans Werner Richter als Erzähler und Zeitzeuge, Netzwerker und Autor*, Carsten Gansel, Werner Nell (eds.), 69–81. Berlin: Erich Schmidt, 2011.

Theweleit, Klaus. *Pocahontas*, vol. 4, *'You give me fever'. Arno Schmidt. Seelandschaft mit Pocahaontas. Die Sexualität schreiben nach WWII*. Frankfurt/Main: Stroemfeld, Roter Stern, 1999.

Toker, Leona. *Documentation. Return from the Archipelago. Narratives of Gulag Survivors*. Bloomington, Indianapolis: Indiana University Press, 2000.

Vogt, Jochen. '*Sie fielen aus Gottes Hand*' zwischen Sozialreportage und Epochenroman. In '*Es sind alles Geschichten aus meinem Leben.' Hans Werner Richter als Erzähler und Zeitzeuge, Netzwerker und Autor*, Carsten Gansel, Werner Nell (eds.), 83–92. Berlin: Erich Schmidt, 2011.
Wingenroth, Carl. Das Jahrhundert der Flüchtlinge. *Außenpolitik* 10:8 (1959): 491–499.

Holt Meyer

Displacement and its Nationalist Totalitarian Compensations in Puschdorf/Pushkino

(Re)Constructing 'Germanness' and 'Russianness' in Cartographic Practices Pertaining to East Prussia in the 1930s and 1940s

> And in this country, if the cobbles in the streets could talk, they would say: Stalin
> *(Henri Barbusse[1])*

0 Hard Reset

A hard reset, otherwise known as a factory reset or a master reset, is a full restoration of an electronic device to its factory settings. It deletes all information stored in the device during the history of its usage, and thus essentially erases history. The spatiotemporal nature of computer storage units and devices (spatial material for storage technologies based on time) makes this analogy or metaphor all the more apt – a storage device devoid of all data produced during the use of the device is in a way reduced to pure space, or at least motionlessness. This can, as we shall see, be said just as well about the political phenomenon I am discussing here: late Stalinism as absolute stagnation, i.e. as the unthinkability of the system experiencing events, much less going through developments.

In this paper I ask the reader to consider the grounds of East Prussia under Soviet rule after World War II, particularly the first years of that rule, as a device subject to hard reset. Precisely the fate of time and its effective dissolution (one might say: time as fate, i.e. as ineluctable teleology) allows us to think of official memory practices in this non-human way, as a "production of history".[2]

For this reason as well, the hard reset is a fitting metaphor for what goes on in the region I am discussing in the time mentioned. The difference is, however, that in this case the 'restored' factory settings were not (re)discovered but rather

1 Henri Barbusse. *Stalin. A new world seen through one man.* New York: International Publishers, 1935, vi.

2 Evgenij Dobrenko. *Stalinist Cinema and the Production of History*. Edinburgh: Edinburgh University Press, 2008.

DOI 10.1515/9783110418750-023

invented, and the apparatus known as "Kaliningradskaja oblast'" 'returns' to a format and settings it actually never had. This reset is thus 'historiographically psychotic', assuming the existence of a non-existent past.

1 Imperial 'Germanness', Antiimperial 'Russianness' and Anachronistic Absurdity

Within a little more than a decade of each other, i.e. in July of 1938 and in the period 1946–1950, the map of what until 1946 was called East Prussia – Ostpreußen, (in Russian Vostočnaja Prussija) – was totally altered, i.e. literally 'manipulated'. Numerous toponyms changed twice, most of them, particularly those which were located on Soviet territory, at least once. In 1938 it was the officialdom of Nazi Germany, to be more specific Adolf Hitler personally[3], who took the initiative and the responsibility. In 1946–1950 it was the USSR enmeshed in the ideological mazes of late Stalinism. Both times, the administrations renamed places whose official cartographic designations were deemed not sufficiently 'national'. In other words, many toponyms were deemed by the Nazis, then the Stalinist Communists, as 'Ungerman' or 'Unrussian', respectively, and eradicated from the map.

One exemplary case is the town of Stallupönen (Lithuanian: Stalupėnai; Polish: Stołupiany) which was 'Germanized' by the Nazi administration into Ebenrode in 1938 and then 'Russianized' into Nesterov by the Soviet authorities after 1945.

Here is an excerpt from a map from the Perthes collection which documents the firm's loyal notation of the Nazi measure of 1938 (note the marking of Stallupönen/Ebenrode about a third of the way down the list):

The two general cases of a wave of 'national renaming' in the region look similar, but are in fact radically different both in terms of the percentage of renamed towns (in "Kaliningradskaja oblast" over 99% in the entire region, and in Ostpreußen of 1938 decidedly less, with regional variations) and with respect to discrepancies between the official roles of nationalism in the respective ideologies. This, in turn, has serious consequences for the analyses of these processes as imperial acts.

3 Hitler's order was executed by East Prussian Gauleiter Erich Koch. See Uwe Neumärker et.al. *„Wolfsschanze": Hitlers Machtzentrale im Zweiten Weltkrieg*. 3d ed. Berlin: Ch. Links Verlag, 2007, 202.

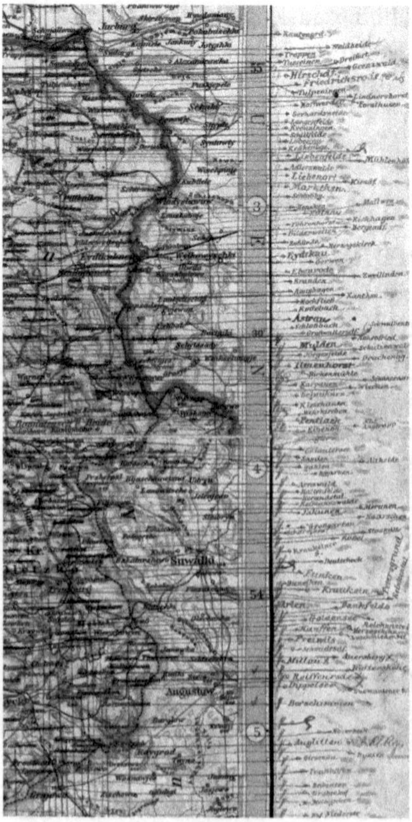

Map 1: Over 100 renamings on the right margin: The cartographic results of the 'Germanization' of East Prussia, with changes in red writing in preparation for the 1939 edition.

For instance, "Ebenrode" and "Nesterov" are two semantically utterly different cases. "Ebenrode" of 1938 is a generic and artificial 'naturalization' ("Ebenrode" means something like "a place completely cleared of trees") with no relation to the original name of Lithuanian origin.

The naming of the Soviet town (or the Soviet naming of the town) after the Red Army Guard colonel Stepan Kuzmich Nesterov, a native of Dobrinsky District of the Lipetsk region of Russia, who fell in the battle to conquer "Ebenrode" on October 20, 1944, also has no relation to the original Polish or Lithuanian name. In contrast to Ebenrode, it is, however, a direct and literal embodiment of socialist realism. It is a telling of the events of October 1944 which led to Nesterov's death in a way which might be summarized in this manner: the history of this town was always already destined to participate in the bright future of Soviet rule, something predetermined by "scientific socialism"[4] itself, and this necessary fate came to fruition in October 1944. The story of Nesterov as the 'positive hero' must thus be told with all the partisanship appropriate for an agent of Soviet (i.e. Russian) progress just as "Great Russia" ("velikaja Rus") was historically always destined to "weld forever" the "unbreakable Union of freeborn Republics" (thus the Soviet national anthem after 1944). Nesterov as the name of a Red Army officer and as the name of a town in Kaliningrad district is at the center of a story which dissolves all time into the absolute significance of this place, and all tales told about the respective present of this place point to this necessary future.

If it were not for the extreme Russian nationalism in this view, there would be nothing special about this particular playing out of 'socialist realism'. The extreme nationalist aspect gives it an inner contradiction which is a peculiarity of chauvinistic and 'anti-cosmopolitan' late Stalinism: of 'Marxism' having 'developed' so long in a particular direction that it became its total(itarian) opposite.

This is a significant contrast to "National Socialism", which of course always vociferously and openly opposed all forms of Marxism. When researching the 'Germanization' of toponyms in the years leading up to the outbreak of World War II, such as that in Sorbia in 1937 or in East Prussia in 1938, there thus appears to be no need to question the 'inner consistency' of the measures with respect to the political program which stood behind them. There are no glaring inner contradictions of the type which emerge when a supposedly Marxist government develops policies which in practice are hard to distinguish from the ones of the Nazis.

The directly and immediately imperial orientation of the German case is indisputable, indeed trivial: we are dealing with a straightforward and openly nationalist and racist policy whose nuances are interesting mainly for historians reconstructing the details of National Socialist rule. There is no Soviet version of genocidally oriented Nazi publications such as "Der Untermensch". At the same time, the deportation of entire peoples such as the Crimean Tartars differs little in its cruelty, its racism and its methods from similar practices of the "Third Reich". Despite this, the fact that these practices are put into place in the name of "anti-fascism" is not simply a question of window dressing. It is a core issue, particularly as it concerns the imperial and its conceptualization of space and time. For this reason, this subject cannot be simply lumped together with that of Nazi atrocities (it seems that Hardt and Negri are suggesting just this).

---

4 Stalin used the term "scientific socialism", claiming it is developed already in Friedrich Engels's "Anti-Dühring", in his essay "Anarchism or Socialism" which was written at the end of 1906 and the beginning of 1907 (official English translation in: J.V. Stalin. *Works*, Vol. 1, November 1901 – April 1907. Foreign Languages Publishing House, Moscow, 1954, 297–372.

Let us look at some of the research on the renaming of northern East Prussian cities. Most recently it was Jurij Kostjashov, who in his 2009 book "The Secret History of Kaliningradskaja oblast"[5] provides a handy chronology of the renaming process and offers a typology of renaming techniques:

Kostjashov[6] notes that many of the new names were connected to the war and its heroes [TYPE 1]. Another common source was what he calls "symbols of the socialist era", referring to names like Sovjetsk, Pravdinsk, Pionersk etc [TYPE 2]. Related to this were names derived from leading figures in officially propagated Russian and Soviet political, literary and cultural history (Chapaevsk, Chernyshevskoe) but also German communists (Tel'manovo for the murdered party leader Ernst Thälmann) [TYPE 3]. One assumes that names associated with Pushkin would be seen by Kostjashov to be in this category of the third type. Other sources were geographical features in the area [TYPE 4], but also towns where the new inhabitants came from in the first place (Vladimirovo, Mordovskoe, Moskovskoe etc.) [TYPE 5]. Without giving any statistical material on its general significance, Kostjashov also underscores what he calls the "russification of toponyms" [7] by assigning names which were similar to the original German ones [TYPE 6]. Kostjashov takes examples from the Eastern region around Sovjetsk (formerly Tilsit), with Petraschen becoming Petrovka, Kraupischken becoming Krupinovka etc. This type is especially interesting, particularly when it is hybridized with the type 3, "leading figures from history" (as in the case of Pushkino and Pushkarevo, the case discussed at the end of this study). Very seldom was a seventh type which Kostjashov views as the most logical: translating or transliterating the German toponym into Russian [TYPE 7]. The actual translation was more common in the Polish southern part of East Prussia, and its blocking in the northern part could be seen as the difference between a hard and a soft reset. There are cases of translations being suggested, such as "Shum" (murmur, noise) for the sea resort "Rauschen", but these were rejected in the end. By explicitly refusing to translate, the authorities confirmed the principle of total erasure.

The topic of renaming was researched thoroughly also by Per Brodersen, who in his book "Die Stadt im Westen" (The City in the West) of 2008 carefully retraces the steps of the Kaliningrad and Moscow administrative bodies from 1945 until June of 1950, when the Politburo finalized the renaming process.[8] Brodersen

---

5 Jurij Kostjashov. *Sekretnaja istorija Kaliningradskoj oblasti*. Kaliningrad: Terra baltika, 2009.

6 Kostjashov, *Sekretnaja istorija*, 247f.

7 Kostjashov, *Sekretnaja istorija*, 276.

8 Per Brodersen. *Die Stadt im Westen: Wie Königsberg Kaliningrad wurde*. Göttingen: Vandenhoeck & Ruprecht, 2008, 71–72.

notes that the smaller the town or village, the stronger was the tendency to let the new inhabitants from the mainland of the USSR to take part in the process.[9] He states that this renaming was a form of taking possession of the respective location. This would apply to the little town Puschdorf-Pushkarevo.

Of particular interest for the issue of restoration and translation is Brodersen's detailed description of the conflict, briefly mentioned by also by Kostjashov, between Russian and Lithuanian positions in a committee which began meeting in February of 1947. The Lithuanian historian and minister of education Povilas Pakarklis insisted that Lithuanian names, particularly those which had been eliminated by the Nazis in a wave of Germanizing renamings in 1938, should be restored.[10] At that time, it was not clear to the Lithuanians what would actually happen to the territory, and some assumed it would eventually become part of the Lithuanian "SSR" anyway. Pakarklis and other Lithuanian politicians were overruled and outvoted at the end of February 1947 by the leadership of the Russian Federal Socialist Republic[11], mainly with the argument that the new inhabitants needed names familiar to them and closer to their own language. Brodersen summarizes the real reason: "without doubt, the leadership of the republic was firmly convinced that the locations of the Kaliningrad territory should contain no reference to their origins" ("Ohne Zweifel war die Republiksführung fest entschlossen, den Orten im Gebiet Kaliningrad keinerlei Referenz an deren Ursprung zu gewähren"[12]), so that also "russified toponyms with Lithuanian roots were unacceptable."[13]

In other words, everything that was not felt to be 'purely Russian', be it a translation, be it the reproduction of a Lithuanian sound (i.e. anything which displayed traces of actual linguistic or cultural history), was implicitly antithetical to the Soviet 'anti-imperial' project which was equated with natural Russian progressiveness. Precisely this was the factory setting not only of Königsberg, but also of the entirety of humanity begging for Russian Soviet salvation and prevented only by "imperialism" from achieving its deep-seated goal.

In the case at hand, this leads to a total erasure of historical roots for the names of an entire territory. The total erasure and renaming surpasses cases such as that of the United States, where a large number of toponyms in lands seized from natives have roots in local native American languages. This is a totality of

9 Brodersen, *Die Stadt*, 62.
10 Brodersen, *Die Stadt*, 65.
11 Brodersen, *Die Stadt*, 70.
12 Brodersen, *Die Stadt*, 70.
13 Brodersen, *Die Stadt*, 70–71.

renaming practically unknown in recorded history. It is connected to an officially supported total amnesia lasting well into the post-Soviet era.

What are the key points for the research of the 1938 case? From the point of view of the chronology of Nazi Germany's aggression, we are dealing with the general 'geopolitical' configuration in the months between the Anschluss of Austria in March 1938 and the "Munich accords" of September 1938.

The specific issues which can be investigated might be:

1. structures of decision making at that particular time in that particular region, e.g. as concerns the relationship between the administration of the Reich and that of Prussia;
2. the political role of figures like Erich Koch in the execution of the plan;
3. the communication between the government and the cartographic institutes, particularly the publisher Perthes in Gotha with respect to the representation of the renaming in maps;
4. the makeup of the commission put into place by the Prussian Ministry for Science, Education and Culture, containing employees of the University and the State Archive of Königsberg,[14] which can shed light on nature of the collaboration of scholars and government in Nazi Germany.

The third point would involve the emergence of practices such as the cartographic recording of all the changes of 1938 for the Stieler's Atlas of 1939. Here is the entirety of the map excerpted above, this being a direct result of the measures initiated personally by Hitler:

The makeup and the modus operandi of the commission which put forth the concrete names, of which there were over one hundred in the Stallupönen-Ebenrode district alone,[15] is clearly of interest. It was thoroughly investigated by Andreas Kossert for the case of the Deutsche Reich's region Masuren which corresponded to the Regierungsbezirk Alleinstein (Stallupönen-Ebenrode was in the Regierungsbezirk Gumbinnen to the northeast of the Allenstein district). The institutional location of the persons involved in the conceptualization and execution of such measures can be compared (it is interesting, for instance, that

14 See Kossert, Andreas. 'Grenzlandpolitik' und Ostforschung an der Peripherie des Reiches. Das ostpreußische Masuren 1919–1945. *Vierteljahresschrift für Zeitgeschichte* 51:2 (2003): 117–146.

15 Eydtkuhnen became Eydtkau, Lengwehnen became Grenzkrug, Kinderweitschen became Kinderhausen, Kryschullen became Narwickau, Nickelnischken became Nickelsfelde, Escherkehmen became Seebach, Jodringkehmen became Sinnhöfen, Jocknen became Jocken, Plimballen became Lehmfelde etc. etc. See Preußisches statistisches Landesamt (ed.). *Gemeindelexikon für den Freistaat Preußen*, vol. 1: Provinz Ostpreußen, Berlin 1931; Statistisches Reichsamt (ed.). *Amtliches Gemeindeverzeichnis des Deutschen Reiches* 1939, 2nd ed. Berlin 1941.

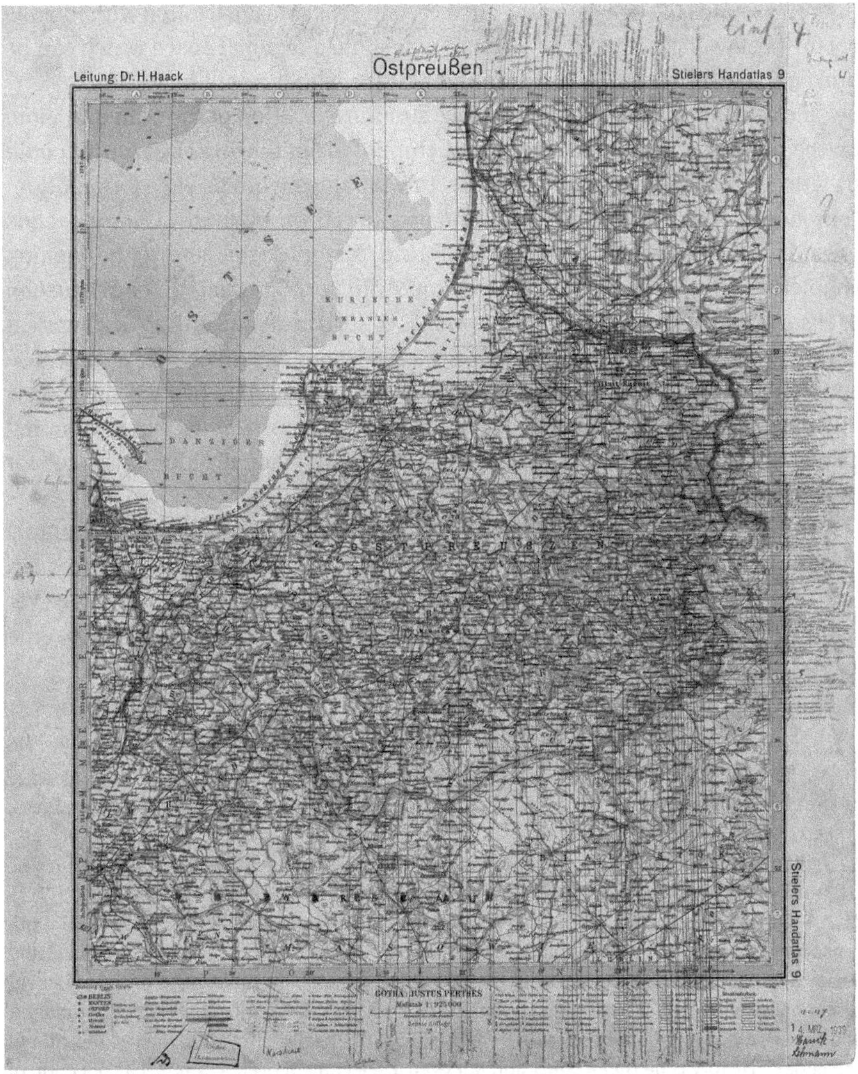

Map 2: The complete cartographic results of the 'Germanization' of East Prussia.

experts for Slavic studies participated). The practices of totally invented narratives with a strong ethnic motivation are in any case analogous.

This, in turn, is significant for the analysis of the cases from the point of view of spatiotemporality, for the toponymic traces of the actual course of history (in the form of the ethnicity of the actual places involved) are replaced with a kind of 'national timelessness' which assigns the status of 'national eternity' to the (literal) grounds of such measures. In a certain sense we are dealing with (details of) the bestowal of ahistorical 'German time' onto a regional space which can be broken down into individual stages of execution with eminent historicity.

If one concentrates less on the stages of the execution of the plan and more on the renaming in one fell swoop, the chronotope of the Soviet narrative could be termed 'zero-chrono-plus-total-tope' (my rephrasing of Epstein's formulation: "chronos is consistently displaced and swallowed up by topos. Chronos tends towards zero",[16] but also of Dobrenko's phrase "history degree zero",[17] the opening chapter of in the 2008 English translation of *Stalinist Cinema and the Production of History*). In the background of the moves of 1938 is an ahistorical Germanness which is reminiscent of metaphysically based mappings of holy lands whose actual recent history is irrelevant compared to the 'big time beyond time'. This 'big time' is assigned authentic reality, in extreme cases viewing 'earthly powers' as negligible (see Katerina Clark's conception of the "sacralization of space" for the Soviet Stalinist case[18]).

Thus, there is in both cases more behind the renaming policies than simply opportune prevarication and the arrogant cynicism of absolute power (though this is the prerequisite). What is behind the seeming prevarication is however quite different in each case. I would like to demonstrate this by taking up some

---

16 Mikhail Epstein, Russo-Soviet Topoi. In *The Landscape of Stalinism the art and ideology of Soviet space*, Evgenij Dobrenko, Eric Naiman (eds.). Seattle: University of Washington Press, 2003, 277.

17 Dobrenko, *Stalinist Cinema*, 1ff. The phrase is obviously a reference to Roland Barthes *Le Degré zéro de l'écriture suivi de Nouveaux essais critiques*. Paris: Éditions du Seuil, 1972 [1953]. Transl. Annette Lavers and Colin Smith. Writing Degree Zero. London: Jonathan Cape 1967. Dobrenko quotes the following formulation from that book: "the teleology common to the Novel and to narrated History is the alienation of the facts." (op.cit., p.7). In the context of my argumentation one might rephrase: "the displacement of the facts". Dobrenko continues the thought, shoring up his view of Stalinism as the "museum of the revolution" (thus the second title of the book): "History is neither a collection nor even a selection of events. Nor is it a collection of interpretations of these events, but rather a system of techniques and optics which produces the events themselves from the past and gives them the very status of events. In other words, history can be conceived as the alienated past [...]". Again I would rephrase: "the displaced past". For the ironic application of the early anti-bourgeois Barthes to a bourgeois pseudo-socialist setting see also Holt Meyer. Spiel(k)arten der Distanz in der ‚begrenzten Souveränität': Bohumil Hrabals Kdo jsem. In *Distanz. Schreibweisen, Entfernungen, Subjektkonstitutionen in der tschechischen und mitteleuropäischen Literatur*. Nora Schmidt, Anna Förster (eds.). Weimar: VDG, 2014, 23–63.

18 Katerina Clark. Socialist Realism and the Sacralizing of Space. In *The Landscape of Stalinism the art and ideology of Soviet space*, Evgenij Dobrenko, Eric Naiman (eds.), Seattle: University of Washington Press, 2003, 3–19.

of Kossert's formulations. Kossert notes that the Nazi officialdom in its renaming policy

> griff [...] auch zu den Mitteln der Neuinterpretation historischer Fakten und bewußter Fälschung. Auf diesem Wege wollte man nicht nur polnischen Gebietsansprüchen den Boden entziehen, sondern auch das leidige Thema polnisch-sprachiger Ethnien in Deutschland ein für alle Mal aus der Welt schaffen.[19]

Kossert's phrasing in the last part is somewhat odd. He refers to the "annoying topic of Polish speaking ethnic groups in Germany". The topic is "annoying" (*leidig*) for the Nazi officials, for they are proponents of a system of social mapping which simply erases inconvenient issues from cartographic consideration and backs this erasure policy up ideologically with a radical privileging of 'pure Germanness' in particular regions. I would suggest that "leidig" would have to be marked somehow as not one's own word, e.g. with quotation marks in order to avoid puzzling ambiguities. The formulations "reinterpretation of historical facts" (*Neuinterpretation historischer Fakten*) and "conscious falsification" (*bewußte[r] Fälschung*) are not inaccurate. It is, however, not unambiguously pointed out in the article that eradicating the "annoying topic" (*das leidige Thema*) of Polish language minorities from the face of the earth (*aus der Welt schaffen*) equates imperial domination of space with a manipulation of time, corresponding consistently with the dissolution of actual history in Nazi ahistorical framings of Germannness. One might speak more in terms of blindness caused by genocidal ideology which first dissolved toponymic words in the caustic acid of extreme nationalist (German) and totalitarian verbal and administrative aggression and then persecuted and/or 'liquidated' those who used and lived those (Polish) words.

1.1 Methodological Challenges and Non-Challenges

Having said this, there is, as I suggested before, hardly anything methodologically challenging in the reconstruction of a set of political practices which is radically nationalist and openly colonial and imperial with respect to its neighbors, particularly to the neighbors in the East who were for the most part ideologically downgraded into – at best – second class humans ("Untermenschen"), if they were considered humans at all (and if not, the consequence is the conventional

19 Kossert, ‚Grenzlandpolitik', 141.

colonial assumption of effectively unpopulated lands waiting to welcome its 'first human settlers'). This is explicit, extreme and by no means complex colonial thinking which does not even make the effort to hide behind 'humanist' considerations – au contraire.

Kossert notes that the scholars, particularly the historians, who worked in the committees who set forth the guidelines for renaming in the Nazi period made use of categories which had already been developed in the Weimar republic:

> Ein klares Freund-Feind-Schema erlaubte – wie Peter Schöttler feststellte – nur zwei Optionen: Staatsraison oder Subversion, Nationalismus oder Defätismus. Diese absurden ideologischen Alternativen waren derart prägend, daß sich ihnen kaum ein deutscher Wissenschaftler entziehen konnte. Was in der Weimarer Republik begann, setzte sich mit veränderten ideologischen Koordinaten nach 1933 fort.[20]

The "absurdity" which Kossert notes here could also be rephrased, for the term is an anachronistic and to a certain degree half-hearted attempt at political correctness. The infusion of the logic of 'for or against' is very much in line with Carl Schmitt's concept of "friendship" dependent on "enmity", whose nature, when properly reconstructed and indeed deconstructed, can be seen to go far beyond extreme and "absurd" simplification.[21]The historical point (that the general style of thinking was already widespread long before the Nazis came to power in 1933) is well taken. What is being described here is the impossibility of a thought of displacement – an either-or situation. That is the configuration which is in search on a methodology of analysis.

As for the Soviet case, the methodological challenge is the linking of thought on empire, particularly that of Hardt and Negri, with the special case of the practices of late Stalinism on the exceptional territory of Königsberg/Kaliningrad with all of its ideological and human displacements.

As was already noted in the introduction to this section, the concept of "displacement" is – not unexpectedly – applied by Hardt and Negri to the area of sovereignty, particularly as concerns the nation-state. When they address the

---

20 Kossert, ‚Grenzlandpolitik', 119.

21 See Carl Schmitt. *The Concept of the Political*. Chicago: University of Chicago Press, 1996. See also Derrida's deconstuctionist reading in Jacques Derrida. *The Politics of Friendship*. London: Verso 1997. This discussion is germane for my topic in that Schmitt's thinking is clearly rooted in the pattern described, be it of Nazi origin or be it traceable further back to contexts of the Weimar Republic. The Soviet practices after the war are, on the other hand the absolute revocation of hospitality based on a radically binary friend-enemy pattern of thinking and thus reproduces the Nazis' manner of thinking instead of questioning its form – much less deconstructing it.

problem of the actual displacement of people and populations, they address it in terms of forced mobility between center and periphery:

"Certainly from the standpoint of many around the world, hybridity, mobility, and difference do not immediately appear as liberatory in themselves. Huge populations see mobility as an aspect of their suffering because they are displaced at an increasing speed in dire circumstances".[22]

The fact of people's being "displaced" at great "speed" and "in dire circumstances" certainly applies to the case of the emergence of Kaliningrad. The "displacement" Hardt and Negri are addressing is, of course, caused by economic necessity and not by measures backed by direct political violence. This means that it is the temporal development of the market economy which causes the spatial displacements in question.

In terms of Hardt and Negri's terminology, we are, in the case of postwar Stalinism, dealing with a model of "national sovereignty" which seems to be nationalist, imperial and above all radically colonial in a completely different way from 'National Socialism'. This is the systematic point I am making here when it comes to the operations of both systems in the part of East Prussia which became the Kaliningrad District (Kaliningradskaja oblast) in the Russian Soviet Federal Socialist Republic. Here I am referring to issues of naming, i.e. to 'toponymic policy' which was made possible by the Red Army's conquest of the territory.

The central aspect is as follows: "National Socialism" is not explicitly based on anything like "socialist realism". This difference is of extreme significance if one accepts the leading role assigned to "socialist realism" by Evgenij Dobrenko and other like minded scholars of Stalinism, particularly late Stalinism. Due to this systematic point, a semiotic slant needs to be taken in order to take account of the specificities of Stalinism. Adding writings edited and written by Evgenij Dobrenko and Eric Naiman into the mix of approaches suited for an analysis of the practices at hand, particularly writings of Dobrenko himself here and in other places, provides that very slant. It is Dobrenko who most accurately and compellingly captures the issue of spatiality in Stalinism. I would go so far as to say that his take on Stalinist totalitarian spatiality not only augments the scope of Empire for the Soviet case, but also enables the correction of vagueness in Hardt and Negri's comments on the specificity of nationalism in Stalinism.

Epstein's already quoted key phrase comes from in the closing essay of that collection: "chronos is consistently displaced and swallowed up by topos".[23] The phrase "displacement of time" seems paradoxical, but it is actually very helpful,

---

22 Michael Hardt, Antonio Negri. *Empire*. Cambridge: Harvard University Press, 2000, 154–155.
23 Epstein, Russo-Soviet Topoi, 289.

since, as I try to argue here, a timeless space ceases to be space at all: space is defined by the ability to pass through it in time. This displacement of time itself is central for my argumentation, and works together with the displacement of sovereignty (Hardt and Negri) and displacement of masses of people (historical fact of the 1940s) to clarify the specificities of the spatiotemporality of the issue at hand.

This view, in turn, allows us also to expand the notion of "cartographic practices", including but also going beyond the simple making of maps. Mapping is more generally the 'partisan notation' of space in the context at hand (partisan not with reference to partisans, but rather in the sense of the 'partisanism' of socialist realism). In this field too we can think of Dobrenko's pioneering thought – it is not Stalinism which creates socialist realism in order to beautify itself, but rather socialist realism which creates Stalinism.

This, in turn, is a way of generally introducing spatiotemporality into Hardt and Negri's account of empire (and Empire[24]), particularly as it concerns its relationship to the "totalitarian", be it "national socialist" or Stalinist "socialist realist". For Hardt and Negri, Nazi Germany is "the ideal type of the transformation of modern sovereignty into national sovereignty and of its articulation in capitalist form". They mention Nazi Germany and "Stalinist Russia" several times in one breath, indeed already in their second mention of the Nazis which I just referred to, and which I now quote in extenso:

> If Nazi Germany is the ideal type of the transformation of modern sovereignty into national sovereignty and of its articulation in capitalist form, then Stalinist Russia is the ideal type of the transformation of popular interest and the cruel logics that follow from it into a project of national modernization, mobilizing for its own purposes the productive forces that yearn for liberation from capitalism.
>
> Here we could analyze the national socialist apotheosis of the modern concept of sovereignty and its transformation into national sovereignty: nothing could more clearly demonstrate the coherence of this passage than the transfer of power from the Prussian monarchy to Hitler's regime, under the good auspices of the German bourgeoisie. This passage, however, is well known, as are the explosive violence of this transfer of power, the exemplary obedience of the German people, their military and civil valor in the service of the nation, and the secondary consequences that we can call, in a kind of intellectual shorthand, Auschwitz (as symbol of the Jewish holocaust) and Buchenwald (as symbol of the extermination of communists, homosexuals, Gypsies, and others). Let us leave this story to other scholars and to the disgrace of history.[25]

24 Empire with a capital "E" refers to the present of the year 2000 as represented by Hardt and Negri.

25 Hardt and Negri, *Empire*, 110.

In reference to the link between the imperial and spatiotemporality at the core of this study, one might say that their concept of time has much of the teleology of orthodox Marxist thinking. The space addressed is that of the nation state ruled by such teleological historiographies (or pseudo-historiographies). Let us now hone in on their approach to Stalinism, or to what they surprisingly call "Stalinist Russia".

1.2 Hardt and Negri's Take on "Stalinist Russia"

> ... then Stalinist Russia is the ideal type of the transformation of popular interest and the cruel logics that follow from it into a project of national modernization ...
> *(Hardt and Negri[26])*

The appearance of the USSR in Hardt and Negri is sparse, indeed sparser than it should be given the global scope of the study. They assume the existence of "Soviet imperialism", but take no particular interest in the special features of this model.

> During the cold war [...], imperialist temptation – or really the ambiguity between protector and dominator – became more intense and more extensive. In other words, protecting countries across the entire world from communism (or, more accurately, Soviet imperialism) became indistinguishable from dominating and exploiting them with imperialist techniques.[27]

The specificities of "Soviet imperialism" in its various phases and in its entirety are not an issue which fits well into the model of analysis put forth by Hardt and Negri.

In another passage, this time making specific reference to temporality and spatiality, they refer globally to "totalitarianism":

> Even though the state of exception and police technologies constitute the solid nucleus and the central element of the new imperial right, however, this new regime has nothing to do with the juridical arts of dictatorship or totalitarianism that in other times and with such great fanfare were so thoroughly described by many (in fact too many!) authors. On the contrary, the rule of law continues to play a central role in the context of the contemporary passage: right remains effective and (precisely by means of the state of exception and police techniques) becomes procedure. This is a radical transformation that reveals the unmediated relationship between power and subjectivities, and hence demonstrates

26 Hardt and Negri. *Empire*, 110.
27 Hardt and Negri, *Empire*, 178.

both the impossibility of 'prior' mediations and the uncontainable temporal variability of the event.[28]

In their section on the "Death Throes of Soviet discipline", they leap over the period considered in my study:

> According to our thesis, then, after the dramatic final years of Stalin's rule and Khrushchev's abortive innovations, Brezhnev's regime imposed a freeze on a productive civil society that had reached a high level of maturity and that, after the enormous mobilizations for war and productivity, was asking for social and political recognition. In the capitalist world, the massive cold war propaganda and the extraordinary ideological machine of falsification and misinformation prevented us from seeing the real developments in Soviet society and the political dialectics that unfolded there. Cold war ideology called that society totalitarian, but in fact it was a society criss-crossed by extremely strong instances of creativity and freedom, just as strong as the rhythms of economic development and cultural modernization. The Soviet Union was better understood not as a totalitarian society but rather as a bureaucratic dictatorship. And only if we leave these distorted definitions behind can we see how political crisis was produced and reproduced in the Soviet Union, to the point finally of burying the regime.[29]

The key passage covering the period in question is contained in a section called "The Totalitarianism of the Nation State" and is unambiguous:

> In the Second World War, Nazi Germany, along with the various European fascisms, stood opposed to socialist Russia. Nations were presented as mystifications of, or stand-ins for, the class subjects in conflict. If Nazi Germany is the ideal type of the transformation of modem sovereignty into national sovereignty and of its articulation in capitalist form, then Stalinist Russia is the ideal type of the transformation of popular interest and the cruel logics that follow from it into a project of national modernization, mobilizing for its own purposes the productive forces that yearn for liberation from capitalism.
>
> [...]
>
> Through Stalin's translation [*of the fact that "nationalism is an ineluctable stage in development" – H.M.*] as nationalism becomes socialist, socialism becomes Russian, and Ivan the Terrible is laid to rest in the tomb beside Lenin. The Communist International is transformed into an assembly of the "fifth column" of Russian national interests. The notion of communist revolution – the deterritorializing specter that had haunted Europe and the world, and that from the Paris Commune to 1917 in Saint Petersburg and to Mao's Long March had managed to bring together deserters, internationalist partisans, striking workers, and cosmopolitan intellectuals – was finally made into a reterritorializing regime of national sovereignty. It is

28 Hardt and Negri, *Empire*, 26.
29 Hardt and Negri, *Empire*, 278.

> a tragic irony that nationalist socialism in Europe came to resemble national socialism. This is not because 'the two extremes meet,' as some liberals would like to think, but because the abstract machine of national sovereignty is at the heart of both.[30]

This argument, which Hardt and Negri view as a welcome opportunity to attack "liberals" for thinking that "the two extremes meet" and to make their point that the "abstract machine of national sovereignty" is equally, indeed one might assume identically, at work in the Nazi and (late) Stalinist case, also contains the implicit claim that as long as a political regime has the status of "reterritorializing" in accordance with "national sovereignty", it doesn't matter whether or not this "reterritorialization" takes place in the name of "Marxism-Leninism", "National Socialism" or some other label. This is certainly debatable.

Particularly from the point of view of spatiotemporality, one might well ask whether a political program which places all of history, particularly that after 1917, under a temporal vector of the successive "liberation" (in violent and dictatorial Bolshevik style) of all territories on the face of the earth from the 'yoke of class domination' can be placed in the same category as one whose idea of "Lebensraum" openly places Germanic national empowerment, posited as unique, at the core of all political activity. It is true that the almost mystical and certainly not rational claim that the Russian nation is at the core of world liberation in postwar Stalinism shares many traits of Nazi ideology, and that the latter also had a diffuse 'anti-capitalist' tendency. But there are still huge differences.

I would claim that the claim that the case of Kaliningrad – as part of the Stalinist literally radical (uprooting) nature of dealing with all peoples, including the Russians – represents a case study in the totality of space over time, since time (in the sense of total teleology) is not real time, but a timelessness which replaces the actual course of history.

2 The Stalinist 'Nationalization' of East Prussian Toponyms as an Imperial hard reset and a Displacement of Time

For the reasons given in my introductory remarks, and also based on my critique of Hardt and Negri's theses that "nationalist socialism in Europe came to resemble national socialism", I am using the case of the renaming of East Prussian cities in 1938 more as a foil than as an independent object of study. It is the 'zero point' from which the Stalinist 'Russification' begins, since the restitution of the

30 Hardt and Negri, *Empire*, 110–112.

place names of the pre-1938 era seems not to have been even considered as an option for the fate of East Prussian toponyms.

This is the case only and especially for Russian Soviet practices, which are thus the central topic of my paper. In the southern part of East Prussia, the Polish governmental "Commission for the Determination of Place Names" (Komisja Ustalania Nazw Miejscowości[31]) did not operate in the same way. The heirs to the Polish-Lithuanian rzeczpospolita did not 'totally reset' the maps as the officials of the Russian Soviet Federative Socialist Republic on the northern side of the border did.

I would thus, again, claim that the work of Stalinist Soviet officials was qualitatively different from that of Nazi officials in the late 30s and Polish officials of the mid 40s, also and particularly with respect to the specific features of the imperial quality of their measures – particularly as concerns the relationship of space and time.

It is in this spirit that Dobrenko and Naiman as editors of the collection *The Landscape of Stalinism: the art and ideology of Soviet space* join Hardt and Negri on the stage of my methodological remarks. The theses of Evgenij Dobrenko's studies on Stalinism of the last fifteen years are of key importance for my argumentation. I am referring Dobrenko's own article "The Art of Social Navigation: The Cultural Topography of the Stalin Era" in the *Landscape of Stalinism*[32] collection, as well as Dobrenko's own monograph on the *Political Economy of Socialist Realism*[33]. As I already indicated, the central idea of the latter book is central for my argumentation: socialist realism creates Stalinism, not the other way around.

Having said that, it is important to make clear the extreme significance of socialist realism for the spatiotemporality of Stalinist thinking, and in particular in its later period (during and after World War II) and the policies connected to this thinking. Together with "partisanism", the "typical", and the "positive hero"[34], it

---

31 See Jun Yoshioka. Imagining Their Lands as Ours: Place Name Changes on Ex-German Territories in Poland after World War II. In *Regions in Central and Eastern Europe: Past and Present*, Hayashi Tadayuki, Hiroshi Fukuda (eds.). 21st Century COE Program Slavic Eurasian Studies 15. Slavic Research Center, Hokkaido University, Sapporo, Japan. 207, 273–288.

32 Evgenij Dobrenko. The Art of Social Navigation: The Cultural Topography of the Stalin Era. In *The Landscape of Stalinism: the art and ideology of Soviet space*, Evgenij Dobrenko, Eric Naiman (eds.). Seattle: University of Washington Press, 2003, 163–200.

33 Evgenij Dobrenko. *Political Economy of Socialist Realism*. New Haven: Yale University Press, 2007.

34 For a recent attempt to account for socialist realism in art in all of its historical and aesthetic complexity see Matthew Brown and Matteo Lanfranconi. *Socialist Realisms. Great Soviet Painting 1920–1970*, New York: Skira 2012.

is, of course, its explicit temporality which interests me the most here: the prescription that every depiction of the present has to display an orientation towards the future, and indeed not just any future, but a future based on (vulgar) Marxist teleology. All of these fundamental principles of socialist realism can be found in the naming of former East Prussian towns after this 'future oriented' persons. A temporality which reduces all time to pure space (for time loses all of the openness which gives it its existence) is in the forefront. The various approaches and objects of the analyses in the collection *Landscape of Stalinism* provide helpful methodological and thematic underpinnings for the spatiotemporal points made up till now.

Katerine Clark's already mentioned underscoring of "sacralization" in Stalinism is in many ways expanded upon by Plamper's analysis of the "spatial poetics of the personality cult" in the in the *Landscape of Stalinism* collection[35]. Clark notes "architecture's central role in Stalinist culture" [36], as a symptom of the general "sacralization of space". She argues convincingly that the "defining features of the Soviet regime and its ideological underpinnings were presented through the discourse of space and architecture" [37] pointing to the widespread metaphor of the "building of communism", and notes the revealing fact that in socialist realist films and novels, "'the leader' and even the most positive of heroes participate in two entirely different orders of time".[38] One might rephrase this in terms of my argumentation in this way: approximating Stalin, which like the *imitatio Christi* must be the goal of all true believers, involved nothing less than exiting time itself.

Plamper expands on these thoughts, pointing out that "visual culture" of the 1930s was "preoccupied with establishing Stalin as the center of representation" and that in the post-war period this had been so deeply ingrained in official visual representation that Stalin is assumed to be there even if he is not shown at all.[39] The "centripetal organization of society" with Stalin at the center of all circles as the "sacrally charged embodiment of the Soviet state".[40]

Dobrenko's own article in the collection, "The Art of Social Navigation: The Cultural Topography of the Stalin Era", makes a huge leap for all discussions of

---

35 Jan Plamper. The Spatial Poetics oft he Personality Cult. In *The Landscape of Stalinism: the art and ideology of Soviet space,* Evgenij Dobrenko, Naiman, Eric (eds.). Seattle: University of Washington Press, 2003, 19–50.

36 Clark, Socialist Realism, 4.

37 Clark, Socialist Realism, 15.

38 Clark, Socialist Realism, 15.

39 Plamper, Spatial Poetics, 43.

40 Plamper, Spatial Poetics, 44.

Stalinist space and thus also Stalinist spatiotemporality. His linking of postage stamps, tourist guides and actual borders and maps (their visualization and their de-visualization), the latter concentrating on the 1947 and 1949 versions of Mikhailov's "Nad kartoi rodiny", ends with a figure of thought which prefigures the argument of his 2007 book (it is the last sentence of the article): "one can, of course, object that nobody ever lived in this world anyway. But this is just as true as saying that it was precisely here that we did live".[41] This is a performance of the "displacement of time" connected to Epstein's 'zero-chronos'.

It is this very ambivalence (living and not living in "this world") which dominates the toponymic work with Kaliningradskaja oblast". In this context it is enlightening to take up the passages from the 1949 version of Mikhailov's "Nad kartoi rodiny" which Dobrenko quotes extensively, especially considering the fact that Kaliningrad makes an appearance here:

> The holy boundary of the Soviet land is unshakable. The red line is firmly etched on the map of the world – from Pechenega to Kaliningrad, from Kaliningrad to Ismail, from Ismail to the Kurile Ridge. This is not only the border of the greatest country in the world / this is a line that lawlessness, tyranny, and oppression cannot pass. [...] Our country has become a mighty power. And these historical shifts have been reflected on the map. [42]

Not only do these formulations elegantly sum up the 'anti-imperialistic imperial': they collect the essential features of the 'red frontier' as a key feature of late Stalinist spatiality and explicitly place Kaliningrad on that "holy boundary". Now that the "country" is a "mighty power", the static spatial depiction in the form of a map is sufficient for the representation of "historical shifts". Time has become a stationary "red line".

The arguments and examined objects which lead up to this striking example are just as important. Particularly informative is Dobrenko's 'brief course on the history of the Soviet stamp', which again culminates in an example from 1949: "Stalin's Plan for Transforming Nature" – a six stamp series which Dobrenko describes as "synthesizing, as it were, the various means of landscape depiction of the previous five years[43], the synthesis being based on the postwar tendency of "generic consolidation of the presentation of center and periphery" (which in terms of this paper can be rephrased as the issue of empire).

We are dealing with something one might indeed call socialist realist cartography: a collection of toponyms populated by 'postitive heroes' representing

41 Dobrenko, Art of Social Navigation, 199.

42 Quoted in Dobrenko, Art of Social Navigation, 192.

43 Dobrenko, Art of Social Navigation, 177.

the “shining future” of the “liberated” that northern part of East Prussia which became “Kaliningrad County”.

3 Absence as Presence: /push/

1949 is not only the year of the stamp series “Stalin’s Plan for Transforming Nature” which is discussed by Dobrenko, and not only the year of Stalin’s official 70th birthday, it is also the almost equally universally and hysterically celebrated 150th anniversary of Pushkin. This anniversary is the time of the emergence of the many Pushkin streets in East Central Europe (also in Erfurt and in Weimar, in the latter case also with the Pushkin bust which stands at the end of the street to this day), since 1948 and 1949 were the years of the establishment of Soviet style “people’s republics” there (e.g. the GDR in the fall of 1949). This can also, from the point of view of the expanding Soviet empire, be also termed, to quote Dobrenko’s words once again, an aspect of the “generic consolidation of the presentation of center and periphery”.

Just as the northern part of East Prussia which became part of the USSR was an exceptional territory, Pushkin’s presence in this territory is both typical and exceptional at the same time. I will be looking at a trace or a series of traces which mark not only his presence, but also his absence. This presence and/as absence is a symptom of a very specific form of Russification in the Baltic region. This Russification manufactures Russian presence not despite, but rather precisely out of its absence. In this case, Pushkin’s absence is not just the circumstance, but rather the material of his presence. This is the basis of my work with the chain of sounds which forms the first part of Pushkin’s name: /push/. This para-morpheme is one of the heroes in my contribution.

I call it a para-morpheme because it functions in this poetical-political context like a morpheme, but in none of the cases is an actual morpheme. It seems to be a carrier of meaning, but it is in fact not. Linguistically – although it coincidentally sounds very much like the English word „push“ – it is a semantically meaningless territory of sound which wanders between languages.

After all that argumentatio finally the narratio: the village (with just under 200 inhabitants) of Pushkarevo in the Cherniakhovsk district and a three dozen kilomenters the west of the town Cherniakhovsk, arrived at that final nomenclature in the period from 1946 to 1950 which I already mentioned at the beginning of my article. The documents in the archives of Moscow[44] and Kaliningrad (city archive) indicate that the name took a detour from "Puschdorf" to "Pushkarevo" by way of "Pushkino". The para-morpheme "push" thus appeared not in two, but in three hypostases (I have not been able to find any reason for this particular name "Pushkarevo" being given to Puschdorf-Pushkino, but it is not impossible that some of its not terribly numerous inhabitants came there from another Pushkarevo, e.g. in the Omsk region in Siberia).

In the short period in which the toponym "Pushkino" existed, the documents further indicated that this naming was taking place "in honor of the Russian poet". In the lists of renamings, a column headed by the word "motivation" (motivirovka) sometimes appeared, and a glance at these lists shows how heterogeneous this "motivations" were even within a very limited geographical radius. For instance, the list of "motivations" corresponding to the new names "Pushkino", "Mezhdurech'e" and "Svetlovka" was, respectively, as follows:

"in honor of the Russian poet"
"located between two rivers"
"upon the request of the workers".[45]

The only common denominator of these three "motivations" was their 'Russianness'. It is symptomatic for the role of history – as erased – that "Pushkino" 'metonymically' neighbored or, to put it another way, was placed into the same paradigm as "Mezhdurech'e" (a pure 'natural' location between rivers) and "Svetlovka" (referring either to "light" or "Svetlov"), and thus in effect relativized the assignment of the name "Pushkin" to an actual historical figure or even to history itself. Pushkin became an 'act of God' like the placement of two rivers or the creation of light. As for light, it is worth remembering that one of the Pushkin lines most often quoted in the late 1940s was "Da zdravstvuet solntse, da skroetsia t'ma" – "long live the sun, let darkness be hidden" – taken from a completely incongruous source in a very early anacreontic verse, i.e. effectively denying any real historical development or context for individual acts, words or works of Pushkin aside from their location in vulgar Marxist teleology, i.e. in timeless space.

The fact that the toponym "Cherniakhovsk" in this region, like "Nesterov" in the region further to the northeast, was derived from the name of a Red Army officer whose blood was shed on East Prussian soil, completes the picture of an historical event which returns to the ground that which always already belonged

44 GARF (State Archive of the Russian Federation) Fond A259, Opis 6, delo 3923.
45 GARF(State Archive of the Russian Federation) Fond A259, Opis 6, delo 3923.

to it: Russianness (a Russian who stands synechdocically for the Russian *Ding an sich* based on a view which insists on making Immanuel Kant a Russian since the Russians briefly conquered East Prussia during Kant's lifetime, allowing him to briefly taste ahistorical 'Russian emancipation').

4 The Hard Reset of Puschdorf

I would like now to touch on the historical conditions of the wandering of "push" between German and Russian toponyms.

For over 540 years, a town in East Prussia carried the para-morpheme "push" in the name "Puschdorf". The morpheme /push/ is in the case of Puschdorf, Pushkino and Pushkarevo (/push-dorf//push-kino/ /push-kar'iovo/) part of toponymic names and words, and thus produces elements of maps. But it is also part of a 'double Russian philology' which produces a fork in the road between language and literature in the process of Russification. We are, in other words, dealing with a hard philological reset or a philologically hard reset.

"Puschdorf": this is the first known and thus the identifying sign for a town at the beginning of the 15th century, to be exact, 605 years ago, in June of 1406, when the brothers Peter and Bathel Kuschenpusch receive 16,5 hectares of land from the 25th Grand Master of the Teutonic Knights, Konrad von Jungingen. The further history of Puschdorf and the towns and villages linked to Norkitten in the five and a half centuries until the middle of the 20th century is not of great relevance to the topic at hand. It is perhaps significant to note that the wave of "Germanizations" which engulfs East Prussia in 1938 leaves the name unaffected, as did most of the towns around Insterburg (in contrast to Stallupönen/Ebenrode/Nesterov). This town, like Stallupönen/Ebenrode, was not exceptional in becoming part of the virtually total population switching, the toponymic 'hard reset' of a territory subject to 'population exchange'.[46]

46 See the term "échanger les peuples" in the title of Catherine Gousseff. *Échanger les peuples. Le déplacement des minorités aux confins polono-soviétiques* (1944–1947). Paris: Fayard, 2015, as well as her formulation "L'URSS, grand maître d'oeuvre de l'échange de population á la fin de la guerre" (327) to sum up the role of the Soviet Union in postwar East Central Europe. I would claim that notion of a "switch" or "exchange" of peoples assumes the point of view of "espace" as opposed to "lieu" in terms of Michel de Certeau (Michel de Certeau. *L'invention du quotidien*. Paris: Gallimard, 1990), i. e. the view from above of the cartographer, for instance. "Displacement", on the other hand, is thought more with the logic of "lieu", of the place on the ground as experienced by the object or victim of the "exchange" policies.

In the renaming process between the start in 1947 and the official completion of the process in Moscow in 1950, the documents indicate that Puschdorf first becomes Pushkino, then Pushkarevo. It is this short life as Pushkino which interests me the most. In the case of (the short-lived) Pushkino, we are dealing with a type of renaming after historical figures such as "Pushkin , Nekrasov or Bagration" which Brodersen describes as "historisierende Variante"[47] and corresponds to Kostjashov's type 3 ("names derived from leading figures in officially propagated Russian and Soviet political, literary and cultural history"). Both scholars use the term "history", but this is "history" in Dobrenko's sense of "a system of techniques and optics which produces the events themselves from the past and gives them the very status of events"[48] and at the same time dissolves real time into teleological stasis in the immobile space of 'Stalinist liberation'.

A "push" is senseless violence in itself. Pushkin himself seems to become this senselessness when he is dissolved into a reference to artillerists ("pushka" in Russian means "cannon" and "pushkar'" means "canoneer") which is an ironically inadvertent expression of truthfulness. It is dissolved history pretending to be an automatic one-way Russian-guaranteed liberation which is actually the result of very complex military-economic activity involving a sizable number of cannons (pushki).

Here again we can bring Dobrenko's trailblazing analysis into the discussion. Dobrenko's three-pronged reconstruction of 'Stalinized space' is a story of violent "transformation". "Stalin's plan for Transforming Nature" in the stamp discourse 1949 is echoed in the 1949 version of Mikhailov's book as a literal "push" of all real temporality beyond the frame of the image containing the supposed map which is not a map in the sense of an image at all, but rather a screen memory of real Russian space planted into the consciousness of the Soviet populace as a replacement for the real maps which are "all located in Stalin's office".[49] This, in turn, is a analogous to the assumption and manufacture of deficient subjectivity in Soviet films which allow only Stalin to be an authentic father, lover and man.[50] Just so, the periphery always points the Moscow, precisely through its absence.

This is especially relevant for the concept of "restoration"[51] which is applied equally and synonymically both to the liberation from capitalism and to the con-

---

47 Brodersen, *Die Stadt*, 63.

48 See above, footnote 17.

49 Dobrenko, Art of Social Navigation, 190.

50 See Lilya Kaganovsky. *How the Soviet Man Was Unmade*. Pittsburgh: University of Pittsburgh Press, 2008.

51 Dobrenko, Art of Social Navigation, 193.

quest of territory from foreign powers which is assumed to have always already been Russian-Soviet, thus making all historical development into pseudo-history and pseudo-temporality, derailed from its true path and waiting to be brought back onto the track of Russian-guided and Russian-guaranteed Stalinist teleology. Conquest as the "restoration of historical justice"[52] – this is precisely the argument which makes northern Ostpreußen always already Russian – erases its status as an historical development and puts history itself into the position of timelessly waiting for things to return the state they were already in the first place. In effect, according to this view, we were all at one point Russian and will at some point in the future return to this 'natural state'.

This is analogous to the "topographic paranoia" [53] which "taboos" space and at the same time encourages hiking within the "red line" of 'liberated humanity' as a unification of anti-spy vigilance and pro-Soviet happiness.[54] The seeming motion of the hiker on the periphery of the empire is nothing but the confirmation of the motionless absolute center. The domination of landscape on the Soviet postage stamp from the 1930s to the 1950s does the same representational work and deceives the viewer into thinking he is seeing actual land while actually presenting a screen memory of space which precludes any possibility of individually moving in real time through this space. This is Stalin's "transformation of nature" described in the spirit of socialist realism (and thus only promised, sketched in an ideal future), congealed into the slogan "Kolkhoz farmers – build ponds and reservoirs!" (kolkhozniki, stroite prudy i vodoemy). This is translated in the English version of Dobrenko's text without the limitation to "Kolkhoz farmers" ("Farmers – build ponds and reservoirs"). This limitation to collective farmers is, however, important, since it is not this any farmer, but rather precisely the worker on the collective farm who can "build" a containment of/for water which fills the essential quality of "self-sufficient closure". This feature is brilliantly deduced by Dobrenko from the Soviet tourism poem "Motherland" written by the "tourist Lev Chernomortsev"[55] the key line of which is "Wherever I may be, I'm always in the motherland". The poem was printed in the immensely popular tourist periodical "Na sushe i na more"[56] (At Land and at Sea) , the second object of Dobrenko's

---

52 Mikhailov 1949, bibliography. 24, quoted in Evgenij Dobrenko, Eric Naiman (eds.). *The Landscape of Stalinism: the art and ideology of Soviet space.* Seattle: University of Washington Press, 2003, 193.
53 Dobrenko, Art of Social Navigation, 180.
54 Dobrenko, Art of Social Navigation, 186.
55 Dobrenko, Art of Social Navigation, 85.
56 Vol. 21, 1935, 17 quoted in Dobrenko, Art of Social Navigation, 186.

studies after Soviet stamps and before the 'screen mnemotechnics' in Mikhailov's best-seller "Map of the Motherland" in its edition of 1949.

All three objects of Dobrenko's study perform the grinding up of time into ahistorical space produced in and by late Stalinist empire. None of this can be found in this form in fascist or Nazi totalitarianisms. This is why Hardt and Negri's lumping together of Nazism and Stalinism as practically identical forms of "national sovereignty" misses the specificity of what is going on in the latter.

I now come back again to Pushkin in this context. When a Soviet postage stamp appears in 1949 in honor of the 150th anniversary of Pushkin's birth, it is symptomatic that the feature of the original portrait by Kiprenskii which links the poet to Scotland is sliced off, 'transforming' the poet from a figure of international potency into one of limited and castrated Russian chauvinist 'Soviet patriotism'. Against this backdrop, we can now also return to the specificities the idea behind of naming a town in the USSR's "new territories" after the "inventor of modern Russian literature" and "the man who built the foundation (osnovopoložnik) for modern Russian language", Pushkin, and of the execution this idea in the case of Puschdorf with the violent para-morpheme "push" at its core. The irony is that the potential of "push" in its final manifestation pushes even the "great poet" Pushkin out of the picture in favor of the artillerist as his truth and the final goal of his poetry. If Brodersen is right and the new Soviet settlers had significant influence on the final naming in smaller towns, i.e. on the switch from Pushkino to Pushkarevo,[57] then were are perhaps dealing with the product of direct ideologization, if it is not the case that the settlers came from another Pushkarevo in Russia. But it is not the final naming as Pushkarevo, but rather the official suggestion of Pushkino which is my concern here.

When Pushkin's name flickers up in the soon-to-be-former Puschdorf and is then extinguished in favor of Pushkarevo, one can read this as a comment on Pushkin himself in Stalinist pseudo-history. Pushkin is here a function of operations which are in themselves senseless and only serve as signifiers whose signified is violently established and maintained. Absolute Bolshevik power legitimates itself and controls space by replacing the memory of real time with ahistorical screen memories. The cannon (pushka) or cannoneer, i.e. artilleryman (pushkar) at the core of Pushkarevo – like the coincidental violent meaning of "push" in English – ironically expresses this very function of the Stalinist Pushkin as a place-keeper of Stalin. This is nicely illustrated by the celebration of the 150th and 70th birthdays of the former and the latter in 1949.

57 Brodersen, *Die Stadt*, 59f.

Conclusion: The Anti-Colonial Imperial Implosion of Time and the "Explosion of Time" 40 Years Later

In Jáchym Topol's 1994 novel *Sestra* ("Sister", translated by Alex Zucker as *City Sister Silver* in 2000)[58], the period after 1989 is linked to the expression "explosion of time". The term is used repatedly in the novel. The period 40 years earlier discussed here, particularly 1949, might be described as the 'implosion of time' (manifesting itself in the replacement of all temporal dynamics with a static architecture representing the place to which all of history always already been directed). The year 1949, on the eve of the final approval of Kaliningrad toponyms in 1950, represents the last gasp of history in the region before it suffocates in the sweet atmosphere of the "bright future" which seems to contain everything necessary for life, but is in fact lacking in oxygen, this unfortunately being the one substance needed for the metabolism of most complex living things.

Be it Pushkino or Pushkarevo, be it Cherniakhovsk or Nesterov, or be it Kaliningrad itself, a place name in this new (and at the same time according to official propaganda extremely old) piece of the "Russian [...] Republic" marks the fact that the victory of the cannons (*pushki*) of the Red Army did not bring a new time, but rather the (teleo)logically necessary and absolutely just "push", the necessary violence, of the Red Army returned the landscape to where it has always has been.

One might, with Hardt and Negri, say that the extreme national hard reset at hand is nothing but the ruthlessness of national sovereignty insisting on its due. But the mercilessly displacing hardness of the Kaliningrad region's Soviet reset (at least in the area of naming, where absolutes are quickly and cheaply attainable, as opposed to actual physical change for which both the funding and the will turned out to be insufficient[59]) cannot be explained solely by this post-Marxist observation of the workings of economy and ideology. The complete shutting down of time, the "degree zero of history" (Dobrenko) which 'dawns' (actually there can be no dawn in timelessness) on the northern part of the former Ostpreußen in the late 1940s, is the effect of a late Stalinist chronotope of pure spatiality which absolutely displaces (i.e. dissolves) time. Its toponyms are symptoms and indices of this chronotope. The same thing can be said about them which Dobrenko says about official Stalinist depictions of reality in film: "one can, of

58 Jáchym Topol. *Sestra*, Prague: Atlantis, 1994; *City Sister Silver*, translated by Alex Zucker, North Haven/CT: Catbird Press, 2000, passim.
59 See Bert Hoppe's article in this volume.

course, object that nobody ever lived in this world anyway. But this is just as true as saying that it was precisely here that we did live".[60]

And having said all this, I would repeat the argument that "here" is a space basically without time, and in further reflections one would have to analyze what the words "ever lived", "is true" and "did live" actually can mean with respect to a time which never was, since time itself was displaced as part of the political program which forcibly displaced millions, including the roughly half a million who wound up becoming citizens of the "most Western" part of Russia[61].

References

Barbusse, Henri. *Stalin. A new world seen through one man*. New York: International Publishers, 1935.

Barthes, Roland. *Le Degré zéro de l'écriture suivi de Nouveaux essais critiques*. Paris: Éditions du Seuil, 1972 [1953].

Brodersen, Per. *Die Stadt im Westen: Wie Königsberg Kaliningrad wurde*. Göttingen: Vandenhoeck & Ruprecht, 2008.

Brown, Matthew and Matteo Lanfranconi. *Socialist Realisms. Great Soviet Painting 1920–1970*. New York: Skira 2012.

Certeau, Michel de. *L'invention du quotidien*. Paris: Gallimard, 1990

Clark, Katerina. Socialist Realism and the Sacralizing of Space. In *The Landscape of Stalinism the art and ideology of Soviet space*, Evgenij Dobrenko, Eric Naiman (eds.), 3–19. Seattle: University of Washington Press, 2003.

Derrida, Jacques. *The Politics of Friendship*. London: Verso, 1997.

Dobrenko, Evgenij. The Art of Social Navigation: The Cultural Topography of the Stalin Era. In *The Landscape of Stalinism: the art and ideology of Soviet space*, Evgenij Dobrenko, Eric Naiman (eds.), 163–200. Seattle: University of Washington Press, 2003.

Dobrenko, Evgenij. *Political Economy of Socialist Realism*. New Haven: Yale University Press, 2007.

Dobrenko, Evgenij. *Stalinist Cinema and the Production of History*. Edinburgh: Edinburgh University Press, 2008.

Dobrenko, Evgenij and Eric Naiman (eds.). *The Landscape of Stalinism: the art and ideology of Soviet space*. Seattle: University of Washington Press, 2003.

Epstein, Mikhail. Russo-Soviet Topoi. In *The Landscape of Stalinism the art and ideology of Soviet space*, Evgenij Dobrenko, Eric Naiman (eds.), 277–306. Seattle: University of Washington Press, 2003.

Gousseff, Catherine. *Échanger les peuples: Le déplacement des minorités aux confins polono-soviétiques* (1944–1947). Paris: Fayard, 2015.

60 Dobrenko, Art of Social Navigation, 199.

61 Models of displacement can also be applied to the 2.5 million East Prussians after the war, particularly those who wound up in West Germany. See Charlton Payne's article in this volume.

Hardt, Michael and Antonio Negri. *Empire*. Cambridge: Harvard University Press, 2000.
Kaganovsky, Lilya. *How the Soviet Man Was Unmade*. Pittsburgh: University of Pittsburgh Press, 2008.
Kossert, Andreas. 'Grenzlandpolitik' und Ostforschung an der Peripherie des Reiches. Das ostpreußische Masuren 1919–1945. *Vierteljahrsschrift für Zeitgeschichte* 51:2 (2003): 117–146.
Kostjashov, Jurij. *Sekretnaja istorija Kaliningradskoj oblasti*. Kaliningrad: Terra baltika, 2009.
Meyer, Holt. Spiel(k)arten der Distanz in der ‚begrenzten Souveränität': Bohumil Hrabals Kdo jsem. In *Distanz. Schreibweisen, Entfernungen, Subjektkonstitutionen in der tschechischen und mitteleuropäischen Literatur*. Nora Schmidt, Anna Förster (eds.), 23–63. Weimar: VDG, 2014.
Mikhailov, Nikolaj. *Nad kartoi rodiny*. Moscow: Molodaia Gvardiia, 2d ed., Moscow: Voennoe izdatel'stvo, 1949.
Neumärker, Uwe et. al. *"Wolfsschanze": Hitlers Machtzentrale im Zweiten Weltkrieg*. Berlin: Ch. Links Verlag, 3. Aufl. 2007 [1999].
Plamper, Jan. The Spatial Poetics of the Personality Cult. In *The Landscape of Stalinism*, Evgenij Dobrenko, Eric Naiman (eds.), 19–50. Seattle: University of Washington Press, 2003.
Schmitt. Carl. *The Concept of the Political*. Chicago: University of Chicago Press, 1996.
Topol, Jáchym. *Sestra*, Prague: Atlantis, 1994; *City Sister Silver*, translated by Alex Zucker, North Haven/CT: Catbird Press, 2000.
Yoshioka, Jun. Imagining Their Lands as Ours: Place Name Changes on Ex-German Territories in Poland after World War II. In *Regions in Central and Eastern Europe: Past and Present*, Hayashi Tadayuki, Hiroshi Fukuda (eds.). 21st Century COE Program Slavic Eurasian Studies 15. 273–288. Sapporo, Japan: Slavic Research Center, Hokkaido University, 2007.

Maps

Map 1 [= detail]: Over 100 renamings on the right margin: The cartographic results of the 'Germanization' of East Prussia, with changes in red writing in preparation for the 1939 edition. (Excerpt from *Stielers Hand-Atlas. 254 Haupt- und Nebenkarten in Kupferstich*, Hermann Haack (ed.), Gotha: Justus Perthes, sheet 9: Ostpreußen, version March 14th, 1939. Forschungsbibliothek Gotha der Universität Erfurt, SPA 2 45 (9)).

Map 2: The complete cartographic results of the 'Germanization' of East Prussia (*Stielers Hand-Atlas. 254 Haupt- und Nebenkarten in Kupferstich*, Hermann Haack (ed.), sheet 9: Ostpreußen, version March 14, 1939. Forschungsbibliothek Gotha der Universität Erfurt, SPA 2 45 (9)).

Bert Hoppe

Der Traum von Klein-Moskau im Westen

Architektonische Visionen und Gestaltungen Kaliningrads als eines imperialen Raums in der Nachkriegszeit

English abstract: The Dream of Little Moscow in the West. Architectural Visions and Designs of Kaliningrad as a Postwar Imperial Space. The resettling and reconstruction of Königsberg as the Kaliningrad was originally conceived of by the new Soviet administration as a radical departure from the past. The space was to completely adapted to the new time. The new municipality was not only to rise like a phoenix from the ashes of the destroyed city, but was also to become the antithesis of the "robber's den of German imperialism". However, both in the postwar public discussion and in the actual city planning, the new Kaliningrad remained much more connected to the old Königsberg as the planners and the party functionaries would have preferred. This was caused not only by purely material and financial limitations, causing the impossibility of building anything. It was also a result of the conceptualization of the empire which depicted that city as its furthest outpost and was perceived by its inhabitants as such.

Zu den Absurditäten der Geschichte Königsbergs-Kaliningrads im 20. Jahrhunderts zählt, wie ähnlich die Stadt und ihre Region in den deutschen und sowjetischen Diskursen in den jeweils zwei Jahrzehnten vor und nach 1945 dargestellt wurden: Dominierendes Element war in beiden Fällen das Bild von der Stadt als „Bollwerk" oder „Vorposten". Der Unterschied lag auf dieser Ebene lediglich darin, gegen wen sich die Abwehr richtete – waren Ostpreußen und Königsberg in der Zwischenkriegszeit zu zentralen Orten des „Volkstumskampfes" gegen die „slawische Flut" stilisiert worden, galten Kaliningrad und das umgebende Gebiet von 1945 an als Verteidigungsposten gegen den deutschen „Drang nach Osten".

Wie stark diese beiden Diskurse aufeinander bezogen waren, hatte sich bereits zwei Tage nach der Kapitulation der deutschen Verteidiger von Königsberg am 9. April 1945 abgezeichnet, als die *Pravda* voller Stolz meldete: „Die deutschen Räuber machten seit je her aus Königsberg einen Vorposten ihrer Eroberungspolitik im Osten Europas. Die Deutschen haben das Territorium Ostpreußens in eine Aufmarschbasis für Eroberungsüberfälle auf slawisches Gebiet verwandelt.

DOI 10.1515/9783110418750-024

[…] Nun hat die Rote Armee dieses Räubernest des deutschen Imperialismus in Königsberg auf ewig liquidiert."[1]

Seit Stalins Neuinterpretation des Lenin'schen Imperialismusbegriffs wurde dieser als Synonym für die Bedrohung der Sowjetunion durch kapitalistische Staaten verwendet, allerdings ursprünglich in erster Linie gegen das britische Empire gerichtet. Seit Mitte der 1930er-Jahre waren aufgrund der geänderten außenpolitischen Lage die Achsenmächte in den Vordergrund der anti-imperialistischen Polemik getreten. Bemerkenswert war in diesem Zusammenhang, dass sich der zeitliche Fokus des sowjetischen Imperialismus-Diskurses immer weiter in die vormoderne (und vorkapitalistische) Vergangenheit verlagerte.

Das lässt sich als Reaktion auf die Propagierung des nationalsozialistischen „Lebensraum"-Konzeptes interpretieren. So hatte Himmler etwa die Stiftkirche zu Quedlinburg zu einer NS-Weihestätte umwandeln lassen, weil dort der deutsche König Heinrich I. begraben lag, den er als Ahnherrn der deutschen Ostsiedlung verehrte; die sowjetische Antwort auf derlei historische Rückgriffe gipfelte konsequenterweise in Sergej Eisenstein Film „Aleksandr Nevskij" von 1938 über die Niederlage des Deutschen Ordens im Jahr 1242. Mit diesem Film zog Eisenstein eine geistige Verbindungslinie von der Eroberungspolitik des Deutschen Ordens zur aggressiven Annexionspolitik Hitlerdeutschlands (Kurz vor dem Abschluss des Hitler-Stalin-Pakts wurde der Streifen aus dem Programm genommen, und Eisenstein inszenierte stattdessen am Moskauer Bolschoj Theater Wagners „Walküre". Erst nach dem deutschen Überfall auf die Sowjetunion am 22. Juni 1941 wurde Eisensteins Film wieder in sowjetischen Kinos gezeigt.)[2]

Der *Pravda*-Artikel vom 9. April 1945 über Königsberg fügte sich also in einen schon bestehenden Diskurs ein und gab zugleich den Ton vor, mit dem in der Folgezeit über die Hauptstadt Ostpreußens in der sowjetischen Öffentlichkeit gesprochen wurde. Aus der Logik dieses Diskurses heraus betrachtet lag es daher nahe, die angenommene Rolle der Stadt beizubehalten und nur die Stoßrichtung umzudrehen. Trotz der ideologischen Kluft, die sich zwischen dem deutsch-nationalen und dem russisch-sowjetischen Diskurs auftat, ergab sich auf diese Weise eine bemerkenswerte strukturelle Parallele – in der Darstellung der Stadt traten deren zivilen Merkmale hinter ihre geostrategische Bedeutung zurück.

---

1 A. Ivanov, Kenigsberg. In *Pravda*, 11. April 1945.

2 Zu den sowjetischen Filmen der Vorkriegs- und Kriegszeit vgl.: Richard Stites, *Russian popular culture entertainment and society since 1900*, Cambridge: Cambridge University Press, 1993, 112–115. Zu Eisensteins Wagnerinszenierung vom Dezember 1939 vgl.: Karl Schlögel. *Moskau – offene Stadt eine europäische Metropole*. Reinbek bei Hamburg: Rowohlt Verlag 1992 [1984], 303.

Das galt auch für das Selbstverständnis, das den neuen Bewohnern Kaliningrads zugedacht wurde. Manche der Resolutionen, die ihnen in der Nachkriegszeit zur Unterzeichnung vorgelegt worden sind, hätten mit geringfügigen Änderungen ebenso gut den Angehörigen der örtlichen Garnision der Roten Armee unterschrieben werden können. Die Arbeiter der Papierfabrik etwa durften sich mit folgenden markigen Worten zu ihrer neuen Heimatstadt bekennen: „Wir, die ersten Sowjetbürger der Stadt Kaliningrad am westlichsten Vorposten der UdSSR, werden mit beharrlicher Arbeit die Verteidigung der Sowjetunion stärken."[3]

Mit derlei Bekenntnissen wurde der „Vorposten" Kaliningrad aber nicht ausschließlich auf seine quasi-militärische Funktion reduziert (ironischerweise ist das russische Wort „forpost" wie zahlreiche andere russischen Militärbegriffe ein Lehnwort aus dem Deutschen). Ebenso wichtig war, dass Stadt und Region mit diesem Begriff als integraler Teil eines viel größeren Ganzen definiert wurden – des sowjetischen Imperiums. Ein Autor der bald nach Kriegsende neugegründeten lokalen *Kaliningradskaja Pravda* widmete daher im Mai 1948 der Düne mit dem westlichsten Grenzpfahl der Sowjetunion an der Küste der Frischen Nehrung einen ganzen Artikel, um von der Weite des Landes zu schwärmen, dem sich die neuen Bewohner des Kaliningrader Gebiets zugehörig fühlen durften: „Und jeden, der hier hinaufsteigt, ergreift das unbeschreibliche Gefühl von Stolz auf die Größe der Heimat, auf die Endlosigkeit ihres Raums – von den weit entfernten Kurilen bis zu diesem Hügel."[4] In diesen Sätzen vom „Vorposten" und von der Düne am Rande des vermeintlich „endlosen" Raums findet sich das eigentümliche Paradoxon des „antiimperialen Imperialismus" gespiegelt: Die Ausdehnung des sowjetischen Imperiums galt in dieser Sichtweise quasi als Akt der „Vorwärtsverteidigung", gerichtet gegen die Herausforderer jenseits der Grenzen, die es ihrerseits auf die Beherrschung der russischen Kernlande absahen.

Hinsichtlich der architektonischen Planungen für die im Sommer 1944 bei zwei Luftangriffen der Royal Air Force und im Frühjahr 1945 im Zuge wochenlangen Artilleriebeschusses und heftiger Straßenkämpfe weitgehend zerstörten Stadt hatte der auf seinen deutschen Gegenpart bezogene Diskurs von Kaliningrad als „Vorposten" dementsprechend eine doppelte Wirkung: Zum einen galten die Bauten aus deutscher Zeit fortan als Zeugnisse einer Jahrhunderte andauernden Besatzung der Region und des aggressiven deutschen „Drangs nach Osten"

---

3 Resolution einer Versammlung von Arbeitern der Kaliningrader Gebietsverwaltung für Zellulose- und Papierindustrie vom 2. Juli 1946, zitiert nach: Per Brodersen. *Die Stadt im Westen. Wie Königsberg Kaliningrad wurde*. Göttingen: Vandenhoeck & Ruprecht, 2008, 173.

4 Il'ja Zhernakov, Na krajnij zapade strany. In *Kaliningradskaja Pravda*, 1. Mai 1948, zitiert nach: ebenda.

und sollten daher nach Möglichkeit aus dem Weichbild der Stadt verschwinden: Die Arbeiterviertel des ehemaligen Königsberg, so konstatierte etwa Kaliningrads erster Chefarchitekt Dimitrij Navalichin Mitte der 1950er-Jahre, zeichneten sich „durch eine äußerste Primitivität der planerischen und architektonischen Lösungen“ aus, die Häuser wirkten wie „trostlose, freudlose Kasernen“, zivile Gebäude hätten nur eine „untergeordnete Bedeutung“ besessen. Das so entstandene Stadtbild des ehemaligen Königsberg trage daher nicht allein die für privatkapitalistische Städte ohnehin typischen (negativen) Merkmale, sondern zudem „die eigenartigen Züge einer deutschen Militärstadt, „die Ideen und Vorstellungen propagiert und bekräftigt, die der Menschheit verhasst sind“.[5]

Zum anderen wurden die konkreten Wiederaufbauplanungen von der Vorstellung geprägt, im neuen Stadtbild müsse sich deutlich ablesen lassen, dass Kaliningrad ein untrennbarer Bestandteil des sowjetischen Imperiums war: Gerade aufgrund der Position dieser Stadt als „Vorposten“ am „äußersten Rand“ des Landes durfte kein Zweifel an ihrer Zugehörigkeit aufkommen.

Die städtebaulichen und architektonischen Lösungen, die die Verantwortlichen in den folgenden Jahren umzusetzen versuchten, offenbarten allerdings schließlich vor allem ein hohes Maß an Unsicherheit, die zu einem paradoxen Ergebnis führte. Während Imperien sich dadurch auszeichnen, dass sie bei aller Dominanz des Zentrums an der Peripherie ein gewisses Maß an kultureller Eigenständigkeit bzw. Andersartigkeit aufweisen, sollte der Wiederaufbau von Kaliningrad eben diese Unterschiede zwischen Zentrum und Peripherie völlig vergessen machen. Einerseits wurde also die Lage der Stadt „ganz im Westen“ des sowjetischen Imperiums bei jeder sich bietenden Gelegenheit in Reden, Resolutionen, Zeitungsartikeln und Gedichten thematisiert und überhöht, andererseits herrschte unter den Machthabern eine große Furcht davor, es könne jemand daraus die falschen Schlussfolgerungen ziehen.

Die Verwirrung über das richtige Verhältnis von Zentrum und Peripherie ist in Kaliningrad ganz bildlich am Erscheinungsbild der Stadt abzulesen – bis heute ist der Wiederaufbau der Innenstadt Stückwerk geblieben, insbesondere der zentrale Bereich der einstigen Altstadt zwischen den Fundamenten des einstigen Ordensschlosses und der Kathedrale auf der Pregelinsel präsentiert sich als eine weite, freie Fläche, durchzogen nur von breiten Verkehrsschneisen, während sich das gesellschaftliche, politische und wirtschaftliche Leben des modernen Kaliningrads überwiegend in den ehemaligen Randbereichen der ostpreußischen Metropole abspielt. Innerhalb des durch die frühere Stadtbefestigung gebildeten

5 Dmitrij Navalichin. *K voprosu rekonstrukcii centra goroda Kaliningrada.* Moskau 1958 (zweite Fassung der Dissertation), Bd. I, 26–33 und 167.

Straßenrings um die Innenstadt besteht bis heute keine wichtige öffentliche Institution, mit der einsamen Ausnahme der Kant-Universität, die im wiedererrichteten Hauptgebäude der Universität Albertina eine Art Außenstelle eingerichtet hat (der eigentliche Campus liegt mehrere Hundert Meter entfernt an der Ausfallstraße zum Flughafen).

Wenngleich die Verlagerung des Zentrums letztlich nicht auf eine bewusst getroffene Entscheidung zurückzuführen ist, sondern in erster Linie Folge eines permanenten Mangels an Ressourcen war, wurde in der Nachkriegszeit zwischenzeitlich tatsächlich erwogen, das alte Königsberg als eine Art neuzeitlichem Karthago wüst fallen zu lassen und die neue sowjetische Stadt abseits neu zu errichten. Ein Text, 1951 in Kaliningrad unter dem Titel „Im Westen gibt es kein Ostpreußen mehr“ in einem „literarisch-künstlerischem“ Sammelband publiziert, gibt eindringlich wieder, wie das offiziell gewünschte Verhältnis zu den Überresten des unterworfenen Königsberg aussehen sollte: „Was für eine Stadt! Die Straßenbahn führt uns durch buckelige und enge Straßen des ehemaligen Königsbergs. Ehemalig deshalb, weil Königsberg tatsächlich eine ehemalige Stadt ist. Sie existiert nicht. Kilometerweit öffnet sich ein unvergessliches Ruinengemälde. Dort stehen, eng aneinander gedrückt, zerstörte leere Häuser, ohne Dächer und Abdeckungen. Sie sind mit Unkraut und Efeu überwuchert. In jeder Ruine sprießt ein kleiner Wald. Seitlich, inmitten der Ruinen, liegt das Schloss der drei Könige – eine Zitadelle, deren massive Mauern von der Zeit und vom Wetter geschwärzt sind. Das alte Königsberg ist eine tote Stadt. Es wieder aufzubauen wäre sinnlos. Einfacher, praktischer ist es, eine neue Stadt zu bauen. Was aber mit der alten Stadt machen! Die Kaliningrader schlagen im Ernst vor, um das zerstörte Königsberg eine Mauer zu ziehen und von Zeit zu Zeit dorthin Anwärter auf die Weltherrschaft zu führen.“[6]

Der Begriff „ehemalig“ hat innerhalb des stalinistischen Diskurses einen besonderen Klang, ist er doch mit den „byvshie ljudi“ verknüpft, den „ehemaligen Leuten“, wie die Angehörigen des Bürgertums und Vertreter der vorsowjetischen Staatsmacht und Bürokratie genannt wurden. Mit diesem Begriff war also weit mehr als nur die bloße Tatsachenfeststellung verbunden, dass das alte Königsberg „nicht mehr existent“ war. Er beinhaltete zugleich die geschichtsmaterialistische Sichtweise, dass sich die „ehemalige Stadt“ Königsberg – ebenso wie die „ehemaligen Leute“ – geschichtlich überlebt hatten und daher nun völlig zu Recht unter Unkraut und Efeu dahindämmerte, gewissermaßen als Friedhof ihrer selbst und der historischen Epoche, die sie verkörperte.

---

6 Na zapade – net bol'she vostochnoj Prussii. Literaturno-chudozhestvennij i obshestvenno-politcheskij sbornik, Kaliningrad 1951, 14.

Zum Zeitpunkt als dieser Aufsatz gedruckt wurde, war allerdings schon die Entscheidung gefallen, doch die gesamte Innenstadt innerhalb des historischen Befestigungsrings wieder aufzubauen, bzw. neu zu errichten. Im Mai 1948 hatte der Chef der Verwaltung für Architektur beim Ministerrat der RSFSR, V. Škvarikov, den Direktor des Moskauer Staatlichen Institutes für die Projektierung von Städten (*Giprogor*), N. Ja. Burlakov, mit der Ausarbeitung eines Generalplans für Kaliningrad beauftragt.[7] Und als der Leiter des *Giprogor*-Projekts, M. R. Naumov, ein knappes Jahr später die Konzeption seines Teams in einem Artikel für die *Kaliningradskaja Pravda* vorstellte, sprach er sich eindeutig gegen die Verlagerung des Zentrums an die Peripherie aus. Er führte für seine Entscheidung jedoch an erster Stelle pragmatische Gründe an: Im Zentrum träfen die Verkehrsverbindungen zwischen den Außenbezirken aufeinander, in diesem Bereich gebe es eine Reihe wiederaufbaufähige Gebäude, und schließlich seien die unterirdischen Versorgungsleitungen noch zu verwenden.

Auffällig war auch, dass der konkrete Wiederaufbauplan für das Zentrum, den die Mitarbeiter von Giprogor entworfen hatten, zwar formal den Forderungen nach einer repräsentativen Umgestaltung entsprach, etwa durch Verbreiterung und Begradigung von Straßen und die Errichtung eines großen Palastes der Sowjets anstelle der Ruine des Schlosses. Gerade dieser Palast der Sowjets aber sollte dieser Planung nach den besonderen Ort der Stadt symbolhaft widerspiegeln: „Das große Gebäude des Sowjetpalastes soll ein Denkmal für Michail Ivanovich Kalinin sein, den großen Aktivisten der Kommunistischen Partei und des sowjetischen Staates. Der Palast soll gekrönt sein von einem hohen, schon von fern sichtbaren Leuchtturm, der den Charakter der Stadt Kaliningrad als Hafenstadt unterstreichen soll. Die Schaffung des künftigen Palastes ist die Aufgabe unserer sowjetischen Architekten, die ein Projekt ausarbeiten sollen, das des neuen Kaliningrads würdig ist."[8]

Bemerkenswerterweise bemühten sich die Moskauer Planer mit solchen Ideen stärker als die in Kaliningrad selbst maßgeblichen Architekten, dem genius loci zumindest in Ansätzen gerecht zu werden, während bei Letzteren der Gedanke der Repräsentativität bei der Frage nach Lage und Gestaltung des neuen Stadtzentrums eindeutig im Vordergrund stand. Der 1948 berufene Chefarchitekt von Kaliningrad, Dimitrij Navalichin, entschied sich in erster Linie gegen dessen Verlagerung der Stadtmitte in westlicher Richtung, weil dies die Realisierung eines „genügend ausdrucksstarken allgemeinen kompositorischen Stadtentwur-

---

7 Gosudarstvennij Archiv Kaliningradskoj Oblasti (GAKO), f. 520, op. 1, d. 8, l. 1, Befehl Nr. 195 des Chefs der Verwaltung für Architektur beim Ministerrat der RSFSR, Shkvarikov.

8 M. R. Naumov, Kakim budet Kaliningrad. In *Kaliningradskaja Pravda*, 30. April 1949.

fes" verhindert hätte[9] – ein schwarzes Loch im geographischen Mittelpunkt des sozialistischen Kaliningrad wollte er verhindern.

Der Leiter der Gebietsverwaltung für Architektur, Dimitrij Tjan, machte sich wiederum in erster Linie Gedanken über die Außenwirkung der neuen Stadt: Für ihn stand bereits Ende 1947 fest, dass sich in der städtebaulichen Gestalt Kaliningrads die Rolle der Stadt „als westlicher Vorposten, als Vorhof der Russländischen Föderation" ausdrücken müsse. „Die architektonische Planung der Stadt soll die Macht unseres großen Staates und die russische Gastfreundlichkeit ausdrücken." Diese widersprüchlich erscheinende Verbindung sollte Assoziationen wecken an das populäre Lied „Shirokaja strana maja rodnaja" („Weit ist mein Heimatland") aus dem Film „Zirkus" von 1936 über eine schwarze Amerikanerin, die vor dem Rassismus in den USA in die Sowjetunion flieht, dort warmherzig aufgenommen wird und schließlich gemeinsam mit ihren neuen Freunden und Genossen an der monumentalen Mai-Parade auf dem Roten Platz teilnimmt.

Den Vorstellungen Tjans zufolge sollte längs durch das alte Zentrum in Nord-Südrichtung ein gigantischer Prospekt durchgebrochen werden: „Für den Bau der Magistrale werden die zerstörten Stadtviertel abgetragen. Sie wird so breit werden, dass gleichzeitig Dutzende Autos nebeneinander herfahren können."[10] Als Vorbild für den vorgesehenen übertrieben breiten Prospekt diente ihm die Vorzeigestraße der Sowjetunion, die Gorkij-Straße in Moskau, deren Umgestaltung eines der Kernprojekte des „Generalplanes für die Rekonstruktion Moskaus" darstellte. Von ursprünglich 18 Metern war sie auf 59 Meter verbreitert worden.[11] Die Nord-Süd-Achse Kaliningrads sollte die Gorkij-Straße in ihrer Breite nun sogar noch übertreffen: Zeitweise wurde eine Breite von 65 Metern ins Auge gefasst.[12] Damit war Tjan im Wortsinne über das Ziel hinausgeschossen: Moskau in einer im Wortsinne am Boden zerstörten Stadt am äußersten Rand des sowjetischen Imperiums übertrumpfen zu wollen, war weder ökonomisch sinnvoll noch politisch statthaft. Es ist vor dem Hintergrund der extrem hierarchisierten politischen Struktur der spätstalinistischen Sowjetunion zwar kaum anzunehmen, dass die entsprechenden Pläne mit dieser Absicht entworfen worden, es gibt jedenfalls keine entsprechenden Äußerungen. Allerdings zeichnet sich hier bereits ein charakteristisches Merkmal der lokalen Wiederaufbauprojekte ab: Auffällig an den

---

9 Dimitrij Navalichin. *K voprosu rekonstrukcii centra goroda Kaliningrada*. Moskau 1955 (erste Fassung der Dissertation), 96–97.

10 Dimitrij Tjan, Sovetskij gorod Kaliningrad. In *Kaliningradskaja Pravda*, 7. November 1947.

11 Vgl. Andrew Day, Building Socialism. The Politics of Soviet Cityscape in the Stalin-era, Dissertation Columbia University New York 1997, 65, Anm. 272. Zum Moskauer Generalplan von 1935 vgl. ebd., 44–48 und 64–70.

12 Vgl. die Abbildung Nr. 11.3 in: Navalichin, *Rekonstrukcii* (1958), Bd. 2.

Planungen sowohl von Tjan wie von Navalichin war die exzessive Orientierung an Moskau. Die sowjetische Hauptstadt galt nicht lediglich als Vorbild in dem Sinne, dass dort die Leitlinien vorexerziert wurden, nach denen sich die städtebauliche Gestalt einer Provinzhauptstadt zu orientieren hatte. Wie noch zu zeigen sein wird, sollte Moskau im Falle Kaliningrads darüber hinaus als ganz konkretes „Vorbild“ bei der Ausgestaltung des städtischen Weichbilds dienen.[13]

Im folgenden Jahrzehnt standen die Entwürfe aus dem Büro des Chefarchitekten der Stadt Kaliningrad fortwährend in Konkurrenz zu den Plänen des Moskauer *Giprogor*, dessen Einfluss als Zentralinstitut deutlich abgenommen hatte, seitdem das Staatliche Komitee für architektonische Angelegenheiten Ende 1949 aufgelöst und durch ein Ministerium für Städtebau ersetzt worden war. Da die Aufgaben dieses neuen Ministeriums mehr darin bestanden, sich um quantitative Erfolge beim Wiederaufbau der sowjetischen Städte zu kümmern als um die städtebauliche Gestaltung,[14] vergrößerte sich der Spielraum der lokalen Chefarchitekten auf dem Gebiet der Gestaltung.

Diesen Spielraum nutzte Navalichin, um die Planung seiner Kollegen aus der Hauptstadt als zu unentschieden zu kritisieren und dieser im Mai 1953 einen wesentlich radikaleren Entwurf entgegenzusetzen. In der Gegenüberstellung seines Projektes mit der Planung von *Giprogor* warf er den Moskauer Architekten noch 1958 vor, zu zaghaft mit den bestehenden Strukturen umgegangen zu sein. Schon der Ansatz der Planung, die deutsche Bebauung und die alten Versorgungsleitungen so weit wie möglich zu erhalten, um so Geld zu sparen, erklärte er für falsch. Die deutsche Bebauung werde völlig überbewertet und führe dazu, dass man sich unnötigerweise von der alten städtebaulichen Gestalt abhängig mache. Mit ihren Entwürfen hätten die Architekten von *Giprogor* deshalb „unfreiwillig einige negative Merkmale der ehemaligen Stadt Königsberg in der sowjetischen Stadt Kaliningrad verstärkt“ und auf diese Weise die weitere Entwicklung Kaliningrads „als moderne sowjetisch-russische Stadt“ behindert.[15] Indirekt forderte er, die Innenstadt als *tabula rasa* zu betrachten: Insgesamt seien die „Möglichkeiten, die in Zusammenhang mit den bestehenden Zerstörungen der Stadt

---

13 Zu den konkreten Bauplanungen der Wiederaufbauphase siehe: Markus Podehl. *Architektura Kaliningrada. Wie aus Königsberg Kaliningrad wurde*. Marburg: Herder-Institut, 2012, v.a. 82–185.

14 Tatsächlich wurden im Gründungsdekret des Ministeriums die bisherigen Leitbegriffe der sowjetischen Architektur „Kunst“ und „Schönheit“ gar nicht erwähnt. Vgl. Ukaz Verkhovnogo Soveta SSSR. Ob obrazovanij obshesojuznogo Ministerstva gorodskogo stroitel'stva. In *Arkhitektura i stroitel'stva* 7 (1949), 1, und Day, Building Socialism, 86–87.

15 Navalichin, *Rekonstrukcii* (1958), Bd. 1, 101, 104.

für die Schaffung eines maximalen Komforts für die Bevölkerung entstanden sind“, viel zu wenig genutzt worden.[16]

Ein besonderes Anliegen war Navalichin schließlich das Haus der Sowjets, dessen Bearbeitung im Generalplan von *Giprogor* in seinen Augen zu kurz gekommen war (von einem „Palast der Sowjets“ war bei ihm forthin nicht mehr die Rede, offenkundig war er sich der hierarchischen Unterordnung der Gebietshauptstadt unter Moskau bewusst). Negativ beurteilte Navalichin vor allem die im Generalplan vorgesehene „mittlere Höhe“ des Hauses der Sowjets: Die Behandlung des „führenden Ensembles“ als „gewöhnliches Gebäude“ ließ es für ihn als Fluchtpunkt von Magistralen völlig ungeeignet erscheinen. Der Generalplan habe weder die Frage nach der „plastischen Verbindung“ des Hauses der Sowjets mit anderen städtischen Ensembles, noch die „symbolisch-künstlerische Gestalt und die Silhouette“ dieses Gebäudes gelöst. All diese Fragen seien beim Wiederaufbau „einer gewöhnlichen sowjetischen Stadt möglicherweise nicht Aufgabe des Generalplanes, doch unter den gegenwärtigen Verhältnissen von Kaliningrad, wo nicht nur Fragen der Organisation gründlich revidiert werden, sondern auch der symbolische Gehalt und die früher bestehende Form, ist es unabdingbar, darauf eine ausführlichere und inhaltsvollere Antwort gerade in diesem Dokument zu finden“.[17]

Was in Navalichins Kritik am Generalplan bereits anklang, entwickelte er in dem Entwurf weiter, den er 1954 gemeinsam mit Architekten des Gebietsplanungsinstituts (*Kaliningradskaja oblastnaja proektnaja kontora – Olproekt*) ausgearbeitet hatte.[18] Kernpunkt seines Planes war das Bemühen, für Kaliningrads Zentrum eine „funktionale, symbolische und kompositorische Einheit“ herzustellen. Es sollten also nicht nur ästhetisch-städtebauliche Aspekte und die Erfordernisse einer modernen Verkehrsinfrastruktur berücksichtigt werden, die Stadt sollte zugleich ein neues ideologisches Gepräge erhalten. Bezeichnenderweise führte Navalichin als Beispiel für eine solche Einheit vor allem Moskauer Siegesmonumente an, an erster Stelle christliche Sakralbauten, bei denen er aber vor allem ihre Bedeutung als nationale Monumente in den Vordergrund rückte: Neben der Grablegungskirche [*cerkov rizpolozhenija*], gebaut 1485 bis 1486 zum Gedächtnis des Sieges von Vasili des Dunklen über die Tataren im Jahr 1451, war dies die Kirche des Seligen Vasili [*cerkov Vasili Blazhennogo*], gebaut 1555 bis 1560 zum Gedächtnis des Sieges von Ivan des Schrecklichen über die Khanate Kazan

---

16 Navalichin, *Rekonstrukcii* (1958), 102.

17 Navalichin, *Rekonstrukcii*, 107–108.

18 Vgl. Razrabotka detal'nich proektov central'nich magistralej. In *Kaliningradskaja Pravda*, 3. Februar 1954.

und Astrachan. Auch in sowjetischen Städten wie Stalingrad seien „eine Reihe von Plätzen und die sie bildenden Gebäude [...] bestimmten Ereignissen gewidmet: Der Platz der Kämpfer, der Platz der Helden, der Park des Sieges sind dazu berufen, das Heldentum und das Pathos des Großen Vaterländischen Krieges auszudrücken.“[19] Der Bezug auf Stalingrad und den Weltkrieg ist hierbei besonders interessant. Das heutige Wolgograd gehörte nach 1945 zu den sogenannten Heldenstädten, die für den Behauptungswillen der „Sowjetmenschen“ gegen die deutschen Invasoren standen – eine vergleichbare Rolle konnte Kaliningrad als einstiges „Räubernest“ des deutschen Imperialismus freilich nicht übernehmen, ihre Zerstörung wurde in der sowjetischen Propaganda ja vielmehr als gerechtes Purgatorium dargestellt. Mit den Bezügen auf Stalingrad wurde nun versucht, das kriegerische Renommee jener Heldenstadt wenigstens teilweise auf Kaliningrad zu übertragen – etwa indem man die wichtigste Straße der Pregel-Stadt in „Stalingrader Prospekt“ umbenannte. In seinen bereits 1950 verfassten und vom Gebietsexekutivkomitee gebilligten Gestaltungsregeln für die Umgestaltung dieser Straße betonte der Chefarchitekt, die Gebäude sollten „den Heroismus und das Pathos unserer Epoche“ wiederspiegeln, damit die rekonstruierte Magistrale „ihres ruhmvollen Namens würdig“ sei.[20] In diesen Zusammenhang gestellt, gewinnt die von Navalichin für Kaliningrad geplante Schaffung eines „ganzen Systems von Ensembles, die reich an neuem symbolischen Gehalt sind“, eine ganz konkrete Aussage. Die möglichst umfassende Umgestaltung der Stadt, die Navalichin im Falle Kaliningrads als besonders dringlich ansah, da das ehemalige Königsberg keine „tiefen russischen Traditionen“ besäße,[21] zielte darauf ab, die ganze Stadt zu einem bewohnten Siegesdenkmal umzugestalten – und der Stadt eben jene Traditionen zu geben.

Die Traditionen, auf die Navalichin hier abhob, beschränkten sich allerdings nicht allein auf die historische Abolge militärischer Siege des Russischen Reichs, in die sich nun der Sieg der Sowjetunion im Zweiten Weltkrieg einreihte. Navalichin wollte mit seinem konkreten Entwurf für den Wiederaufbau des Stadtzentrums auch die enge Verbundenheit Kaliningrads mit der sowjetischen Hauptstadt visualisieren. Nicht nur das Erscheinungsbild einzelner Gebäude und Straßenzüge sollte den Betrachter an die russische Hauptstadt erinnern, der gesamte Stadtgrundriss Kaliningrads sollte dem von Moskau angepasst werden. Als Grundstruktur beabsichtigte der Chefarchitekt den bestehenden ehemaligen Befestigungsring um die Innenstadt zu einer äußeren Ringstraße auszubauen,

19 Navalichin, *Rekonstrukcii* (1958), Bd. 1, 162–163.

20 Zitiert nach A. Lapin, Glavnaja Magistral. In *Kaliningradskaja Pravda*, 21. April 1953.

21 Navalichin, *Rekonstrukcii* (1958), Bd. 1, 164.

die durch eine neu zu schaffende innere Ringstraße („Dzerzhinskij-Straße") ergänzt werden sollte. Diese innere Ringstraße war in Kaliningrad als Verkehrsader überflüssig – tatsächlich ist auch nur ein kurzer Abschnitt von knapp dreihundert Metern verwirklicht worden –, sie ergab lediglich als Pendant zum Moskauer Boulevardring einen Sinn und sollte wie dieser einen grünen Mittelstreifen erhalten, auf dem man promenieren konnte. Ebenfalls nach Moskauer Vorbild wollte Navalichin die bisherigen, durch das Zentrum laufenden Hauptstraßen radikal begradigen und durch weitere Radialstraßen ergänzen, die auf das Haus der Sowjets zulaufen sollten, das weiterhin anstelle des Schlosses vorgesehen war. Und wie in Moskau sollten schließlich an den Kreuzungen der äußeren Ringstraße mit den Radialstraßen eine Reihe von Türmen und Hochhäusern errichtet werden.[22] Das zentrale Hochhaus in Kaliningrad sollte die gleiche Funktion wie der Palast der Sowjets in Moskau übernehmen: Anstelle einer mächtigen Kathedrale wie in Köln, einem Schloss wie in Berlin oder einem Rathaus wie Hamburg sollten in Moskau und nun in Kaliningrad Gebäude der Partei den symbolträchtigen Platz in der Mitte der Stadt einnehmen. Auf derlei historische Bezüge wies der Chefarchitekt wenige Jahre später in seiner Dissertation mehrfach hin. Während die Struktur des neuen Straßennetzes und die Platzierung der geplanten städtebaulichen Dominanten in Kaliningrad also auf geradezu absurd exakte Weise dem Moskauer Vorbild folgen sollten, wich Navalichin bei dem konkreten Gestaltungsvorschlag für das Hauses der Sowjets in Kaliningrad in einem entscheidenden Punkt ab: In seinen Entwürfen ist es nicht nach dem Vorbild des damals noch geplanten Palastes der Sowjets gestaltet, sondern nach dem des neuen Gebäudes der Moskauer Lomonossov-Universität, die nach den Plänen von Lev Rudnev 1947–1949 südwestlich vom alten Moskauer Zentrum gebaut worden war. Eine Skizze und ein Modell von Navalichins Zentrumsplanungen zeigen das Haus der Sowjets als eine verkleinerte Kopie dieses Universitätsgebäudes. Der Grund für diese Abweichung bei der ansonsten so genauen Übernahme so vieler Details des Moskauer Stadtbildes ist nicht bekannt, dürfte aber wohl in der hierarchischen Symbolik des Bauwerkes liegen: Während etwa einzelne US-Bundesstaaten wie Arkansas, Kentucky oder Idaho für ihr Parlamentsgebäude bewusst den Stil und die Struktur des Kapitols in Washington, DC übernahmen und damit ihre Eigenständigkeit betont haben, wäre ein ähnliches Vorgehen in der spätstalinistischen Sowjetunion als gefährliche Anmaßung begriffen worden.

In Navalichins Entwürfen für den Wiederaufbau von Kaliningrad findet sich aber noch eine weitere bemerkenswerte Abweichung. Seine städtebauliche Kon-

22 Ebenda, 175–176 und 183–185. Zu der Moskauer Hochhausplanung vgl.: V.K. Oltarshevskij, *Stroitel'stvo vysotnikh zdanii*. Moskau 1953.

zeption für die Innenstadt bewegte sich im Rahmen des zeitgenössischen Planungswesens, das von breiten Straßen, Hochhäusern und dem für Imperien allgemein als angemessen angesehenen klassizistischem Baustil geprägt war, bei der Neugestaltung der wichtigsten Gebäude des wenig zerstörten Platz des Sieges und des Staliningrader Prospekts westlich der Innenstadt – wo sich bis heute das Zentrum Kaliningrads befindet – zu einem deutlich anderen Formenrepertoire. Neben Gebäude mit Säulenportiken und ausladende Kranzgesimsen wie dem in Anlehnung an das Bolschoj Theater wiederaufgebauten Stadttheater setzte der Chefarchitekt in seinen Entwürfen Türme, die in ihrer Formensprache auf spätmittelalterliche Vorbilder zurückgriffen, konkret: auf die Türme des Moskauer Kreml. Ein Gutachter des Kollegiums der Verwaltung für Architektur in Moskau erkannte diese architektonischen Zitate sofort, kritisierte den Entwurf ungeachtet dessen trotzdem. Er bemängelte, dass Navalichin den noch aus deutscher Zeit stammenden Nordbahnhof mit einem „schwerförmigen, zu großen und unmaßstäblichen Turm“ ergänzen wollte, der ganz offensichtlich dem Vorbild des ikonischen Spasskaja Turms am Roten Platz folgen solle. Doch werde das „ohnehin schon so finstere Bauwerk“ des Bahnhofs durch diese Ergänzung nur noch weiter beschwert. Offenkundig, so spottete der Gutachter, sei Navalichin durch den von ihm selbst kritisierten Genius loci des früheren Königsbergs negativ beeinflusst worden. Er regte an, den Turm nicht als „Festungsturm“ zu gestalten, sondern als „leichtes, ziviles“ Bauwerk.[23]

Damit hatte er die eigentliche Intention des Chefarchitekten Kaliningrads allerdings gründlich missverstanden. Leichte, „zivile“ Türme hätten der neuen Rolle der Stadt widersprochen, als „Vorposten“ zu dienen. Die Türme hatten darüber hinaus eine doppelte Funktion zu erfüllen: Sie sollten nicht nur auf die Bedeutung der Stadt als Abwehrbastion gegen Gefahren aus dem Westen verweisen, sondern Kaliningrad zudem auch gedanklich von der Küste weit in das Landesinnere versetzen, indem sie mit ihrer Architektur die Zugehörigkeit der Stadt zum alten, vorpetrinischen Kern Russlands propagierten. Diese Absicht lässt sich bei Navalichin bis in die spezifische Verwendung des Begriffs der „russischen Traditionen“ hinein verfolgen, die er mit seinen Entwürfen in Kaliningrad einpflanzen wollte. Navalichin hatte hierbei offensichtlich nicht die Traditionen des multiethnischen Russländischen Reichs im Blick – denn anstelle des auf den Staat bezogenen Begriffs „rossijskij – russländisch“ nutzte er in diesem Zusammenhang stets das auf die Ethnie abzielende Wort „russkij – russisch“.

23 GAKO, f. 520, op. 1, d. 8, l. 13–16, Expertise von Ja. A. Kornfel'd zum Schema der vorrangigen Bebauung und zum Projekt der Bebauung des Stalingrader Prospektes vom 2. Dezember 1949.

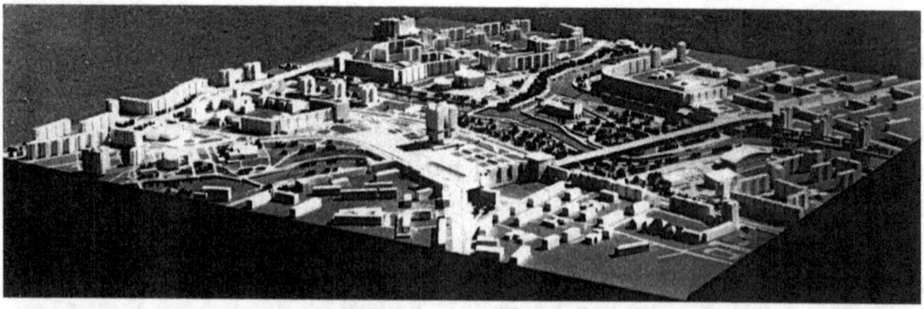

Abb. 1: 1972 fiel in einem städtebaulichen Wettbewerb die Entscheidung auf den Entwurf des örtlichen Stadtplanungsinstituts, der in den nächsten anderthalb Jahrzehnten realisiert wurde: Das Stadtzentrum wurde als eine offene Parklandschaft ausgebildet, die von bis zu zehnstöckigen Wohnriegeln umstellt und von zwei Autostraßen durchschnitten wurde und in der sich die Ruine des Doms und die Dominante des neuen Hauses der Sowjets erheben.

Kaliningrad war, wie die Beispiele Voronezh oder Stalingrad zeigen, nicht die einzige Stadt, die nach den Prinzipien des Moskauer Generalplans umgestaltet werden sollte,[24] doch gab es keine sowjetische Stadt, in der vergleichbar radikale Planungen entworfen wurden, die letztlich auf die Schaffung eines städtebaulichen Klons abzielten. Realisiert wurde von all diesen hochfliegenden Projekten kaum etwas. Der Wiederaufbau des Theaters am Stalingrader Prospekt (heute „Friedensprospekt") zog sich 14 Jahre lang hin und wurde erst 1960 abgeschlossen; es blieb bis in die 1970er Jahre das einzige repräsentative öffentliche Gebäude dieser Größe in der ganzen Stadt. Das einzige geschlossene architektonische „Ensemble", das in Kaliningrad während der sowjetischen Periode tatsächlich entstand, war die „Ausstellung der Landwirtschaft" des Kaliningrader Gebiets, ein Miniaturversion der gleichnamigen Unionsausstellung in Moskau, die sich zu einer Art sowjetischem Disneyland entwickelte. In Kaliningrad hatte es nur für fünf Pavillons und einen kleinen Triumphbogen als Eingang zum Gelände gereicht, wie die ausschweifenden Pläne für die gesamte Innenstadt hatte man aber auch hier versucht, die Bezüge zum Moskauer Vorbild überdeutlich herauszustreichen und bis in Details wie die Spitze des zentralen Pavillons hinein auszuformulieren.

Die alte Innenstadt blieb bis in die späten 1950er-Jahre eine Wüstenei, in die sich von ihren Rändern her einzelne Häuserreihen entlang des Lenin-Prospektes hineintasteten, der die Innenstadt in Nord-Südrichtung als Hauptverkehrsachse

24 Siehe dazu: Alexej Tarchanow, Sergej Kawtaradse. *Stalinistische Architektur*. München: Klinkhardt & Biermann, 1992, 97–115; Day, *Building Socialism*, 70–110.

vom Hauptbahnhof zum provisorischen neuen Zentrum am Stalingrader Prospekt durchzieht. Der Begriff „Wüstenei" ist dabei ganz wörtlich zu nehmen: Luftaufnahmen aus jener Zeit zeigen hohe Trümmerberge, hinter denen die Straßenbahnen halb verschwinden, dazwischen einzelne Ruinen bedeutender Gebäude wie des Schlosses, des Universitätsgebäudes, des Hauptpostamtes oder des Doms. Und nachdem die Stadtverwaltung von 1960 an fast alle Ruinen einebnen und die Trümmerberge abtragen ließ, wurde die Leere nur umso offensichtlicher: Im Zentrum Kaliningrad tat sich nun eine mit Rasen bedeckte Ebene von etwa einem Kilometer Durchmesser auf, aus der in der Mitte die ausgebrannte frühere Kathedrale emporragte. Es ist nicht gelungen, das städtebauliche Geflecht jemals wiederherzustellen. Das Zentrum dieser Stadt am Rande des sowjetischen Imperiums blieb selbst Peripherie – bis das Imperium in seine Einzelteile zerfiel.

Quellen

Staatsarchiv des Kaliningrader Gebiets – Gosudarstvennij Archiv Kaliningradskoj Oblasti (GAKO)

GAKO, f. 520, op. 1, d. 8, l. 1, Befehl Nr. 195 des Chefs der Verwaltung für Architektur beim Ministerrat der RSFSR, Shkvarikov.

GAKO, f. 520, op. 1, d. 8, l. 13–16, Expertise von Ja. A. Kornfel'd zum Schema der vorrangigen Bebauung und zum Projekt der Bebauung des Stalingrader Prospektes vom 2. Dezember 1949.

Literaturverzeichnis

Brodersen, Per. *Die Stadt im Westen. Wie Königsberg Kaliningrad wurde*. Göttingen: Vandenhoeck & Ruprecht, 2008.

Day, Andrew. *Building Socialism. The Politics of Soviet Cityscape in the Stalin-era*. Dissertation Columbia University, 1998.

Ivanov, A. Kenigsberg. In *Pravda*, 11. April 1945.

Lapin, A. Glavnaja Magistral. In *Kaliningradskaja Pravda*, 21. April 1953.

Na zapade – net bol'she vostochnoj Prussii. Literaturno-chudozhestvennij i obshestvenno-politcheskij sbornik, Kaliningrad 1951.

Naumov, M. R. Kakim budet Kaliningrad. In *Kaliningradskaja Pravda*, 30. April 1949.

Navalichin, Dmitrij. *K voprosu rekonstrukcii centra goroda Kaliningrada*, Moskau 1955 (erste Fassung der Dissertation), Moskau 1958 (zweite Fassung der Dissertation).

Oltarshevskij, V. K. *Stroitel´stvo vysotnikh zdanii*, Moskau 1953.

Podehl, Markus. *Architektura Kaliningrada. Wie aus Königsberg Kaliningrad wurde*. Marburg: Herder-Institut, 2012.

Razrabotka detal'nich proektov central'nich magistralej. In *Kaliningradskaja Pravda*, 3. Februar 1954.

Schlögel, Karl. Moskau – offene Stadt eine europäische Metropole, Reinbek bei Hamburg. Rowohlt-Verlag, 1992 [1984].

Stites, Richard. *Russian popular culture entertainment and society since 1900*. Cambridge: Cambridge University Press, 1993.

Tarchanow, Alexej und Sergej Kawtaradse. *Stalinistische Architektur*. München: Klinkhardt & Biermann, 1992.

Tjan, Dimitij. Sovetskij gorod Kaliningrad. In *Kaliningradskaja Pravda*, 7. November 1947.

Ukaz Verkhovnogo Soveta SSSR. Ob obrazovanij obshesojuznogo Ministerstva gorodskogo stroitel'stva. In *Arkhitektura i stroitel'stva* 7 (1949): 1.

Abbildungsnachweis

Abb. 1: 1972 fiel in einem städtebaulichen Wettbewerb die Entscheidung auf den Entwurf des örtlichen Stadtplanungsinstituts, der in den nächsten anderthalb Jahrzehnten realisiert wurde: Das Stadtzentrum wurde als eine offene Parklandschaft ausgebildet, die von bis zu zehnstöckigen Wohnriegeln umstellt und von zwei Autostraßen durchschnitten wurde und in der sich die Ruine des Doms und die Dominante des neuen Hauses der Sowjets erheben. (Quelle: Privatsammlung Jurij Vaganov, Architekt)

About the authors

Asbach, Olaf

Olaf Asbach is Professor for Political Science and Heisenberg Chair 'Europe and Modernity' at the Institute for Political Science, University of Hamburg, Germany. Research areas: History of Political Ideas; State, Democracy and (Inter-)National Law in the Modern World System. His publications include: *Der moderne Staat und ‚le doux commerce'. Politik, Ökonomie und internationale Beziehungen im politischen Denken der Aufklärung*. Baden-Baden: Nomos, 2014; *Europa – Vom Mythos zur Imagined Community? Zur historischen Semantik 'Europas' von der Antike bis ins 17. Jahrhundert*. Hannover: Wehrhahn, 2011; Dynamics of Conflict and Illusions of Law: Making War and Thinking Peace in the Modern International System. In *War, the State and International Law in Seventeenth Century Europe*, O. Asbach and P. Schröder (eds.), 249–266. Farnham: Routledge, 2010.

Banerjee-Dube, Ishita

Ishita Banerjee-Dube is Professor of History at the Centre for Asian and African Studies, El Colegio de México, Mexico City, and a member of the National System of Researchers (SNI), Mexico, where she holds the highest rank. Her 4 authored books include: *A History of Modern India*. Cambridge, Delhi and New York: Cambridge University Press, 2015 *and Religion, Law and Power*. London: Anthem Press, 2007. *Cooking Cultures*. Cambridge, Delhi and New York: Cambridge University Press, 2016 is the most recent of her 10 edited volumes. She edits the series "Hinduism" with De Gruyter Open and has contributed articles and book chapters to international journals and anthologies such as Current Sociology; Studies in History; and Subaltern Studies. She also acts as a referee for several important journals and publishing houses.

De Groot, Joanna

Joanna de Groot is a Senior Lecturer in the History Department at the University of York UK. Her main interests are in gendered approaches to histories [a] of societies in the Middle East, especially Iran, since c.1800; [b] of imperial and colonial relationships and cultures since c.1700; [c] of historical practices and theories. Recent publications include: *Empire and history writing in Britain 1750–2010*. Manchester: Manchester University Press, 2013; 'Hybridising past, present, and future': reflections on the "sexology" of R. F. Burton. In *Sex, knowledge and receptions of the past*, K. Fisher and R. Langland (eds.), 135–159. Oxford: Oxford University Press, 2016; The bureaucrat, the mulla, and the maverick intellectual "at home": domestic narratives of patriarchy, masculinity and modernity in Iran c.1880–1980. *Gender and History* 27:3 (2015): 791–811.

Dorsch, Sebastian

Sebastian Dorsch, after having studied Latin American History at Passau, Erfurt and Mexico City he earned his PhD in Erfurt. As postdoctoral research fellow he currently works on a study of

São Paulo and spatio-temporal forms of appropriating and representing the world. In 2010 he co-founded the research unit Erfurter RaumZeit-Forschung (ERZ, Erfurt Spatio-Temporal Studies) and organized several ERZ-workshops and publications. Selected relevant publications: „Translokale Wissensakteure": Ein Debattenvorschlag zu Wissens- und Globalgeschichtsschreibung. *Zeitschrift für Geschichtswissenschaft* (to be published in 09/2016); Emil Göldis ‚spezifisch brasilianisches Prisma' gegen französische Kolonialansprüche in den Guyanas. Wissen produzieren, lokalisieren und kartographieren (1880er bis 1900er Jahre). *Berichte zur Wissenschaftsgeschichte* (forthcoming); „Brasilien" zwischen Insel-Topos und „Neuer Welt": Ein raum-zeitlicher Versuch über Weltbilder und Welterfahrungen im frühen Anthropozän. In *Welt-Anschauungen. Interdisziplinäre Perspektiven auf die Ordnungen des Globalen*, Olaf Breidbach etc. (eds.), 15–41. Stuttgart: Wissenschaftliche Verlagsgesellschaft, 2015.

Dube, Saurabh

Saurabh Dube (PhD, Cantab) is Research Professor in History, Centre for Asian and African Studies, El Colegio de México and also holds the highest rank in the National System of Researchers (SNI), México. He has been a fellow of the John Simon Guggenheim Memorial Foundation, New York; the Institute of Advanced Study, University of Warwick; the Indian Institute of Advanced Study, Shimla; and the Stellenbosch Institute of Advanced Study, South Africa. Apart from more than one hundred and twenty journal articles and book chapters, Dube's authored books include *Untouchable Pasts*, Albany and New York: State University of New York Press, 1998 (Sage, 2001); *Stitches on Time*, Durham: Duke University Press, 2004; (Oxford University Press, 2004); *Subjects of Modernity*, Manchester and New York: Manchester University Press, 2017. He has also written a quintet in historical anthropology in the Spanish language published by El Colegio de México, and edited fifteen volumes on subjects of colony and postcolony, nation and modernity, and history and anthropology. Dube has also been visiting professor several times at institutions such as Cornell University and the Johns Hopkins University.

Fischer, Robert

Robert Fischer, historian of Latin American and North American History, studied at University of Erfurt and UNAM, Mexico City. Currently working as a fellow researcher at the German Historical Institute in Washington, D.C. Selected relevant works: *Sex im Grenzbereich. Sexualität, Devianz und Regulierung in den US-amerikanischen Grenzstädten Ciudad Juárez/El Paso, 1880–1950*. Unpublished Dissertation: University of Erfurt, 2016; Pinturas de Castas. Verbildlichung von *race* und *sex* im kolonialen Mexiko. In *race&sex. Eine Geschichte der Neuzeit. 49 Schlüsseltexte aus vier Jahrhunderten neu gelesen*, Jürgen Martschukat und Olaf Stieglitz (eds.), 365–374. Berlin: Neofelis, 2016; Mobility and Morality at the Border. A Lefebvrian Spatio-Temporal Analysis in Early Twentieth-Century Ciudad Juárez and El Paso. *Historical Social Research* 38:3 (2013): 176–196.

Frischmann, Bärbel

Bärbel Frischmann is Professor for History of Philosophy in the Department of Philosophy of the University of Erfurt. Her main research interests are the history of philosophy, cultural philosophy and social philosophy. Selected relevant publications: *Vom transzendentalen zum frühromantischen Idealismus. J. G. Fichte und Fr. Schlegel.* Paderborn etc.: Verlag Ferdinand Schöningh, 2005; Philosophie der Anerkennung. Historische und aktuelle Bezüge. In *Differenz und Dialog. Anerkennung als Strategie der Konfliktbewältigung?*, Vera Flocke, Holger Schoneville (eds.), 15–35. Berlin: Berliner Wissenschafts-Verlag, 2011; Frischmann, Bärbel (ed.). *Grenzziehungen und Grenzüberwindungen. Philosophische und interdisziplinäre Zugänge.* Hannover: Wehrhahn Verlag, 2014.

Holtorf, Christian

Christian Holtorf, Professor of Science Studies and Science Communications at the University of Applied Sciences in Coburg. In 2013 he published the book *Der erste Draht zur neuen Welt. Die Verlegung des trantlantischen Telegrafenkabels*, Göttingen: Wallstein. [The First Wire to the New World. The laying of the transatlantic telegraph cable]. Currently he is pursuing a research project entitled 'Imaging the Arctic. Towards a history of knowledge concerning space and time during the 19th and 20th centuries. Selected relevant publications: Der Nordpol – eine Erzählung. In *Erzählung und Geltung. Wissenschaft zwischen Autorschaft und Autorität*, Safia Azzouni, Stefan Böschen, Carsten Reinhardt (eds.), 133–155. Weilerswist: Velbrück Wissenschaft, 2015; Das offene Polarmeer. Ein Bilddiskurs im 19. Jahrhundert. In *Bilder in historischen Diskursen*, Franz K. Eder, Oliver Kühschelm, Christina Linsboth (eds.), 145–172. Wiesbaden: VS Verlag für Sozialwissenschaften, 2014; Die Südsee im Norden. Technik und Transzendenz in Narrativen der Arktisforschung des 19. Jahrhunderts. In *Technik und Transzendenz. Zum Verhältnis von Technik, Religion und Gesellschaft*, Katharina Neumeister, Peggy Renger-Berka, Christian Schwarke (eds.), 181–208. Stuttgart: Kohlhammer, 2012.

Hoppe, Bert

Bert Hoppe worked from 1999 to 2004 as an assistant professor (wissenschaftlicher Mitarbeiter) at the Humboldt University Berlin (Department of History). From 2004 until 2005 he was a co-worker of the German-Russian Forum Berlin. Between 2005 and 2011 he worked as freelancer at the Institute of Contemporary History Berlin/Munich. From 2011 to 2014 he was a nonfiction editor of the publishing house Rowohlt Berlin. Since 2015 he is Subject Librarian in the Department Central and Eastern Europe of the Friedrich Ebert Foundation Berlin. Selected relevant publications: *Auf den Trümmern von Königsberg. Kaliningrad 1946–1970*, München: Oldenbourg-Verlag 2000; *In Stalins Gefolgschaft. Moskau und die KPD 1928–1933*, München: Oldenbourg-Verlag 2007; als Bearbeiter: *Die Verfolgung und Ermordung der europäischen Juden durch das nationalsozialistische Deutschland.* Vol. 7 and 8: Besetzte sowjetische Gebiete, München: Oldenburg Verlag, 2011 and 2015.

Kugele, Jens

Jens Kugele studied Comparative Religious Studies, Jewish History and Political Science at Georgetown University and at the LMU Munich where he completed his MA. He received his PhD in Literature and Cultural Studies from Georgetown University, Washington DC. After two years as Visiting Assistant Professor in the Study of Religion Program back at LMU Munich, he joined the International Center for the Study of Culture (GCSC) in Gießen, funded by the German Excellence Initiative. Here, he is one of the research center's Principal Investigators and has been a member of the center's Executive Board since September 2014. He is co-editor and co-founder of the peer-reviewed academic journal "On_Culture". His research interests include intersections of history of religion with literary and cultural history, German-Jewish history, constructions of belonging, memory and sacred space.

Mackenthun, Gesa

Gesa Mackenthun is Professor of American Studies at Rostock University. Her publications include *Fictions of the Black Atlantic in American Foundational Literature*, London: Routledge, 2004 and *Metaphors of Dispossession. American Beginnings and the Translation of Empire, 1492–1637*, Norman: University of Oklahoma Press, 1997. She is series editor of "Cultural Encounters and the Discourses of Scholarship" at Waxmann (Münster). Her current research deals with nineteenth-century imperial travel and archaeology and the scientific construction of American antiquity.

Maier, Harry O.

Harry O. Maier is Professor of New Testament and Early Christian Studies at Vancouver School of Theology, Vancouver, Canada and a Fellow of the Max Weber Center for Advanced Cultural and Social Studies at the University of Erfurt. He is also a Fellow of the Alexander von Humboldt Foundation. He publishes widely across a variety of fields, usually with a focus on the inter-disciplinary study of early Christianity from the point of view of social sciences and social history. Research with a focus on spatial geography has appeared in several publications. The most notable of these are his recent monographs, *Picturing Paul in Empire: Imperial Image, Text and Persuasion in Colossians, Ephesians and the Pastoral Epistles*. New York: Bloomsbury, 2013; as well as a number of recently published essays including From Material Place to Imagined Space: Emergent Christian Community as Thirdspace in the Shepherd of Hermas. In *Christian Communities in the Second Century: Between Ideal and Reality*, Joseph Verheyden (eds.), 66–89. Tübingen: Mohr Siebeck, 2015 and Soja's Thirdspace, Foucault's Heterotopia and de Certeau's Practice: Space-time and Social Geography in Emergent Christianity. *Historical Social Research* 38:3 (2013): 76–92.

Meyer, Holt

Holt Meyer is Professor of Slavic Literatures at the University of Erfurt since 2001. Organized the international conference "Imperial Traces" in Erfurt in 2008. Selected relevant publications: Spiel(k)arten der Distanz in der ‚begrenzten Souveränität': Bohumil Hrabals Kdo jsem. In *Distanz. Schreibweisen, Entfernungen, Subjektkonstitutionen in der tschechischen und mitteleuropäischen Literatur*, Nora Schmidt , Anna Förster (eds.), 23–63. Weimar: VDG 2014; '... and qua criminal he is of an imperfectly formed mind...': Transylvania and Ireland as 'Criminally Deficient Territories' and the Terrain of the Sponsa Christi as Accomplice of/in Bram Stoker's

Dracula. In *Crime and Nation. Political and Cultural Mappings of Criminality in New and Traditional Media*, Immacolata Amodeo, Eva Erdmann (eds.), 143–161. Trier: Wissenschaftlicher Verlag Trier, 2009; Die vergebliche (Über)Gabe des Geheimagenten. Trauer, Spionage und Wertschöpfung im 2. Kapitel von Nabokovs Dar [Die Gabe]. In *Gabe und Opfer in der russischen Literatur und Kultur der Moderne*, Rainer Grübel, Gun-Britt Kohler (eds.), 414–442. Oldenburg: BIS-Verlag der Carl-von-Ossietzky-Universität, 2006.

Michael, Bernardo

Bernardo Michael is a professor of history at Messiah College in the United States. His research has largely focused on statemaking and space on the Anglo-Gorkha borderlands. He has a number of publications on this subject, including his 2012 book *Statemaking and Territory: Lessons from the Anglo-Gorkha War (1814–1816)*. London, New York and Delhi: Anthem Press, 2012. He is currently working on the life of the Anglican educator and activist Charles F. Andrews (1871–1940). His research interests also include world history, the history of cartography, interdisciplinary history, and studies on diversity, equity, and inclusion. Over the past ten years he has also worked as an administrator, directing the Center for Public Humanities and was a Special Assistant to the President and Provost for Diversity Affairs.

Millán, Elizabeth

Elizabeth Millán is Professor of Philosophy at DePaul University (Chicago). She works on aesthetics, German Idealism/Romanticism and Latin American Philosophy. Before joining the faculty at DePaul, she worked at the Universidad Simón Bolívar in Venezuela (1997–98). In 2004–05 she was awarded an Alexander von Humboldt Fellowship and spent the year in Leipzig, doing research on Alexander von Humboldt, focusing upon his view of nature and his presentation of America. Her publications include: *Friedrich Schlegel and the Emergence of Romantic Philosophy*, Albany: SUNY, 2007; with Bärbel Frischmann, *Das neue Licht der Frühromantik/The New Light of German Romanticism*, Paderborn: Schöningh Verlag, 2008. In 2011, Millán co-edited (with John Smith) a special volume of the *Goethe Yearbook* on Goethe and German Idealism. In 2014, she was guest editor, with Hugo Moreno, of a special volume on *Latin American Aesthetics* for *Symposium: Canadian Journal of Continental Philosophy*, Vol. 18, No. 2, Fall 2014.

Niedermeier, Silvan

Silvan Niedermeier is assistant professor (wissenschaftlicher Mitarbeiter) of history at the University of Erfurt. He received his PhD from the University of Erfurt in 2011 and worked as a postdoctoral fellow at the Graduate School Cultural Encounters and the Discourses of Scholarship at the University of Rostock in 2011 and 2012. Among his most important publications are: *Rassismus und Bürgerrechte. Polizeifolter im Süden der USA 1930–1955*. Hamburg: Hamburger Edition, 2014; *Violence and Visibility in Modern History*. New York: Palgrave Macmillan 2013 (edited together with Jürgen Martschukat); Imperial Narratives: Reading U.S. soldiers' photo albums from the Philippine American War. *Rethinking History. The Journal of Theory and Practice*, 18:1 (2014), 28–49.

Payne, Charlton

Charlton Payne is Visiting Assistant Professor in the German Department at UC Berkeley. He is completing the book On the Trail of Refugees: Documentality and Narrative in Twentieth and Twenty-First Century German-Language Literature and Culture. He was awarded his PhD in German Literature from UCLA in 2007, was a doctoral fellow in the DFG Research Group "The Figure of the Third" as well as a researcher in the Cluster of Excellence "Cultural Foundations of Integration" in Konstanz. From 2011 to 2013, he was a postdoctoral fellow at the University of Erfurt. During 2014–2015, he was Fellow of the Alfried Krupp Institute for Advanced Study in Greifswald. Publications include The Epic Imaginary. Political Power and Its Legitimations in Eighteenth-Century German Literature, and essays on cosmopolitanism in Goethe and Wieland, hospitality and asylum in Kleist, narratives about the assimilation of Vertriebene in Treichel and Hein, and articles on refugee voice and archive in the writing of Saša Stanišić. Forthcoming publications include a special issue of symplokē (co-edited with Jesper Gulddal) on "Passports" as well as the edited volume Flüchtlinge in der deutschsprachigen Gegenwartsliteratur.

Rau, Susanne

Susanne Rau, M.A. in History, Romance Languages and Literatures and Philosophy in 1997, PhD in medieval and modern History in 2001 at Hamburg University, Habilitation in 2008 at Dresden University. Since 2009, she is professor of spatial history and cultures at Erfurt University, Germany; since 2011 regular visiting professor at the ENS Lyon and the CIHAM, France, and, since 2014, Vice President for Research at Erfurt University. Regarding urban and spatial history, she is member of several advisory boards of academic journals ("Histoire urbaine", "Informationen zur modernen Stadtgeschichte", prospectively "Moderne Stadtgeschichte", and "Historische Zeitschrift") and member of the editorial board of the Encyclopedia of Early Modern History (1450–1850). Her last publications in this field are: *Räume. Konzepte – Praktiken – Nutzungen*. Frankfurt: Campus, 2013; *Räume der Stadt. Eine Geschichte Lyons 1300–1800*. Frankfurt: Campus, 2014; Imaginierte Räume. Zur Geschichte nicht-realisierter Stadtplanungsprojekte. *Informationen zur modernen Stadtgeschichte* 2 (2015): 129–147.

Schmid, Alfred

Alfred Schmid, PhD in Basel 2003. Currently working as co-editor of unpublished lectures of Jacob Burckhardt on the Ancient Orient and on Rome and serving as lecturer in Ancient History at the University of Basel. Publications include: *Die Geburt des Historischen aus dem Geiste der Politik. Zur Bedeutung frühgriechischer Geschichtsschreibung, mit einem Seitenblick auf Altchina*. Basel: Schwabe Verlag, 2016; Das ptolemäische Weltbild. In *Mensch und Raum von der Antike bis zur Gegenwart* (= Colloquium Rauricum IX), Antonio Loprieno (ed.), 127–149. München and Leipzig: Saur Verlag, 2006; *Augustus und die Macht der Sterne. Die antike Astrologie und die Etablierung der Monarchie in Rom*. Köln and Weimar: Verlag Böhlau, 2005.

Schmidtke, Alrun

Alrun Schmidtke is a doctoral candidate at Humboldt-Universität zu Berlin. After studying the history of science at the Technische Universität and Humboldt-Universität zu Berlin she received a Herzog Ernst scholarship from the Fritz Thyssen Foundation at the University of Erfurt for her research on cartographic publishing at Perthes. She has worked as research fellow at the chair for the history of science at Humboldt-Universität zu Berlin and is now an affiliate member of the graduate school "Cultures of Decision-Making" at the University of Münster. Her PhD thesis, which is funded by the Gerda Henkel Foundation, focusses on scientists as publishing advisors for Springer, Wiley and others up until the 1960s.

Schmolinsky, Sabine

Sabine Schmolinsky received her PhD in German philology at the Ludwig Maximilian University in Munich and her 'Habilitation' in Medieval History and Auxiliary Sciences of History at the Helmut-Schmidt-University (HSU), Hamburg. Professor of Medieval History at the University of Erfurt since 2009, after having worked as an Acting Professor of Medieval History at the HSU in 2007/08. Member of the Erfurt Spatio-Temporal Studies Group. Selected relevant publications: Dialogue Situations. Considerations on Self-Identification in the Middle Ages. In *Forms of Individuality and Literacy in the Medieval and Early Modern Periods*, Franz-Josef Arlinghaus (ed.), 303–317. Turnhout: Brepols, 2015; The Production of Future. Chronotope and Agency in the Middle Ages. *Historical Social Research/Historische Sozialforschung* (Special Issue: Space/Time Practices and the Production of Space and Time) 38:3 (2013): 93–104; *Sich schreiben in der Welt des Mittelalters. Begriffe und Konturen einer mediävistischen Selbstzeugnisforschung*, Bochum: Verlag Dr. Dieter Winkler, 2012.

Schröder, Iris

Iris Schröder is vice director of the Gotha Research Center on Knowledge Cultures in Modern History, where she is currently building up a research group on 19–20th centuries' knowledge cultures; in addition she holds the chair for Global History at Erfurt University. Her main research interests draw on the intersections of history of science, history of knowledge, and media history, treating especially issues of space, geography and cartography, the visual and the material world. Moreover, her work has focussed on the history of international organizations. In terms of historiography she is interested in combining European and Global history from the late 18th to the 20th century onward. Selected relevant publications: *Das Wissen von der ganzen Welt: Globale Geographien und räumliche Ordnungen Afrikas und Europas, 1790er–1860er Jahre*. Paderborn: Schöningh, 2011; Disziplinen. Zum Wandel der Wissensordnungen im 19. Jahrhundert. In *Von Käfern, Märkten und Menschen: Kolonialismus und Wissen in der Moderne*, Rebekka Habermas, Alexandra Przyrembel, (eds.), 148–161. Göttingen: Vandenhoeck & Ruprecht, 2013; as well as a volume coedited with Sabine Höhler: *Welt-Räume. Geographie, Geschichte und Globalisierung seit 1900*, Frankfurt am Main: Campus, 2005.

Waldner, Katharina

Katharina Waldner is Professor of Religious Studies at the University of Erfurt since 2009. She studied Classics and Archaeology at Zurich and Berlin (FU) and was assistant professor in Classics and Religious Studies at Berlin, Munich, and Erfurt. Her research fields include: ancient Greek religion, mystery cults and individuality, Early Christianity in its cultural context. Selected relevant publications: Letters and Messengers: the Construction of Christian Space in the Roman Empire in the Epistles of Ignatius of Antioch. In *The Changing face of Judaism, Christianity, and Other Greco-Roman Religion*, Ian H. Henderson and Gerbern S. Oegema (eds.), 72–86. Güthersloh: Güthersloher Verlagshaus, 2006. Aitiologie und Religion in der griechisch-römischen Antike. In *Von Ursachen sprechen: Eine aitiologische Spurensuche*, Christiane Reitz, Anke Walter (eds.), 25–57. Hildesheim u.a.: Olms, 2014.

Walsh, Stephen A.

Stephen A. Walsh has been a Christoph Martin Wieland Postdoctoral Fellow at the Gotha Research Center of the University of Erfurt since 2016. After receiving his PhD from Harvard University, he also held research positions at the European University Institute (Florence) and the Smithsonian Institution (Washington DC). His publications include Liberalism at High Latitudes: The Politics of Polar Exploration in the Habsburg Monarchy. *Austrian History Yearbook* 47 (2016): 89–106 and On Slippery Ice: Discovery, Imperium & the Austro-Hungarian North Polar Expedition. In *Expedition into Empire: Exploratory Journeys and the Making of the Modern World*, Martin Thomas (ed.), 148–170. New York and London: Routledge, 2015.

Index of persons

Index of places and spaces